Translational Nanomedicine

Edited by
Robert A. Meyers

Translational Nanomedicine

Edited by
Robert A. Meyers

Editor

Dr. Robert A. Meyers
Editor-in-Chief
Ramtech Limited
34896 Staccato St.
Palm Desert, CA 92211
United States

Cover

Steps of antitumor activity of a redox-sensitive DOX/GO/PEG hybrid. For details see chapter 4, figure 8.

Library of Congress Card No.: applied for

British Library Cataloguing-in-Publication Data
A catalogue record for this book is available from the British Library.

Bibliographic information published by the Deutsche Nationalbibliothek
The Deutsche Nationalbibliothek lists this publication in the Deutsche Nationalbibliografie; detailed bibliographic data are available on the Internet at <http://dnb.d-nb.de>.

Print ISBN: 978-3-527-33789-7
ePDF ISBN: 978-3-527-68431-1
ePub ISBN: 978-3-527-68428-1

Cover Design Adam Design, Weinheim, Germany

Typesetting SPi Global, Chennai, India

Printing and Binding Markono Print Media Pte Ltd, Singapore

Printed on acid-free paper

10 9 8 7 6 5 4 3 2 1

Contents

Preface

Our book is aimed at providing students, professors, physicians and research scientists at universities, research laboratories, hospitals and drug companies with the latest developments in the very broad and fertile field of nanomedicine. Nanomedicine research and development utilizes the unique nanoscale properties of particles and certain molecules for diagnosis, delivery, sensing and actuation of treatment of diseases. These nanostructures very in size from 1 to 100 nm. Such structures have unique abilities for delivery of drugs to and into diseased cells allowing for less dosage and side effects, as well as unique abilities in diagnostics because of their distinct optical, magnetic and structural properties. Ultimately it may be possible to develop nanorobots and nanodevices that even more effectively may be applied in both diagnosis and repair of diseased cells and tissues. All of these approaches and advances are covered in our book.

Our book is divided into five sections describing the advances in Nanomedical Laboratory Techniques, Nanoscale Devices, Pharmaceuticals Delivery, Cancer Treatment and Tissue Engineering.

The Nanomedical Laboratory Techniques section includes recent approaches using microfluidic technology to accelerate the clinical translation of nanomedicine devices and drug complexes. The microfluidic technologies constitute a novel platform capable of replacing the entire nanomedicine production process in a scalable manner. Fabrication and utilization of nanocrystals or quantum dots for more accurate identification of diseases as well as drug carriers is covered. Special attention is given to clearing of nanoparticles from the body after use as well as any possible toxicity.

The Nanoscale Devices section ranges from DNA origami used to design and fabricate an autonomous, logic-guided DNA origami nanorobot, which can be programmed to transport molecular payloads between selected points of origin and target, and can be loaded with a variety of cargoes including small molecules, drugs, proteins, and small (<30–35 nm) nanoparticles; to synthetic gene circuits that allow the engineering of biological embedded computing devices; as well as electrodes that can easily be miniaturized to micron size, or even to nanometer size. Graphene is currently a "shining star" in nanomedicine on account of its good biocompatibility, low cytotoxicity, and ease of functionalization. The unique applications of graphene and its derivatives in biosensing, bioimaging, therapeutics, and genetic engineering are discussed.

The Pharmaceuticals Delivery section includes reviews of carbon nanotubes (CNTs) that can be applied as versatile biopharmaceutical delivery systems due to their high drug-loading capacity, excellent cell-penetrating ability, and customizable surface chemistry. CNTs synthesis, functionalization, and application is presented including transport of peptides and proteins, antibodies and nucleic acids (e.g. siRNA) into cells for therapy. Many other nano-delivery systems are covered including cholesterol, and a wide range of nanoparticle conjugates. In addition, a major problem in application of NPs systems is addressed and this is that the first major barrier that NPs encounter after entering the body is the innate immune system. Before reaching the target sites, NPs are readily cleared by macrophages in the liver and spleen. To overcome the rapid body clearance and minimize the immune recognition of NPs using different chemical or physical surface modifications, the mechanisms of NP interaction with the immune system is addressed in detail as are methods to overcome this barrier.

The Cancer Treatment Section applies NPs and NP derivatives to the diagnosis and treatment of various types of cancer. Nanomedicine is a major potential effector of the theranostics or specific targeted therapy approach. More specifically, the goal is to unite diagnostic and therapeutic individualized treatment to provide diagnosis, drug delivery and treatment response monitoring. Gold nanoparticles (AuNPs) and iron oxide nanoparticles (IONPs) conjugated to antibodies, silica-based NPs and others are discussed. Brain cancer therapy has become a huge challenge compared with peripheral cancers because of the physiological characteristic of the brain-blood barrier (BBB), which prevents most therapeutic drugs from reaching the cancer tissues. For years, efforts have been made in nanotechnology, especially employing nanoparticles (NPs), to overcome this limitation. This section covers research aimed at focusing on three elements, namely functionalization, targeting, and imaging, to fabricate smart nanoparticles that are capable of crossing the BBB, can respond to multiple internal or external stimuli, and deliver therapeutic or diagnostic agents to cancer cells through systemic administration. We also review approaches to utilizing RNA interference (RNAi), a ubiquitous cellular pathway of post-transcriptional gene regulation, that provides an intriguing tool for an innovative rational cancer drug design. Among the endogenous mediators of RNAi, microRNAs (miRNAs) represent the most important class of small RNAs whose global dysregulation is a typical feature of human tumors. Harnessing of the RNAi machinery by using small, synthetic RNAs that target, or mimic endogenous miRNAs offers the opportunity to reach virtually any gene and pathway relevant to tumor maintenance.

The Tissue Engineering Section focuses on the use of cellular and material-based therapies aimed at targeted tissue regeneration caused by traumatic, degenerative, and genetic disorders. Current treatments for bone injuries and defects that will not spontaneously heal employ replacement rather than regeneration, which is accompanied by long-term complications. In this chapter, attention is focused on state-of-the-art research techniques and materials that are aimed at alleviating these complications by inducing bone-healing rather than bone-substitution; specifically, the use of nanoscale materials and surface modifications in order to closely mimic the microenvironment of bone. Today, these nanoscale technologies are coming to the forefront in medicine because of their biocompatibility, tissue-specificity, and integration and ability to act as therapeutic carriers. Nanoparticles and nanotubes investigated or proposed include nanoscale ceramics, titanium nanotubes, graphene and graphene oxide, carbon nanotubes (CNTs), and iron oxide as well as derivatives. Techniques for producing bionanomaterials and structures are reviewed including 3D printing, Micropatterning, Nanopatterning, and Lithography, as well as Chemical Etching and Vapor Deposition.

Palm Desert, California, June 2019

Robert A. Meyers
Editor-in-Chief

Part I
Laboratory Techniques

Translational Nanomedicine. First Edition. Edited by Robert A. Meyers.

1 Microfluidics in Nanomedicine

YongTae Kim[1] *and Robert Langer*[2]
[1]*Georgia Institute of Technology, George W. Woodruff School of Mechanical Engineering, Wallace H. Coulter Department of Biomedical Engineering, Institute for Electronics and Nanotechnology, Parker H. Petit Institute for Bioengineering and Bioscience, 345 Ferst Drive, Atlanta, GA 30318, USA*
[2]*Massachusetts Institute of Technology, Department of Chemical Engineering, Harvard-MIT Division of Health Sciences and Technology, David H. Koch Institute for Integrative Cancer Research, 500 Main Street, Cambridge, MA 02139, USA*

Translational Nanomedicine. First Edition. Edited by Robert A. Meyers.

Keywords

Microfluidics
The science and technology that involves the manipulation of nanoscale amounts of fluids in microscale fluidic channels for applications that include chemical synthesis, and biological analysis and engineering.

Nanotechnology
The manipulation of matter on atomic and molecular scales.

Nanomedicine
The medical application of nanotechnology for the advanced diagnosis, treatment and prevention of a number of diseases.

Biomimetic microsystem
A microscale device that mimics biological systems and is used to probe complex human problems.

Clinical translation
Clinical translation involves the application of discoveries made in the laboratory to diagnostic tools, medicines, procedures, policies and education, in order to improve the health of individuals and the community.

Nanomedicine is the medical application of nanotechnology for the treatment and prevention of major ailments, including cancer and cardiovascular diseases. Despite the progress and potential of nanomedicines, many such materials fail to reach clinical trials due to critical challenges that include poor reproducibility in high-volume production that have led to failure in animal studies and clinical trials. Recent approaches using microfluidic technology have provided emerging platforms with great potential to accelerate the clinical translation of nanomedicine. Microfluidic technologies for nanomedicine development are reviewed in this chapter, together with a detailed discussion of microfluidic assembly, characterization and evaluation of nanomedicine, and a description of current challenges and future prospects.

1 Introduction

Nanomedicine is the medical application of nanotechnology that uses engineered nanomaterials for the robust delivery of therapeutic and diagnostic agents in the advanced treatment of many diseases, including cancer [1–3], atherosclerosis [4–6], diabetes [7–9], pulmonary diseases [10, 11] and disorders of the central nervous system [12, 13]. One key advantage of nanomedicine is the ability to deliver poorly water-soluble drugs [14–16] or plasma-sensitive nucleic acids (e.g., small interfering (si)RNA [17, 18]) into the circulation with enhanced stability. Nanomedicine is also capable of providing contrast agents for different imaging modalities and the targeting of specific sites for the delivery of drugs and/or genes [19–23]. Engineered nanomaterials, developed as particulates that are widely referred to as nanoparticles (NPs), have been formulated using a variety of materials that includes lipids, polymers, inorganic nanocrystals, carbon nanotubes, proteins, and DNA origami [24–36]. The ultimate goal of nanomedicine is to achieve a robust, targeted delivery of complex assemblies that contain sufficient amounts of multiple therapeutic and diagnostic agents for highly localized drug release, but with no adverse side effects [37, 38], and a reliable detection of any site-specific therapeutic response [39, 40].

1.1 Nanomedicine Development

Typical nanomedicine development processes for the clinical translation include benchtop syntheses, characterizations, *in-vitro* evaluations, *in-vivo* evaluations with animal models, and scaled-up production in readiness for clinical trials. Although, previously, several NPs have been reported as superior platforms, many are still far from their first stages of patient clinical trials due to several critical challenges [41, 42]. Such challenges mainly result from batch-to-batch variations of NPs produced in the benchtop synthesis process, and from insignificant outcomes in the *in-vitro* evaluation process under physiologically irrelevant conditions. These limitations ultimately lead to highly variable results in the *in-vivo* evaluation, or to failure in clinical trials. In order to address these challenges, the following methodologies need to be established in the nanomedicine development process:

- Nanomedicine needs to be continuously produced in a high-throughput fashion. The large-scale, continuous production of nanomedicines will allow a robust supply of highly reproducible materials for the *in-vitro* and *in-vivo* evaluation stages and clinical trials, ultimately increasing the success rate in clinical trials.
- Nanomedicines synthesized using large-scale, continuous production methods also need to be characterized in a high-throughput manner. Rapid characterization will create an efficient production cycle for an optimized nanomedicine via feedback loops between the synthesis and characterization stages.
- The *in-vitro* evaluation of nanomedicine must be conducted in more physiologically relevant environments. Highly repeatable results obtained from these biomimetic conditions will allow the obviation of a number of simple screening experiments in animal studies, not only saving costly animal models but also accelerating the clinical translation.

1.2 Microfluidics Technology

Microfluidics technology provides highly compatible platforms to create a new nanomedicine development pipelines that include the required methodologies introduced above. Basically, microfluidics presents a number of useful capabilities to manipulate very small quantities of samples, and to detect substances with a high resolution for a wide range of applications, including chemical syntheses [43, 44] and biological analysis [45, 46]. More importantly, the adaptability of microfluidics allows its integration with many other technologies, such as micro/nanofabrication, electronics, and feedback control systems [47–52]. Recently, microfluidic platforms integrated with control systems and advanced microfabrication technologies have been used to address the critical challenges in nanomedicine [53–57]. For example, the continuous synthesis of NPs in microfluidics has demonstrated a versatility to produce a variety of NPs with different sizes, shapes, and surface compositions [58, 59]. Several advances have recently been made in the label-free detection, characterization and identification of single NPs [60]. The confluence of microfluidics and biomimetic design has enabled the creation of physiologically relevant microenvironments for the evaluation of drug candidates [61–63]. The key microfluidic technologies in nanomedicine, including microfluidic assembly, and the characterization and evaluation of nanomedicines, are discussed in the following sections (see Fig. 1), and their current challenges and future research directions are highlighted.

2
Microfluidic Assembly of Nanomedicines

The bulk synthesis of NPs typically has strong dependencies on nonstandard multistep processes which are time-consuming, difficult to scale up, and depend heavily on specific synthetic

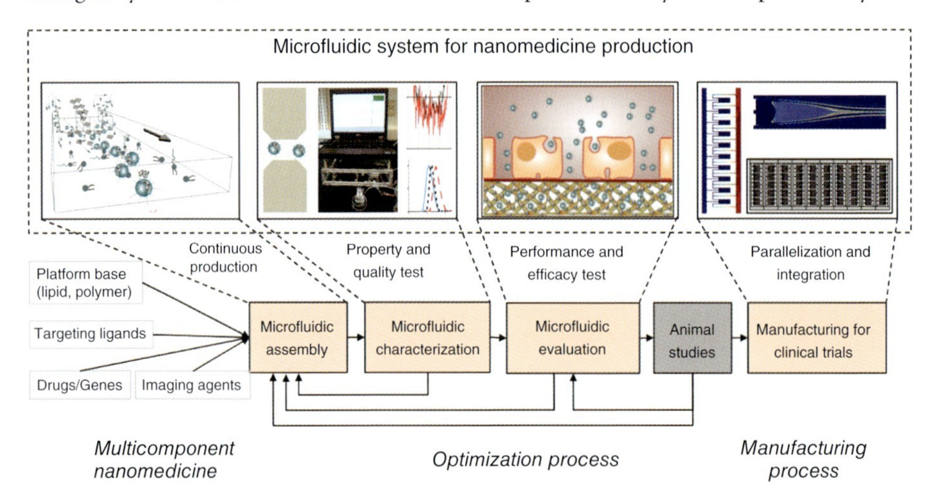

Fig. 1 A new nanomedicine development pipeline using microfluidic systems. First, a designed nanomedicine with multiple precursor components is continuously assembled through controlled strong mixing patterns in the Microfluidic Assembly stage, and the properties of the nanomedicine produced are identified at the Microfluidic Characterization stage. Only if those properties meet the nanomedicine design criteria will the performance and efficacy of a selected nanomedicine be evaluated in *in-vitro* biomimetic microsystems that recapitulate the structure and function of human organs in the Microfluidic Evaluation stage. If the targeting, therapeutic, and imaging efficacies are satisfactory in the *in-vitro* model system, the nanomedicine will then be validated with animal models. All nanomedicine candidates that are unsuccessful at the above stage will be reformulated in the Microfluidic Assembly stage and go through the iterative processes. If successful in animal models, the selected nanomedicine will then be manufactured through parallelized microfluidic platforms. The pressure and flow patterns in the integrated microfluidic system are regulated by high-precision control systems.

conditions in the laboratory. This reliance of NPs on such nonstandard multistep processes inevitably causes high batch-to-batch variations in their physico-chemical properties [64–69]. Batch size is also subject to custom protocols that vary among laboratories, leading to difficulties in screening and identifying optimal NP physico-chemical characteristics for enhanced drug delivery. Furthermore, the introduction and combination of multiple materials for creating multicomponent NPs compromises the expected functionality of the individual elements. This is largely because of an inability to precisely control the continuous assembly process in various conventional bulk syntheses that involve the macroscopic mixing of precursor solutions [58, 70]. As the micrometer- and nanometer-scale interactions of precursors will direct the characteristics of NPs, it is essential that their composition is fine-tuned in order to attain the anticipated functionalities of multicomponent NP assemblies. In general, the central challenge for the synthesis of multicomponent NPs is to establish large-scale and continuous manufacturing methodologies with high reproducibility.

Amphiphilic blocks self-assemble spontaneously into NPs through size-dependent formation mechanisms and on timescales governed by diffusion [71]. The physico-chemical properties of NPs are, at least in part, determined by the timescales at which the multiple solutions mix in the system [72], as well as the thermodynamic characteristics of the block polymers [73]. Thus, a mixing timescale that is longer than the characteristic time for chemical chain formation will result in an uncontrolled aggregation due to incomplete solvent change. Conversely, a complete solvent change through shorter mixing times in rapid precipitations can result in stable assembly kinetics that lead to the production of homogeneous NPs [74]. One critical difference between conventional bulk synthesis and microfluidic assembly is the mixing time, which occurs on the order of seconds in bulk synthesis and contrasts with those in the millisecond and microsecond range in microfluidic assembly [75]. This shorter mixing time results in more homogeneous NPs by reducing the aggregation of precursors, leading to high reproducibility, which in turn prevents the subsequent thermal and mechanical agitation needed in conventional bulk synthesis for NP homogenization. Therefore, a precise control of microfluidic flow patterns with tunable characteristic mixing times will offer a better understanding of the effect of the mixing time on NP reproducibility and homogeneity.

Microfluidic technologies have demonstrated a better control over effective mixing of the precursor solutions for assembling a range of NP types (Fig. 2a) when compared to conventional bulk methods, due to the larger contact surface areas given per unit volume of fluid in microfluidics [58, 70]. For example, typical laminar flows in microfluidics enabled the controlled syntheses of several NPs (Fig. 2b), including liposomes [76–78] and polymeric NPs [75, 79, 80], with a narrower size distribution compared to those of conventional bulk synthesis. Under laminar flow conditions at a low Reynolds number (~1),[1] mixing occurs only through diffusion across the interface between two miscible fluids moving next to each other in viscous flows. Unfortunately, NP synthesis by diffusive mixing does not allow for the development

[1] The Reynold's number is a dimensionless number that provides a ratio of inertial to viscous forces to quantify the relative importance of these two types of force for given flow conditions.

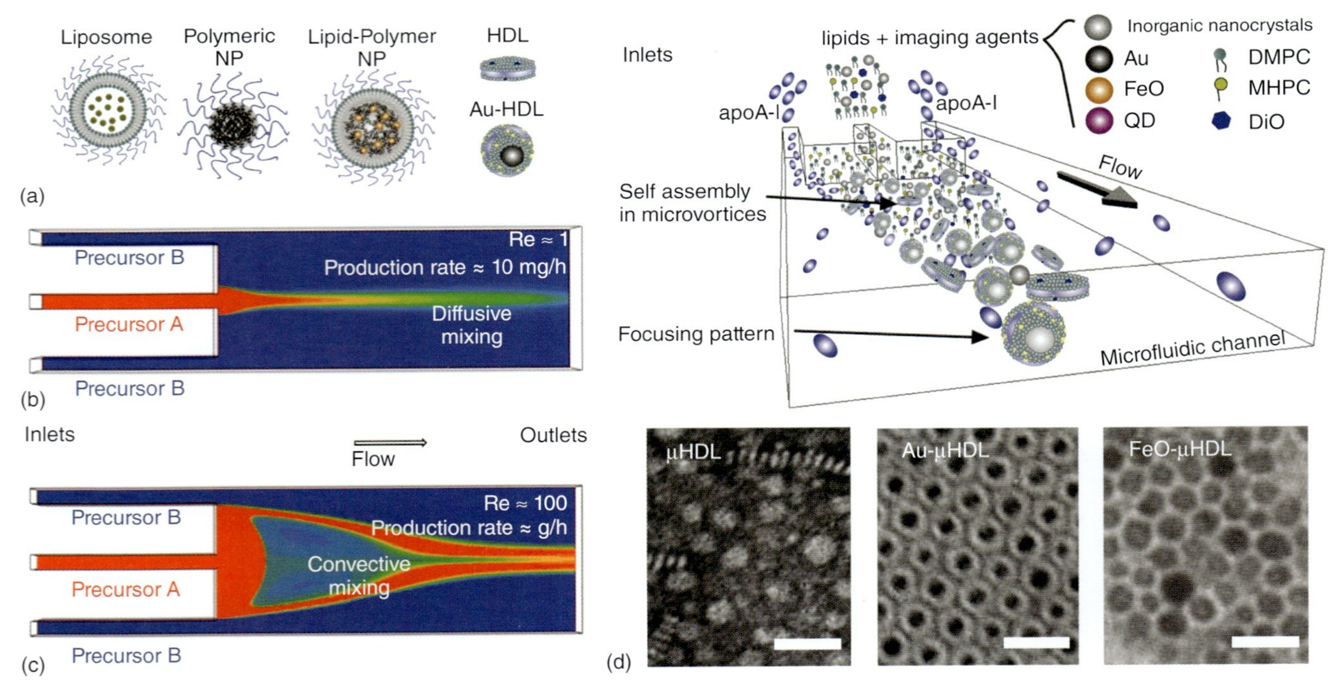

Fig. 2 (a) Representative nanomedicine types (liposome, polymeric NP, lipid-polymer NP, and high-density lipoprotein (HDL)) that have been assembled by microfluidic technology; (b) A microfluidic channel that creates diffusive mixing across laminar flow interfaces at Re ~ 1; (c) A microfluidic channel that creates convective mixing between two precursors at Re ~ 100; (d) Microfluidic reconstitution of HDL (referred to as ■ μHDL) with inorganic nanocrystals, such as gold (Au) and iron oxide (FeO), using controlled microvortices. The transmission electron microscopy images show μHDL, Au-μHDL, and FeO-μHDL (scale bars = 20 nm). Reproduced with permission from Ref [100].

of materials such as lipid–polymer hybrid NPs [81], which require a strong mixing of solutions in the aqueous and organic phases. Lipid–polymer hybrid NPs have shown a higher drug loading within the polymer core and a slower drug release due to the lipid shell when compared to pure polymeric NPs [82, 83]. Furthermore, the diffusive mixing required is difficult to scale up and ultimately leads to a limited controllability of the precursor mixing time, thereby restricting NP homogeneity and leading to a low-throughput production of NPs.

One approach to facilitate precursor mixing (i.e., shortening the mixing times) is to use convective mixing, thereby increasing the interfacial surface area between fluids and reducing the diffusion length scales. Whereas, conventional microfluidic systems exploit easy-to-control flow patterns, which are strictly laminar at low Re values (~1), an increase in Re ($10 < Re < 30$) generates complex flow patterns under a variety of geometric conditions of the microfluidic channel, such as local microvortices and flow separation due to an increase in inertial forces [84–86]. In order to implement convective mixing in microfluidic devices, microfluidic platforms have been designed for the rapid mixing of fluids using relatively higher inertial forces in localized regions with moderate Re values ($10 < Re < 100$) [87–93]. Furthermore, microvortices have demonstrated the ability to rapidly manipulate, sort, and excite particles in microfluidics [94–97]. Recently, a new generation of three-dimensional (3D) focusing patterns in a simple, single-layer microfluidic channel has allowed the development of a pattern-controllable microvortex platform (Fig. 2c). This device has been used for the highly reproducible synthesis of lipid–polymer hybrid NPs with multiple drugs and imaging agents [98, 99], and multifunctional high-density lipoprotein-derived NPs [100] with high productivity (up to $1\,g\,h^{-1}$) (Fig. 2d).

3
Microfluidic Characterization of Nanomedicines

The most important properties of NPs to be characterized before probing their interaction with biological systems are size, shape, surface chemistry/charge, and stability. The development of novel NP characterization tools will impact heavily on nanomedicine, as the lack of characterization standards and quality control tools for NPs has inhibited their clinical adoption to date. One practical obstacle to the clinical-scale commercialization of NPs is an inability to certify the stability of formulations, as even small property variations will have significant effects on *in-vivo* distribution, causing unpredictable therapy outcomes. Recently, several studies have been conducted on NP quality evaluation in microfluidics. For example, a rapid liposome quality assessment in microfluidics allows for quantitative results on liposome formulation composition and stability using dielectric spectroscopy and multivariate data analysis methods [101]. Instantaneous immobilization by ultrarapid cooling in microfluidics reveals the formation of nonequilibrium liposomes in detail [102]. Yet, in spite of recent advances in the label-free characterization of single NPs [60], it remains difficult to effectively integrate a high-throughput microfluidic technology capable of detecting NPs (or their motion) with currently available characterization equipment that includes dynamic light scattering (DLS) [103], transmission electron microscopy

(TEM) [104], atomic force microscopy (AFM) [105], Auger electron spectroscopy (AES) [106], nuclear magnetic resonance (NMR) [107], and flow cytometry [108].

Meanwhile, recent advances in nanofabrication and microfluidics have allowed for the development of high-throughput devices capable of characterizing NP properties, including size and surface charge. The most common electrical technique is to probe impedance changes in nanowire-embedded microfluidic and nanofluidic

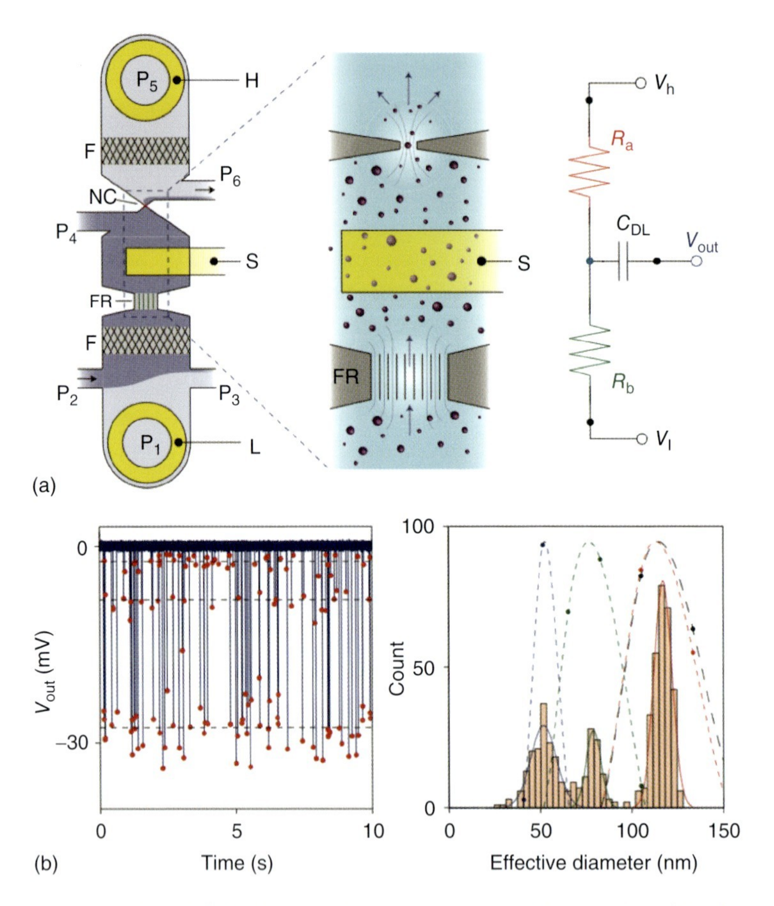

Fig. 3 (a) Overall device layout (left) depicting the electrical and fluidic components: external voltage bias electrodes (H, L) and sensing electrode (S); embedded filters (F); fluid resistor (FR); nanoconstriction (NC); pressure-regulated fluidic ports (P_1~P_6). The nanoparticle (NP) suspension enters at P2 and exits at P6. A detailed image of the dashed box area in (a) (middle) shows the key sensing parts. While NPs flow in the direction of the arrows, changes in the electrical potential in the NC are detected by the electrode S. Electrical circuit expression of the device (right): a constant bias voltage V_h (V_l); Resistors R_a and R_b represent the resistance of the nanoconstriction and the fluidic resistor, respectively; (b) An example analysis of a NP mixture with polydispersity. Left: Output voltage over time for a mixture of NPs of different diameters. Events marked with red circles cluster around three values of V (horizontal dashed black lines). Right: Histogram of effective diameters (40 s measurements). Reproduced with permission from Ref. [117].

channels, and to detect any perturbation in the local electrical properties of the nanowires in response to disturbance by NP solutions. For example, nanowire field-effect transistors allowed for the real-time sensitive detection of label-free molecules [109–111]. A combination of these technologies with nanoscale mechanical systems offers real-time, high-precision, single-molecule/NP/cell detectors, such as advanced mass spectrometry [112] and microfluidic and nanofluidic channel resonators [113–115].

With recent advances in the fundamental physical chemistry of nanoscale pore sensors, several pore-based sensors have been developed in the nanoscale range, offering a rapid and specific, yet simple, biosensing strategy with an improved measurement sensitivity over a wide particle size range [116]. An example of this is a high-throughput microfluidic analyzer that has been developed to detect and characterize unlabeled NPs in a multicomponent mixture at a rate of 500 000 particles per second (Fig. 3) [117]. In this case, a real-time single-nucleotide detection of a model G487A mutation (which is responsible for glucose-6-phosphate dehydrogenase deficiency) was achieved by leveraging the *in-situ* reaction-monitoring capability of the nanopore platform [118]. Tunable pore sensors, which can elastically adjust the size of their pore, have been used to count and detect the size and concentration of smaller NPs compared to other techniques such as flow cytometry [119]. This platform also allows for a simultaneous extraction of the size and zeta-potential of NPs from their charge density, under electrophoretic forces [120]. NP translocation was also detected using a pressure-reversal technique through a cone-shaped nanopore membrane [121]. Today, these approaches, all of which employ size-tunable pore sensors, are starting to provide a better understanding of the fundamental behavior of NPs, as well as a high-throughput characterization of their properties.

4
Microfluidic Evaluation of Nanomedicines

Nanomedicines needs to be nontoxic, biodegradable, sufficiently stable to be delivered to targeted sites, and to have a superior therapeutic advantage over the free drug [122, 123]. Conventionally, nanomedicine evaluation has been made in static cell culture plates, but unfortunately this neglects the important effects of flowing conditions and subsequent transport phenomena on the microenvironment. In contrast to static conditions, flowing conditions assist in the homogeneous distribution of NPs with no gravitational sedimentation, which is similar to the physiological conditions encountered *in vivo*. For example, microfluidic approaches have been used to measure the cytotoxicity of quantum dots (QDs) in a flowing condition [124, 125] and to examine the stability of multicomponent NPs across a laminar flow interface [126]. Compared to tests conducted in conventional static plates, these approaches have provided a more accurate approximation of nanomedicine performance *in vivo*. Microfluidic approaches were also used to evaluate the selective binding of NPs to cells while varying the fluid shear stress, the targeting ligand concentration, receptor expression on target cells, and NP size [127, 128]. The targeted delivery of a nanomedicine represents a powerful technology for the development of safer and more effective therapeutics compared to systemic delivery by nontargeted formulations [83]. Indeed, such approaches show that the targeting performance of an NP can be examined under more physiologically relevant conditions, which is preferential to examining a wide array of cell–particle interactions prior to *in-vivo* experiments. The accurate detection of biomarkers also holds significant promise for "personalized" cancer diagnostics, with more physiologically relevant 3D platforms having been developed for identifying and validating ubiquitous biomarkers [129].

4.1 Mimicking Physiological Environments

Today, an increasing number of engineered NPs requires a reliable high-throughput screening methodology with more physiologically accurate conditions. Whereas, microfluidic approaches have demonstrated the potential to closely approximate physiological environments, current preclinical studies on drug candidates mostly rely on costly and highly variable animal models, mainly because existing cell culture models fail to recapitulate the organ-level pathophysiology of humans. This lack of accurate predictive models highlights the need for better approaches to mimic the structure and function of cells, tissues and organs, as well as the dynamically changing environments *in vivo*. Recently, the evolution of microfluidics has witnessed the integration of *in-vitro* cellular approaches onto chips, which allow real-time, *in-vitro* microscopic observations to be made as well as an evaluation of cell function [130]. But, in order to probe the targeting, therapeutic and diagnostic efficacy of NPs in spatially and temporally regulated environments, it is important first to examine how the NPs interact with cells, tissues and organisms under more physiologically realistic conditions [131, 132]. Consequently, the microfluidic approaches for replicating organ-level structure and function will be discussed at this point (Fig. 4a) [133], and current applications and potentials for the *in-vitro* evaluation of nanomedicines highlighted.

4.2 Endothelial Cell Systems

The vascular endothelium is a crucial target for therapeutic intervention in pathological processes that include inflammation, atherosclerosis, and thrombosis. Endothelial cells exist under dynamically changing mechanical stresses that are generated by blood flow patterns.

Fig. 4 (a) Schematic of microengineered biomimetic systems with spatiotemporal control over physiological effectors, including mechanical cues, chemical factors, electrical signals, multilayered platform with 3D scaffold; (b) Schematic depiction of a biomimetic model that mimics the function of the blood–brain barrier. b.END3 brain endothelial cells and C8-D1A astrocyte cells are cultured on either side of a porous membrane between two microfluidic flow chambers; (c) Schematic of a microfluidic model that mimics the permeable endothelium in artery-surrounding microvessels. Permeability is detected using microelectrodes embedded in the chip. The fluorescent image shows the disrupted endothelial connections. Adherens junctions are shown in green, and nuclei in blue (scale bar = 20 μm). Schematic and TEM image of nanoparticles used for NP translocation studies in the chip (scale bar = 100 nm); (d) Schematic of a lung-on-a-chip device showing IL-2-induced pulmonary edema (scale bar in contrast image = 200 μm). The graph shows barrier permeability in response to IL-2, with and without cyclic strain. Error bars indicate SEM. The fluorescent images show that immunostaining of epithelial occulidin (green) and vascular endothelial cadherin (VE-cadherin; red) with 10% strain with and without IL-2 (scale bars = 30 μm). Reproduced with permission from Refs [133, 143, 149, 152].

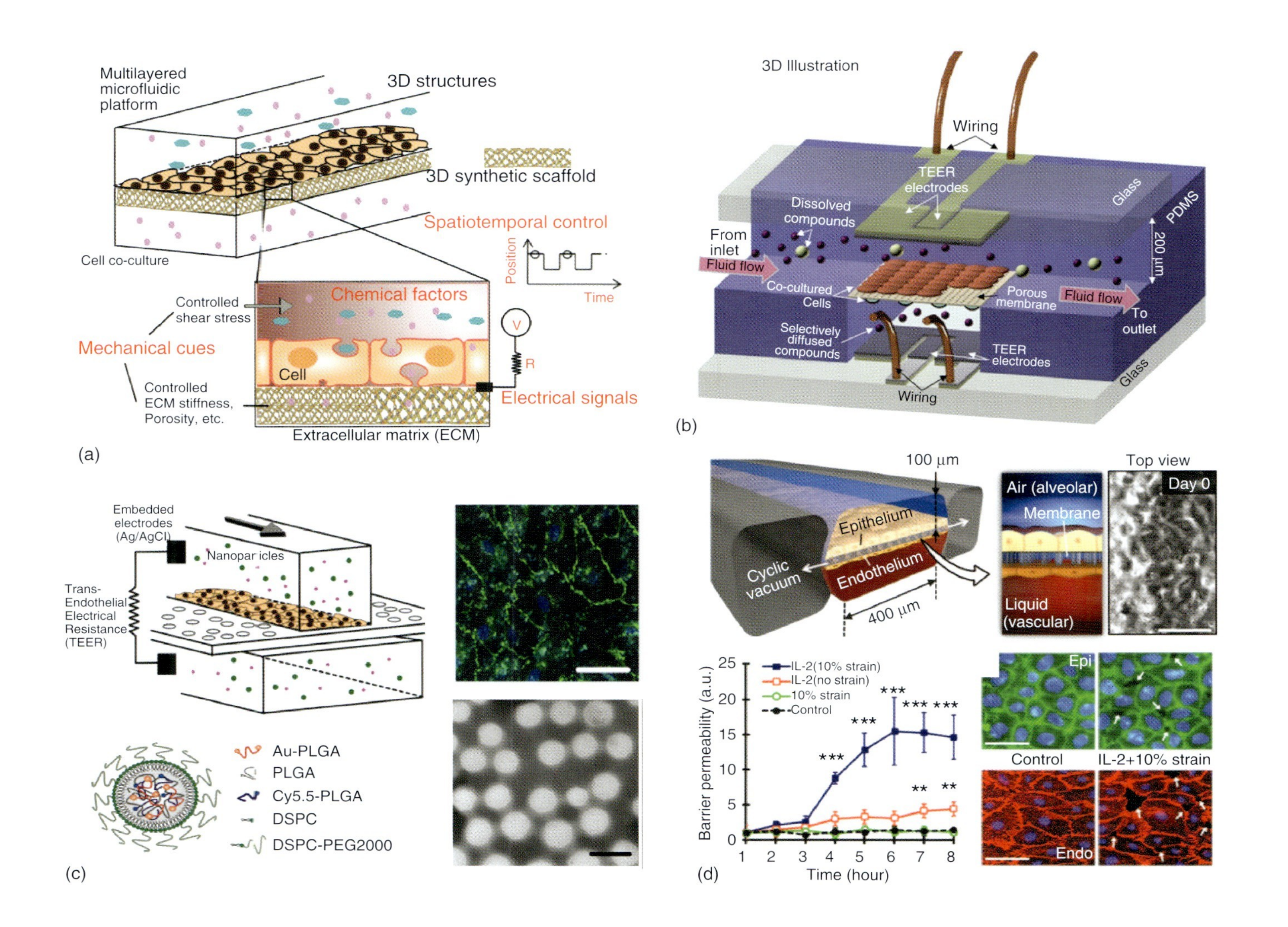
Multilayered microfluidic platform
3D structures
3D synthetic scaffold
Cell co-culture
Spatiotemporal control
Position
Time
Chemical factors
Controlled shear stress
Mechanical cues
Cell
Controlled ECM stiffness, Porosity, etc.
V
R
Electrical signals
Extracellular matrix (ECM)
(a)
3D Illustration
Wiring
TEER electrodes
Glass
PDMS
Dissolved compounds
From inlet
Fluid flow
200 µm
Co-cultured Cells
Porous membrane
Fluid flow
To outlet
Selectively diffused compounds
TEER electrodes
Glass
Wiring
(b)
Embedded electrodes (Ag/AgCl)
Nanopar icles
Trans-Endothelial Electrical Resistance (TEER)
Au-PLGA
PLGA
Cy5.5-PLGA
DSPC
DSPC-PEG2000
(c)
100 µm
Top view
Epithelium
Air (alveolar)
Membrane
Day 0
Cyclic vacuum
Endothelium
400 µm
Liquid (vascular)
IL-2(10% strain)
IL-2(no strain)
10% strain
Control
Barrier permeability (a.u.)
Time (hour)
Epi
Control
IL-2+10% strain
Endo
(d)

Yet, endothelial cell monolayers cultured in conventional multiwell plates fail to reproduce the complex architecture of a vascular network *in vivo* and thus fail to capture the relationship between shear stress experienced by the cells and the local concentration of the drug used. Recent developments in microengineered vascular systems have shown the potential for evaluating nanomedicines under physiologically realistic conditions. For example, replicating the structure and function of blood vessels *in vitro* can be helpful for investigating NP behavior and interaction in and around the targeted sites [134–138]. An accurate reconstitution of the geometric configuration of natural blood vessels is also important, as the interactive effects between blood flow and drug concentration were not captured by a rectangular channel coated with endothelial cells. Rather, a branching network with tubular channels was constructed in order to reproduce these effects [139]. In addition, microvessels supported by the extracellular matrix (ECM), when patterned in a tubular structure, establish the endothelial monolayer, maintain permeability, and are not prone to delamination (which was relatively common in the rectangular channels). The rectangular channels were also very susceptible to delamination that disrupted local permeability, due mainly to the poor connections between the endothelial cells and ECM at the sharp corners [140, 141].

Several additional methods to reproduce microvessels accommodating multiple cells (endothelial cells, pericytes, and astrocytes) have allowed the development of microfluidic models of the blood–brain barrier (BBB) (Fig. 4b) [142–148], as well as for endothelial dysfunction and permeability control in atherosclerosis (Fig. 4c) [149–151].

4.3 "Organ-On-A-Chip"

In combining microfabrication techniques with tissue engineering, the "lung-on-a-chip" device offered a novel *in-vitro* approach to drug screening by mimicking the mechanical and biochemical activities of the human lung (Fig. 4d) [152]. For example, a recent study using this device revealed that mechanical strains associated with physiological breathing movements play an essential role in the development of the increased vascular leakage which leads to pulmonary edema, and that circulating immune cells are not necessary for this disease to develop. The same studies also led to the identification of potential new therapeutics, including angiopoietin-1 (Ang-1) and a new transient receptor potential vanilloid 4 (TRPV4) ion channel inhibitor (GSK2193874), which might prevent the severe toxicity associated with interleukin (IL)-2 therapy. An *in-vitro* model of the intestine has also been developed, together with its crucial microbial symbionts [153, 154]. Whereas, previous *in-vitro* models of intestinal function depend on the use of epithelial cell lines (e.g., Caco-2 cells) which create polarized epithelial monolayers but fail to mimic human intestinal functions for drug development. The recent development of a "gut-on-a-chip" device recreated the gut microenvironment with low shear stress (0.02 dyne cm^{-2}) with cyclic strain (10%; 0.15 Hz) that mimicked physiological peristaltic motions. This precise regulation allowed for an increased exposure of the intestinal surface area and a robust 3D intestinal villi morphogenesis, which mimicked the enhanced cytochrome isoform-based drug-metabolizing activity and the absorptive efficiency of the human intestine.

4.4 Renal Toxicity and Hepatotoxicity

A further use of biomimetic microfluidic platforms for nanomedicine is to evaluate renal toxicity and hepatotoxicity in preclinical studies. Renal excretion represents a clearance pathway for the removal of molecules from vascular compartments, during which time the circulating NPs enter the glomerular capillary and undergo a size-dependent filtration. Those NPs smaller than the pore size of glomerular filtration (~5 nm) can be filtered to enter the proximal tubule, where the brush border of the epithelial cells is negatively charged. As a consequence, positively charged NPs are readily resorbed from the luminal space compared to the negatively charged NPs. The recent development of a microfluidic device lined by human kidney epithelial cells that can be exposed to a shear rate demonstrated a significant increase in albumin transport, glucose reabsorption and brush border alkaline phosphatase activity, all of which are crucial functions of the human kidney proximal tubule [155]. This approach also confirmed that cisplatin toxicity and Pgp efflux transporter activity detected on-chip more closely mimicked the *in-vivo* responses than those obtained with cells maintained on conventional culture plates.

It should be noted that any NPs which are not cleared via the kidney are excreted via the hepatobiliary system. The hepatocytes, which are referred to as potential sites for toxicity, play an important role in liver clearance through endocytosis and the enzymatic breakdown of NPs. As NPs between 10 and 20 nm in size are efficiently eliminated via the liver, any NPs designed in this size range must be modified in order to avoid their prolonged retention in the liver as they undergo excretion. Recently, microfluidic devices have been developed that reconstitute the function of the hepatocytes; for example, a 3D hepatocyte chip has been fabricated for *in-vitro* drug toxicity examinations aimed at predicting drug hepatotoxicity *in-vivo* [156]. This device allowed for the controlled delivery of multiple drug doses to functional primary hepatocytes, while an incorporated concentration gradient generator created *in vitro* dose-dependent drug responses in order to predict *in-vivo* hepatotoxicity.

4.5 Live Tissue Explants

The introduction of *ex-vivo* live tissue explants into microfluidics may provide additional physiological conditions to be investigated. For example, embryonic tissues excised from live frog embryos were used to examine dynamic responses to time-varying chemical stimuli (Fig. 5a) [157]. Carcinoma tumor biopsies were also introduced into a reproducible glass microfluidics system to study the tumor environment, thus offering a preclinical model for the creation of "personalized" treatment regimens [158]. The culturing of brain tissue slices on transistor arrays fabricated on silicon chips may also become a novel platform for neurophysiological and pharmacological studies. Typical microfluidics and semiconductor technologies can be integrated to produce high-resolution, planar transistor arrays for mimicking neuronal structures in long-term studies of topographic mapping [159], and also for mapping evoked extracellular field potentials in organotypic brain slices of rat hippocampus [160].

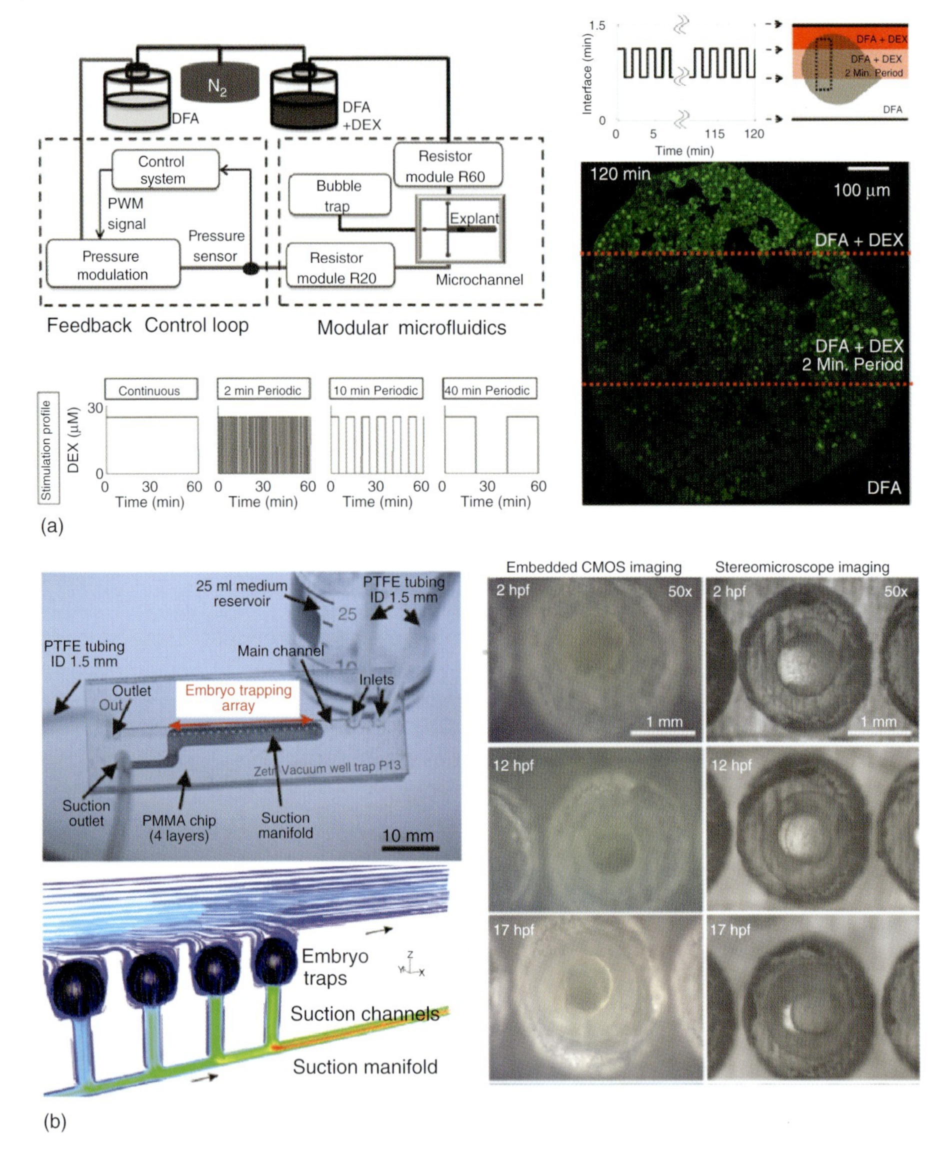

Fig. 5 (a) Schematic of feedback control system that allows for long-term culture of an embryonic tissue excised from a live frog embryo within a microfluidic channel to examine dynamic responses to spatially and temporally varying chemical stimuli. The fluorescent image show the distinct localized responses of an embryonic tissue to dynamic stimuli in a single tissue explant for 2 h, showing that stimulation with localized bursts versus continuous stimulation can result in highly distinct responses. This platform can be used to investigate the cell intercommunication in response to localized drug stimulation for toxicity tests; (b) Schematic of microfluidics-based chip integrating embedded electronic interfaces, and CFD simulation for the flow stream line prediction. Embryos cultured in the device that allowed for immobilization, culture, and treatment of developing zebrafish embryos for toxicity tests. Reproduced with permission from Refs [157, 167].

4.6 Intact Organisms

Small multicellular organisms such as nematodes, fruit flies, clawed frogs and zebrafish, allow for toxicological screening in the normal physiological environments of intact organisms, providing substantial advantages over cell lines and extracted tissues [161–163]. While the fully automated analysis of these model systems in a high-throughput manner remains challenging, the application of microfluidics to these model organisms has demonstrated the ability to handle multicellular organisms in an efficient manner and to precisely manipulate the local conditions to allow for the assessment and imaging of these small organisms [163–166]. For example, manipulating small organisms, such as the worm *Caenorhabditis elegans* , allowed the observation of neuronal responses in order to correlate the activity of sensory neurons with the worm's behavior *in vivo* [161]. Moreover, the integration of embedded electronic interfaces with microfluidic chip-based technologies allowed for the automatic immobilization, culture, and treatment of developing zebrafish embryos during fish embryo toxicity (FET) biotests (Fig. 5b) [167].

To capture the interactions between multiple organs on microfluidic chips would potentially enable a more accurate model of how organs function and interact with one another for potential drug development applications [168, 169]. For example, the combination of a mathematical model and a multiorgan approach provided a novel platform with improved predictability for testing the toxicity of an anticancer drug, 5-fluorouracil, in a pharmacokinetics-based manner [170]. These approaches can help to achieve a better insight into the mechanisms of action of drug candidates, perhaps leading to patient-specific therapies in the future [171]. While multiorgan chips have the potential to simulate human body functions for patient-specific point-of-care devices and therapies, the inherent complexity of each organ itself hinders the development of reliable "human-on-a-chip" model systems. For example, the practical challenges include an optimization of organ size, the control of fluid volumes, the maintenance of coupled organ systems, and the development of a universal blood substitute [172]. The key question for building multiorgan systems is how to simplify the organ complexity without losing physiological accuracy.

5 Challenges and Opportunities

A new nanomedicine development pipeline using microfluidics technologies includes microfluidic assembly, characterization, evaluation, and the manufacture of nanomedicines. These technologies will allow the robust supply of highly reproducible nanomedicines to the entire development process and thereby increase the success rates in clinical trials. In addition to the stages discussed above, microfluidics technology for nanomedicine manufacture is key to the successful translation of a nanomedicine from the laboratory to the clinic. A long-term vision for the manufacture of nanomedicines is to create reliable, continuous and scalable assembly methodologies for a variety of multifunctional NPs with high reproducibility, yield, and homogeneity [173]. The development of these assembly methods requires microfluidic approaches to allow for an efficient and strong mixing of precursors, modular methods for incorporating multicomponents (e.g., therapeutic compounds, imaging agents, targeting ligands, etc.) into multifunctional NPs, and automated control systems for the large-scale integration and parallelization of microfluidic modules [117, 174]. The ability to integrate microfluidics with dynamics, control, and more complex microfabrication techniques opens the door for high-throughput, automated manufacturing.

A current challenge of nanomedicine manufacture using microfluidics is to optimize and maximize microfluidic devices with tunable mixing flow patterns that are applicable to the synthesis of a wide range of nanomedicine types, without losing the physico-chemical properties of the designed nanomedicine [175]. The key technologies required for the development of these nanomedicine manufacturing techniques are computational fluid dynamics to allow simulation of mixing flow patterns in microfluidic devices, highly reliable microfabrication techniques capable of integrating microscale pumps, valves and detecting sensors [176], and high-precision control systems that regulate parallelization and automation capability [177–180].

While the quantities of NPs synthesized by microfluidic devices are often in the microgram to milligram range, the parallelization of microfluidic channels has the potential to scale-up the synthesis by several orders of magnitude to a clinical scale of grams to kilograms. With parallel and stackable microfluidic systems, gram to kilogram scales of NPs could be prepared with the same properties as those prepared at the bench scale, as long as a precise control of either flow rates or pressure in the microfluidic platform is achieved. Pressure control is far better than flow control for controlling the flow rate into a microfluidic network because the flow rate is proportional to the inlet pressure, which can be easily measured for high bandwidth feedback control [181]. To maintain a precise control over fluid pressure with the potential to scale-up production, it is necessary to isolate the pressure-regulating mechanism from the fluid reservoir and the microfluidic device, so that larger reservoir volumes and diverse microfluidic devices can be used independently and integrated as needed. Three important features should be considered for robust and reliable parallelization:

- Fouled modules must be easily replaced or disconnected from other systems.
- Unexpected disturbances due to air bubble formation in microfluidic devices need to be compensated, as this increases the hydraulic resistance between neighboring devices in the parallelized network, leading to an imprecise regulation of the entire system.
- Bridging or networking channels that connect microfluidic modules need to be well designed with minimal secondary flows, which may affect the main bulk flow streams, leading to chip-to-chip variations.

While the microfluidic assembly of nanomedicine has demonstrated much progress using several platforms and various mixing patterns, practical development has been significantly constrained due to a lack of tools capable of detecting, characterizing, and analyzing NPs in a high-throughput manner. Although single NP detectors and characterization tools using microfluidics and nanofabrication technologies have been demonstrated, these are still limited to specific solutions and NP types and need to be generalized in order to function with multicomponent NPs. In addition, there is a need for a technique allowing the easy preparation of highly concentrated NP samples to be developed, including the purification of toxic solvents via either separation or filtration. Furthermore, while many approaches have shown the promise of combining optical systems with microfluidics [60], the precise and reliable control of light delivery to targeted areas in microfluidic platforms remains an active topic of research and engineering. A combination of microfluidics and optics – termed optofluidics – has demonstrated a synergistic effect for new capabilities in several applications including lens, colloidal suspensions, and flow cytometry [182–185]. In addition, optofluidics technology could be employed to incorporate the microfluidic characterization capabilities into NP production platforms in a high-throughput fashion.

Reliability of biomimetic microsystems that mimic the structure and function of human organs for nanomedicine evaluation is crucial. Current challenges include the reliability of long-term cultivations of multiple cell types [186], as well as real-time monitoring of nanomedicines, cellular response to nanomedicines, and critical chemical cues (e.g., reactive oxygen stress) in 3D microenvironments [187–189]. In addition, the development of synthetic biomaterials remains a critical topic of research for physiological accuracy and niches for specific cells, tissues and organs. Furthermore, the development of *in vitro* model systems that can accurately replicate the structure and function of *in-vivo* systems necessitates a precise 3D control of dynamically changing properties, such as mechanical properties of the ECM, at a scale comparable to human cells, tissues and organs [132, 172, 190].

6
Concluding Remarks

Microfluidics in nanomedicine has demonstrated the ability to overcome critical issues with conventional approaches used for nanomedicine development. When combined with advanced nanofabrication, synthetic biomaterials and high-precision control systems, microfluidic technologies constitute a novel platform capable of replacing the entire nanomedicine production process in a scalable manner. Although microfluidics as applied to nanomedicine is still in its infancy, it will surely continue to expand to provide innovative systems at industrially relevant scales in the near future.

Acknowledgments

The authors thank the members of their laboratories for participating in stimulating discussions of these investigations.

References

1. Peer, D., Karp, J.M., Hong, S., Farokhzad, O.C., *et al.* (2007) Nanocarriers as an emerging platform for cancer therapy. *Nat. Nanotechnol.* **2** (12), 751–760.
2. Namiki, Y., Fuchigami, T., Tada, N., Kawamura, R., *et al.* (2011) Nanomedicine for cancer: lipid-based nanostructures for drug delivery and monitoring. *Acc. Chem. Res.* **44** (10), 1080–1093.
3. Seigneuric, R., Markey, L., Nuyten, D.S., Dubernet, C., *et al.* (2010) From nanotechnology to nanomedicine: applications to cancer research. *Curr. Mol. Med.* **10** (7), 640–652.
4. Lobatto, M.E., Fuster, V., Fayad, Z.A., Mulder, W.J. (2011) Perspectives and opportunities for nanomedicine in the management of atherosclerosis. *Nat. Rev. Drug Discovery* **10** (11), 835–852.

5. Psarros, C., Lee, R., Margaritis, M., Antoniades, C. (2012) Nanomedicine for the prevention, treatment and imaging of atherosclerosis. *Nanomedicine* **8** Suppl. 1, S59–S68.
6. Schiener, M., Hossann, M., Viola J.R., Ortega-Gomez, A., *et al.* (2014) Nanomedicine-based strategies for treatment of atherosclerosis. *Trends Mol. Med.* **20** (5), 271–281.
7. Sung, H.W., Sonaje, K., Feng, S.S. (2011) Nanomedicine for diabetes treatment. *Nanomedicine* **6** (8), 1297–1300.
8. Pickup, J.C., Zhi, Z.L., Khan, F., Saxl, T., *et al.* (2008) Nanomedicine and its potential in diabetes research and practice. *Diabetes Metab. Res. Rev.* **24** (8), 604–610.
9. Sonaje, K., Lin, K.J., Wey, S.P., Lin, C.K., *et al.* (2010) Biodistribution, pharmacodynamics and pharmacokinetics of insulin analogues in a rat model: oral delivery using pH-responsive nanoparticles vs. subcutaneous injection. *Biomaterials* **31** (26), 6849–6858.
10. Taratula, O., Kuzmov, A., Shah, M., Garbuzenko, O.B., *et al.* (2013) Nanostructured lipid carriers as multifunctional nanomedicine platform for pulmonary co-delivery of anticancer drugs and siRNA. *J. Controlled Release* **171** (3), 349–357.
11. Mansour, H.M., Rhee, Y.S., Wu, X. (2009) Nanomedicine in pulmonary delivery. *Int. J. Nanomed.* **4**, 299–319.
12. Muldoon, L.L., Tratnyek, P.G., Jacobs, P.M., Doolittle, N.D., *et al.* (2006) Imaging and nanomedicine for diagnosis and therapy in the central nervous system: report of the 11th Annual Blood-Brain Barrier Disruption Consortium meeting. *Am. J. Neuroradiol.* **27** (3), 715–721.
13. Sharma, H.S., Sharma, A. (2011) New strategies for CNS injury and repair using stem cells, nanomedicine, neurotrophic factors and novel neuroprotective agents. *Expert Rev. Neurother.* **11** (8), 1121–1124.
14. Morgen, M., Lu, G.W., Du, D., Stehle, R., *et al.* Targeted delivery of a poorly water-soluble compound to hair follicles using polymeric nanoparticle suspensions. *Int. J. Pharm.* 2011, **416** (1), 314–322.
15. Sigfridsson, K., Bjorkman, J.A., Skantze, P., Zachrisson, H. (2011) Usefulness of a nanoparticle formulation to investigate some hemodynamic parameters of a poorly soluble compound. *J. Pharm. Sci.* **100** (6), 2194–2202.
16. Jia, L. (2005) Nanoparticle formulation increases oral bioavailability of poorly soluble drugs: approaches experimental evidences and theory. *Curr. Nanosci.* **1** (3), 237–243.
17. Gao, W., Xiao, Z., Radovic-Moreno, A., Shi, J., *et al.* (2010) Progress in siRNA delivery using multifunctional nanoparticles. *Methods Mol. Biol.* **629**, 53–67.
18. Leuschner, F., Dutta, P., Gorbatov, R., Novobrantseva, T.I., *et al.* (2011) Therapeutic siRNA silencing in inflammatory monocytes in mice. *Nat. Biotechnol.* **29** (11), 1005–1010.
19. Soppimath, K.S., Aminabhavi, T.M., Kulkarni, A.R., Rudzinski, W.E. (2001) Biodegradable polymeric nanoparticles as drug delivery devices. *J. Controlled Release* **70** (1-2), 1–20.
20. Chen, X., Schluesener, H.J. (2008) Nanosilver: a nanoproduct in medical application. *Toxicol. Lett.* **176** (1), 1–12.
21. Huang, X., Jain, P.K., El-Sayed, I.H., El-Sayed, M.A. (2007) Gold nanoparticles: interesting optical properties and recent applications in cancer diagnostics and therapy. *Nanomedicine* **2** (5), 681–693.
22. Sanvicens, N., Marco, M.P. (2008) Multifunctional nanoparticles–properties and prospects for their use in human medicine. *Trends Biotechnol.* **26** (8), 425–433.
23. Jain, K.K. (2003) Nanodiagnostics: application of nanotechnology in molecular diagnostics. *Expert Rev. Mol. Diagn.* **3** (2), 153–161.
24. Torchilin, V.P. (2005) Recent advances with liposomes as pharmaceutical carriers. *Nat. Rev. Drug Discovery* **4** (2), 145–160.
25. Gu, F., Zhang, L., Teply, B.A., Mann, N., *et al.* (2008) Precise engineering of targeted nanoparticles by using self-assembled biointegrated block copolymers. *Proc. Natl Acad. Sci. USA* **105** (7), 2586–2591.
26. Salvador-Morales, C., Zhang, L., Langer, R., Farokhzad, O.C. (2010) Immunocompatibility properties of lipid-polymer hybrid nanoparticles with heterogeneous surface functional groups. *Biomaterials* **30** (12), 2231–2240.
27. Shi, J., Xiao, Z., Votruba, A.R., Vilos, C., *et al.* (2011) Differentially charged hollow core/shell lipid-polymer-lipid hybrid

nanoparticles for small interfering RNA delivery. *Angew. Chem. Int. Ed.* **50** (31), 7027–7031.

28. Bianco, A., Kostarelos, K., Prato, M. (2005) Applications of carbon nanotubes in drug delivery. *Curr. Opin. Chem. Biol.* **9** (6), 674–679.
29. Huang, X., El-Sayed, I.H., Qian, W., El-Sayed, M.A. (2006) Cancer cell imaging and photothermal therapy in the near-infrared region by using gold nanorods. *J. Am. Chem. Soc.* **128** (6), 2115–2120.
30. Gupta, A.K., Gupta, M. (2005) Synthesis and surface engineering of iron oxide nanoparticles for biomedical applications. *Biomaterials* **26** (18), 3995–4021.
31. Mulder, W.J., Koole, R., Brandwijk, R.J., Storm, G., *et al.* (2006) Quantum dots with a paramagnetic coating as a bimodal molecular imaging probe. *Nano Lett.* **6** (1), 1–6.
32. Kratz, F. (2008) Albumin as a drug carrier: design of prodrugs, drug conjugates and nanoparticles. *J. Controlled Release* **132** (3), 171–183.
33. Skajaa, T., Cormode, D.P., Jarzyna, P.A., Delshad, A., *et al.* (2011) The biological properties of iron oxide core high-density lipoprotein in experimental atherosclerosis. *Biomaterials* **32** (1), 206–213.
34. Basta, T., Wu, H.J., Morphew, M.K., Lee, J., *et al.* (2014) Self-assembled lipid and membrane protein polyhedral nanoparticles. *Proc. Natl Acad. Sci. USA* **111** (2), 670–674.
35. Maune, H.T., Han, S.P., Barish, R.D., Bockrath, M., *et al.* (2010) Self-assembly of carbon nanotubes into two-dimensional geometries using DNA origami templates. *Nat. Nanotechnol.* **5** (1), 61–66.
36. Pal, S., Deng, Z., Ding, B., Yan, H., *et al.* (2010) DNA-origami-directed self-assembly of discrete silver-nanoparticle architectures. *Angew. Chem. Int. Ed.* **49** (15), 2700–2704.
37. Farokhzad, O.C., Cheng, J., Teply, B.A., Sherifi, I., *et al.* (2006) Targeted nanoparticle-aptamer bioconjugates for cancer chemotherapy in vivo. *Proc. Natl Acad. Sci. USA* **103** (16), 6315–6320.
38. Xiao, Z., Levy-Nissenbaum, E., Alexis, F., Luptak, A., *et al.* (2012) Engineering of targeted nanoparticles for cancer therapy using internalizing aptamers isolated by cell-uptake selection. *ACS Nano* **6** (1), 696–704.
39. Gianella, A., Jarzyna, P.A., Mani, V., Ramachandran, S., *et al.* (2011) Multifunctional nanoemulsion platform for imaging guided therapy evaluated in experimental cancer. *ACS Nano* **5** (6), 4422–4433.
40. Lobatto, M.E., Fayad, Z.A., Silvera, S., Vucic, E., *et al.* (2010) Multimodal clinical imaging to longitudinally assess a nanomedical anti-inflammatory treatment in experimental atherosclerosis. *Mol. Pharm.* **7** (6), 2020–2029.
41. Venditto, V.J., Szoka, F.C., Jr (2013) Cancer nanomedicines: so many papers and so few drugs! *Adv. Drug Delivery Rev.* **65** (1), 80–88.
42. Duncan, R., Gaspar, R. (2011) Nanomedicine(s) under the microscope. *Mol. Pharm.* **8** (6), 2101–2141.
43. Olofsson, J., Bridle, H., Sinclair, J., Granfeldt, D., *et al.* (2005) A chemical waveform synthesizer. *Proc. Natl Acad. Sci. USA* **102** (23), 8097–8102.
44. Duraiswamy, S., Khan, S.A. (2010) Plasmonic nanoshell synthesis in microfluidic composite foams. *Nano Lett.* **10** (9), 3757–3763.
45. El-Ali, J., Sorger, P.K., Jensen, K.F. (2006) Cells on chips. *Nature* **442** (7101), 403–411.
46. Schimek, K., Busek, M., Brincker, S., Groth, B., *et al.* (2013) Integrating biological vasculature into a multi-organ-chip microsystem. *Lab Chip* **13** (18), 3588–3598.
47. Andersson, H., van den Berg, A. (2004) Microfabrication and microfluidics for tissue engineering: state of the art and future opportunities. *Lab Chip* **4** (2), 98–103.
48. Minteer, S.D., Moore, C.M. (2006) Overview of advances in microfluidics and microfabrication, in: *Methods in Molecular Biology*, vol. **321**, Humana Press, pp. 1–2.
49. Ozaydin-Ince, G., Coclite, A.M., Gleason, K.K. (2012) CVD of polymeric thin films: applications in sensors, biotechnology, microelectronics/organic electronics, microfluidics, MEMS, composites and membranes. *Rep. Progr. Phys. Phys. Soc.* **75** (1), 016501.
50. Shih, S.C., Fobel, R., Kumar, P., Wheeler, A.R. (2011) A feedback control system for

high-fidelity digital microfluidics. *Lab Chip* **11** (3), 535–540.
51. Welch, D., Christen, J.B. (2014) Real-time feedback control of pH within microfluidics using integrated sensing and actuation. *Lab Chip* **14** (6), 1191–1197.
52. Prohm, C., Stark, H. (2014) Feedback control of inertial microfluidics using axial control forces. *Lab Chip* **14** (12), 2115–2123.
53. Valencia, P.M., Farokhzad, O.C., Karnik, R., Langer, R. (2012) Microfluidic technologies for accelerating the clinical translation of nanoparticles. *Nat. Nanotechnol.* **7** (10), 623–629.
54. Capretto, L., Carugo, D., Mazzitelli, S., Nastruzzi, C. (2013) Microfluidic and lab-on-a-chip preparation routes for organic nanoparticles and vesicular systems for nanomedicine applications. *Adv. Drug Delivery Rev.* **65** (11-12), 1496–1532.
55. Hashimoto, M., Tong, R., Kohane, D.S. (2013) Microdevices for nanomedicine. *Mol. Pharm.* **10** (6), 2127–2144.
56. Bhise, N.S., Ribas, J., Manoharan, V., Zhang, Y.S., *et al.* (2014) Organ-on-a-chip platforms for studying drug delivery systems. *J. Controlled Release* **190**, 82–93
57. Lee, J.B., Sung, J.H. (2013) Organ-on-a-chip technology and microfluidic whole-body models for pharmacokinetic drug toxicity screening. *Biotechnol. J.* **8** (11), 1258–1266.
58. Song, Y., Hormes, J., Kumar, C.S. Microfluidic synthesis of nanomaterials. *Small*, 2008 **4** (6), 698–711.
59. Marre, S., Jensen, K.F. (2010) Synthesis of micro and nanostructures in microfluidic systems. *Chem. Soc. Rev.* **39** (3), 1183–1202.
60. Yurt, A., Daaboul, G.G., Connor, J.H., Goldberg, B.B., Ünlü, M.S. (2012) Single nanoparticle detectors for biological applications. *Nanoscale* **4** (3), 715–726.
61. Neeves, K.B., Onasoga, A.A., Wufsus, A.R. (2013) The use of microfluidics in hemostasis: clinical diagnostics and biomimetic models of vascular injury. *Curr. Opin. Hematol.* **20** (5), 417–423.
62. Kuo, C.T., Chiang, C.L., Chang, C.H., Liu, H.K., *et al.* (2014) Modeling of cancer metastasis and drug resistance via biomimetic nano-cilia and microfluidics. *Biomaterials* **35** (5), 1562–1571.
63. Domachuk, P., Tsioris, K., Omenetto, F.G., Kaplan, D.L. (2010) Bio-microfluidics: biomaterials and biomimetic designs. *Adv. Mater.* **22** (2), 249–260.
64. Jonas, A. (1986) Synthetic substrates of lecithin: cholesterol acyltransferase. *J. Lipid Res.* **27** (7), 689–698.
65. Mieszawska, A.J., Mulder, W.J., Fayad, Z.A., Cormode, D.P. (2013) Multifunctional gold nanoparticles for diagnosis and therapy of disease. *Mol. Pharm.* **10** (3), 831–847.
66. Cormode, D.P., Skajaa, T., van Schooneveld, M.M., Koole, R., *et al.* (2008) Nanocrystal core high-density lipoproteins: a multimodality contrast agent platform. *Nano Lett.* **8** (11), 3715–3723.
67. Chorny, M., Fishbein, I., Danenberg, H.D., Golomb, G. (2002) Lipophilic drug loaded nanospheres prepared by nanoprecipitation: effect of formulation variables on size, drug recovery and release kinetics. *J. Controlled Release* **83** (3), 389–400.
68. Boehm, A.L., Martinon, I., Zerrouk, R., Rump, E., *et al.* (2003) Nanoprecipitation technique for the encapsulation of agrochemical active ingredients. *J. Microencapsul.* **20** (4), 433–441.
69. Betancourt, T., Brown, B., Brannon-Peppas, L. (2007) Doxorubicin-loaded PLGA nanoparticles by nanoprecipitation: preparation, characterization and in vitro evaluation. *Nanomedicine* **2** (2), 219–232.
70. Medina-Sanchez, M., Miserere, S., Merkoci, A. (2012) Nanomaterials and lab-on-a-chip technologies. *Lab Chip* **12** (11), 1932–1943.
71. Johnson, B.K., Prud'homme, R.K. (2003) Mechanism for rapid self-assembly of block copolymer nanoparticles. *Phys. Rev. Lett.* **91** (11), 118302.
72. Capretto, L., Cheng, W., Carugo, D., Katsamenis, O.L., *et al.* (2012) Mechanism of co-nanoprecipitation of organic actives and block copolymers in a microfluidic environment. *Nanotechnology* **23** (37), 375602.
73. Zhu, Z. (2013) Effects of amphiphilic diblock copolymer on drug nanoparticle formation and stability. *Biomaterials* **34** (38), 10238–10248.
74. Chen, T., Hynninen, A.P., Prud'homme, R.K., Kevrekidis, I.G., *et al.* (2008) Coarse-grained simulations of rapid assembly kinetics for polystyrene-b-poly(ethylene oxide) copolymers in aqueous solutions. *J. Phys. Chem. B* **112** (51), 16357–16366.

75. Karnik, R., Gu, F., Basto, P., Cannizzaro, C., *et al.* (2008) Microfluidic platform for controlled synthesis of polymeric nanoparticles. *Nano Lett.* **8** (9), 2906–2912.
76. van Swaay, D., deMello, A. (2013) Microfluidic methods for forming liposomes. *Lab Chip* **13** (5), 752–767.
77. Jahn, A., Stavis, S.M., Hong, J.S., Vreeland, W.N., DeVoe, D.L., Gaitan, M. Microfluidic mixing and the formation of nanoscale lipid vesicles. *ACS Nano* 2010 **4** (4), 2077–2087.
78. Hong, J.S., Stavis, S.M., DePaoli Lacerda, S.H., Locascio, L.E., *et al.* (2010) Microfluidic directed self-assembly of liposome-hydrogel hybrid nanoparticles. *Langmuir* **26** (13), 11581–11588.
79. Kolishetti, N., Dhar, S., Valencia, P.M., Lin, L.Q., *et al.* (2010) Engineering of self-assembled nanoparticle platform for precisely controlled combination drug therapy. *Proc. Natl Acad. Sci. USA* **107** (42), 17939–17944.
80. Rhee, M., Valencia, P.M., Rodriguez, M.I., Langer, R., *et al.* (2011) Synthesis of size-tunable polymeric nanoparticles enabled by 3D hydrodynamic flow focusing in single-layer microchannels. *Adv. Mater.* **23** (12), H79–H83.
81. Tan, S., Li, X., Guo, Y., Zhang, Z. (2013) Lipid-enveloped hybrid nanoparticles for drug delivery. *Nanoscale* **5** (3), 860–872.
82. Zhang, L., Chan, J.M., Gu, F.X., Rhee, J.-W., *et al.* (2008) Self-assembled lipid-polymer hybrid nanoparticles: a robust drug delivery platform. *ACS Nano* **2** (8), 1696–1702.
83. Shi, J., Xiao, Z., Kamaly, N., Farokhzad, O.C. (2011) Self-assembled targeted nanoparticles: evolution of technologies and bench to bedside translation. *Acc. Chem. Res.* **44** (10), 1123–1134.
84. Cheng, C.M., Kim, Y., Yang, J.M., Leuba, S.H., *et al.*, (2009) Dynamics of individual polymers using microfluidic based microcurvilinear flow. *Lab Chip* **9** (16), 2339–2347.
85. Kim, Y., Joshi, S.D., Davidson, L.A., LeDuc, P.R., *et al.*, (2011) Dynamic control of 3D chemical profiles with a single 2D microfluidic platform. *Lab Chip* **11** (13), 2182–2188.
86. Kim, Y., Pekkan, K., Messner, W.C., Leduc, P.R. (2010) Three-dimensional chemical profile manipulation using two-dimensional autonomous microfluidic control. *J. Am. Chem. Soc.* **132** (4), 1339–1347.
87. deMello, A.J. (2006) Control and detection of chemical reactions in microfluidic systems. *Nature* **442** (7101), 394–402.
88. Lee, M.G., Choi, S., Park, J.K. (2009) Three-dimensional hydrodynamic focusing with a single sheath flow in a single-layer microfluidic device. *Lab Chip* **9** (21), 3155–3160.
89. Mao, X.L., Lin, S.C.S., Dong, C., Huang, T.J. (2009) Single-layer planar on-chip flow cytometer using microfluidic drifting based three-dimensional (3D) hydrodynamic focusing. *Lab Chip* **9** (11), 1583–1589.
90. Chang, C.C., Huang, Z.X., Yang, R.J. (2007) Three-dimensional hydrodynamic focusing in two-layer polydimethylsiloxane (PDMS) microchannels. *J. Micromech. Microeng.* **17** (8), 1479–1486.
91. Nguyen, N.T., Wu, Z.G. (2005) Micromixers – a review. *J. Micromech. Microeng.* **15** (2), R1–R16.
92. Valencia, P.M., Basto, P.A., Zhang, L., Rhee, M., *et al.* (2010) Single-step assembly of homogenous lipid-polymeric and lipid-quantum dot nanoparticles enabled by microfluidic rapid mixing. *ACS Nano* **4** (3), 1671–1679.
93. Fang, R.H., Chen, K.N., Aryal, S., Hu, C.M., *et al.* (2012) Large-scale synthesis of lipid-polymer hybrid nanoparticles using a multi-inlet vortex reactor. *Langmuir* **28** (39), 13824–13829.
94. Liu, S.J., Wei, H.H., Hwang, S.H., Chang, H.C. (2010) Dynamic particle trapping, release, and sorting by microvortices on a substrate. *Phys. Rev. E Stat. Nonlin. Soft Matter Phys.* **82** (2 Pt 2), 026308.
95. Stott, S.L., Hsu, C.H., Tsukrov, D.I., Yu, M., *et al.* (2010) Isolation of circulating tumor cells using a microvortex-generating herringbone-chip. *Proc. Natl Acad. Sci. USA* **107** (43), 18392–18397.
96. Hsu, C.H., Di Carlo, D., Chen, C., Irimia, D., *et al.* (2008) Microvortex for focusing, guiding and sorting of particles. *Lab Chip* **8** (12), 2128–2134.
97. Shelby, J.P., Lim, D.S.W., Kuo, J.S., Chiu, D.T. (2003) High radial acceleration in microvortices. *Nature* **425** (6953), 38.
98. Kim, Y., Lee, C.B., Ma, M., Mulder, W.J., *et al.* (2012) Mass production and size control of lipid-polymer hybrid nanoparticles through controlled microvortices. *Nano Lett.* **12** (7), 3587–3591.

99. Mieszawska, A.J., Kim, Y., Gianella, A., van Rooy, I., *et al.* (2013) Synthesis of polymer-lipid nanoparticles for image-guided delivery of dual modality therapy. *Bioconjugate Chem.* **24** (9), 1429–1434.
100. Kim, Y., Fay, F., Cormode, D.P., Sanchez-Gaytan, B.L., *et al.* (2013) Single step reconstitution of multifunctional high-density lipoprotein-derived nanomaterials using microfluidics. *ACS Nano* **7** (11), 9975–9983.
101. Birnbaumer, G., Kupcu, S., Jungreuthmayer, C., Richter, L., *et al.* (2011) Rapid liposome quality assessment using a lab-on-a-chip. *Lab Chip* **11** (16), 2753–2762.
102. Jahn, A., Lucas, F., Wepf, R.A., Dittrich, P.S. (2013) Freezing continuous-flow self-assembly in a microfluidic device: toward imaging of liposome formation. *Langmuir* **29** (5), 1717–1723.
103. Hinterwirth, H., Wiedmer, S.K., Moilanen, M., Lehner, A., *et al.* (2013) Comparative method evaluation for size and size-distribution analysis of gold nanoparticles. *J. Sep. Sci.* **36** (17), 2952–2961.
104. Ungureanu, C., Kroes, R., Petersen, W., Groothuis, T.A., *et al.* (2011) Light interactions with gold nanorods and cells: implications for photothermal nanotherapeutics. *Nano Lett.* **11** (5), 1887–1894.
105. Ramachandran, S., Lal, R. (2010) Scope of atomic force microscopy in the advancement of nanomedicine. *Indian J. Exp. Biol.* **48** (10), 1020–1036.
106. Wang, Y., Wu, Q., Sui, K., Chen, X.X., *et al.* (2013) A quantitative study of exocytosis of titanium dioxide nanoparticles from neural stem cells. *Nanoscale* **5** (11), 4737–4743.
107. Vasanthakumar, S., Ahamed, H.N., Saha, R.N. (2014) Nanomedicine I: in vitro and in vivo evaluation of paclitaxel loaded poly-(epsilon-caprolactone), poly (DL-lactide-*co*-glycolide) and poly (DL-lactic acid) matrix nanoparticles in Wistar rats. *Eur. J. Drug Metab. Pharmacokinet.*, in press.
108. Zhu, G., Zheng, J., Song, E., Donovan, M., *et al.* (2013) Self-assembled, aptamer-tethered DNA nanotrains for targeted transport of molecular drugs in cancer theranostics. *Proc. Natl Acad. Sci. USA* **110** (20), 7998–8003.
109. Stern, E., Wagner, R., Sigworth, F.J., Breaker, R., *et al.* (2007) Importance of the Debye screening length on nanowire field effect transistor sensors. *Nano Lett.* **7** (11), 3405–3409.
110. Patolsky, F., Zheng, G., Hayden, O., Lakadamyali, M., *et al.* (2004) Electrical detection of single viruses. *Proc. Natl Acad. Sci. USA* **101** (39), 14017–14022.
111. Sridhar, M., Xu, D., Kang, Y., Hmelo, A.B., *et al.* (2008) Experimental characterization of a metal-oxide-semiconductor field-effect transistor-based Coulter counter. *J. Appl. Phys.* **103** (10), 104701–10470110.
112. Naik, A.K., Hanay, M.S., Hiebert, W.K., Feng, X.L., *et al.* (2009) Towards single-molecule nanomechanical mass spectrometry. *Nat. Nanotechnol.* **4** (7), 445–450.
113. Burg, T.P., Godin, M., Knudsen, S.M., Shen, W., *et al.* (2007) Weighing of biomolecules, single cells and single nanoparticles in fluid. *Nature* **446** (7139), 1066–1069.
114. Lee, J., Shen, W., Payer, K., Burg, T.P., *et al.* (2010) Toward attogram mass measurements in solution with suspended nanochannel resonators. *Nano Lett.* **10** (7), 2537–2542.
115. Lee, J., Chunara, R., Shen, W., Payer, K., *et al.* (2011) Suspended microchannel resonators with piezoresistive sensors. *Lab Chip* **11** (4), 645–651.
116. Kozak, D., Anderson, W., Vogel, R., Trau, M. (2011) Advances in resistive pulse sensors: devices bridging the void between molecular and microscopic detection. *Nano Today* **6** (5), 531–545.
117. Fraikin, J.L., Teesalu, T., McKenney, C.M., Ruoslahti, E., *et al.* (2011) A high-throughput label-free nanoparticle analyser. *Nat. Nanotechnol.* **6** (5), 308–313.
118. Ang, Y.S., Yung, L.Y. (2012) Rapid and label-free single-nucleotide discrimination via an integrative nanoparticle-nanopore approach. *ACS Nano* **6** (10), 8815–8823.
119. Roberts, G.S., Yu, S., Zeng, Q., Chan, L.C., *et al.* (2012) Tunable pores for measuring concentrations of synthetic and biological nanoparticle dispersions. *Biosens. Bioelectron.* **31** (1), 17–25.
120. Kozak, D., Anderson, W., Vogel, R., Chen, S., *et al.* (2012) Simultaneous size and zeta-potential measurements of individual nanoparticles in dispersion using size-tunable pore sensors. *ACS Nano* **6** (8), 6990–6997.

121. Lan, W.J., White, H.S. (2012) Diffusional motion of a particle translocating through a nanopore. *ACS Nano* **6** (2), 1757–1765.
122. Love, S.A., Maurer-Jones, M.A., Thompson, J.W., Lin, Y.S., *et al.* (2012) Assessing nanoparticle toxicity. *Annu. Rev. Anal. Chem.* **5**, 181–205.
123. Kim, S.T., Saha, K., Kim, C., Rotello, V.M. (2013) The role of surface functionality in determining nanoparticle cytotoxicity. *Acc. Chem. Res.* **46** (3), 681–691.
124. Wu, J., Chen, Q., Liu, W., Zhang, Y. (2012) Cytotoxicity of quantum dots assay on a microfluidic 3D-culture device based on modeling diffusion process between blood vessels and tissues. *Lab Chip* **12** (18), 3474–3480.
125. Mahto, S.K., Yoon, T.H., Rhee, S.W. (2010) A new perspective on in vitro assessment method for evaluating quantum dot toxicity by using microfluidics technology. *Biomicrofluidics* **4** (3), 034111.
126. Ozturk, S., Hassan, Y.A., Ugaz, V.M. (2012) A simple microfluidic probe of nanoparticle suspension stability. *Lab Chip* **12** (18), 3467–3473.
127. Farokhzad, O.C., Khademhosseini, A., Jon, S., Hermmann, A., *et al.* (2005) Microfluidic system for studying the interaction of nanoparticles and microparticles with cells. *Anal. Chem.* **77** (17), 5453–5459.
128. Kusunose, J., Zhang, H., Gagnon, M.K., Pan, T., *et al.* (2013) Microfluidic system for facilitated quantification of nanoparticle accumulation to cells under laminar flow. *Ann. Biomed. Eng.* **41** (1), 89–99.
129. Lai, Y., Asthana, A., Kisaalita, W.S. (2011) Biomarkers for simplifying HTS 3D cell culture platforms for drug discovery: the case for cytokines. *Drug Discovery Today* **16** (7-8), 293–297.
130. Hsieh, C.C., Huang, S.B., Wu, P.C., Shieh, D.B., *et al.* (2009) A microfluidic cell culture platform for real-time cellular imaging. *Biomed. Microdevices* **11** (4), 903–913.
131. Zhang, X.Q., Xu, X., Bertrand, N., Pridgen, E., *et al.* (2012) Interactions of nanomaterials and biological systems: implications to personalized nanomedicine. *Adv. Drug Delivery Rev.* **64** (13), 1363–1384.
132. Huh, D., Hamilton, G.A., Ingber, D.E. (2011) From 3D cell culture to organs-on-chips. *Trends Cell Biol.* **21** (12), 745–754.
133. Sei, Y., Justus, K., LeDuc, P., Kim, Y. (2014) Engineering living systems on chips: from cells to human on chips. *Microfluid. Nanofluid.* **16** (5), 907–920.
134. Zheng, Y., Chen, J., Craven, M., Choi, N.W., *et al.* In vitro microvessels for the study of angiogenesis and thrombosis. *Proc. Natl Acad. Sci. USA* 2012 **109** (24), 9342–9347.
135. Borenstein, J.T., Tupper, M.M., Mack, P.J., Weinberg, E.J., *et al.* (2010) Functional endothelialized microvascular networks with circular cross-sections in a tissue culture substrate. *Biomed. Microdevices* **12** (1), 71–79.
136. Srigunapalan, S., Lam, C., Wheeler, A.R., Simmons, C.A. (2011) A microfluidic membrane device to mimic critical components of the vascular microenvironment. *Biomicrofluidics* **5** (1), 13409.
137. Shin, Y., Jeon, J.S., Han, S., Jung, G.-S., *et al.* (2011) In vitro 3D collective sprouting angiogenesis under orchestrated ANG-1 and VEGF gradients. *Lab Chip* **11** (13), 2175–2181.
138. Gunther, A., Yasotharan, S., Vagaon, A., Lochovsky, C., *et al.* (2010) A microfluidic platform for probing small artery structure and function. *Lab Chip* **10** (18), 2341–2349.
139. Zhang, B., Peticone, C., Murthy, S.K., Radisic, M. (2013) A standalone perfusion platform for drug testing and target validation in micro-vessel networks. *Biomicrofluidics* **7** (4), 44125.
140. Esch, M.B., Post, D.J., Shuler, M.L., Stokol, T. (2011) Characterization of in vitro endothelial linings grown within microfluidic channels. *Tissue Eng. Part A* **17** (23-24), 2965–2971.
141. Wong, K.H., Truslow, J.G., Khankhel, A.H., Chan, K.L., *et al.* (2013) Artificial lymphatic drainage systems for vascularized microfluidic scaffolds. *J. Biomed. Mater. Res. Part A* **101** (8), 2181–2190.
142. Cucullo, L., Marchi, N., Hossain, M., Janigro, D. (2011) A dynamic in vitro BBB model for the study of immune cell trafficking into the central nervous system. *J. Cereb. Blood Flow Metab.* **31** (2), 767–777.
143. Booth, R., Kim, H. (2012) Characterization of a microfluidic in vitro model of the blood-brain barrier (muBBB). *Lab Chip* **12** (10), 1784–1792.
144. Culot, M., Lundquist, S., Vanuxeem, D., Nion, S., *et al.* (2008) An in vitro

blood-brain barrier model for high throughput (HTS) toxicological screening. *Toxicol. in Vitro* **22** (3), 799–811.

145. Cucullo, L., McAllister, M.S., Kight, K., Krizanac-Bengez, L., *et al.* (2002) A new dynamic in vitro model for the multidimensional study of astrocyte-endothelial cell interactions at the blood-brain barrier. *Brain Res.* **951** (2), 243–254.
146. Cucullo, L., Hossain, M., Tierney, W., Janigro, D. (2013) A new dynamic in vitro modular capillaries-venules modular system: cerebrovascular physiology in a box. *BMC Neurosci.* **14**, 18.
147. Prabhakarpandian, B., Shen, M.C., Nichols, J.B., Mills, I.R., *et al.* (2013) SyM-BBB: a microfluidic blood brain barrier model. *Lab Chip* **13** (6), 1093–1101.
148. Parkinson, F.E., Friesen, J., Krizanac-Bengez, L., Janigro, D. (2003) Use of a three-dimensional in vitro model of the rat blood-brain barrier to assay nucleoside efflux from brain. *Brain Res.* **980** (2), 233–241.
149. Kim, Y., Lobatto, M.E., Kawahara, T., Lee Chung, B., *et al.* (2014) Probing nanoparticle translocation across the permeable endothelium in experimental atherosclerosis. *Proc. Natl Acad. Sci. USA* **111** (3), 1078–1083.
150. Estrada, R., Giridharan, G.A., Nguyen, M.D., Prabhu, S.D., *et al.* (2011) Microfluidic endothelial cell culture model to replicate disturbed flow conditions seen in atherosclerosis susceptible regions. *Biomicrofluidics* **5** (3), 32006–3200611.
151. Polk, B.J. Stelzenmuller, A., Mijares, G., MacCrehan, W., Gaitan, M. (2006) Ag/AgCl microelectrodes with improved stability for microfluidics. *Sens. Actuators, B* **114**, 239–247.
152. Huh, D., Leslie, D.C., Matthews, B.D., Fraser, J.P., *et al.* (2012) A human disease model of drug toxicity-induced pulmonary edema in a lung-on-a-chip microdevice. *Sci. Transl. Med.* **4** (159), 159ra47.
153. Kim, H.J., Ingber, D.E. (2013) Gut-on-a-chip microenvironment induces human intestinal cells to undergo villus differentiation. *Integr. Biol.* **5** (9), 1130–1140.
154. Kim, H.J., Huh, D., Hamilton, G., Ingber, D.E. (2012) Human gut-on-a-chip inhabited by microbial flora that experiences intestinal peristalsis-like motions and flow. *Lab Chip* **12** (12), 2165–2174.
155. Jang, K.J., Mehr, A.P., Hamilton, G.A., McPartlin, L.A., *et al.* (2013) Human kidney proximal tubule-on-a-chip for drug transport and nephrotoxicity assessment. *Integr. Biol.* **5** (9), 1119–1129.
156. Toh, Y.C., Lim, T.C., Tai, D., Xiao, G., *et al.* (2009) A microfluidic 3D hepatocyte chip for drug toxicity testing. *Lab Chip* **9** (14), 2026–2035.
157. Kim, Y., Joshi, S.D., Messner, W.C., LeDuc, P.R. (2011) Detection of dynamic spatiotemporal response to periodic chemical stimulation in a *Xenopus* embryonic tissue. *PLoS ONE* **6** (1), e14624.
158. Hattersley, S.M., Sylvester, D.C., Dyer, C.E., Stafford, N.D., *et al.* (2012) A microfluidic system for testing the responses of head and neck squamous cell carcinoma tissue biopsies to treatment with chemotherapy drugs. *Ann. Biomed. Eng.* **40** (6), 1277–1288.
159. Hutzler, M., Fromherz, P. (2004) Silicon chip with capacitors and transistors for interfacing organotypic brain slice of rat hippocampus. *Eur. J. Neurosci.* **19** (8), 2231–2238.
160. Besl, B., Fromherz, P. (2002) Transistor array with an organotypic brain slice: field potential records and synaptic currents. *Eur. J. Neurosci.* **15** (6), 999–1005.
161. Chronis, N., Zimmer, M., Bargmann, C.I. (2007) Microfluidics for in vivo imaging of neuronal and behavioral activity in *Caenorhabditis elegans*. *Nat. Methods* **4** (9), 727–731.
162. George, S., Xia, T., Rallo, R., Zhao, Y., *et al.* (2011) Use of a high-throughput screening approach coupled with in vivo zebrafish embryo screening to develop hazard ranking for engineered nanomaterials. *ACS Nano* **5** (3), 1805–1817.
163. Jimenez, A.M., Roche, M., Pinot, M., Panizza, P. (2011) Towards high throughput production of artificial egg oocytes using microfluidics. *Lab Chip* **11** (3), 429–434.
164. Crane, M.M., Chung, K., Stirman, J., Lu, H. (2010) Microfluidics-enabled phenotyping, imaging, and screening of multicellular organisms. *Lab Chip* **10** (12), 1509–1517.
165. Wlodkowic, D., Khoshmanesh, K., Akagi, J., Williams, D.E., *et al.* (2011) Wormometry-on-a-chip: innovative technologies for in

situ analysis of small multicellular organisms. *Cytometry, Part A* **79** (10), 799–813.
166. Shi, W., Wen, H., Lin, B., Qin, J. (2011) Microfluidic platform for the study of *Caenorhabditis elegans. Top. Curr. Chem.* **304**, 323–338.
167. Wang, K.I., Salcic, Z., Yeh, J., Akagi, J., *et al.* (2013) Toward embedded laboratory automation for smart lab-on-a-chip embryo arrays. *Biosens. Bioelectron.* **48**, 188–196.
168. Moraes, C., Mehta, G., Lesher-Perez, S.C., Takayama, S. (2012) Organs-on-a-chip: a focus on compartmentalized microdevices. *Ann. Biomed. Eng.* **40** (6), 1211–1227.
169. Sung, J.H., Esch, M.B., Prot, J.M., Long, C.J., *et al.* (2013) Microfabricated mammalian organ systems and their integration into models of whole animals and humans. *Lab Chip* **13** (7), 1201–1212.
170. Sung, J.H., Kam, C., Shuler, M.L. (2011) A microfluidic device for a pharmacokinetic-pharmacodynamic (PK-PD) model on a chip. *Lab Chip* **10** (4), 446–455.
171. Williamson, A., Singh, S., Fernekorn, U., Schober, A. (2013) The future of the patient-specific body-on-a-chip. *Lab Chip* **13** (18), 3471–3480.
172. Wikswo, J.P., Block, III, F.E., Cliffel, D.E., Goodwin, C.R., *et al.* (2013) Engineering challenges for instrumenting and controlling integrated organ-on-chip systems. *IEEE Trans. Biomed. Eng.* **60** (3), 682–690.
173. Olcum, S., Cermak, N., Wasserman, S.C., Christine, K.S., *et al.* (2014) Weighing nanoparticles in solution at the attogram scale. *Proc. Natl Acad. Sci. USA* **111** (4), 1310–1315.
174. Yang, M., Sun, S., Kostov, Y., Rasooly, A. (2011) A simple 96 well microfluidic chip combined with visual and densitometry detection for resource-poor point of care testing. *Sens. Actuators, B* **153** (1), 176–181.
175. Biswas, S., Miller, J.T., Li, Y., Nandakumar, K. (2012) Developing a millifluidic platform for the synthesis of ultrasmall nanoclusters: ultrasmall copper nanoclusters as a case study. *Small* **8** (5), 687–698.
176. Hong, J.W., Quake, S.R. (2005) Integrated nanoliter systems. *Nat. Biotechnol.* **21** (10), 1179–1183.
177. Kobayashi, I., Mukataka, S., Nakajima, M. (2005) Novel asymmetric through-hole array microfabricated on a silicon plate for formulating monodisperse emulsions. *Langmuir* **21** (17), 7629–7632.
178. Kobayashi, I., Mukataka, S., Nakajima, M. (2005) Effects of type and physical properties of oil phase on oil-in-water emulsion droplet formation in straight-through microchannel emulsification, experimental and CFD studies. *Langmuir* **21** (13), 5722–5730.
179. Nisisako, T., Torii, T. (2008) Microfluidic large-scale integration on a chip for mass production of monodisperse droplets and particles. *Lab Chip* **8** (2), 287–293.
180. Li, W., Greener, J., Voicu, D., Kumacheva, E. (2009) Multiple modular microfluidic (M3) reactors for the synthesis of polymer particles. *Lab Chip* **9** (18), 2715–2721.
181. Kim, Y., LeDuc, P., Messner, W. (2013) Modeling and control of a nonlinear mechanism for high performance microfluidic systems. *IEEE Trans. Control Syst. Technol.* **21** (1), 203–211.
182. Psaltis, D., Quake, S.R., Yang, C.H. (2006) Developing optofluidic technology through the fusion of microfluidics and optics. *Nature* **442** (7101), 381–386.
183. Mao, X., Waldeisen, J.R., Juluri, B.K., Huang, T.J. (2007) Hydrodynamically tunable optofluidic cylindrical microlens. *Lab Chip* **7** (10), 1303–1308.
184. Yang, A.H., Erickson, D. (2010) Optofluidic ring resonator switch for optical particle transport. *Lab Chip* **10** (6), 769–774.
185. Song, C.L., Luong, T.D., Kong, T.F., Nguyen, N.T. (2011) Disposable flow cytometer with high efficiency in particle counting and sizing using an optofluidic lens. *Opt. Lett.* **36** (5), 657–659.
186. Ziolkowska, K., Stelmachowska, A., Kwapiszewski, R., Chudy, M. (2013) Long-term three-dimensional cell culture and anticancer drug activity evaluation in a microfluidic chip. *Biosens. Bioelectron.* **40** (1), 68–74.
187. Lii, J., Hsu, W.J., Parsa, H., Das, A. (2008) Real-time microfluidic system for studying mammalian cells in 3D microenvironments. *Anal. Chem.* **80** (10), 3640–3647.

188. Cheah, L.T., Dou, Y.H., Seymour, A.M., Dyer, C.E., *et al.* (2010) Microfluidic perfusion system for maintaining viable heart tissue with real-time electrochemical monitoring of reactive oxygen species. *Lab Chip* **10** (20), 2720–2726.
189. Richter, L., Charwat, V., Jungreuthmayer, C., Bellutti, F. (2011) Monitoring cellular stress responses to nanoparticles using a lab-on-a-chip. *Lab Chip* **11** (15), 2551–2560.
190. Ahmad, A.A., Wang, Y., Gracz, A.D., Sims, C.E. (2014) Optimization of 3-D organotypic primary colonic cultures for organ-on-chip applications. *J. Biol. Eng.* **8**, 9.

2
Quantum Dots for Biomedical Delivery Applications

Abolfazl Akbarzadeh[*,1,3], *Sedigheh Fekri Aval*[1,2,4], *Roghayeh Sheervalilou*[4,5], *Leila Fekri*[6], *Nosratollah Zarghami*[1,2], *and Mozhdeh Mohammadian*[7]

[1] *Tabriz University of Medical Sciences, Drug Applied Research Center, Daneshgah Street, Tabriz 51656-65811, Iran*
[2] *Tabriz University of Medical Sciences, Department of Medical Biotechnology, School of Advanced Medical Sciences, Golgasht Ave, Azadi Street, Tabriz 51666-14766, Iran*
[3] *Tabriz University of Medical Sciences, Department of Medical Nanotechnology, Faculty of Advanced Medical Sciences, Golgasht Ave, Azadi Street, Tabriz 51666-14766, Iran*
[4] *Tabriz University of Medical Sciences, Advanced Medical Sciences Research Center, Student's Research Committee, Golgasht Ave, Azadi Street, Tabriz 51666-14766, Iran*
[5] *Tabriz University of Medical Sciences, Department of Molecular Medicine, Faculty of Advanced Medical Sciences, Golgasht Ave, Azadi Street, Tabriz 51666-14766, Iran*
[6] *Islamic Azad University, Plasma Physics Research Center, Science and Research Branch, Shahid Sattari Highway - University Square, Shohadaye Hesarak Street, Tehran 1477893855, Iran*
[7] *Mazandaran University of Medical Sciences, Amol Faculty of Paramedical Sciences, Valiasr Street, Sari 48178-44718, Iran*

Translational Nanomedicine. First Edition. Edited by Robert A. Meyers.

Keywords

Lithography
The most common method of quantum dot fabrication.

Quantum dot (QD)
A nanocrystal made from semiconductor materials, exhibiting unique optical properties.

Detection
Identification process of a sign, using tools such as QDs, to provide additional information and superior results.

Nanocrystals
A crystalline material with size in the nanometer range.

SiRNA
Small interfering RNA is a powerful tool for knocking down or gene silencing in most cells.

Quantum dots (QDs) are novel, photostable fluorescent semiconductor nanocrystals. The understanding of quantum-confined electrons in very small particles forms the basis for an understanding of the exceptional features described in this chapter. Lithographically defined QDs, epitaxially self-assembled QDs, and colloidal QDs are methods of QD synthesis. In this chapter, an outline is provided of the unique characteristics of QDs, such as their wide excitation spectra, narrow symmetrical emission spectra, tunable emission, superior brightness, and photostability. These properties have led to QDs becoming an ideal platform for many beneficial applications in a variety of scientific and medical fields.

1 Introduction

During the past decade, new approaches have been devised that involve the ability to fabricate, characterize and manipulate artificial structures, the features of which may be controlled at the nanometer level. These embrace areas of research as diverse as physics, engineering, chemistry, materials sciences, pharmaceuticals, and biomedicine. This rapid progress has led to an expansion of possible extensive research at the nanoscale, which has in turn led to the fabrication of materials with exceptional properties. In the case of very small crystals of nanometer dimensions – termed nanocrystals – assumptions of translational symmetry and infinite sizes of crystals are no longer valid, and consequently these systems cannot be described using the same model as can be applied to a bulk solid [1, 2]. The electronic structure of a nanocrystal should be intermediate between the discrete levels of an atomic system and the band structure of a bulk solid. Nanocrystals have demonstrated discrete energy level structures and narrow transition line widths, and it is because of these discrete energy levels that such structures are referred to as quantum dots (QDs). The density of these structures is much greater, and their spacing is smaller than for the corresponding levels of one atom or a small atomic cluster [3, 4]. The highest occupied atomic levels and lowest unoccupied levels of the atomic species interact to form the valence band and conduction band of the QD, respectively. One of the most exciting interfaces of QDs is their high efficiency in biomedical applications, and this makes them ideal donor fluorophores – a property which has led to a revolution in detection processes [5, 6]. QDs also provide a versatile platform for the design and engineering of nanoparticle-based drug delivery (NDD) vehicles [7–9].

2 Properties of QDs

QDs exhibit unique optical properties due to a combination of their material band gap energy and quantum phenomena. Colloidal semiconductor QDs can have a very narrow symmetrical intense distribution, at specific wavelengths, ranging from the ultraviolet to the infra-red [10].

Although the emission peak from a single colloidal QD can be less than 0.1 meV wide, this emission energy shifts randomly over time, and such behavior is referred to as either "spectral jumps" or "spectral diffusion" [11]. The fluorescence emission from a single QD exhibits a dramatic on/off behavior, which is referred to as "blinking"; this is another spectroscopic property that distinguishes colloidal from self-assembled QDs [12, 13].

3 Lithographically Defined QDs

Lithographically defined QDs are formed by isolating a small region of a two-dimensional (2D) system. Such two-dimensional electron systems (2DESs) can be found in metal oxide-semiconductor field-effect transistors, or in so-called semiconductor heterostructures [14, 15]. Heterostructures are composed of several thin layers of semiconductor materials grown on top of each other, using the technique of molecular beam epitaxy (MBE). The layer sequence can be chosen in such a way that all free charge carriers are confined to a thin slice of the crystal, forming essentially a 2DES. A superstructure derived from the periodic repetition of this sequence of layers is also termed a "multiple quantum well." Electrons will be repelled by the electric field of the electrodes, such that the region of the 2DES below the electrodes will be depleted of electrons; this charge-depleted region will then behave like an insulator. By applying an electric field with metal electrodes of an appropriate shape, it is possible to create an island of charges insulated from the remainder of the 2DES, and if the island within the 2DES is small enough it will behave as a QD. In the vertical geometry, a small pillar of the

2DES can be isolated by etching away the heterostructure around it, and in such an arrangement the charge carriers will again be confined in all three dimensions. To date, most of the electron transport measurements performed on QDs have employed the two types of QD in the form just described. The lateral arrangement offers a relatively high degree of freedom for the design of the structure, as this is determined by the choice of electrode geometry. In addition, it is possible to fabricate and study "artificial molecules" [16–19] composed of several QDs linked together. When using the vertical arrangement, structures with very few electrons can be fabricated [20].

At present, a variety of ongoing studies are being undertaken to investigate many-body phenomena in these QD systems. Relevant examples include studies of the Kondo effect [14, 15, 21] and the design and control of coherent quantum states, with the ultimate goal being to perform quantum information processing. One remarkable advantage of lithographically defined QDs is that their electrical connection to the "macro-world" is straightforward. Moreover, the manufacturing processes are similar to those used in chip fabrication and, at least in principle, such structures could be embedded within conventional electronic circuits. The geometry of these QDs is determined lithographically, and is limited to the usual size and resolution limits of lithographic techniques. However, even when using electron beam lithography to fabricate the QDs it is not possible to tailor their size with nanometer precision. Typically, QDs fabricated lithographically are >9 nm in size, and consequently only relatively low lateral confining energies can be achieved.

4
Colloidal QDs

Colloidal QDs can be chemically synthesized using wet chemistry, and are free-standing nanoparticles (NPs) or nanocrystals grown in solution [22]. Colloidal QDs represent only a subgroup of a broader class of materials that can be synthesized at the nanoscale level using wet chemical methods, where the reaction chamber contains a liquid mixture of compounds that control both nucleation and growth. In a general synthesis of QDs in solution, each of the atomic species that will form part of the nanocrystals is introduced into the reactor in the form of a precursor. (A precursor is a molecule or complex containing one or more of the atomic species required to grow the nanocrystals.) When the precursors are introduced into the reaction flask they decompose, forming new reactive species (the monomers) that will in turn cause nucleation and growth of the nanocrystals. The energy required to decompose the precursors is provided by the liquid in the reactor, either by thermal collisions, by a chemical reaction between the liquid medium and the precursors, or by a combination of these two mechanisms [23].

The controlling growth of colloidal nanocrystals is the presence of mobile molecular species to provide access for the addition of monomer units in the reactor, broadly termed "surfactants" [24]. Suitable surfactants include alkyl thiols, phosphines, phosphine oxides, phosphates, phosphonates, amides or amines, carboxylic acids, and QDs of nitrogen-containing aromatics. If the growth of nanocrystals is carried out at high temperatures (e.g., at 200–400 °C), the surfactant molecules must be stable under such conditions if they are to serve as suitable candidates for controlling the growth. At low temperatures or, more generally, when growth ceases, the surfactants will be more strongly bound to the surface of the nanocrystals such that the latter are soluble in a wide range of solvents. The coating process allows for a greater synthetic flexibility, in that it can be exchanged for another coating of organic molecules with different functional groups or polarity. The surfactants can also be temporarily removed such that an epitaxial layer of another material with different electronic, optical or magnetic properties can be grown on the initial nanocrystal [25, 26].

An excellent control of the size and shape of QDs is possible by controlling the mixture of surfactant molecules present during their generation and growth [23, 27, 28]. Colloidal nanocrystals, when dispersed in solution, are not bound to any solid support, as is the case for the other two QD systems described above. Rather, colloidal nanocrystals can be produced in large quantities in a reaction flask, and later transferred to any desired substrate or object. It is, for example, possible to coat the nanocrystal surfaces with biological molecules such as proteins or oligonucleotides. The fact that many biological molecules perform tasks of molecular recognition with extremely high efficiency means that the ligand molecules can bind with very high specificity to certain receptor molecules, similar to a "key-and-lock" system. Notably, if a colloidal QD is tagged with ligand molecules it can bind specifically to all of the positions where a receptor molecule is present, and in this way it has been possible to create small groupings of colloidal QDs mediated by molecular recognition [29–31], and also to label the specific compartments of a cell with different types of QD [32–34]. Although colloidal QDs are rather difficult to connect electrically, a few electron transport experiments have been reported in which nanocrystals were used as the active material in devices that behave as single-electron transistors (Fig. 1) [35, 36].

5
QDs Application in Optics

Many applications of QDs can be found in optics. As with the more general case of atoms or molecules, QDs can be excited optically, and regardless of the nature of the excitation they may emit photons as they relax from an excited state to ground state. Based on these properties, QDs may be used as lasing media, as single-photon sources, as optically addressable charge storage devices, or as fluorescent labels. Self-assembled QDs incorporated into the active layer of a quantum well laser can significantly improve the operation characteristics of the laser, due to the zero-dimensional density of states of QDs. In QD lasers, the threshold current density is reduced, its temperature stability is improved, and the differential gain is increased. The absorption operation of a QD laser structure was first demonstrated as early as 1994 [37, 38], since when the lasing characteristics have been improved by a better control of the growth of the self-organized QD layers, such that QD lasers are now approaching commercialization [39]. Optical gain and stimulated emission have also been observed from CdSe and CdS colloidal nanocrystal QDs [40]. On the basis of these results, it is conceivable that optical devices could also be built using the self-assembly of colloidal QDs.

In the past, QDs have been used not only as conventional laser sources but also as "non-classical" light sources. Photons emitted from thermal light sources have

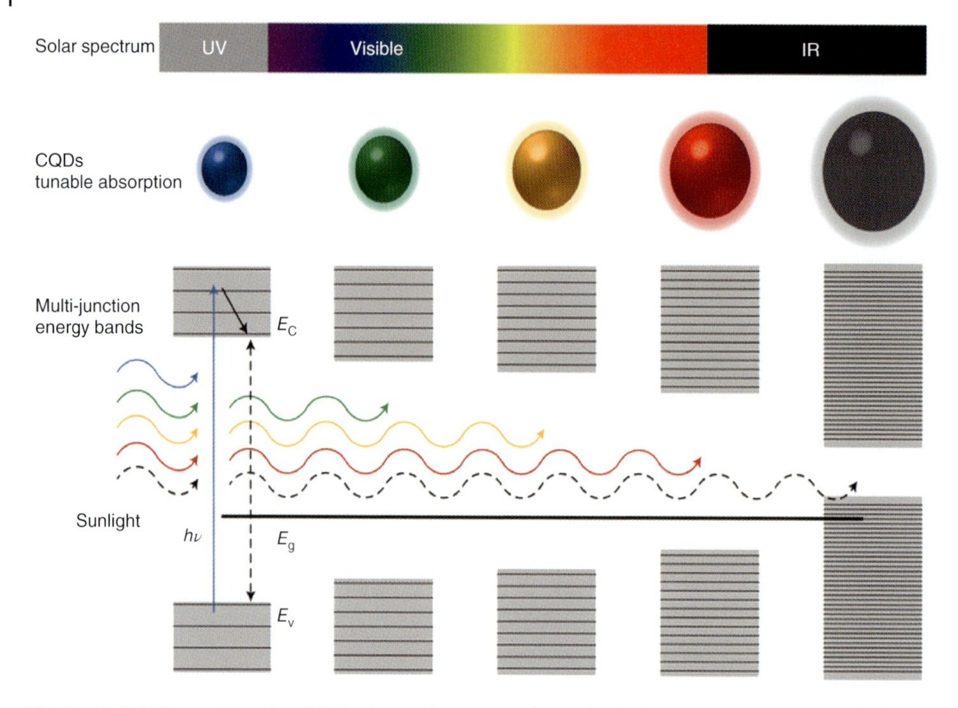

Fig. 1 Colloidal quantum dot (CQD) photovoltaics, size-dependent absorption enables CQDs to be tuned to absorb, sequentially, the constituent bands making up the Sun's broad spectrum, paving the way for the construction of multijunction solar cells that overcome the energy loss.

characteristic statistical correlations. In arrival time measurements, it is found that they tend to "bunch" together (super-poissonian counting statistics), but for applications in quantum information processing it would be desirable to emit single photons one at a time (sub-poissonian counting statistics, or anti-bunching). During the past few years it has been possible to demonstrate the first prototypes of such single-photon sources based on single QDs [41–43]. Self-assembled QDs have also been discussed as the basis of an all-optical storage device, whereby excitons would be generated optically and the electrons and holes stored separately in coupled QD pairs [44]. By applying an electric field, the electron and hole could then be forced to recombine and to generate a photon that would provide an optical read-out. Colloidal QDs have also been used in the development of light-emitting diodes (LEDs) [45, 46], in which colloidal QDs are incorporated into a thin film of a conducting polymer. Colloidal QDs have also been used in the fabrication of photovoltaic devices [47, 48].

Chemically synthesized QDs fluoresce in the visible range with a wavelength that is tunable by the size of the colloids. The possibility of controlling the onset of absorption and the color of fluorescence by tailoring the size of colloidal QDs makes them interesting objects for the labeling of biological structures [32, 33], specifically as a new class of fluorescent biomarkers. Such tenability, combined with extremely reduced photobleaching, would make colloidal QDs an interesting alternative to conventional fluorescent molecules [42]. The possible biological applications of fluorescent colloidal QDs are discussed in detail elsewhere in this encyclopedia.

6
Transport (Electrical) Properties of QDs

Electron transport through ultrasmall structures such as QDs is governed by charge and energy quantization effects. Charge quantization comes into play for structures with an extremely small capacitance. The capacitance of a nanostructure which, roughly speaking, is proportional to its typical linear dimension, may become so small that the energy required to charge the structure with one additional charge carrier (electron, hole, cooper pair) would exceed the thermal energy available. In this case, charge transport through the structure would be blocked – an effect of QDs which has appropriately been termed "Coulomb blockade" (CB).

The CB effect can be exploited to manipulate single electrons within nanostructures. In contrast to bulk structures, charge carriers within a QD are only allowed to occupy discrete energy levels, as in the case of electrons within an atom. In metallic nanostructures, CB can also occur without "quantum aspects," as the energy level spacing in these structures is usually too small to be observable. In fact, energy quantization is the main reason for using the name "quantum dot," and differentiates this from most of the metallic nanostructures. A brief history and fundamentals of the CB effect are provided in the following section [49]. Spectral diffusion is likely related to the local environment of the QDs, which creates rapidly fluctuating electric fields that may perturb the energy levels of the system. Similarly, spectral diffusion can also be observed in single organic fluorophores [50]. Conversely, self-assembled QDs embedded in a matrix do not exhibit spectral jumps, because their local environment does not change with time. In self-assembled QDs, multiexciton states can be observed and studied at high pumping power, but these have never been observed in colloidal QDs. The absence of multiexcitons in single colloidal QDs is believed to be correlated with the fluorescence intermittence observed in these systems [51]. The fluorescence emission from a single QD exhibits a dramatic on/off behavior that is referred to as "blinking," and is another spectroscopic feature that distinguishes colloidal from self-assembled QDs [12, 13].

7
Application of QDs in Diagnostics

The accurate identification of diseases requires a precise detection. The high extinction coefficient of QDs, in addition to their long lifetime and broad absorption spectra that may lead to Stokes shifts, makes them more appropriate for diagnostic purposes than organic fluorophores. The size-dependent properties of QDs such as CdSe provide the possibility for multiplex diagnosis (Fig. 2), with application as either donors or acceptors. The fluorescent properties of QDs make them ideal as donor fluorophores in hybridization-mediated Förster resonance energy transfer (FRET) [53–56], and they are also exceptionally well suited to *in-situ* hybridization, immunohistochemistry, and immunoassay. As QDs are more resistant to degradation than are other optical imaging probes, this allows them to be used for cell tracking over longer periods of time.

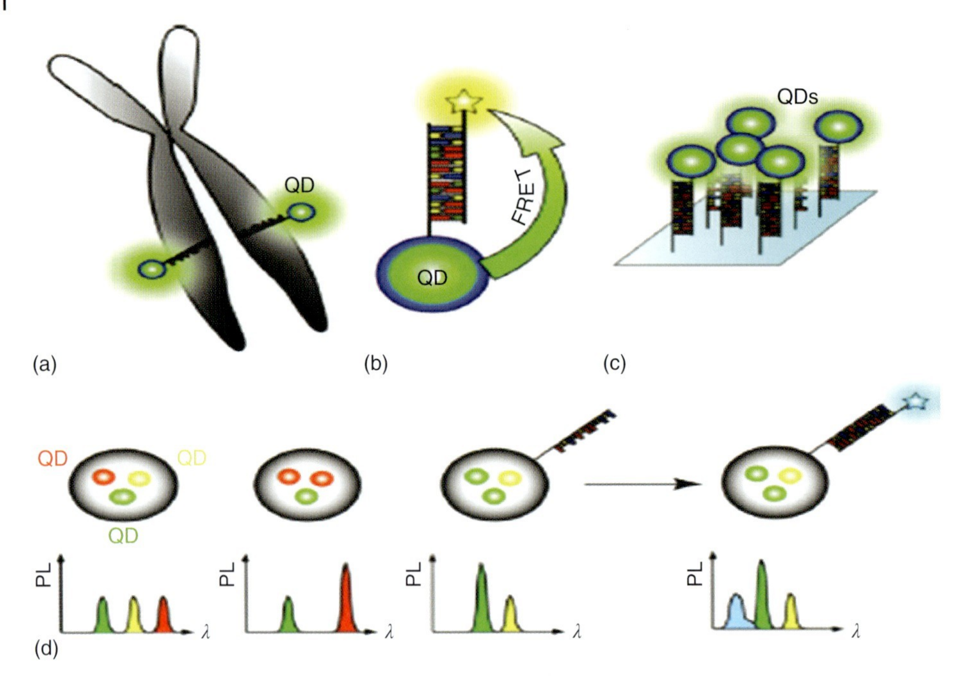

Fig. 2 Applications of quantum dots (QDs) in nucleic acid diagnostics. (a) QDs as labels for FISH; (b) QDs as donors in hybridization-mediated fluorescence resonance energy transfer (FRET); (c) QDs as labels for microarray/solid-phase hybridization; (d) Encapsulated combinations of QDs for spectral bar-coding [52].

8
Nucleic Acid Detection

Fluorescence *in-situ* hybridization (FISH) is an optical nucleic acid diagnostic method that employs QDs, with FISH-based QDs having been used to detect and localize target DNA sequences on chromosomes by hybridization with a specific QD-labeled probe DNA fragment. The QD-labeled probes were used to identify mutations in a Y-chromosome-specific sequence in human sperm, using *in-situ* hybridization. Subsequently, streptavidin-conjugated QDs were used successfully to quantify Fourier transform infra-red (FTIR) signals in human, mouse, and plants [53, 54, 57, 58]. In another study, QDs were shown to be photostable and to have twofold brightness compared to Texas Red and fluorescein dyes in the HER2 locus of breast cancer cells [57].

QD-based molecular beacons (MBs) were fabricated specifically to detect β-lactamase genes that were located in pUC18 and which were responsible for antibiotic resistance in *Escherichia coli* DH5 [59]. The comparison between QD-based and Cy3-based FISH showed, despite brightness and less photobleaching, the background signal to be higher for the QD-based system than for the Cy3-based counterpart [60]. An additional application of QDs to probe DNA is that of FRET, the process of which depends on the extent of spectral overlap between two particles and involves the transfer of fluorescence energy from a donor to an acceptor whenever the distance is typically 1–10 nm [61–63]. When QDs are used as donor particles, there is a limitation in the acceptor particles. The FRET efficiency can be improved with increasing particle as acceptors around a central QD. The limitation of QDs as an acceptor particle is inefficient FRET, as a result of the broad spectrum, that results in a small ratio of excited-state donors to ground-state QD acceptors. Donors with long excited-state lifetimes and pulsed excitation can be used to resolve this problem [60]. QD-based FRET can be applied to photochromic switching, photodynamic medical therapy, pH, and ion sensing and the sensing of enzymatic activity [64–70]. A new class of QD bioconjugates used in bioluminescence resonance energy transfer (BRET). When donors are not excited optically, self-illuminating QDs are suitable acceptors in BRET experiments for deep tissue imaging [71]. This application is indebted to the exceptional optical features of QDs (e.g., superior brightness and photostability, tunable emission, multiplexing) as well as the high sensitivity of bioluminescence imaging [72]. Recently, a BRET-based sensing system was described that was capable of detecting a nucleic acid target within 5 min, with high sensitivity and selectivity. This system as a result of adjacent binding of oligonucleotide probes labeled with Renilla luciferase (Rluc) and QD on the nucleic acid target [73]. The electrochemiluminescent resonance energy transfer (ECRET) process takes place between luminol and QDs. Resonance energy transfer (RET) with electrochemiluminescence (ECL) between CdS QDs and tris (2,2′-bipyridine)ruthenium($^{2+}$) ion ($Ru(bpy)_3{}^{2+}$) and between the ECL of 4-aminobutyl-*N*-ethylisoluminol (ABEI-luminol)/H_2O_2 and QDs. Compared to FRET and BRET, ECRET demonstrated an even better controllability, a higher sensitivity, and a wider dynamic range [74–76].

9 Application of QDs in Drug Treatments

Semiconductor particles act as a model platform in shaping the intricate design criteria for the engineering of NDD vehicles, and can also be considered as potential drug-delivery vehicles [77]. The small size, versatile surface chemistry, high photostability and brightness, large Stokes shift, sensitivity to microenvironment, electron-dense inorganic core (and many other properties, as noted above) enable QDs to be used as a platform for nanocarrier design (Fig. 3).

One major limitation of using QDs is the presence of toxic components such as cadmium or lead, although layers of organic and/or inorganic polymers (e.g., poly(ethylene glycol) or SiO_2) can be applied in solution form to minimize any possible cytotoxic effects [79–81]. The coated particles sizes typically range from 5 to 20 nm; particles less than 5 nm in size are quickly cleared via renal filtration, while larger particles may be taken up by the reticuloendothelial system before reaching targeted sites, while achieving also a limited penetration into solid tissues [82, 78]. Studies with QDs have provided a systematic assessment of appropriate measurements for efficient delivery; notably, the high surface-to-volume ratio of QDs makes it possible to link multiple functionalities on the particles while maintaining the overall size within an optimal range [78].

Variations in transfection efficiency, delivery-induced cytotoxicity and "off-target" effects at high concentrations of small interfering RNAs (siRNAs) can confound the interpretation of functional

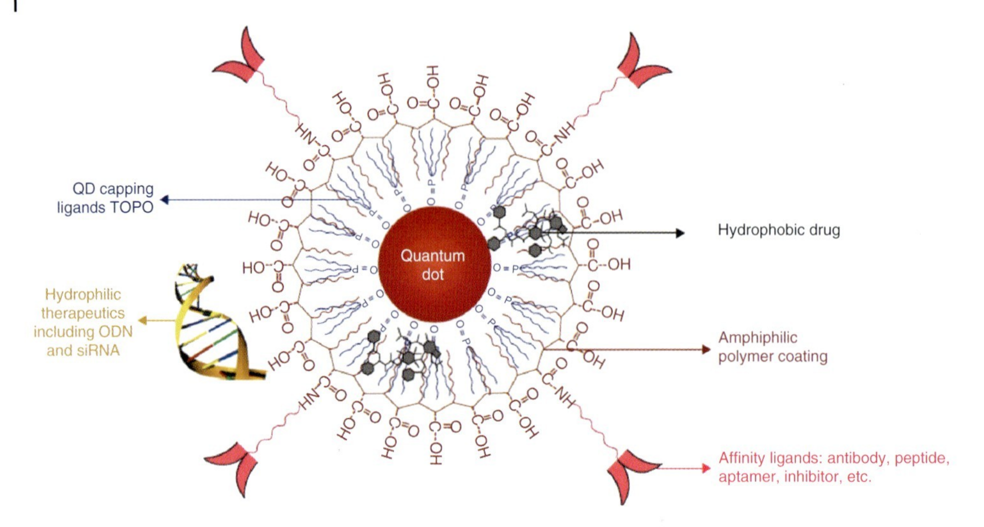

Fig. 3 The high surface-to-volume ratio of QDs makes it possible to link multiple functionalities with optimal size [78]. ODN: oligodeoxynucleotide; TOPO: tri-*n*-octylphosphine oxide.

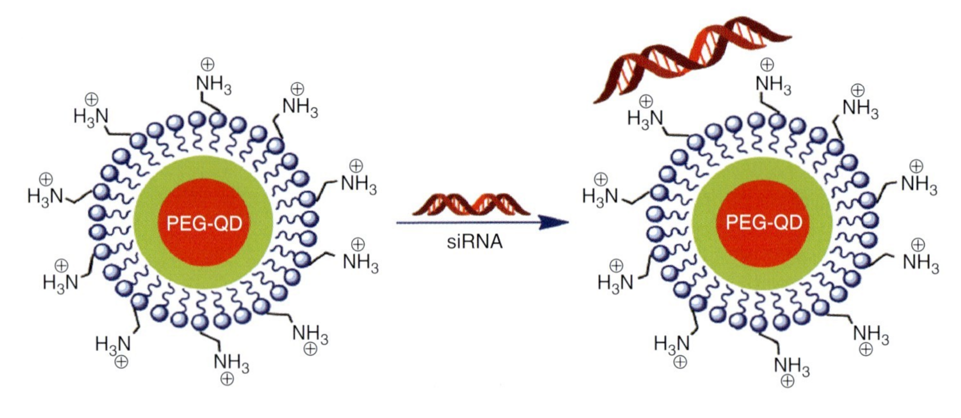

Fig. 4 Formation of QD–poly(ethylene glycol) (PEG)/siRNA complexes by electrostatic interaction.

studies. An unmodified siRNA combination with semiconductor QDs as multicolor biological probes was used to address this problem [83]. Recently, CdSe/ZnS fluorescent QDs, conjugated with amino-polyethylene glycol, were used to deliver siRNAs targeting β-secretase (BACE1) in order to achieve a high transfection efficiency of siRNAs and a reduction in β-amyloid (Aβ) levels in nerve cells (Fig. 4) [84].

Additionally, when mucin1 aptamer-doxorubicin by pH-responsive quantum dot (QD-MUC1-DOX) was applied for the chemotherapy of ovarian cancer, the system showed a preferential targeting of ovarian cancer cells while efficiently releasing doxorubicin at acidic pH; as a consequence, the anticancer efficacy of doxorubicin in ovarian cancer cells was enhanced compared to that of the free drug [85].

The results of *in-vivo* studies of human prostate cancer growing in nude mice have indicated that the QD probes accumulate at tumors, due both to an enhanced permeability and retention at tumor sites, and also by the binding of antibodies to cancer-specific cell-surface biomarkers [86]. Cancer photodynamic therapy is another application of QDs, which also have potential as photosensitizers since irradiation with ultraviolet light can induce the generation of reactive oxygen species (ROS) such as hydroxyl, superoxide radical, cytotoxic single oxygen (O_2), and toxic heavy-metal ions [87, 88]. In one of these studies, fluorescent CdTe and CdSe QDs, with red/brown, brown, or deep brown (close to black) colors, were seen to effectively convert light energy into heat both *in vitro* and *in vivo* when activated by 671 nm laser irradiation. In order to monitor QD retention at the tumor sites, the levels of Cd atoms in the tumors were monitored using inductively coupled plasma atomic emission spectroscopy following the injection of QDs [89]. In the field of immunotherapy, dual QD-magnetic imaging probes were used to trace dendritic cell migration to the lymph nodes in mice, using two-photon optical imaging and magnetic resonance imaging [90].

Acknowledgments

The authors thank the Department of Medical Nanotechnology, Faculty of Advanced Medical Science of Tabriz University for the support provided. These studies were funded by a 2014 Drug Applied Research Center Tabriz University of Medical Sciences Grant.

References

1. Atkins, P.W. (1986) *Physical Chemistry*, 3rd edn. Oxford University Press: Oxford.
2. Karplus, M., Porter, R.N. (1970) *Atoms and Molecules*, 1st edn. W.A. Benjamin, Inc., New York.
3. Kittel, C. (1989) *Einführung in die Festkörperphysik*, 8th edn. R. Oldenbourg Verlag, München, Wien.
4. Ashcroft, N.W., Mermin, N.D. (1976) *Solid State Physics*, Saunders College, Philadelphia.
5. Abrams, B.L., Wilcoxon, J.P. (2005) *Crit. Rev. Solid State Mater. Sci.*, **30**, 153–182.
6. Burda, C., Chen, X., Narayanan, R., El-Sayed, M.A. (2005) *Chem. Rev.*, **105**, 1025–1102.
7. Delehanty, J.B., Boeneman, K., Bradburne, C.E, Robertson, K., Medintz, I.L. (2009) *Expert Opin. Drug Deliv.*, **6** (10), 1091–1112.
8. Bharali, D.J., Mousa, S.A. (2011) *Pharmacol. Ther.*, **128** (2), 324–335.
9. Barreto, J.A., O'Malley, W., Kubeil, M., Graham, B., Stephan, H., Spiccia, L. (2011) *Adv. Mater.*, **23** (12), H18–H40.
10. Samir, T.M, Mansour, M.M., Kazmierczak, S.C., Azzazy, H.M. (2012) *Nanomedicine*, **7** (11), 1755–1769.
11. Empedocles, S., Bawendi, M. (1999) *Acc. Chem. Res.*, **32**, 389–396.
12. Efros, A.L., Rosen, M. (1997) *Phys. Rev. Lett.*, **78**, 1110–1113.
13. Kuno, M., Fromm, D.P., Hamann, H.F., Gallagher, A., Nesbitt, D.J. (2000) *J. Chem. Phys.*, **112**, 3117–3120.
14. Davies, J.H. (1998) *The Physics of Low-Dimensional Semiconductors*, Cambridge University Press, Cambridge.
15. Ando, T., Fowler, A. B., Stern, F. (1982) *Rev. Mod. Phys.* , **54**, 437–672.
16. Führer, A., Lüscher, S., Ihn, T., Heinzel, T., Ensslin, K., Wegscheider, W., Bichler, M. (2001) *Nature*, **413**, 822–825.
17. Kemerink, M., Molenkamp, L.W. (1994) *Appl. Phys. Lett.*, **65**, 1012.
18. Waugh, F.R., Berry, M.J., Mar, D.J., Westervelt, R.M., Campman, K.L., Gossard, A.C. (1995) *Phys. Rev. Lett.*, **75**, 705.
19. Bayer, M., Gutbrod, T., Reithmaier, J.P., Forchel, A., Reinecke, T.L., Knipp, P.A., Dremin, A.A., Kulakovskii, V.D. (1998) *Phys. Rev. Lett.*, **81**, 2582–2585.
20. Tarucha, S., Austing, D.G., Honda, T., Hage, R.J., Kouwenhoven, L.P. (1996) *Phys. Rev. Lett.*, **77**, 3613.
21. Frost, J.E.F., Hasko, D.G., Peacock, D.C., Ritchie, D.A., Jones, G.A. (1988) *J. Phys. C*, **21**, L209.
22. Cho, A.Y. (1999) *J. Cryst. Growth*, **202**, 1–7.

23. Elbaum, R., Vega, S., Hodes, G. (2001) *Chem. Mater.*, **13**, 2272.
24. Zhao, H., Douglas, E.P., Harrison, B.S., Schanze, K.S. (2001) *Langmuir*, **17**, 8428.
25. Zhao, H., Douglas, E.P. (2002) *Chem. Mater.*, **14**, 1418.
26. Pinna, N., Weiss, K., Sackongehl, H., Vogel, W., Urban, J., Pileni, M.P. (2001) *Langmuir*, **17**, 7982.
27. Farmer, S.C., Patten, T.E. (2001) *Chem. Mater.*, **13**, 3920.
28. Wang, Y.A., Li, J.J., Chen, H., Peng, X. (2002) *J. Am. Chem. Soc.*, **124**, 2293.
29. Nandakumar, P., Vijayan, C., Murti, Y.V.G.S. (2002) *J. Appl. Phys.*, **91**, 1509.
30. Ispasoiu, R.G., Jin, Y., Lee, J., Papadimitrakopoulos, F., Goodson, T., III, (2002) *Nano Lett.*, **2**, 127.
31. Cumberland, S.L., Hanif, K.M., Javier, A., Khitrov, G.A., Strouse, G.F., Woessner, S.M., Yun, C.S. (2002) *Chem. Mater.*, **14**, 1576.
32. Henglein, A. (1995) *Ber. Bunsen Ges. Phys. Chem.*, **99**, 903.
33. Weller, H., Eychmüller, A. (1995) *Adv. Photochem.*, **20**, 165.
34. Dubertret, B., Skourides, P., Norris, D.J., Noireaux, V., Brivanlou, A.H., Libchaber, A. (2002) *Science*, **298**, 1759–1762.
35. Shirasaki, Y., Supran, G.J., Bawendi, M.G., Bulović, V. (2013) *Nat. Photonics*, **7**, 13–23.
36. Klein, D.L., Roth, R., Lim, A.K.L., Alivisatos, A.P., McEuen, P.L. (1997) *Nature*, **389**, 699–701.
37. Ledentsov, N.N., Ustinov, V.M., Egorov, A.Y., Zhukov, A.E., Maksimov, M.V., Tabatadze, I.G., Kop'ev, P.S. (1994) *Semiconductors*, **28**, 832.
38. Kirstaedter, N., Ledentsov, N.N., Grundmann, M., Bimberg, D., Ustinov, V.M., Ruvimov, S.S., Maximov, M.V., Kop'ev, P.S., Alferov, Z.I., Richter, U., Werner, P., Gösele, U., Heydenreich, J., (1994) *Electron. Lett*, **30**, 1416.
39. Bimberg, D., Grundmann, M., Heinrichsdorff, F., Ledentsov, N.N., Ustinov, V.M., Zhukov, A.E., Kovsh, A.R., Maximov, M.V., Shernyakov, Y.M., Volovik, B.V., Tsatsul'nikov, A.F., Kop'ev, P.S., Alferov, Z.I. (2000) *Thin Solid Films*, **367**, 235.
40. Klimov, V.I., Mikhailovsky, A.A., Xu, S., Malko, A., Hollingsworth, J.A., Leatherdale, C.A., Eisler, H.J., Bawendi, M.G. (2000) *Science*, **290**, 314–317.
41. Michler, P., Imamoglu, A., Kiraz, A., Becher, C., Mason, M.D., Carson, P.J., Strouse, G.F., Buratto, S.K., Schoenfeld, W.V., Petroff, P.M. (2002) *Phys. Status Solidi B-Basic Res.*, **229**, 399–405.
42. Zwiller, V., Blom, H., Jonsson, P., Panev, N., Jeppesen, S., Tsegaye, T., Goobar, E., Pistol, M.E., Samuelson, L., Björk, G. (2001) *Appl. Phys. Lett.*, **78**, 2476.
43. Santori, C., Fattal, D., Vuckovic, J., Solomon, G.S., Yamamoto, Y. (2002) *Nature*, **419**, 594.
44. Lundstrom, T., Schoenfeld, W., Lee, H., Petroff, P.M. (1999) *Science*, **286**, 2312.
45. Colvin, V.L., Schlamp, M.C., Alivisatos, A.P. (1994) *Nature*, **370**, 354–357.
46. Dabbousi, B.O., Bawendi, M.G., Onotsuka, O., Rubner, M.F. (1995) *Appl. Phys. Lett.*, **66**, 1316.
47. Huynh, W.U., Dittmer, J.J., Alivisatos, A.P. (2002) *Science*, **295**, 2425–2427.
48. Huynh, W.U., Peng, X., Alivisatos, A.P. (1999) *Adv. Mater.*, **11**, 923–927.
49. Giaever, I., Zeller, H.R. (1968) *Phys. Rev. Lett.*, **20**, 1504.
50. Basche, T.J. (1998) *Luminescence*, **76–77**, 263–269.
51. Nirmal, M., Dabbousi, B.O., Bawendi, M.G., Macklin, J.J., Trautman, J.K., Harris, T.D., Brus, L.E. (1996) *Nature*, **383**, 802–804.
52. Algar, W.R., Massey, M., Krull, U.J. (2009) *Trends Anal. Chem.*, **28** (3), 292–306.
53. Ma, L., Wu, S.M., Huang, J., Ding, Y., Pang, D.W., Li, L. (2008) *Chromosoma*, **117**, 181–187.
54. Bentolila, L.A., Weiss, S. (2006) *Cell Biochem. Biophys.*, **45**, 59–70.
55. Chan, P., Yuen, T., Ruf, F., Gonzalez-Maeso, J., Sealfon, S.C. (2005) *Nucleic Acids Res.*, **33**, e161.
56. Pathak, S., Choi, S.K., Arnheim, N., Thompson, M.E. (2001) *J. Am. Chem. Soc.*, **123**, 4103.
57. Xiao, Y., Barker, P.E. (2004) *Nucleic Acids Res.*, **32**, e28.
58. Bentolila, L.A, Ebenstein, Y., Weiss, S. (2009) *J. Nucl. Med.*, **50**, 493–496.
59. Wua, S.M., Tiana, Z.Q., Zhang, Z.L., Huanga, B.H., Jianga, P., Xieb, Z.X., Panga, D.W. (2010) *Biosens. Bioelectron.*, **26**, 491–496.
60. Jin, Z., Hildebrandt, N. (2012) *Trends Biotechnol.*, **30**, 7.
61. Selvin, P.R (2000) *Nat. Struct. Biol.*, **7**, 730–734.

62. Medintz, I.L., Mattoussi, H. (2009) *Phys. Chem.*, **11**, 17–45.
63. Iqbal, A., Arslan, S., Okumus, B., Wilson, T.J., Giraud, D.G., Norman, T., Ha, D.M., Lilley, J. (2008) *Proc. Natl Acad. Sci. USA*, **105** (32), 11176.
64. Medintz, I.L., Trammell, S.A., Mattoussi, H., Mauro, J.M. (2004) *J. Am. Chem. Soc.*, **126**, 30–31.
65. Bakalova, R., Ohba, H., Zhelev, Z., Ishikawa, M., Baba, Y. (2004) *Nat. Biotechnol.*, **22**, 1360–1361.
66. Samia, A.C.S., Chen, X., Burda, C. (2003) *J. Am. Chem. Soc.*, **125**, 15736–15737.
67. Chen, Y., Rosenzweig, Z. (2002) *Anal. Chem.*, **74**, 5132–5138.
68. Snee, P.T., Somers, R.C., Nair, G., Zimmer, J.P., Bawendi, M.G., Nocera, D.G. (2006) *J. Am. Chem. Soc.*, **128**, 13320–13321.
69. Medintz, I.L., Clapp, A.R., Brunel, F.M., Tiefenbrunn, T., Uyeda, H.T., Chang, E.L., Deschamps, J.R., Dawson, P.E., Mattoussi, H. (2006) *Nat. Mater.*, **5**, 581–589.
70. Rotello, V.M. (2008) *ACS Nano*, **2**, 4–6.
71. So, M.K., Xu, C., Loening, A.M., Gambhir, S.S., Rao, J. (2006) *Nat. Biotechnol.*, **24**, 339–343.
72. Yu, W.W., Chang, E., Drezek, R., Colvin, V.L. (2006) *Biochem. Biophys. Res. Commun.*, **348**, 781–786.
73. Kumar, M., Zhang, D., Broyles, D., Deo, S.K. (2011) *Biosens. Bioelectron.*, **30**, 133–139.
74. Sun, L., Chu, H., Yan, J., Tu, Y. (2012) *Electrochem. Commun.*, **17**, 88–91.
75. Wu, M., Shi, H., Xu, J., Chen, H. (2011) *J. Chem. Soc., Chem. Commun.*, **47**, 7752.
76. Li, L., Li, M., Sun, Y., Jin, W. (2011) *J. Chem. Soc., Chem. Commun.*, **47**, 8292.
77. Zrazhevskiy, P., Sena, M., Gao, X. (2011) *Chem. Soc. Rev.*, **39** (11), 4326–4354.
78. Qi, L., Gao, X. (2008) *Drug Deliv.*, **5** (3), 263–267.
79. Medintz, I.L., Uyeda, H.T., Goldman, E.R., Mattoussi, H. (2005) *Nat. Mater.*, **4**, 435–446.
80. Clarke, S.J., Hollmann, C.A., Zhang, Z., Suffern, D., Bradforth, S.E., Dimitrijevic, N.M. (2006) *Nat. Mater.*, **5**, 409–417.
81. Hsieh, S.C., Wang, F.F., Hung, S.C., Chen, Y., Wang, Y.J. (2006) *J. Biomed. Mater. Res.*, **79B**, 95–101.
82. Soo Choi, H., Liu, W., Misra, P. (2007) *Nat. Biotechnol.*, **25** (10), 1165–1170.
83. Chen, A.A., Derfus, A.M., Khetani, S.R., Bhatia, S.N. (2005) *Nucleic Acids Res.*, **33** (22), e190.
84. Li, S., Liu, Z., Ji, F., Xiao, Z., Wang, M., Peng, Y., Zhang, Y., Liu, L., Liang, Z., Li, F. (2012) *Mol. Ther. Nucleic Acids*, **1**, e20.
85. Savla, R., Taratula, O., Garbuzenko, O., Minko, T. (2011) *J. Controlled Release*, **153**, 16–22.
86. Gao, X., Cui, Y., Levenson, R.M., Chung, L.W., Nie, S. (2004) *Nat. Biotechnol.*, **22** (8), 969–976.
87. Bakalova, R., Ohba, H., Zhelev, Z., Ishikawa, M., Baba, Y. (2004) *Nat. Biotechnol.*, **22** (11), 1360e1.
88. Azzazy, H.M., Mansour, M.M., Kazmierczak, S.C. (2007) *Clin. Biochem.*, **40** (13–14), 917e27.
89. Chu, M., Pan, X., Zhang, D., Wu, Q., Peng, J., Hai, W. (2012) *Biomaterials*, **33**, 7071–7083.
90. Mackay, P.S., Kremers, G.J., Kobukai, S., Cobb, J.G., Kuley, A., Rosenthal, S.J., Koktysh, D.S., Gore, J.C., Pham, W. (2011) *Nanomedicine*, **7** (4), 489–496.

Part II
Devices

Translational Nanomedicine. First Edition. Edited by Robert A. Meyers.

3
DNA Origami Nanorobots

Ido Bachelet
Bar Ilan University, Faculty of Life Sciences and The Nano Center, Bldg 206, Rm B337, Ramat Gan, 52900 Israel

Translational Nanomedicine. First Edition. Edited by Robert A. Meyers.

Keywords

DNA
Deoxyribonucleic acid.

Origami
The traditional Japanese art of paper folding.

Nanorobotics
The field of study concerned with designing, building and programming robots at the nanoscale (1 nanometer = 10^{-9} meter).

Nanobiotechnology
The study of biological phenomena and entities at the nanometer scale.

Nanomechanical engineering
The design of machines at the nanometer scale.

Synthetic biology
The synthesis of new biological or genetic constructs not found in Nature, using natural biological or genetic parts of design principles.

Robots augment the ability to automate the perception and control of reality. During the past century, robotics has become a multidisciplinary field, where insights from physics, biology, engineering and design have been integrated to drive an evolution of robotic devices for diverse tasks, from industry and mass production to scientific research and space exploration. Recent advances in the design and fabrication of nanoscale machines have enabled the introduction of man-made robots into the biology of living organisms, a realm that has so far remained essentially inaccessible to them. In this chapter this relatively new field will be reviewed, starting with the roots that led to its existence and its enablement. How robotics can revolutionize medicine and therapeutics, as they are currently known, will also be highlighted.

1 Introduction to Robotics

Difficulties in actually defining a robotare rather surprising considering how essential robots have become in terms of humankind's technology, society and culture during the past century. Robots can be defined in the broadest sense as automata linking the perception and processing of environmental information to the performance of defined tasks. However, this definition seems to encapsulate many different entities, from living organisms and single cells to machines and virtual agents living in cyberspace, to the stage where this definition itself is pointless. It is therefore perhaps better to define robotics in terms of its components, development, and implementations.

1.1 A Brief History of Robotics

In contrast to robots, the more general term machine can refer to a device that simplifies task performance or problem solving (for the purposes of this discussion the thermodynamic definition of work, which is done by *engines*, can be neglected). As such, machines were probably used as early as 2.6 million years ago by hominines, in the form of simple stone tools. These later evolved into models designed for specific tasks (cutting, piercing, sewing, etc.) made from various materials. The discovery, later in documented history, of basic machines such as the wheel, the lever, and the inclined plane, spawned new generations of machines for more elaborate and demanding tasks, enabling human technology, economy, and society to be scaled up.

However, these machines were still just brainless tools that required a human user for their operation. The knowledge that paved the way to robotics in its modern sense appeared during the Renaissance, and involved various branches of mechanical kinematics (the study of motion), applied physics, and engineering. Chronometry (the measurement of time) enabled the generation of automata that could perform series of defined tasks along a timeline, and simple energy sources such as a spring could ensure their sustained activity. In parallel, inventions such as the thermometer and barometer (seventeenth century), as well as advances in analytical chemistry, enabled the introduction of sensing of environmental information into machines. The industrial age (late nineteenth century) already saw elaborate machinery, powered by high-energy sources (steam, fuel, electricity), which could perform faster, stronger, and more accurately than any human. Some of these machines were operated by instructions encoded on hardware such as punched cards (e.g., the Jacquard loom, 1801).

Of particular interest is the emergence of the special category of machines known today as *computers*, as machines capable of carrying out a diverse set of tasks defined by a *program* specifying the mapping of input to an output. The most well-known example of an early computer was the difference engine, designed by Charles Babbage, which was an astoundingly ingenious mechanical device for automating polynomial calculations. Several decades later, Alan Turing will have outlined the first modern computer, the universal Turing Machine, which is still the most powerful model of computation known to exist. By the mid-twentieth century, Turing, Von-Neumann, and others had already designed advanced electronic computer architectures, setting the stage for the future fields of programming languages, algorithms, and artificial intelligence.

What allows the integration of all the separate components seen so far – basic mechanical devices, sensors, a computer, a program, a clock, and so on – is the framework of *cybernetics*, coined by Norbert Wiener as "... the study of control and communication," in machines, be they natural or artificial. Cybernetics is the wiring of machine components in a way that achieves control mechanisms such as feedback and timing, in order to optimize the function of the machine.

It can now be seen that robots may be defined based on these components, as devices linking sensing to actuation by information processing, to perform tasks. Later, it will become clear how

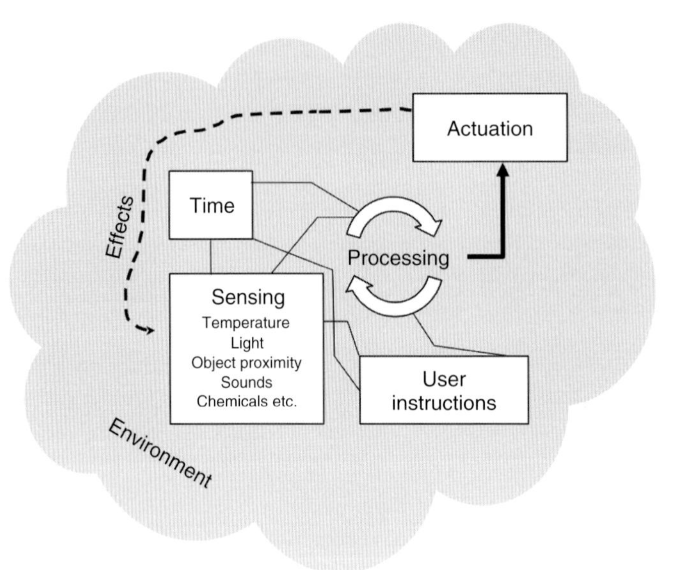

Fig. 1 Schematic abstraction of a robot. The robot is a machine linking the sensing of environmental information to an actuation of some kind, via processing or computation. The framework of cybernetics defines the ways in which the various components connect with each other to achieve control and optimization (feedback, timing, etc.). The effects resulting from the robot's activity can be measured again as environmental information. User instructions can be integrated into the processor (middle) or at any peripheral location (sensors, clock, etc.)

this definition is suitable to define robots operating at the nanoscale with a biological system as its nonlinear, highly dynamic environment (Fig. 1).

1.2 Robotic Control of Molecules

Robots augment the ability to automate the perception and control of reality. Modern industry, scientific research, manufacturing and more, all obligate the utilization of robots to cut costs, improve efficiency, and optimize performance. However, one realm is still largely inaccessible to robots, namely the biology of living organisms. Biological systems consist of small components such as molecules, molecular assemblies and single cells, which interact on size and time scales that it have not, until very recently, been capable of exploration. Moreover, biological interaction networks are among the most complex structures studied in Nature. Hence, to control the location and timing of the activity of molecules is a central requirement for biology.

Every cellular process, such as growth, metabolism and signaling, is the result of molecular interactions – whether between enzymes and substrates or between receptors and ligands – that are exquisitely regulated in time and space. In contrast, the ability to mimic this level of arbitrary control at the molecular scale is exceedingly limited. This is particularly significant for the field of pharmacology and drug design. The ideal drug will act only at its designated target with the correct timing. Unfortunately, however, most drugs currently in use have an enormous range of adverse effects due to a lack of correct spatiotemporal control [1–4]. Indeed, whilst the currently available therapeutic arsenal includes very effective molecules, a lack of ability to correctly control them hinders their safe and effective use.

During the past four decades, remarkable progress has been made in the area of drug control. Currently, an improved spatial control of therapeutics is achieved mainly by conjugating a therapeutic molecule to a target-specific carrier such as an antibody or an interleukin [5]. An alternative approach may involve the local implantation of a therapeutic molecule, but this is not always applicable. The temporal control of drug therapy is mostly achieved by embedding the drug in a matrix that hinders and/or prolongs its dissolution and subsequent diffusion, thereby causing a sustained and more stable distribution of the drug [6]. Various elaborations of this approach have been proposed, including matrices comprised of multiple phases, which can release the drug in stepwise fashion [7]. Other matrices can be triggered externally by physical means, such as infrared laser or ultrasound [8], although even these modifications will not always improve the drug's performance [9].

While clearly representing improvements over conventional, unmodified drug administration (which currently accounts for most therapeutics used), many desired features are not enabled by these techniques. Examples of such features include: (i) reversing the availability of a molecule at will, for example by sequestering or shielding it reversibly; (ii) coordinating the activity of two or more molecules by turning one molecule off as the other is turned on, and vice versa, such that they do not collide or compete with each other; or (iii) activating a molecule by a complex combination of biological conditions, for example, a target cell expressing molecules A, B, and C but not D and not E. Cells routinely exhibit such features, and the application of an equivalent level of expertise in therapeutics would lead to a paradigm shift in how drugs are used and designed. Moreover, it would directly lead to significant improvements in the safety and efficacy of many drugs in use today.

Cells can be thought of as nanoscale computers, which generate precise outputs based on molecular inputs [10]. This analogy holds a potential key to solve the challenge just described. Unlike the fields of pharmacology and drug design, computers have experienced a meteoric progress since the mid-twentieth century, which is described by Moore's law. In fact, computers readily enable high levels of arbitrary control over processes, much like cells do at the molecular scale. A computer capable of reading biomolecular inputs, and linking its output to drug release, could arguably solve the challenge of drug control. Until now, however, it has not been possible to implement computing capabilities to the nanoscale, due mainly to the uncertainty of how such computers should be designed and programmed in the first place. Moreover, it has not been clear how computers could interface with living cells.

2 DNA Nanotechnology

2.1 Computing

During the past two decades it has become clear that computers could be built from molecules, of which DNA would be a natural substrate for universal computing. The tractability of Watson–Crick pairing enables the programming of self-assembling DNA interactions for a diverse set of computing paradigms such as tiling [11], neural networks [12], and cellular automata [13]. In 1994, Adleman demonstrated a solution to the Hamiltonian path problem, a nondeterministic polynomial time (NP)-complete problem, using DNA strands that represented the vertices and paths of a seven-node graph [14]. The strand set representing the specific graph in question was mixed in a tube and allowed to self-assemble into a longer strand that would encode the solution, effectively emulating massively parallel computing as the strands were allowed to generate every possible solution. DNA computing has since been applied to a variety of computational problems [11, 12, 14–22], although its scalability has been questioned [23], as the complexity of solvable problems is limited by the amount of DNA required to carry out the computation process.

The seamless integration of molecular computing into a biological system was first demonstrated by Benenson, who described DNA computers that could recognize gene expression patterns in target cells and activate a desired genetic response in a logical fashion [24, 25]. However, responding exclusively to nucleic acids as inputs ostensibly limits the types of cues sensed by the system; moreover, such a computer would have to be delivered into cells or genetically encoded in order to function. Although nucleic acids are natural inputs for a DNA-based system, the introduction of aptamers (short nucleic acid sequences selected to recognize specific epitopes [26]) made possible the sensing of virtually any type of biological molecule, outside the cell, by nucleic acid-based sensors.

On the other hand, DNA–DNA interactions have provided a robust platform for molecular computing and logic. Two DNA strands competing on the same complementary strand can produce programmable kinetics, which led to the development of systems based on strand displacement reactions [27–29], such as a polymerase chain reaction (PCR) performed without temperature changes [30]. Strand displacement reactions were recently shown to be scalable, and have been utilized in the construction of neural networks capable of complex computations [12, 31, 32]. DNA strand displacement kinetics has also enabled the design of strands capable of "walking" along a DNA track made from complementary strands, with some of these actually referred to as "*robots*" [33–37]. Another implementation was in the construction of self-replicating devices [38] and a bacteria-inspired motor [39].

2.2 Nanofabrication

The tractability of Watson–Crick pairing enables the use of DNA as a programmable building block for molecular self-assembly applications. In a seminal report in 1982, Seeman first proposed the concept of fabricating two-dimensional (2D) and three-dimensional (3D) lattices from DNA [40], which could enable the parallel construction of nanoscale objects with nanometer-scale features and accuracy [17]. This technology was further elaborated by the introduction of DNA origami (see below), which enabled the relatively simple fabrication of arbitrary 2D and 3D shapes from folded DNA. Using DNA in such a way enables the integration of function into structure. As noted above and as will be discussed below, DNA molecules can be designed to carry out molecular computing [12, 14, 31], selectable molecular recognition [26], enzyme-free logic circuits [27], catalytic activity [41], and mechanical motion [42], while still being usable as a genetic code. This unique versatility makes DNA suitable for the design of advanced self-assembling nanoscale machines. Integrating the components of positional control, information encoding, and computing in DNA has produced a variety of applications, from DNA assembly lines to controllable nanocontainers [36, 43–45].

An interesting branch of DNA computing lies at the interface between DNA nanofabrication and computing by tiling. Hao Wang, who investigated this problem during the early 1960s, proposed that computing by tiling is universal, in the sense that any computation can be carried out by filling a surface with tiles. Winfree and others have used the DNA junction principles and basic shapes outlined by Seeman (double crossover molecules; DX molecules [46–48]) to construct 2D tiles capable of self-assembling, or crystallizing, on a surface to generate such computational patterns and processes [11]. Such 2D DNA tiles can serve as information bearing seeds for self-assembly into defined structures or computational outcomes [13, 49, 50], whereas such algorithmically patterned surfaces can be used as templates to array other materials with nanoscale precision [51].

By using the same principles of DNA junction design and construction [52, 53], larger repetitive structures in the 2D and 3D realms were later created, such as surfaces made by three-, four-, five-, and six-arm junctions of DNA [54] (according to the principles laid by Seeman, eight arms is the maximum rank of a junction using just the four canonical bases A, G, C, and T [18]). Introducing geometric twists into the junction enables multiple junctions to self-assemble into space-filling polyhedra with programmable properties [55, 56].

Interestingly, Aldaye and colleagues have recently shown that DNA nanostructures can be genetically encoded to array intracellular proteins in a certain structure to reprogram biochemical reactions [57].

2.3 Nucleic Acid Enzymes

The astonishing demonstrations by Altman and Cech that RNA molecules can have a catalytic function [58, 59], and their discoveries that synthetic nucleic acids can be selected and evolved *in vitro* to exhibit such functions [60–63], unleashed a new and very powerful technology. A wide range of enzymatic activities has already been selected from nucleic acid pools, and it is conceivable that in the near future, catalytic nucleic acids ("ribozymes") could be designed and synthesized *de novo* without selection.

DNA is less common than RNA in ribozymes, since RNA is more structurally and functionally diverse than DNA; however, the obvious advantages of DNA – namely the chemical stability and synthesis costs – could lead to improved ways of making diverse functions from DNA. Still, RNA catalysts can be integrated into a DNA machine using standard techniques of DNA nanofabrication [64].

2.4 Nucleic Acid Sensors

In 1990, the discovery was made independently in two laboratories that short, single-stranded nucleic acids – aptamers – could be selected from a random pool of sequences to bind to a molecular target of choice [26, 65]. This binding is mediated by the aptamer assuming a certain 3D structure, enabling it to access its target and hydrogen bond with it. This technology garnered high interest as a potential strategy to block undesired functions or proteins in vivo (AIDS, cancer growth factors, etc.), and several aptamers are currently in advanced stages of clinical trials as therapeutic agents.

Aptamers can be used for the detection of very low amounts of a chosen analyte, by their integration into a mechanical sensor. Such sensors, which often are referred to as "*beacons*," usually consist of a DNA/RNA aptamer bound to a partial complementary of itself forming a hairpin, while one end of the molecule is labeled with a fluorophore and the other with a dark quencher. Binding of the analyte by the aptamer displaces the complementary strand and allows a fluorescent signal to be released and detected. This concept of an aptamer-based mechanical sensor is analogous to the "riboswitch," a regulatory region of RNA molecules, which exerts its function by binding to certain target molecules in the cell. An interesting application based on this concept is that of genetically encoded sensors for various molecules in the cell.

Aptamers can also be used as sensors or mechanical gates for DNA nanorobotics, as will be described below [66].

2.5
DNA Actuators and Motors

It has been seen already that strand displacement reactions can be utilized in the design of "walkers" – DNA strands that are displaced and reattached to a complementary DNA track along a certain path, be it linear or circular. Walkers have been utilized for carrying payloads between points of origin and destination to drive chemical syntheses or a more elaborate assembly of large payloads [35, 45]. Additionally, strand displacement was used to drive a polymerization motor inspired by bacterial locomotion [39].

These elegant designs all have a central drawback in an applied context; that is, the reactions are not autonomous but require the sequential addition of displacing strands (termed "*fuel*" [67]) at every step, and removal of the double-stranded displacement products (termed "*waste*"). Since, without the respective addition or removal of fuel and waste, the reaction would rapidly be halted, these actuators are not suitable for use in therapeutic applications unless the fuel strands were to be a product of a disease process, and there existed an intrinsic mechanism for waste removal.

During the decades following Watson and Crick's discovery of the structure of DNA, much information was accumulated regarding the mechanical properties of DNA. Single molecule methods have enabled studies of DNA molecules in single strands and in molecular motors [68, 69]. Properties such as Young's modulus (on the order of 0.3–1.0 GPa, much like a stiff plastic), persistence length (ca. 50 nm), and the ability of DNA to shift between alternative structures driven by environmental changes, can be integrated into the design of nanoscale motors and actuators which do not rely on strand displacement, in which kinetics is dominated by sequence. Mao and colleagues have demonstrated an actuating device based on the transition of DNA from the B to Z structure upon addition of hexaminecobalt [70], a reagent favoring the left-handed folding of DNA. Several examples of mechanical DNA "tweezers," powered by various stimuli and capable of holding a molecule and releasing it on demand, have also been described [71–73].

2.6 Miscellaneous Functions of Nucleic Acids

Based on the above-mentioned reports, it might be assumed that it would be possible to select a DNA or RNA sequence that exhibited any desired property. The property space of nucleic acids, while ostensibly large, is still essentially obscure. For example, the kinetics and specificity of aptamer binding are still not well understood, and whether there is a common structure or sequence motifs in catalytically active sequences is unknown. Nevertheless, some interesting directions have already been highlighted, such as RNA mimics of green fluorescent protein (GFP) [74].

2.7 DNA Origami

DNA origami – the folding of a single DNA strand to yield a diverse array of shapes – was first demonstrated by Rothemund [19]. This technique (termed "*scaffolded*" DNA origami to distinguish it from other methods of DNA nanofabrication by self-assembly), makes use of a "scaffold" DNA strand that is typically several thousands of bases long. Directing the folding of the scaffold strand to the desired shape is achieved by hundreds of short (typically 15 to 60 bases long) oligonucleotides, called "*staple*" strands, which hybridize to distinct regions in the scaffold strand and induce crossovers that keep the entire structure solid. The folding reaction is carried out by mixing the scaffold and staple strands and subjecting the mixture to a temperature annealing ramp, which enables the structure to fold properly without being caught in local energy minima. Scaffolded DNA origami is remarkably robust and reproducible, and allows the fabrication of an astonishing variety of shapes with arbitrary features and geometries [75–79].

The physics of DNA origami folding is still largely unclear in light of the remarkable robustness of this system to generate very complex structures both in 2D and 3D forms. The temperature annealing ramp is used as an error correction mechanism, enabling the system to achieve its designed global minimum efficiently; however, the role of mechanical entropy in DNA origami folding is not clear. Scaffolded DNA origami objects seem to be very tolerant to mismatches and omissions of some strands [19], and can fold correctly most of the time, given that the object has been properly designed, the buffer contains a sufficient magnesium concentration, and that the staple strands are in sufficient excess over the scaffold strand.

Scaffolded DNA origami has several advantages over traditional DNA nanofabrication methods in the context of the serial manufacture of DNA devices. Other DNA nanotechnology techniques rely on a mixture of short oligonucleotides interacting with each other, which makes the structure very sensitive to stoichiometry and strand–strand competition. The way around this is usually to break up the synthesis into steps, where the product of each step is isolated on a solid phase or by some other means. In contrast, scaffolded DNA origami is very tolerant to staple strand concentration variations, and usually folds correctly as long as each staple is given a chance to interact at least once with the scaffold, making synthesis significantly simpler with higher yields. Second, it has been shown recently by Shih and colleagues, that scaffolded DNA origami can be folded from a double-stranded scaffold [80], making it feasible to encode both scaffold and staple strands on a large plasmid for scaling-up purposes.

Tab. 1 Nucleic acid design tools

Tool	*Use*	*Web site*
MFOLD	Predict the folding of RNA/DNA into secondary structures	http://mfold.rna.albany.edu/
NUPACK	Predict the folding of DNA/RNA into secondary structures	http://www.nupack.org/
UNIQUIMER-3D	DNA nanostructure design	http://ihome.ust.hk/~keymix/uniquimer 3D/index.htm
TIAMAT	DNA nanostructure design	http://yanlab.asu.edu/Resources.html
SARSE	DNA nanostructure design	http://cdna.au.dk/software/
CADNANO	Scaffolded DNA origami design	http://cadnano.org
CANDO	Mechanical prediction of DNA origami object properties	http://cando-dna-origami.org/
NANOENGINEER-1	Graphical editor for DNA nanostructures	http://www.nanoengineer-1.com/content/

Recently, Dietz and colleagues have elucidated folding as a phase transition phenomenon [81], where nothing or very little happens until a certain, critical temperature is reached which depends on the folding shape. At the critical temperature, during a potentially very short period, most of the object folds correctly and remains unchanged until the end of the process. This discovery has significant implications regarding the future serial manufacture of DNA origami devices for therapeutic applications (this will be discussed below).

2.8 Computer-Aided Design (CAD) Tools for DNA Nanostructures

Computer-aided design (CAD) tools for the automated design and visualization of DNA objects have led to the promotion and improvement of DNA nanofabrication techniques. NUPACK and MFOLD are web servers for the structural prediction of nucleic acid systems [82, 83]. SEQUIN was the first algorithm to enable the optimal design of DNA junctions [84], and was followed by UNIQUIMER 3D, for the complete design of latticed DNA structures [85]. TIAMAT, SARSE, and NANOENGINEER-1 each provide excellent user interfaces for the editing of DNA structures. Douglas and colleagues developed caDNAno, an open source CAD tool for the rapid design of scaffolded DNA origami structures [86]. caDNAno has also been integrated into Autodesk Maya, a 3D design and animation software with superb visualization capabilities. CANDO is a finite element-based analyzer for caDNAno-designed DNA origami structures, which predicts fluctuations and rigidity [87]. International conferences (such as DNA and FNANO) and friendly competitions (http://biomod.net) have helped to spread this knowledge base and to standardize DNA design and nanofabrication. Some of these design tools, along with the web sites where they are available for download or online work, are listed in Table 1.

2.9 Designing Scaffolded Origami in caDNAno: A Guided Tour

In this section, the aim is briefly to explain the process of designing a DNA origami shape in caDNAno, mostly because this is

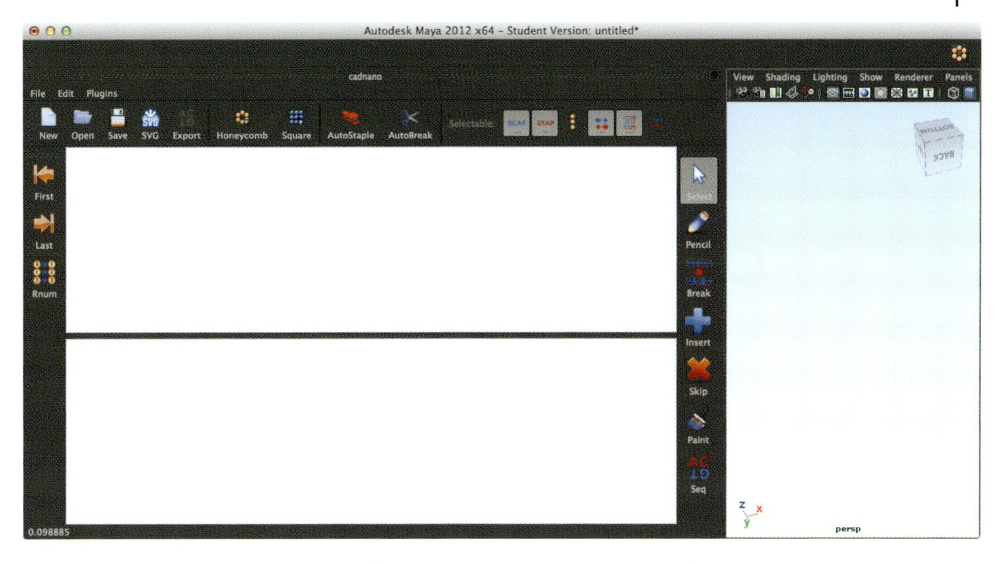

Fig. 2 The caDNAno design interface. The lattice panel is at the top left; the editing panel is at the bottom left; and the 3D visualization panel is on the right.

a highly versatile tool with a user-friendly interface, which was used to design the prototypical nanorobot described later in the chapter.

The explanation will be provided in the form of a list of instructions.

1) Download and install Autodesk Maya (student edition) and the most recent version of caDNAno, according to the instructions on the caDNAno web site (http://cadnano.org).
2) The design interface has three panels (Fig. 2):
   a. Lattice panel on the top left
   b. Editing panel on the bottom left
   c. A 3D graphical visualization panel on the right.
3) Choose, from the top toolbar, the type of lattice you wish to design the shape on. Two types are available, a honeycomb lattice or a square lattice; both are valid for the design of origami shapes. The honeycomb lattice is inherently twist-corrected, while shapes designed on the square lattice are inherently twisted, although this can be corrected during design [88].
4) On the lattice panel, draw the section of the shape. Each circle in the lattice represents a DNA double helix, viewed from the side. Choose helices by clicking on them once. Helices are numbered automatically. For convenience purposes, it is best to start with a helix that is numbered as “0” and not “1.” If the first helix is numbered “1,” simply undo and choose an adjacent helix (Fig. 3).
   Drawing the entire section could be done by clicking one helix at a time, or alternatively by holding the mouse button down and moving the cursor from helix to helix in a continuous path.
5) The first click will tag the helix yellow. A second click will tag the helix orange and open it for editing in the editing panel. Note that helix

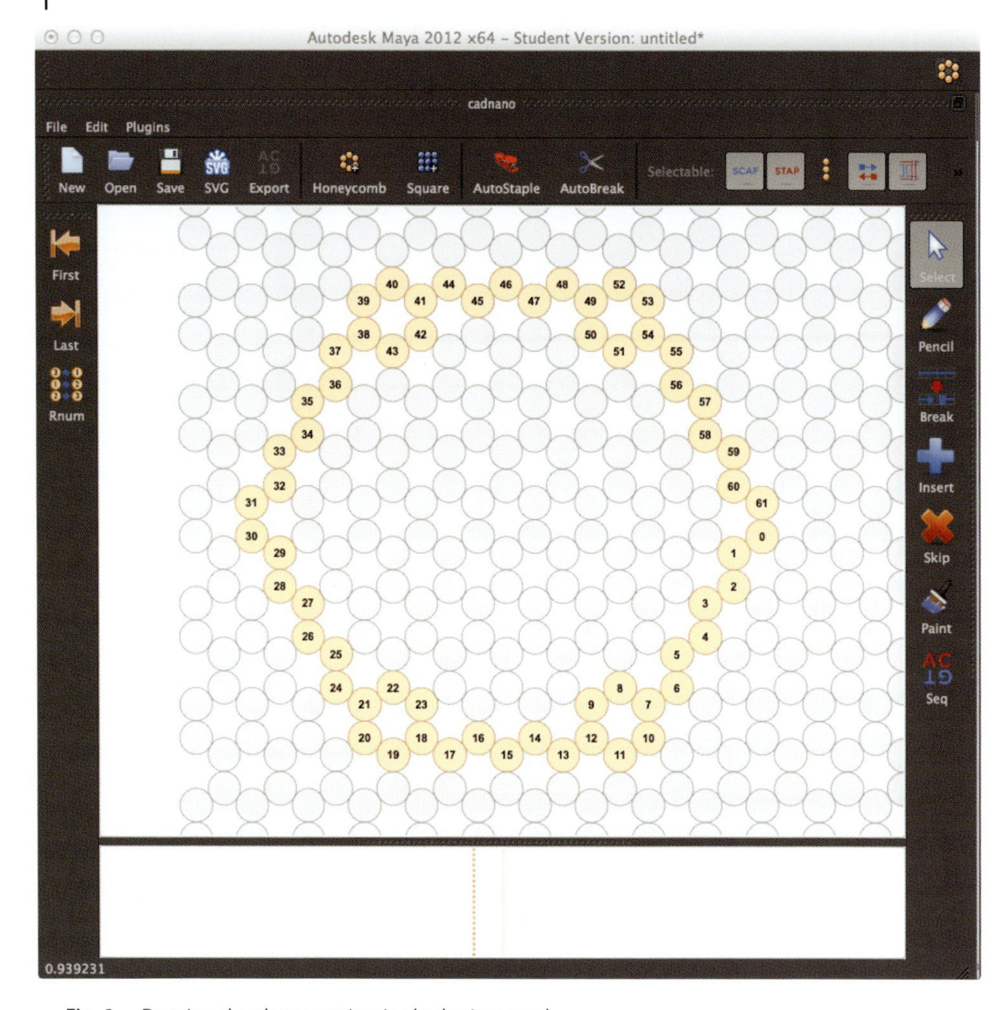

Fig. 3 Drawing the shape section in the lattice panel.

numbering in the editing panel is consistent with numbering in the lattice panel.

Note that while carrying out step (5), the scaffold strand path (blue) has been outlined automatically by caDNAno. If this is satisfactory, continue to step (6); otherwise, one undo action will erase the automatically chosen path, leaving scaffold primers, which can be connected manually as desired.

In addition, the 3D shape will form in real-time as you carry out these actions on the right panel (Fig. 4).

6) The editing panel grid shows each double helix as two rows of square cells. Each row is a strand, with each cell representing a base. Note that the scaffold strand occupies only one row in each double helix, and that the scaffold alternates between the parallel and antiparallel rows in adjacent helices.

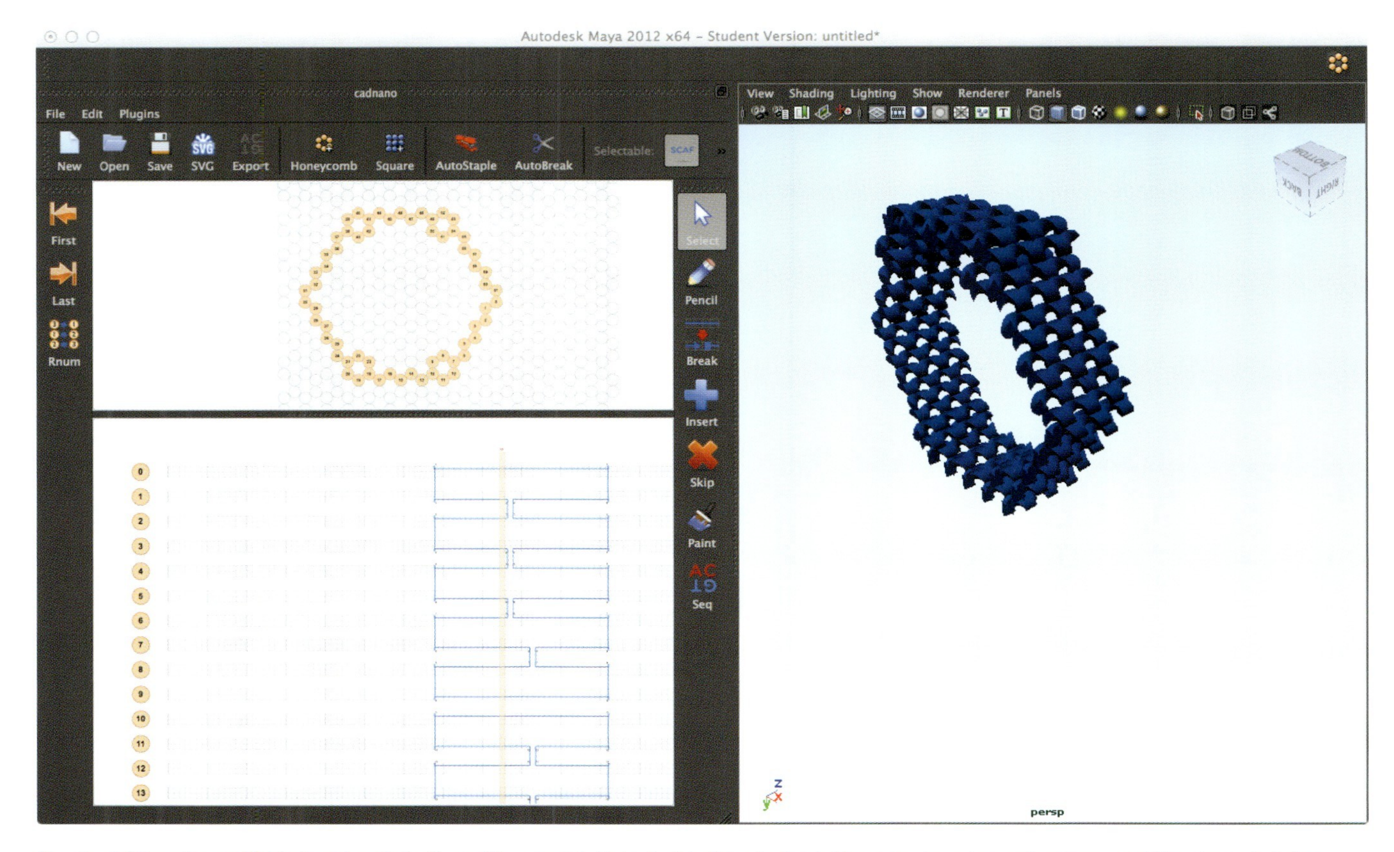

Fig. 4 Editing the scaffold strand path in the editing panel. Note that helices in the lattice panel are tagged orange, and the shape is being formed in real time on the 3D visualization panel on the right.

The grid size is a default value. To extend the grid, click the gray arrowheads on top of helix 0 in the editing panel, and insert the desired extension (in multiples of 21 bases, 21 being two complete turns of the DNA double helix).

7) The scaffold strand path can be extended by selecting parts of it and dragging them to the desired direction. The features to be selected can be chosen in the toolbox under "selectables," allowing the selection of only scaffold, staples, crossovers, or strands. This is particularly useful when the shape is very dense and several features are densely overlaid on each other.
8) To manually introduce a crossover between strands in adjacent helices, look for the locations marked as allowing crossover to take place. In these locations, bases in the two adjacent helices face directly each other, making it possible for a strand to cross from helix to helix without deformation. The allowed locations are marked with little bridge icons (Fig. 5). Simply click an icon to form a crossover. To delete a crossover, select the crossover and delete.
9) Once the scaffold strand path has been drawn, click "Autostaple." This orders caDNAno to staple the entire shape automatically. However, some of the resulting staples will require further editing – these staples will appear as thick lines, in contrast to valid staples which will appear as thin lines.
10) The staples can be edited manually, by extending them (if they are too short), deleting them (if they are unnecessary), or introducing breaks (if they are too long or circular). The latter is carried out by choosing the "Break" tool in the side toolbar, and clicking the point in which the break is to be inserted.
    Alternatively, staples can be automatically broken. For this, click "Autobreak." A dialog box will open asking for values for four parameters:
    a. Target length: optimal length for staples after breaking (default value is 49).
    b. Min length: minimal staple length (default value is 15).
    c. Max length: maximal staple length (default value is 60).
    d. Min distance to crossover: the minimal distance a staple will travel before crossing over to an adjacent helix (default value is 3).

    For now, leave the default values unchanged. With practice and experience, one can get a better sense of how these parameters are transformed to the design.
11) After the shape has been stapled and edited (Fig. 6), all that is left is to assign a scaffold strand sequence to it. For this, click the "Seq" tool in the side toolbar, and click at a point within the scaffold strand where you wish sequence assignment to begin. For practice purposes, this can be any point along the scaffold strand.
12) In the dialog box, choose the correct scaffold DNA from the list. Alternatively, click "custom" and paste a sequence of choice. Note, however, that the sequence chosen must not be shorter than required for completely filling the designed scaffold strand path. You will be notified in such a case to choose a different scaffold sequence.

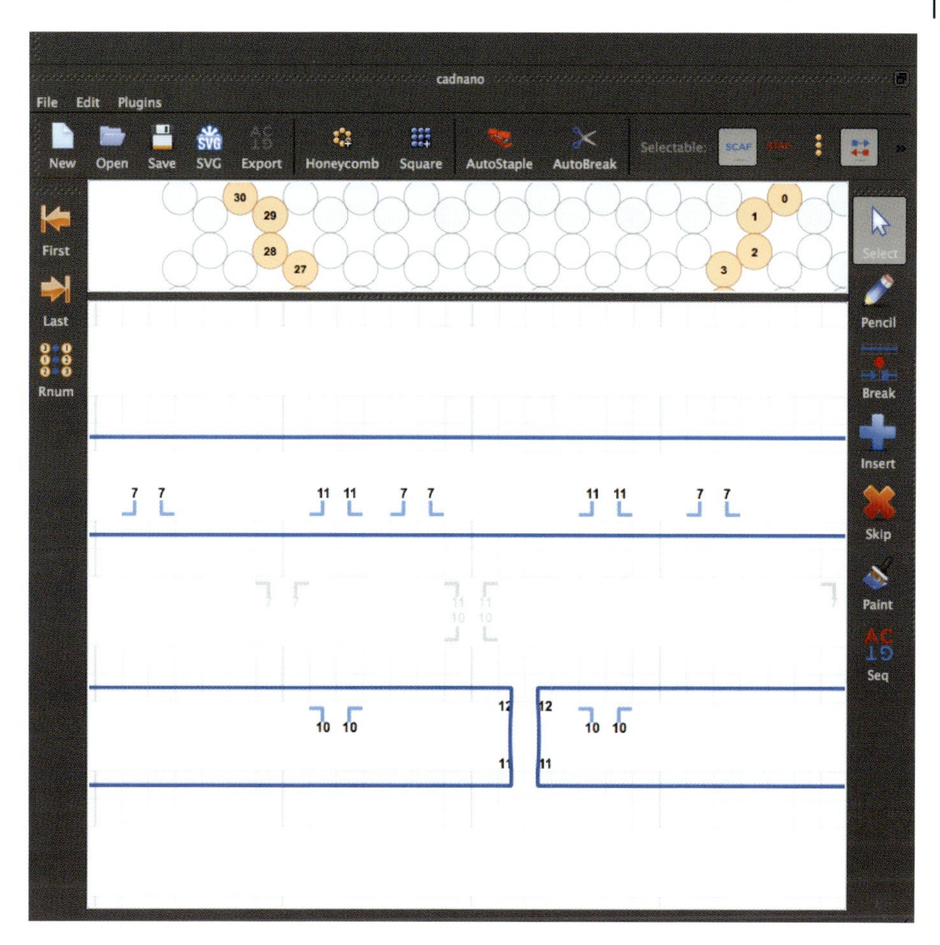

Fig. 5 Introducing crossovers between helices. The bridge icons mark the locations where crossovers are technically allowed. Numbers above and below the icons denote the number of helix to which the strand will cross over.

13) After choosing scaffold sequence, the sequence itself will appear in the editing panel. Make sure the entire shape is assigned with sequence. Regions of scaffold strand that are not assigned indicate the shape is inconsistent, for example, that parts of the scaffold strand form "islands." Go over the entire design again to eliminate such incidents, and assign sequence again to ensure the problem is solved.

14) From the top toolbar, click "Export," which will save the staple strand sequences in. CSV (comma-separated values) format. This can be later opened using a spreadsheet editor.

15) In the spreadsheet editor, sequences of "?????" indicate that this region of the staple is not hybridized with any scaffold. This is fine, and sometimes designed are based on leaving ssDNA tails of staples extending from the shape. These sequences need to be assigned manually.

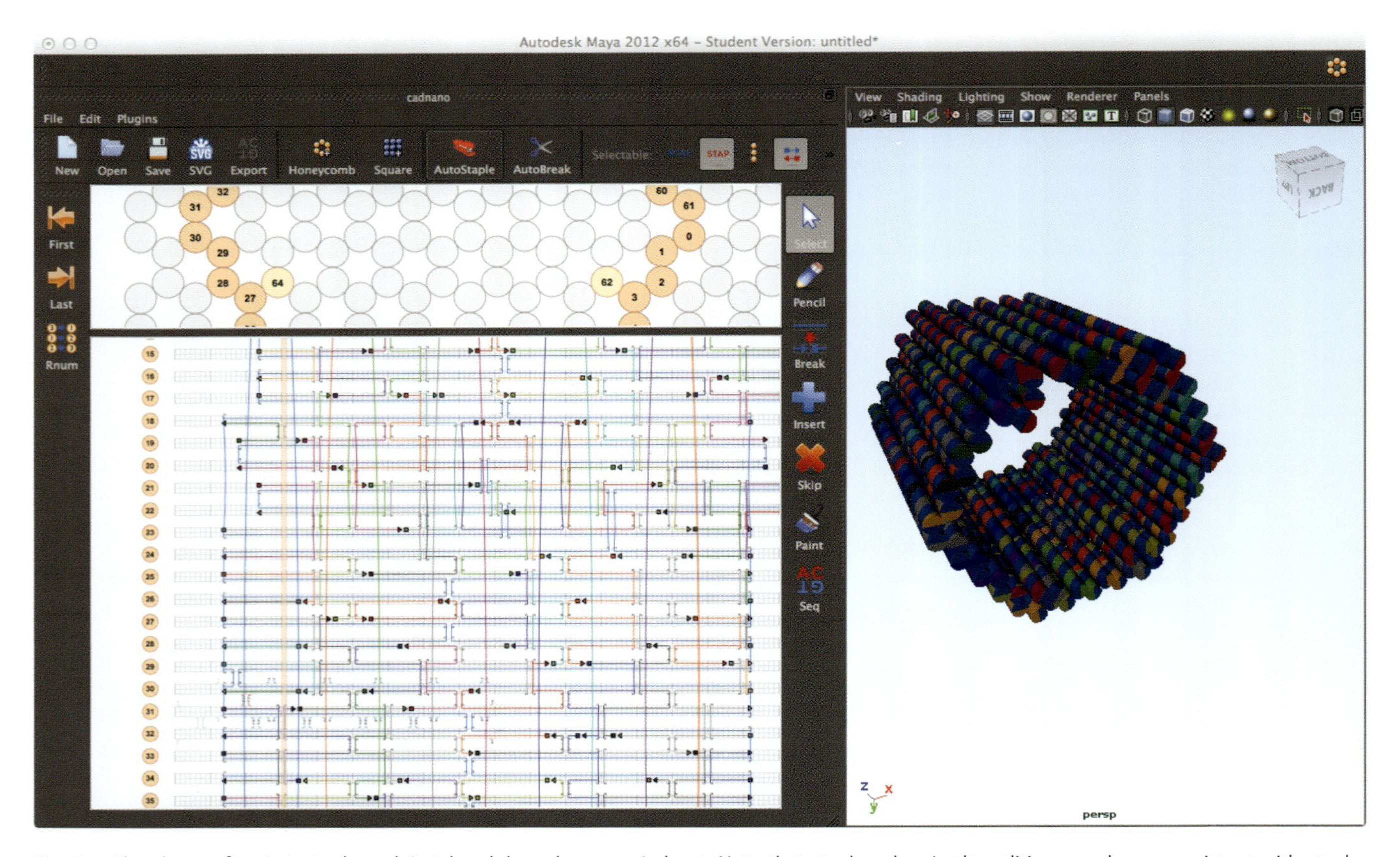

Fig. 6 The shape after Autostaple and Autobreak have been carried out. Note that staple colors in the editing panels are consistent with staple colors in the 3D panel on the right, making it easier to locate them if needed.

3 DNA Nanorobotics

In this section, the ways in which different components can be connected to construct a DNA nanorobot will be discussed, and the best way determined to utilize such a nanorobot for therapeutic applications.

3.1 Case Study: A DNA Nanorobot

Recently, DNA origami was used to design and fabricate an autonomous, logic-guided DNA origami nanorobot, which can be programmed to transport molecular payloads between selected points of origin and target [66].

The nanorobot resembles a hexagonal clamshell open at both ends, with two sides of the clamshell revolving around two single-stranded DNA (ssDNA) axes (Fig. 7). On the opposite side, a gate consisting of two double-stranded DNA (dsDNA) arms controls the nanorobot state. When the arms are in dsDNA configuration, the two halves of the clamshell are held locked; however, when these duplexes unzip and open, the nanorobot is free to entropically open, exposing its internal side. This side can be loaded with a variety of cargoes including small molecules, drugs, proteins, and small (<30–35 nm) nanoparticles. The number, stoichiometry, position, and order of payloads can be carefully planned.

Each DNA arm in the gate is made from a DNA or RNA aptamer, which is designed to sense an input molecule of choice, hybridized to a partially complementary strand. In the presence of the input molecule, the aptamer system switches to an equilibrium between the aptamer–complementary strand complex and the aptamer–molecule complex. The rate at which this occurs is difficult to model, because the system is governed by the thermodynamics of the DNA duplex on the one hand and by the mechanics of the nanorobot structure on the other hand. The hybridization strength between the aptamer and its complementary strand governs a trade-off between specificity and sensitivity; fewer mismatches would lead to a nanorobot that could take a long time (on the order of many hours) and high concentrations of the input to open, but is very specific to the chosen input. Alternatively, more mismatches would make a more sensitive and open-prone nanorobot, which might have a higher chance of a false-positive activation.

Programming of the nanorobot is achieved by choosing the appropriate aptamer components of the gate from the aptamer pool, and assembling them on the nanorobot chassis, in turn setting the nanorobot to be activated when it encounters the predefined signature of molecules and biological conditions recognized as correct input by the gate. On activation, the nanorobot undergoes a drastic conformational change and exposes its cargo, which was previously sequestered, enabling it to interface with the point of destination, for example, a tumor cell or a target tissue. The nanorobot can also revert to its inactive state in the absence of the required inputs, making its cargo concealed again, as demonstrated by dynamic light-scattering experiments.

In various demonstrations, nanorobots loaded with cancer therapeutics were capable of targeting specific tumor cells extremely selectively, and inducing growth arrest and apoptosis. The nanorobots were also capable of scavenging a bacterial protein from dilute solutions and carrying them to T cells, inducing their costimulation specifically for bacteria and

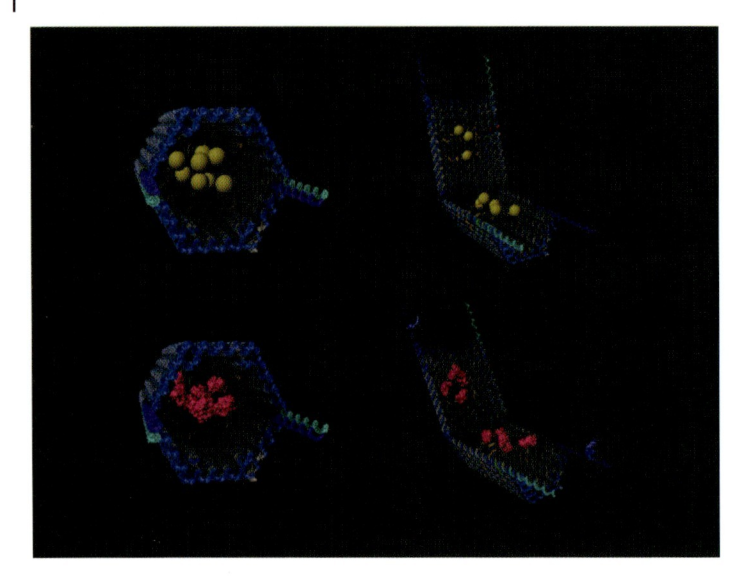

Fig. 7 A molecular model of the prototypical DNA nanorobot. The nanorobot resembles a hexagonal clamshell open at both ends, with two sides of the clamshell revolving around two ssDNA axes (not seen here). On the opposite side (shown here as the front side), a gate consisting of two dsDNA arms controls the nanorobot state: when the arms are in dsDNA configuration, the two halves of the clamshell are held locked. However, when these duplexes unzip and open, the nanorobot is free to entropically open, exposing its internal side. This side can be loaded with a variety of cargoes including small molecules, drugs, proteins (pink), and small (<30–35° nm) nanoparticles (yellow). The number, stoichiometry, position and order of the payloads can be carefully planned.

mimicking a primitive mode of antigen presentation.

An important predecessor to this nanorobot, and a landmark in the field of DNA nanorobotics, was the DNA origami cube described by Andersen and colleagues [44].

3.2 Sensing the Environment

The nanorobot gate in the abovementioned demonstration consisted of two aptamers – one for each input – such that the nanorobot requires both inputs molecules to open. These input molecules are not necessarily proteins. Some types of input that a DNA or RNA sequence can sense are as follows:

1) Proteins or small molecules (e.g., cancer markers, immune mediators, neurotransmitters, hormones, ATP, bacterial lipopolysaccharides) – by aptamers.
2) Bacteria – by a DNA sequence containing a restriction site for one of the enzymes produced by the bacterium of choice. It is important to note, however, that such a design suffers from the obvious drawback of being a one-time mechanism.
3) Nucleic acid sequences (e.g., from viruses) – by designing a gate in which one of the arms is displaced by the target sequence.

4) Temperature – by carefully designing a gate with the appropriate melting temperature (although the design would have to take into account salt concentrations and other factors that modulate the melting temperature).
5) DNA-binding proteins (e.g., transcription factors) – by a DNA sequence containing the cognate binding site of the protein.

In addition to these natural signals readable from the environment, it might be desirable to introduce user-generated signals that would enable the nanorobots to be operate from outside the body. Three examples include:

1) Ultraviolet (UV) light – by DNA gates containing UV-cleavable spacers or UV-sensitive bases. Most UV wavelengths are too energetic to penetrate tissues, and rather are absorbed so as to cause tissue damage; in contrast, infrared (IR) light can safely penetrate tissues to a certain depth, but is not energetic enough to alter DNA. A potential solution to this is to use upconverting nanoparticles [89], that sum up the energy from IR photons to produce UV photons. This design, too, would typically be a one-time mechanism.
2) Magnetic fields – using magnetite nanoparticles tethered to the gate in such a way that in a certain magnetic field the particles would repel each other, leading to an opening of the gate.
3) Radiofrequency – Hamad-Schifferli and colleagues have shown that DNA hybridization can be electronically remote-controlled by attaching the DNA to a metal nanocrystal antenna. Placing this construct in an electromagnetic field at a certain frequency induced heating of the metal and subsequent melting of the DNA duplex [90].

Of course, natural and user-generated signals could be integrated in one nanorobot for purposes of better control and increased safety.

3.3 Information Processing and Logic Types

A central part of a robot is the processor, which processes the inputs received from the sensors and generates a logical output. In this example, the nanorobot gate which serves as the processor is composed of two sensors connected to the chassis in such a way that sensing two inputs emulates a logical AND gate controlling the nanorobot state.

It might be helpful to consider the prototypical nanorobot described here as analogous to a combination lock, in which a series of cams with symbols (usually numerals) etched onto them can freely rotate relative to each other. Thus, the system may be at any of p^n states, where p is the range of symbols (e.g., 0–9) and n is the number of cams. As only one combination of symbols will allow the lock to be opened, the lock functions much like a mechanical-logical n-input AND gate, with inputs selectable from space p. In the same way, each of the arms in the nanorobot can function as a cam in a combination lock, typically with a binary input (the molecule can be either present or absent). In this demonstration, the nanorobot has only two arms, which is precisely equivalent to an electronic two-input AND gate; however, an additional aptamer-complementary system can hypothetically be linked serially to the first system comprising the arm, with or without certain spacing in between them to ensure that the two systems are truly independent.

In the present demonstration, the native state of each arm (in the absence of its cognate input molecule) is closed. However, it should be straightforward to design also arms that are open in the native state, and close in the presence of the input. Such an arm could be based on a hairpin structure, from which binding of the input molecule releases a segment that attaches to the nanorobot chassis and links both halves of the clamshell. This design can be adjusted to sense the other types of input listed above.

Although the design discussed here is merely a prototype, and is not necessarily optimal, its mechanical framework makes it simple to design logic. The concepts discussed here can also be used with other types of device. As an example, consider a DNA polyhedron of the category described by Mao and colleagues [55], in which segments of the junction motifs that build the shape are replaced with aptamers or other DNA-based sensors. Such a device might be a polyhedron that sequestered a drug but disintegrated in the presence of the input, whether a bacterial or tumor enzyme was degrading the sensor sequence. This simple device could respond to a single type of input; however, to raise its logic capacity would require a drastic change in its fabrication strategy, which should be considered. Mao's polyhedra are typically built using *algorithmic self-assembly*, in which the number of parts required to build a structure is larger than the number of *types* of different parts. The self-assembly of scaffolded DNA origami, for example, is not algorithmic as each building block is unique and has a specific position in the folded shape. The integration of additional sensors would almost certainly require additional types of building block, which in turn would mean that the polyhedron could not be constructed elegantly from a single junction motif. However, this could be accomplished with a careful design of the structure, taking into consideration the potential consequences for a correct assembly of the shape.

In this regard it is important also to mention the DNA computers described by Benenson and colleagues [24, 25, 91]. These constructs are not mechanical devices per se; rather, they can be thought of as pieces of code. Indeed, it might be imagined that such a construct could be used as the logical processor of a mechanical DNA nanorobot, similar to those discussed here.

3.4 Collective Behaviors

Natural and artificial systems of many independent agents demonstrate the ability to collectively perform many complex tasks – for example, construction (Turner, J.S., *American Entomologist*, 51, 36–38, 2005 [1]; Petersen, K. *et al.*, *Robotics: Science & Systems VII*, 2011), search [92], and locomotion [93] (Murata, S. and Kurokawa, H., *IEEE Robotics & Automation Magazine*, March 2007) – with greater speed, efficiency, or effectiveness than could a single agent alone. Direct and indirect coordination methods allow agents to collaborate to share information and to adapt their activity to fit the situation.

While most such systems rely on capabilities well beyond those of nanorobotics, some systems could be realized and implemented in real DNA nanorobots for therapeutic applications. For example, DNA strand displacement cascades similar to those demonstrated by Qian and colleagues [12, 31] can form the basis for a population of nanorobots interacting with each other, and operating in a defined sequence. An additional system might be of nanorobots building physical structures in one, two, or three dimensions in response to an input molecule or condition.

DNA nanorobots exhibiting collective behaviors could be extremely valuable in therapeutic applications. For example, they could carry out an autonomous excision of an abnormal tissue by collectively defining the target tissue, enzymatically cutting it from its environment, and then cooperating to carry it to a different location, possibly close to the skin, where it would be simpler to remove. Another example is that of nanorobots which can coordinate their activity to enable multiple drugs to function in parallel, without actually colliding with each other and generating adverse effects.

While this situation is still speculative, the next few years could witness the first prototypes of DNA nanorobots capable of exhibiting collective behaviors – a concept that has so far only been discussed in simulations and *in-silico* models.

3.5 Synthetic Genetic Circuits and DNA Nanorobotics

During the past decade, several research groups have described new types of engineered construct that could program a cell to express genes in a predefined, complex fashion. For example, synthetic circuits have been designed to function as oscillators [94], as toggle switches [95] and counters [96], and multicellular systems have also been described [97]. A comprehensive discussion of these constructs and their engineering and properties is beyond the scope of this chapter, but details can be found elsewhere [98].

While such synthetic circuits can operate only in the context of cells, they can nevertheless be coupled to the synthesis of a secreted signal to which the nanorobots outside the cell can respond. This would enable nanorobots to coordinate their activity, to oscillate, and to perform essentially as an extracellular avatar governed by the genetic circuit's product and following its dynamics. Clearly, this could open up exciting possibilities for controlling nanorobots in therapeutic settings.

4 Challenges of Applying DNA Nanorobots to Therapeutics

DNA nanorobotics could improve medicine and therapeutics with novel capabilities that are completely unachievable using current technologies. However, several technical challenges could hinder the use of DNA nanorobots as therapeutic devices *in vivo*.

DNA is not an optimal material for constructing therapeutic devices for two main reasons. First, it is rapidly degraded by nucleases in the serum and tissues, most notably by DNase I. Although the activity of DNase I is highly variable in the population owing to several mechanisms [99], it is very efficient and DNA origami has been shown to survive for only very short periods of time under DNase I treatment [87]. Efficient extracellular DNA digestion serves an immune function, as DNA floating freely in the serum can be most likely associated with pathogens, particularly viruses. Therefore, the use of DNA nanorobots as therapeutic devices requires, primarily, a good method of overcoming nuclease susceptibility.

Second, as noted above, DNA can be interpreted as a pathogenic component and consequently can exert a potent immune response. The mechanism responsible for this recognition of DNA is Toll-like receptor 9 (TLR9), an intracellular receptor that binds nonmethylated CpG DNA [100].

However, it is not only exogenous DNA that is immunogenic; if DNA that has been spilled from necrotic or damaged cells is not cleared efficiently, then autoimmune diseases such as systemic lupus erythematosus (SLE) might erupt [101, 102]. Clearly, the inherent immunogenicity of DNA is yet another obstacle that could hinder the use of DNA devices in therapeutics.

Nonetheless, some useful insights can be brought from orthogonal fields in which the delivery of systemic DNA is also a primary goal. Such fields include gene therapy and DNA vaccines. In these contexts, methods devised to increase the survivability of DNA in the serum and tissues could (hypothetically) be adapted and used for DNA origami nanorobots. For example, DNA origami structures could be embedded in nanoparticles or liposomes, coated with polyethylene glycol (PEG), or designed to exclude labile or immunogenic motifs, although it is still not clear whether the nanorobots could be modified and maintain their functionality.

Finally, it is still not appreciated how efficiently DNA origami can enter cells. "Professional" phagocytes such as macrophages and dendritic cells can uptake DNA origami efficiently [103], thus highlighting a potential use for DNA origami structures as intracellular delivery vehicles. On the other hand, certain DNA structures such as the nanorobot described here can enter cells extremely slowly, which suggests that this is a cell-specific phenomenon. Nevertheless, DNA origami could be functionalized with cell entry-promoting molecules such as positively charged peptides or polymers (e.g., compound 48/80). Alternatively, DNA origami could be decorated with ligands to receptors undergoing massive endocytosis and recycling, such as receptors for immunoglobulin A and oxidized low-density lipoprotein (LDL).

Despite these major challenges, it is highly likely that DNA nanotechnology will enter the arena of therapeutic devices within the next few years. Moreover, the nanorobots described in this chapter, along with other technologies, represent good candidate platforms [66, 103, 104].

5
Summary and Conclusions

During a 30-year period of research into DNA nanotechnology, insights regarding the structure, topology and mechanics of DNA have been translated into fascinating nanoscale devices, whether computers, motors, and/or architectures. However, it is only just being realized how these components could be integrated to create an entirely new generation of autonomous robots that would be capable of reading from and writing to the biochemistry and physiology of a living organism, in using drugs, molecular surgical tools, endogenous hormones and growth factors to improve the outcome in any case. Although several challenges clearly need to be addressed before this development can take place, it is safe to assume that the next five years will witness the first successful implementations of DNA nanorobotics in therapeutics and biomedical applications. Once successful, DNA nanorobotics could revolutionize the paradigms in a diversity of fields ranging from drug discovery and design to surgical procedures, regenerative medicine, and aging.

References

1. Cooper, W.O., Habel, L.A., Sox, C.M., Chan, K.A. *et al.* (2011) ADHD drugs and serious cardiovascular events in children and young adults. *N. Engl. J. Med.*, **365** (20), 1896–1904.

2. Ray, W.A., Murray, K.T., Hall, K., Arbogast, P.G. *et al.* (2012) Azithromycin and the risk of cardiovascular death. *N. Engl. J. Med.*, **366** (20), 1881–1890.
3. Jukema, J.W., Cannon, C.P., de Craen, A.J., Westendorp, R.G. *et al.* (2012) The controversies of statin therapy: weighing the evidence. *J. Am. Coll. Cardiol.*, **60**, 875–881.
4. Grant, L.M., Rockey, D.C. (2012) Drug-induced liver injury. *Curr. Opin. Gastroenterol.*, **28** (3), 198–202.
5. Govindan, S.V., Goldenberg, D.M. (2012) Designing immunoconjugates for cancer therapy. *Expert Opin. Biol. Ther.*, **12** (7), 873–890.
6. Moses, M.A., Brem, H., Langer, R. (2003) Advancing the field of drug delivery: taking aim at cancer. *Cancer Cell*, **4** (5), 337–341.
7. Sengupta, S., Eavarone, D., Capila, I., Zhao, G. *et al.* (2005) Temporal targeting of tumour cells and neovasculature with a nanoscale delivery system. *Nature*, **436** (7050), 568–572.
8. Schroeder, A., Honen, R., Turjeman, K., Gabizon, A. *et al.* (2009) Ultrasound triggered release of cisplatin from liposomes in murine tumors. *J. Control. Release*, **137** (1), 63–68.
9. Estey, E. (2012) Treatment of AML: resurrection for gemtuzumab ozogamicin? *Lancet*, **379** (9825), 1468–1469.
10. Karr, J.R., Sanghvi, J.C., Macklin, D.N., Gutschow, M.V. *et al.* (2012) A whole-cell computational model predicts phenotype from genotype. *Cell*, **150** (2), 389–401.
11. Winfree, E., Liu, F., Wenzler, L.A., Seeman, N.C. (1998) Design and self-assembly of two-dimensional DNA crystals. *Nature*, **394** (6693), 539–544.
12. Qian, L., Winfree, E., Bruck, J. (2011) Neural network computation with DNA strand displacement cascades. *Nature*, **475** (7356), 368–372.
13. Rothemund, P.W., Papadakis, N., Winfree, E. (2004) Algorithmic self-assembly of DNA Sierpinski triangles. *PLoS Biol.*, **2** (12), e424.
14. Adleman, L.M. (1994) Molecular computation of solutions to combinatorial problems. *Science*, **266** (5187), 1021–1024.
15. Braich, R.S., Chelyapov, N., Johnson, C., Rothemund, P.W. *et al.* (2002) Solution of a 20-variable 3-SAT problem on a DNA computer. *Science*, **296** (5567), 499–502.
16. Yeh, C.W., Chu, C.P., Wu, K.R. (2006) Molecular solutions to the binary integer programming problem based on DNA computation. *Biosystems*, **83** (1), 56–66.
17. Chen, J.H., Seeman, N.C. (1991) Synthesis from DNA of a molecule with the connectivity of a cube. *Nature*, **350** (6319), 631–633.
18. Seeman, N.C. (1982) Nucleic acid junctions and lattices. *J. Theor. Biol.*, **99** (2), 237–247.
19. Rothemund, P.W. (2006) Folding DNA to create nanoscale shapes and patterns. *Nature*, **440** (7082), 297–302.
20. Adar, R., Benenson, Y., Linshiz, G., Rosner, A. *et al.* (2004) Stochastic computing with biomolecular automata. *Proc. Natl Acad. Sci. USA*, **101** (27), 9960–9965.
21. Chang, W.L., Ho, M.S., Guo, M. (2004) Molecular solutions for the subset-sum problem on DNA-based supercomputing. *Biosystems*, **73** (2), 117–130.
22. Yin, Z., Zhang, F., Xu, J. (2002) A Chinese postman problem based on DNA computing. *J. Chem. Inf. Comput. Sci.*, **42** (2), 222–224.
23. Linial, M., Linial, N. (1995) On the potential of molecular computing. *Science*, **268** (5210), 481; author reply 483–484.
24. Xie, Z., Wroblewska, L., Prochazka, L., Prochazka, L. *et al.* (2011) Multi-input RNAi-based logic circuit for identification of specific cancer cells. *Science*, **333** (6047), 1307–1311.
25. Benenson, Y., Gil, B., Ben-Dor, U., Adar, R. *et al.* (2004) An autonomous molecular computer for logical control of gene expression. *Nature*, **429** (6990), 423–429.
26. Ellington, A.D., Szostak, J.W. (1990) In vitro selection of RNA molecules that bind specific ligands. *Nature*, **346** (6287), 818–822.
27. Seelig, G., Soloveichik, D., Zhang, D.Y., Winfree, E. (2006) Enzyme-free nucleic acid logic circuits. *Science*, **314** (5805), 1585–1588.
28. Zhang, D.Y., Seelig, G. (2011) Dynamic DNA nanotechnology using strand-displacement reactions. *Nat. Chem.*, **3** (2), 103–113.
29. Zhang, D.Y., Winfree, E. (2009) Control of DNA strand displacement kinetics using

toehold exchange. *J. Am. Chem. Soc.*, **131** (47), 17303–17314.
30. Yi, J., Zhang, W., Zhang, D.Y. (2006) Molecular Zipper: a fluorescent probe for real-time isothermal DNA amplification. *Nucleic Acids Res.*, **34** (11), e81.
31. Qian, L., Winfree, E. (2011) Scaling up digital circuit computation with DNA strand displacement cascades. *Science*, **332** (6034), 1196–1201.
32. Qian, L., Winfree, E. (2011) A simple DNA gate motif for synthesizing large-scale circuits. *J. R. Soc. Interface*, **8** (62), 1281–1297.
33. Bath, J., Green, S.J., Allen, K.E., Turberfield, A.J. (2009) Mechanism for a directional, processive, and reversible DNA motor. *Small*, **5** (13), 1513–1516.
34. Bath, J., Green, S.J., Turberfield, A.J. (2005) A free-running DNA motor powered by a nicking enzyme. *Angew. Chem. Int. Ed.*, **44** (28), 4358–4361.
35. He, Y., Liu, D.R. (2010) Autonomous multistep organic synthesis in a single isothermal solution mediated by a DNA walker. *Nat. Nanotechnol.*, **5** (11), 778–782.
36. Lund, K., Manzo, A.J., Dabby, N., Michelotti, N. *et al.* (2010) Molecular robots guided by prescriptive landscapes. *Nature*, **465** (7295), 206–210.
37. Wickham, S.F., Bath, J., Katsuda, Y., Endo, M. *et al.* (2012) A DNA-based molecular motor that can navigate a network of tracks. *Nat. Nanotechnol.*, **7** (3), 169–173.
38. Wang, T., Sha, R., Dreyfus, R., Leunissen, M.E. *et al.* (2011) Self-replication of information-bearing nanoscale patterns. *Nature*, **478** (7368), 225–228.
39. Venkataraman, S., Dirks, R.M., Rothemund, P.W., Winfree, E. *et al.* (2007) An autonomous polymerization motor powered by DNA hybridization. *Nat. Nanotechnol.*, **2** (8), 490–494.
40. Seeman, N.C. (1982) Nucleic acid junctions and lattices. *J. Theor. Biol.*, **99** (2), 237–247.
41. Breaker, R.R., Joyce, G.F. (1994) A DNA enzyme that cleaves RNA. *Chem. Biol.*, **1** (4), 223–229.
42. Masoud, R., Tsukanov, R., Tomov, T.E., Plavner, N. *et al.* (2012) Studying the structural dynamics of bipedal DNA motors with single-molecule fluorescence spectroscopy. *ACS Nano*, **6** (7), 6272–6283.
43. Yin, P., Yan, H., Daniell, X.G., Turberfield, A.J. *et al.* (2004) A unidirectional DNA walker that moves autonomously along a track. *Angew. Chem.*, **43** (37), 4906–4911.
44. Andersen, E.S., Dong, M., Nielsen, M.M., Jahn, K. *et al.* (2009) Self-assembly of a nanoscale DNA box with a controllable lid. *Nature*, **459** (7243), 73–76.
45. Gu, H., Chao, J., Xiao, S.J., Seeman, N.C. (2010) A proximity-based programmable DNA nanoscale assembly line. *Nature*, **465** (7295), 202–205.
46. Sa-Ardyen, P., Vologodskii, A.V., Seeman, N.C. (2003) The flexibility of DNA double crossover molecules. *Biophys. J.*, **84** (6), 3829–3837.
47. Fu, T.J., Kemper, B., Seeman, N.C. (1994) Cleavage of double-crossover molecules by T4 endonuclease VII. *Biochemistry*, **33** (13), 3896–3905.
48. Fu, T.J., Seeman, N.C. (1993) DNA double-crossover molecules. *Biochemistry*, **32** (13), 3211–3220.
49. Barish, R.D., Rothemund, P.W., Winfree, E. (2005) Two computational primitives for algorithmic self-assembly: copying and counting. *Nano Lett.*, **5** (12), 2586–2592.
50. Barish, R.D., Schulman, R., Rothemund, P.W., Winfree, E. (2009) An information-bearing seed for nucleating algorithmic self-assembly. *Proc. Natl Acad. Sci. USA*, **106** (15), 6054–6059.
51. Maune, H.T., Han, S.P., Barish, R.D., Bockrath, M. *et al.* (2010) Self-assembly of carbon nanotubes into two-dimensional geometries using DNA origami templates. *Nat. Nanotechnol.*, **5** (1), 61–66.
52. He, Y., Han, S.P., Barish, R.D., Mao, C. (2006) Highly connected two-dimensional crystals of DNA six-point-stars. *J. Am. Chem. Soc.*, **128** (50), 15978–15979.
53. Tian, Y., He, Y., Ribbe, A.E., Mao, C. (2006) Preparation of branched structures with long DNA duplex arms. *Org. Biomol. Chem.*, **4** (18), 3404–3405.
54. Sun, X., Hyeon Ko, S., Zhang, C., Ribbe, A.E. *et al.* (2009) Surface-mediated DNA self-assembly. *J. Am. Chem. Soc.*, **131** (37), 13248–13249.
55. He, Y., Ye, T., Su, M., Ribbe, A.E. *et al.* (2008) Hierarchical self-assembly of DNA into symmetric supramolecular polyhedra. *Nature*, **452** (7184), 198–201.

56. Zhang, C., Tian, C., Li, X., Qian, H. (2012) Reversibly switching the surface porosity of a DNA tetrahedron. *J. Am. Chem. Soc.*, **134** (29), 11998–12001.
57. Delebecque, C.J., Lindner, A.B., Silver, P.A., Aldaye, F.A. (2011) Organization of intracellular reactions with rationally designed RNA assemblies. *Science*, **333** (6041), 470–474.
58. Furdon, P.J., Guerrier-Takada, C., Altman, S. (1983) A G43 to U43 mutation in *E. coli* tRNAtyrsu3+ which affects processing by RNase P. *Nucleic Acids Res.*, **11** (5), 1491–1505.
59. Bass, B.L., Cech, T.R. (1984) Specific interaction between the self-splicing RNA of *Tetrahymena* and its guanosine substrate: implications for biological catalysis by RNA. *Nature*, **308** (5962), 820–826.
60. Zhang, B., Cech, T.R. (1997) Peptide bond formation by in vitro selected ribozymes. *Nature*, **390** (6655), 96–100.
61. Guerrier-Takada, C., Altman, S. (1984) Catalytic activity of an RNA molecule prepared by transcription in vitro. *Science*, **223** (4633), 285–286.
62. Bartel, D.P., Szostak, J.W. (1993) Isolation of new ribozymes from a large pool of random sequences [see comment]. *Science*, **261** (5127), 1411–1418.
63. Ekland, E.H., Szostak, J.W., Bartel, D.P. (1995) Structurally complex and highly active RNA ligases derived from random RNA sequences. *Science*, **269** (5222), 364–370.
64. Ko, S.H., Su, M., Zhang, C., Ribbe, A.E. (2010) Synergistic self-assembly of RNA and DNA molecules. *Nat. Chem.*, **2** (12), 1050–1055.
65. Tuerk, C., MacDougal, S., Gold, L. (1992) RNA pseudoknots that inhibit human immunodeficiency virus type 1 reverse transcriptase. *Proc. Natl Acad. Sci. USA*, **89** (15), 6988–6992.
66. Douglas, S.M., Bachelet, I., Church, G.M. (2012) A logic-gated nanorobot for targeted transport of molecular payloads. *Science*, **335** (6070), 831–834.
67. Yurke, B., Turberfield, A.J., Mills, A.P., Simmel, F.C. (2000) A DNA-fuelled molecular machine made of DNA. *Nature*, **406** (6796), 605–608.
68. Rief, M., Clausen-Schaumann, H., Gaub, H.E. (1999) Sequence-dependent mechanics of single DNA molecules. *Nat. Struct. Biol.*, **6** (4), 346–349.
69. Seidel, R., Dekker, C. (2007) Single-molecule studies of nucleic acid motors. *Curr. Opin. Struct. Biol.*, **17** (1), 80–86.
70. Mao, C., Sun, W., Shen, Z., Seeman, N.C. (1999) A nanomechanical device based on the B-Z transition of DNA. *Nature*, **397** (6715), 144–146.
71. Cheng, C.M., Lee, Y.J., Wang, W.T., Hsu, C.T. *et al.* (2011) Determining the binding mode and binding affinity constant of tyrosine kinase inhibitor PD153035 to DNA using optical tweezers. *Biochem. Biophys. Res. Commun.*, **404** (1), 297–301.
72. Song, G., Chen, M., Chen, C., Wang, C. *et al.* (2010) Design of proton-fueled tweezers for controlled, multi-function DNA-based molecular device. *Biochimie*, **92** (2), 121–127.
73. Han, X., Zhou, Z., Yang, F., Deng, Z. (2008) Catch and release: DNA tweezers that can capture, hold, and release an object under control. *J. Am. Chem. Soc.*, **130** (44), 14414–14415.
74. Paige, J.S., Wu, K.Y., Jaffrey, S.R. (2011) RNA mimics of green fluorescent protein. *Science*, **333** (6042), 642–646.
75. Dietz, H., Douglas, S.M., Shih, W.M. (2009) Folding DNA into twisted and curved nanoscale shapes. *Science*, **325** (5941), 725–730.
76. Douglas, S.M., Dietz, H., Liedl, T., Högberg, B. *et al.* (2009) Self-assembly of DNA into nanoscale three-dimensional shapes. *Nature*, **459** (7245), 414–418.
77. Han, D., Pal, S., Nangreave, J., Deng, Z. *et al.* (2011) DNA origami with complex curvatures in three-dimensional space. *Science*, **332** (6027), 342–346.
78. Yin, P., Hariadi, R.F., Sahu, S., Choi, H.M. *et al.* (2008) Programming DNA tube circumferences. *Science*, **321** (5890), 824–826.
79. Wei, B., Dai, M., Yin, P. (2012) Complex shapes self-assembled from single-stranded DNA tiles. *Nature*, **485** (7400), 623–626.
80. Hogberg, B., Liedl, T., Shih, W.M. (2009) Folding DNA origami from a double-stranded source of scaffold. *J. Am. Chem. Soc.*, **131** (26), 9154–9155.
81. Sobczak, J.P., Martin, T.G., Gerling, T., Dietz, H. (2012) Rapid folding of DNA into nanoscale shapes at constant temperature. *Science*, **338** (6113), 1458–1461.

82. Zuker, M. (2003) Mfold web server for nucleic acid folding and hybridization prediction. *Nucleic Acids Res.*, **31** (13), 3406–3415.
83. Zadeh, J.N., Steenberg, C.D., Bois, J.S., Wolfe, B.R. *et al.* (2011) NUPACK: analysis and design of nucleic acid systems. *J. Comput. Chem.*, **32** (1), 170–173.
84. Seeman, N.C. (1990) De novo design of sequences for nucleic acid structural engineering. *J. Biomol. Struct. Dyn.*, **8** (3), 573–581.
85. Zhu, J., Wei, B., Yuan, Y., Mi, Y. (2009) UNIQUIMER 3D a software system for structural DNA nanotechnology design, analysis and evaluation. *Nucleic Acids Res.*, **37** (7), 2164–2175.
86. Douglas, S.M., Marblestone, A.H., Teerapittayanon, S., Vazquez, A. *et al.* (2009) Rapid prototyping of 3D DNA-origami shapes with caDNAno. *Nucleic Acids Res.*, **37** (15), 5001–5006.
87. Castro, C.E., Kilchherr, F., Kim, D.N., Shiao, E.L. *et al.* (2011) A primer to scaffolded DNA origami. *Nat. Methods*, **8** (3), 221–229.
88. Ke, Y., Douglas, S.M., Liu, M., Sharma, J. *et al.* (2009) Multilayer DNA origami packed on a square lattice. *J. Am. Chem. Soc.*, **131** (43), 15903–15908.
89. Wang, M., Abbineni, G., Clevenger, A., Mao, C. *et al.* (2011) Upconversion nanoparticles: synthesis, surface modification and biological applications. *Nanomedicine*, **7** (6), 710–729.
90. Hamad-Schifferli, K., Schwartz, J.J., Santos, A.T., Zhang, S. (2002) Remote electronic control of DNA hybridization through inductive coupling to an attached metal nanocrystal antenna. *Nature*, **415** (6868), 152–155.
91. Benenson, Y., Paz-Elizur, T., Adar, R., Keinan, E. *et al.* (2001) Programmable and autonomous computing machine made of biomolecules. *Nature*, **414** (6862), 430–434.
92. Wood, R., Nagpal, R., Wei, G.Y. (2013) Flight of the robobees. *Sci. Am.*, **308** (3), 60–65.
93. Bonner, J.T. (2009) *The Social Amoebae: The Biology of Cellular Slime Molds*, Princeton University Press, Princeton, vol. ix.
94. Elowitz, M.B., Leibler, S. (2000) A synthetic oscillatory network of transcriptional regulators. *Nature*, **403** (6767), 335–338.
95. Gardner, T.S., Cantor, C.R., Collins, J.J. (2000) Construction of a genetic toggle switch in *Escherichia coli*. *Nature*, **403** (6767), 339–342.
96. Friedland, A.E., Lu, T.K., Wang, X., Shi, D. *et al.* (2009) Synthetic gene networks that count. *Science*, **324** (5931), 1199–1202.
97. Basu, S., Gerchman, Y., Collins, C.H., Arnold, F.H. *et al.* (2005) A synthetic multicellular system for programmed pattern formation. *Nature*, **434** (7037), 1130–1134.
98. Slusarczyk, A.L., Lin, A., Weiss, R. (2012) Foundations for the design and implementation of synthetic genetic circuits. *Nat. Rev. Genet.*, **13** (6), 406–420.
99. Yasuda, T., Iida, R., Ueki, M., Tsukahara, T. *et al.* (2004) A novel 56-bp variable tandem repeat polymorphism in the human deoxyribonuclease I gene and its population data. *Leg. Med. (Tokyo)*, **6** (4), 242–245.
100. Hemmi, H., Takeuchi, O., Kawai, T., Kaisho, T. *et al.* (2000) A Toll-like receptor recognizes bacterial DNA. *Nature*, **408** (6813), 740–745.
101. Tan, E.M. (2012) Autoantibodies, autoimmune disease, and the birth of immune diagnostics. *J. Clin. Invest.*, **122** (11), 3835–3836.
102. Christensen, S.R., Kashgarian, M., Alexopoulou, L., Flavell, R.A. *et al.* (2005) Toll-like receptor 9 controls anti-DNA autoantibody production in murine lupus. *J. Exp. Med.*, **202** (2), 321–331.
103. Schuller, V.J., Heidegger, S., Sandholzer, N., Nickels, P.C. *et al.* (2011) Cellular immunostimulation by CpG-sequence-coated DNA origami structures. *ACS Nano*, **5** (12), 9696–9702.
104. Lee, H., Lytton-Jean, A.K., Chen, Y., Love, K.T. *et al.* (2012) Molecularly self-assembled nucleic acid nanoparticles for targeted in vivo siRNA delivery. *Nat. Nanotechnol.*, **7** (6), 389–393.

4
Graphene and Graphene Derivatives in Biosensing, Imaging, Therapeutics, and Genetic Engineering

Kim Truc Nguyen[1] *and Yanli Zhao*[2]
[1]*Nanyang Technological University, Division of Chemistry and Biological Chemistry, School of Physical and Mathematical Sciences, 21 Nanyang Link, 637371 Singapore*
[2]*Nanyang Technological University, School of Materials Science and Engineering, 50 Nanyang Avenue, 639798 Singapore*

Translational Nanomedicine. First Edition. Edited by Robert A. Meyers.

Keywords

Graphene
A crystalline allotrope of carbon where the atoms arrange into a two-dimensional structure. Graphene is comprised of carbon atoms in their sp^2 configuration, densely packed into a single atom-thick layer that make graphene the thinnest but strongest material in the universe.

Graphene derivative
A material consisting of graphene or its oxidized form – graphene oxide (GO) or reduced graphene oxide (rGO). A graphene derivative can be obtained by covalent and noncovalent modification on graphene, rGO, or GO surfaces.

Molecular imaging
The twenty-first century medical imaging revolution that integrates molecular biology and *in-vivo* imaging. Molecular imaging allows visualization at the cellular level with specific targets, which dramatically enhances the sensitivity and resolution of the obtained image. This cutting-edge imaging technique has great potential in the early detection (pre-symptom) of life-threatening conditions such as cancer and neurological and cardiovascular diseases.

Theranostics
The combination of diagnostics and therapeutics. This is one of the most common combinations in multifunctional biomaterials, and enables the simultaneous monitoring of treatment and feedback to effectively determine the efficacy of the drug used.

Genetic engineering
Genetic engineering (also called genetic modification) involves direct manipulation of an organism's DNA. By employing biotechnology such as molecular cloning, DNA from the host organism is isolated and modified – that is, specific genes are inserted, replaced, or removed from the origin DNA. The new DNA is then inserted into the host organism to develop desired properties.

Graphene is currently a "shining star" in nanomedicine on account of its good biocompatibility, low cytotoxicity, and ease of functionalization. The versatile nature of modifiable graphene leads to a novel horizon with promising application opportunities in a wide range of technologies and markets. The unique applications of graphene and its derivatives in biosensing, bioimaging, therapeutics, and genetic engineering are discussed in this chapter.

1 Introduction

Graphene, as a potential biomaterial in the twenty-first century medical revolution, is notable for its intrinsic outstanding physico-chemical properties. Currently, graphene is the thinnest – but strongest – material in the universe due to its two-dimensional (2D) sp^2 carbon network [1, 2]. Although the electronic structure of graphene was first studied in 1947, on a theoretical basis, by P. R. Wallace [3], the monolayer of graphene was actually obtained for the first time 57 years later when, in 2004, Geim and Novoselov published their results in *Science* describing the fabrication, identification and characterization of single-layer graphene [4]. Since then, research into graphene has become a "hot" topic, a fact recognized by the exponential increase in the number of publications on the subject each year. Unfortunately, the inert chemical properties of graphene, as well as difficulties with its dispersal in aqueous solvents, have limited its practical applications in biological systems. The biofunctions of graphene can be enhanced by either modifying its inert surface or using alternative graphene derivatives. For example, graphene oxide (GO) and reduced graphene oxide (rGO) are two major graphene derivatives which not only compensate for some of the drawbacks of graphene but also bring about new properties [5–12]. One of the major advantages of GO derivatives is their ability to be easily dispersed in aqueous solution, thus broadening their use in biological applications [13, 14]. rGO, which can be considered as a "transitional material" between graphene and GO, offers both aqueous solubility and a partial recovery of the conjugate system in carbon network, making it a promising candidate for large-scale synthesis in electrochemical sensing applications. The low cytotoxicity of graphene derivatives was demonstrated both *in vitro* and *in vivo*, and this encouraged their further development for biomedical applications. For instance, when GO was coated with chitosan the material demonstrated a significant reduction in hemolytic activity compared to bare graphene sheets [15]. Subsequently, *in-vitro* investigations using neuronal pheochromocytoma (PC12) cells and human alveolar epithelial (A549) cells confirmed the benign properties of graphene derivatives at the cellular level [16, 17]. Moreover, when long-term biodistribution and toxicology studies of polymer-coated graphene were carried out in mice over a period of 3 months, at a dose level of 20 mg kg^{-1}, no appreciable toxicity was demonstrated [18].

The exceptional properties of graphene, along with its potential applications, rely on the following attributes:

1) Graphene is a unique nanoscale material: all sp^2 carbon atoms in graphene are situated at the exposed surface, providing ample opportunities for interactions with biological matter of different sizes and shapes [19–28]. The high electron mobility (10 000 cm^2 V^{-1} s^{-1}) [29, 30] makes graphene the best candidate for facilitating electron transportation. This property is in contrast to conventional metal nanowires, where only a relatively small fraction of the atoms are at the surface while, at the same time, almost all of the current (if this property is exploited) flows in bulk. As an example, a biosensor fabricated from a graphene-based material is expected to be significantly more sensitive than a silicon nanowire-based sensor [31–38].

2) The carbon layer structure of graphene is robust and inert, and can serve as a platform onto which small drug molecules or genes can be attached via various interactions. For example, $\pi-\pi$ stacking interaction is one of the most common interactions for the attachment of drugs and genes.
3) Through oxygen-containing groups on the GO layer, for example, epoxide, –OH, and –COOH, organic synthesis principles can be employed to couple a wide range of biomolecules onto graphene. This approach has been used to prepare graphene-based hybrids with various biomaterial components that can be used for bioimaging and photothermal therapy (PTT) [20–22, 24, 25].

The above-described attributes – often in combination – allow a variety of application opportunities, leading to a rapid development of multifunctional biomaterials based on graphene and its derivatives. In this chapter, some important approaches that have been developed for interfacing graphene-based nanomaterials with biomedicine and biotechnology are highlighted. Notably, four main applications of graphene-based materials in the field of theranostics, including biosensing, bioimaging, therapeutics and genetic engineering, will be discussed.

2
Biosensing

A biosensor normally consists of two parts: (i) the transducer or detector element; and (ii) an electronic system (Fig. 1). The transducer or detector element is fabricated to detect a corresponding biomarker, and is usually comprised of a bioreceptor, such as single-stranded DNA (ssDNA), and antibody or an enzyme, and the electrical interface that might include either a conventional electrode or field-effect transistor (FET).

The performance of a biosensor is normally evaluated based on the following criteria: (i) sensitivity; (ii) selectivity; (iii) stability; and (iv) detection limit. Due to the high active surface area, as well as the ease of introduction of the bioreceptor, graphene and its derivatives have been widely used in the fabrication of electrical interfaces in order to enhance the performance of transducers. The biocompatible properties of graphene represent further advantages brought to the graphene-based hybrids for *in-situ* biosensing. In the following section, attention will be focused on how graphene-based materials can enhance the performance of transducers and bring about new biosensor evolution in medical applications.

2.1
Modification of Bioreceptor on Graphene by Covalent and Noncovalent Interactions

Graphene-based hybrid materials possess several properties that dramatically enhance the performance of the transducer in terms of sensitivity, selectivity, and detection range. First, graphene-modified electrical interfaces normally have an enhanced active surface area compared to conventional electrodes made from graphite. Moreover, due to the 2D structure of graphene the entire carbon network is exposed to the outer environment, thus providing more sites at which the analyte can anchor. The high edge-to-base ratio in graphene also offers more active sites, and this was reported to accelerate electron transfer between the electrode and the analytes [39, 40]. To date, hundreds of

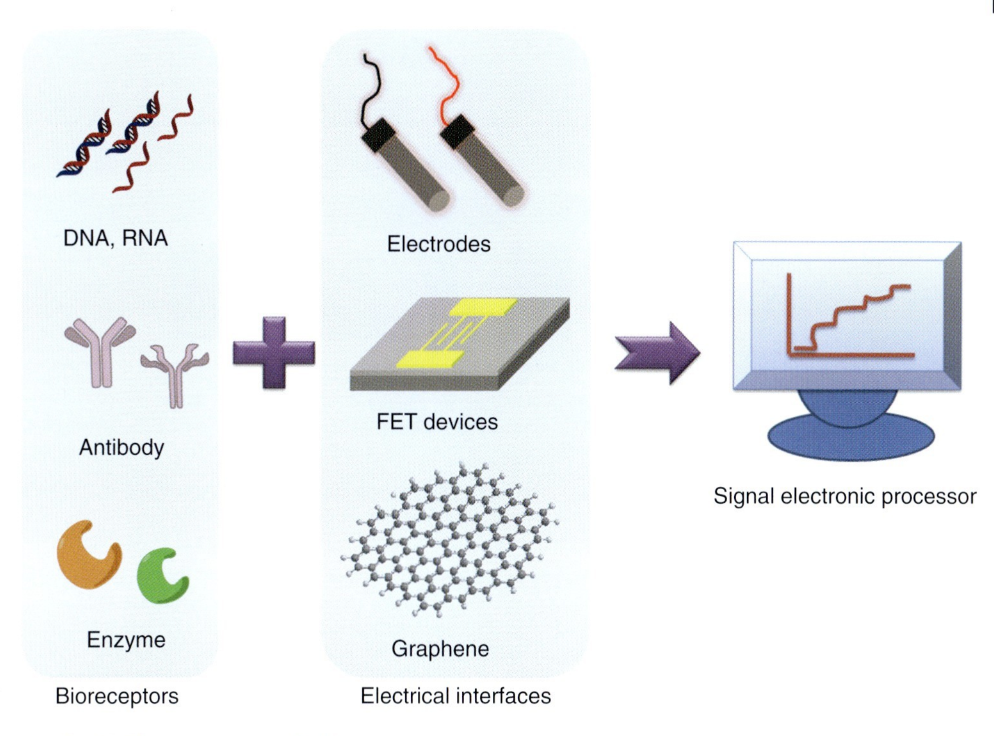

Fig. 1 The components of a biosensor.

graphene-based sensors have been investigated, many of which have shown potential applications in biological areas. The use of different types of supramolecular interaction, such as hydrophobic interactions between proteins and graphene derivatives, aromatic $\pi-\pi$ stacking interactions, and recognition between an antibody and an antigen, is the governing construction strategy for the preparation of cutting-edge biosensors.

For example, a sensitive biosensor to detect the cancer biomarker prostate-specific antigen (PSA), with a detection limit of 0.08 ng ml^{-1}, was prepared by conjugating graphene sheets with antibodies and bovine serum albumin (BSA) through the succinimidyl ester and $\pi-\pi$ stacking interaction provided by 1-pyrenebutanoic acid, respectively [41]. In this case, the protein detection strategy was based on antibody–antigen binding, while the change in the conductivity of graphene film corresponded to the surface attachment of biomolecules around the percolation threshold of the graphene film.

In another example, graphene was used as a platform for the immobilization of the mediator thionine (TH), horseradish peroxidase (HRP), and a secondary PSA antibody (Ab2). The resultant graphene/TH/HRP-Ab2 hybrid was employed as a label for ultrasensitive electrochemical biosensors [42]. This application was effective because: (i) the high surface-to-volume ratio of the graphene sheets allowed the immobilization of a high level of primary antibody (Ab1), Ab2, TH and HRP; and (ii) the good electrical conductivity of graphene improved electron transfer among TH, HRP, H_2O_2, and the electrode. The immunosensor

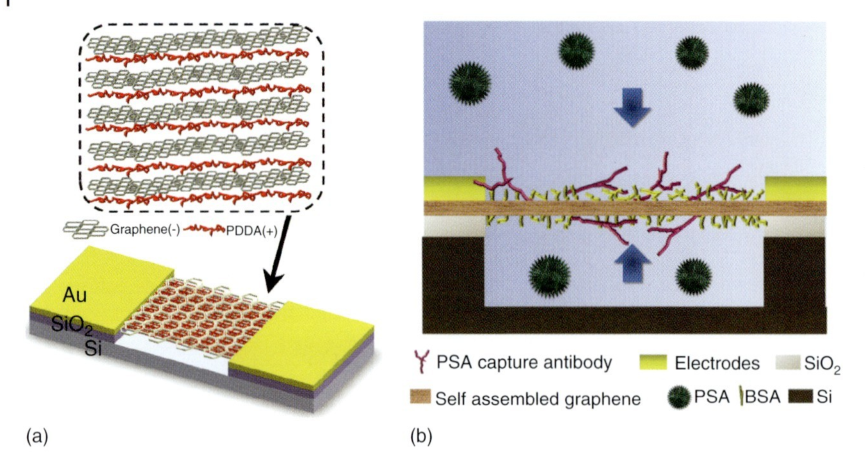

Fig. 2 (a) Schematic representation of suspended LBL self-assembled graphene sensor, where self-assembled graphene serves as the conducting channel bridging the electrodes of a biosensor; (b) Schematic illustration of immunoreaction between PSA-capture antibodies and target protein PSA. When the graphene biosensor modified by capture antibodies encounters the PSA solution, the immunoreaction takes place at both sides of the suspended graphene, and the conductance of graphene changes due to the absorption of PSA on the graphene surface [43]. Reprinted with permission from Ref. [43]; © 2012, Elsevier.

produced demonstrated an effective performance and displayed a wide range of linear response (from 0.002 to 10 ng ml^{-1}), a low detection limit (1 pg ml^{-1}), and also good reproducibility, selectivity, and stability. The transducer, which was based on suspended layer-by-layer (LBL) self-assembled graphene, also exhibited an ultrahigh sensitivity for cancer sensing [43]. For example, the suspended graphene sensors, when functionalized with specific anti-PSA antibodies as bioreceptors, were capable of detecting PSA levels down to 0.4 fg ml^{-1}, which was at least one order of magnitude lower than for unsuspended devices based on conductance shift curves (Fig. 2).

The emergence of novel, label-free detection technologies has dramatically transformed the biosensor field. Conventionally used tagging methods often involve labeling reagents that may have a harmful influence on the bioactivity of the tagging target through misfolding, reduced mobility and, in some cases, incompatibility with live cells. Alternatively, label-free technique offers *in-situ* detection with live cells, which can generate more physiologically relevant data at the cellular level. The label-free detection technique relies on the modification of a bioreceptor, which can specifically recognize the targeted analyte on the electrical interfaces. More recently, the label-free detection of cells or bacteria has attracted much attention, and several of these biosensors have been developed. One such example is a biosensor based on a chitosan/reduced graphene composite that was fabricated using a controllable electrodeposition technique for the selective detection of marine pathogenic sulfate-reducing bacteria (SRB) [44]. With this biosensor, a linear relationship was obtained between charge-transfer resistance and SRB concentration over a range of 18 to 18×10^7 cfu ml^{-1}. However, whilst the impedimetric biosensor provided a distinct response to SRB, it had no obvious response to *Vibrio anguillarum*, and showed a high selectivity for detecting the pathogen.

Graphene-based hybrids have also been used for the label-free detection of cancer cells. In this case, a transducer capable of realizing label-free cancer cell detection was prepared by functionalizing the aptamer AS1411 onto a graphene surface [45]. Intriguingly, when AS1411 formed a stable G-quadruplex structure it showed a high binding affinity to nucleolin, a protein that is overexpressed in cancer cells. This resulted in a graphene-based aptasensor that could be used to perform a targeted detection of as few as 1000 cancer cells.

In addition to the conventional signal-amplifying and processing methods, changes in the color of dyes in response to redox stimuli have also been employed in biosensors. Typically, hybrids created to detect biomarkers produce not only electrochemical but also optical signals; for example, a colorimetric immunoassay as a potential point-of-care tool for clinical diagnosis of the cancer biomarker PSA was recently developed based on the intrinsic peroxidase activity of GO [46]. In the presence of PSA, an immunocomplex was formed between the label for the immunoassay (GO/Ab2) and the magnetic bead (Ab1). Separation of the immunocomplex by applying an external magnetic field led to different amounts of GO/Ab2 in solution being mixed with hydroquinone and H_2O_2 solution, such that color changes (to brown) were displayed following oxidation. In this way, different concentrations of PSA could be determined directly from the various shades of brown coloration produced.

A second example of optical signal production is the highly sensitive and selective fluorescent aptasensor used to detect Mucin 1 (MUC1), a well-known tumor biomarker. In this case, GO was used as a quenching agent to reduce the fluorescence of single-stranded, dye-labeled MUC1-specific aptamer [47]. By using this aptasensor, MUC1 can be detected over a wide range of 0.04 to 10 mM, with a detection limit of 28 nM and a good selectivity. These parameters were verified in a real sample application by determining a series of concentrations of MUC1 which had been added to a 2% serum-containing buffer solution.

In short, several types of biosensor, including electrochemical and colorimetric, have been developed based on the intrinsic properties of graphene – notably the fast electron-transfer kinetics and the promotion of signal amplification – in order to achieve a high detection sensitivity, a wide detection range, good selectivity, and a low detection limit. A new type of graphene hybrid capable of further enhancing the performance of the transducer is described in the following section.

2.2 Advantage of Graphene/Nanoparticle Hybrids in Enhancing Transducer Performance

The modification of electronic interfaces with graphene has produced a major impact on electrochemical biosensing applications. Unfortunately, biological species such as enzymes will normally have poor contacts between their active sites and the electrode surface, on account of the embedded nature of the active site within a thick, insulating protein shell. Consequently, any nanoparticles (NPs) present on the surface of graphene will play an important role as electron transfer mediators between the redox center in proteins and the graphene layers [48]. Furthermore, metal nanoparticles (MNPs) can help to increase the sensitivity of the sensing electrodes by lowering the over-potentials. Several types

Tab. 1 Various graphene/nanoparticle hybrid-based electrodes for electrochemical sensing applications.

Hybrid	*Biomarker*	*Sensitivity* *(μA mM^{-1} cm^{-2})*	*Detection limit* *(S/N = 3)*	*Linear range*	*Reference*
PDDA@AuNP-graphene/MWCNT	Glucose	29.72	4.8 μM	5–175 μM	[60]
Graphene/PtNi NPs	Glucose	20.42	—	Up to 35 mM	[61]
Graphene/Pd	Glucose	—	1 μM	10 μM–5 mM	[49]
GO/Ag/SiO_2	Glucose	—	4 μM	0.1–260 mM	[62]
Chitosan-graphene/PdNPs	Glucose	31.2	0.2 μM	1 μM–1 mM	[63]
rGO/Au/PdNPs	Glucose	266.6	6.9 μM	Up to 3.5 mM	[55]
CVD graphene/AuNPs	Glucose	—	—	—	[53]
Graphene paper/PtAu-MnO_2	Glucose	58.54	0.02 mM	0.1–30 mM	[64]
Cu_2O wrapped by graphene	Glucose	—	3.3 μM	0.3–3.3 mM	[65]
Graphene/AuNPs/GOD/chitosan	Glucose	99.5	180 μM	2–10 mM at −0.2 V	[66]
	—	—	—	2–14 mM at 0.5 V	
Graphene/Pt	AA	0.3457	0.15 μM	0.15–34.4 μM	[51]
	DA	0.9695	0.03 μM	0.03–8.13 μM	
	UA	0.4119	0.05 μM	0.05–11.85 μM	
GS/Fe_3O_4@Au-S-Fc	DA	—	0.1 μM	0.5–50 μM	[67]
	UA	—	0.2 μM	1–300 μM	
	AC	—	0.05 μM	0.3–250 μM	
Graphene/AuNPs/anti-CEA antibody	Carcinoembryonic antigen	—	0.01 ng ml^{-1}	0.05–350 ng ml^{-1}	[68]
Tyr-AuNP/PASE-GO/SPE	Catechol	—	0.024 μM	0.083–23 μM	[69]
Graphene/DA-Fe_3O_4-Fc-Ab_2	Cancer biomarker	—	2 pg ml^{-1}	0.01–40 ng ml^{-1}	[70]
Graphene/Au-Fe_3O_4	Prostate specific antigen	—	5 pg ml^{-1}	0.01–10 ng ml^{-1}	[57]
Graphene/PDDA-AuNPs	Uric acid	0.10308	0.1 μM	0.5–20 μM	[71]
Graphene/AuNP-TiO_2	ProGRP	—	3 pg ml^{-1}	10–500 pg ml^{-1}	[58]
rGO/hollow CoPt	Thrombin	—	0.34 pM	1–50 000 pM	[72]
Graphene/chitosan-PtNPs/AuNPs	Erythromycin	—	0.023 μM	0.07–90 μM	[73]

Graphene/AuNPs	Cholesterol	0.00314	0.05 μM	0.05–0.35 mM	[54]
TiO_2/graphene/Pt–Pd/AuNPs	Cholesterol	—	0.017 μM	0.05–590 μM	[74]
GO/AgNPs	Tryptophan	—	2 nM	—	[52]
Cu_2O wrapped by graphene	H_2O_2	—	20.8 μM	0.3–7.8 mM	[65]
GO/antibody-AuNRs	Transferrin	—	—	0.0375–40 μg ml^{-1}	[75]
G-chitosan/Fc-S/AuNPs	Rutin	—	10 nM	0.04–100 μM	[76]
rGO/AgPd NPs	Ractopamine	—	1.52 pg ml^{-1}	0.01–100 ng ml^{-1}	[56]
rGO/AgPd NPs	Salbutamol	—	1.44 pg ml^{-1}	—	[77]
	Clenbuterol	—	1.38 pg ml^{-1}	—	
Graphene/AuNPs/PANi	DNA	—	2.11 pM	10–1000 pM	[78]
Single layer GO/AuNPs	DNA	—	100 fM	—	[79]
Graphene-thionine/Ag@Fe_3O_4 NPs	Kanamycin	—	15 pg ml^{-1}	0.05–16 ng ml^{-1}	[80]
rGO/Pt	Oxalic acid	—	10 μM	0.1–15 mM	[50]
AuNPs/GO/ITO	Dopamine	62.7	0.06 μM	—	[81]
Cells on rGO/AuNPs	Pheochromocytoma cell (PC-12)	—	5×10^3 cells ml^{-1}	1.6 × 104 to 1.6 × 107 cells ml^{-1}	[82]

GOD = glucose oxidase; ITO = indium tin oxide; MWCNT = multiwalled carbon nanotube; NP = nanoparticle; PANi = polyaniline; S/N = signal-to-noise ratio; SPE = screen-printed electrode.

of NP on graphene sheets for electrochemical sensing applications are detailed in Table 1. Single MNPs, such as Pd [49], Pt [50, 51], Ag [52], and Au [53, 54] NPs, as well as bimetallic NPs such as Au-Pd [55], Ag-Pd [56], Au-Fe_3O_4 [57], and Au-TiO_2 [58], were successfully incorporated onto the graphene sheet, and their usefulness validated for sensing various types of biomarker. The simultaneous detection of ascorbic acid (AA), dopamine (DA) and uric acid (UA) by means of cyclic voltammetry and differential pulse voltammetry (DPV) was achieved by modifying the electrical interface with a hybrid of size-selective Pt NPs (mean diameter 1.7 nm) and graphene sheet. By using optimized, size-selected Pt NPs the electrochemical potential difference between the three analytes was widened, which enabled the concurrent detection [51]. Moreover, the immobilization of bimetallic nanoparticles on graphene led to the formation of a double signal amplification platform [59]. The hybrid of ferrocence thiolate-stabilized Fe_3O_4@Au NPs with graphene sheet showed a wider linear range for all the three analytes, and a higher detection limit compared to the aforementioned hybrids (see Table 1).

Glucose sensing is another common application of graphene/NP hybrids. In fact, various graphene/NP hybrid modified electronic interfaces have been used to

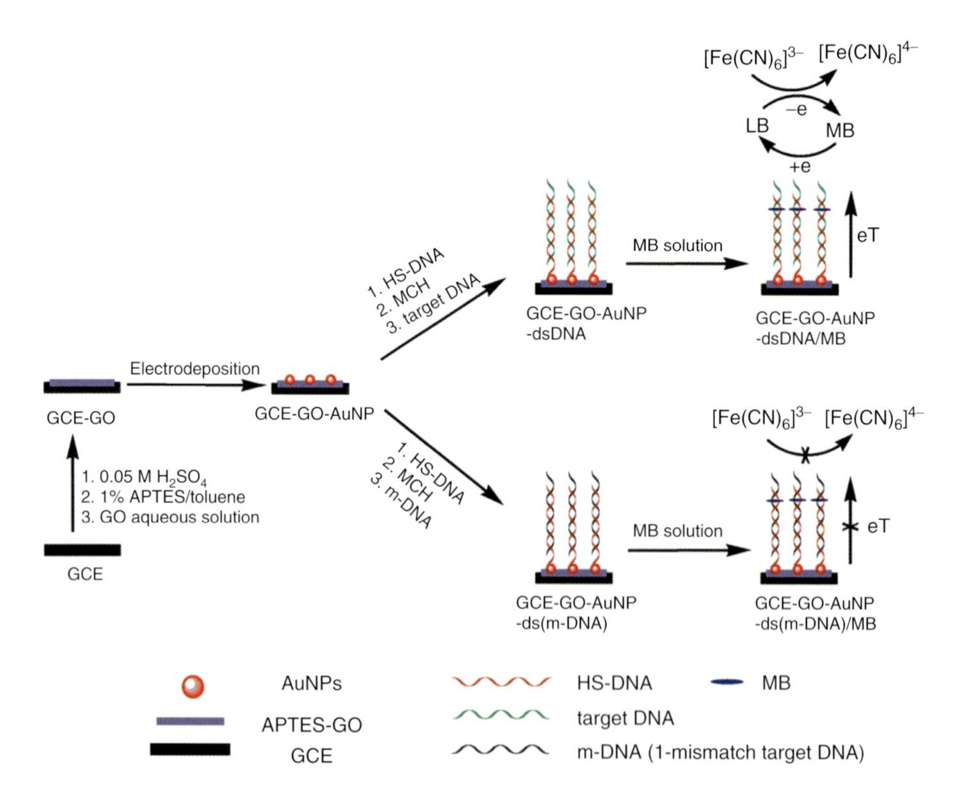

Fig. 3 Schematic representation of the fabrication of the GO/AuNP hybrid-based electrode and its application to DNA detection. Reprinted with permission from Ref. [79]; © The Royal Society of Chemistry.

enhance glucose detection; these include graphene/Pd [49], with a wide linear range of 10 μM to 5 mM, chitosan-graphene/Pd NPs [72], with a low detection limit of 0.2 μM and a signal-to-noise ratio of 3, and rGO/Au/Pd NPs [55], with a high sensitivity of 266.6 $\mu A\,mM^{-1}\,cm^{-2}$. The presence of NPs in the hybrids also had a major impact on DNA detection systems, as the surface of a graphene/gold nanoparticle (AuNP) electrode could easily be modified using thiolated-DNA (probe DNA) through the strong Au-S bonding (Fig. 3) [79]. When the target DNA had become completely hybridized with the probe DNA, the electron would be transferred from the electrode surface to methylene blue in order to catalyze $[Fe(CN)_6]^{3-}$, thus showing a reduction peak in the DPV measurements [79]. The electron transfer could be terminated if the target DNA was mismatched, or if the GO surface was modified by the probe DNA through noncovalent bonding. When using this hybrid-based electrode, the target DNA detection limit was as low as 100 fM. It is noteworthy that the same electrodeposition technique was used for decorating AuNPs onto GO onto the electronic interfaces, and this was initially thought to represent a promising method for their large-scale fabrication.

Based on the same working mechanism, other multicomponent systems have been developed for large biospecies detection, such as PSA [57], thrombin [72], and transferrin [75]. Recently, GO/NP hybrid-modified electronic interfaces have paved the way for the *in-situ* detection of living cells, providing a simple, rapid, label-free, and cell-based sensor for probing the toxicity of cells. For instance, when live PC12 cells were deposited on top of a GO/AuNP-modified glassy carbon electrode using a simple drop-casting method (Fig. 4) [82], the electrochemical behavior of the live cells was characterized using cyclic voltammetry (at different scan rates), DPV and electrochemical impedance spectra (EIS).

3
Bioimaging

Conventional imaging methods currently used to examine and monitor the efficacy of treatments normally rely on anatomic imaging, which reveals internal structures hidden by the skin and tissues, and allows abnormalities in the human body to be visualized and identified. Anatomic imaging techniques used worldwide include X-radiography, magnetic resonance imaging (MRI), medical ultrasonography or ultrasound, endoscopy, elastography, thermography, and positron emission tomography (PET). When identifying lesions, the "normal" database of a species is compared with the sample under inspection, with any differences being visualized only with the naked eye. Clearly, this approach may cause problems in the early detection of life-threatening conditions such as cancer, as well as neurological and cardiovascular diseases. Consequently, the twenty-first century medical imaging revolution, whereby molecular biology is integrated with *in-vivo* imaging, is paving the way for the effective early detection of such diseases. Today, molecular imaging allows changes to be visualized at the cellular level and with specific targets, thus dramatically enhancing the sensitivity and resolution of the images obtained. Moreover, this novel imaging technique employs interdisciplinary approaches for molecular diagnostics, molecular imaging and molecular therapy, leading to a "new age" of medicine.

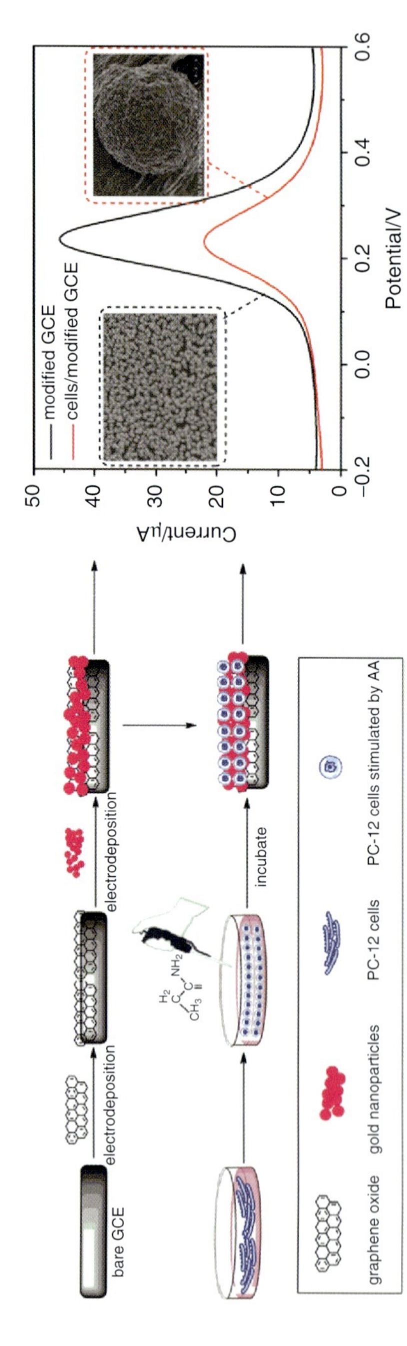

Fig. 4 Schematic representation of the preparation of cell-based electrochemical sensor [82]. Reprinted with permission from Ref. [82]; © 2013, Elsevier.

Based on its good biocompatibility and easy biofunctionalization for biomolecular recognition, graphene shows great potential as a molecular imaging contrast agent and, indeed, various combinations of graphene with biospecies have proved to be effective for *in-vivo*, noninvasive imaging. Recent advances in the use of graphene hybrids for different types of bioimaging are discussed in the following sections.

3.1 Optical Imaging: NIR Imaging, Fluorescence, Multiphoton Excitation Imaging

3.1.1 NIR Imaging

The intrinsic optical properties of graphene in the visible and near-infrared (NIR) light ranges have led to its use for bioimaging, especially of live cells. To date, much effort has been expended in developing graphene derivatives as fluorescent probes for intracellular imaging, both *in vitro* and *in vivo*. For example, the photoluminescence (PL) of poly(ethylene glycol) (PEG)-grafted GO was used for live-cell imaging in the NIR range, with minimal background interference [83]. In this case, biodegradable polymers were selected as linkage groups for loading fluorescent species onto the surfaces of graphene derivatives. Details of some NIR imaging contrast agents are provided in Sect. 4.5.

3.1.2 Fluorescence Imaging

The fluorescence properties of a graphene sheet emerge when its lateral dimension is reduced to only a few nanometers. Strongly fluorescent graphene quantum dots (GQDs) with a lateral size range of 1.5 to 5 nm were prepared using a one-step solvothermal method, whereby the PL quantum yield was as high as 11.4% [84]. The GQDs exhibited a high stability compared to other fluorescent dyes, and could be dispersed in most polar solvents, without further chemical modification. The well-crystallized GQDs exhibited a reversible, pH-responsive green PL, and became saturated at high GQD concentrations. In addition, the PL of GQDs could be tailored by varying QD size [85]. Owing to their unique PL, low cytotoxicity, high solubility and good biocompatibility, GQDs may serve as excellent contrast agents and probes for molecular imaging applications. Moreover, as they are efficiently endocytosed by HeLa cells, they may also serve as a fluorescent nanoprobe for cellular imaging. Consequently, GQDs with a strong yellow emission were prepared via the electrochemical exfoliation of graphite, followed by room-temperature reduction with hydrazine [86]. A bright yellow luminescence with 14% quantum yield confirmed that the nanoscale graphene had been modified with abundant phthalhydrazide-like groups, with hydrazide groups at the edges. Additional studies with stem cells revealed the harmless effects of GQDs on the viability, proliferation and differentiation capacity of stem cells, thereby justifying the valuable application of this novel material in long-term stem cell imaging in order to explore the native development of cells.

When fluorescent NPs are anchored onto the graphene surface, the hybrids obtained can also be used as fluorescent probes for biomedical targeting and imaging. Thus, a new hybrid was generated by decorating GO sheets with zinc-doped $AgInS_2$ NPs. The PEGylated hybrid exhibited four different emitting colors, including red, orange, green, and yellow [87]. Cellular fluorescent imaging on a NIH/3T3 cell line using this hybrid was successful, and demonstrated the feasibility of applying the hybrid to biomedical imaging. Another good example is the biocompatible, sandwich-like fluorescent bis-(8-hydroxyquinoline) cadmium (CdQ2)/GO hybrid. Fluorescent CdQ2 nanorods, prepared by complexation between 8-hydroxyquinoline and Cd^{2+}, were used to construct the sandwich-like structure with GO in water–ethanol solution [88]. The CdQ2/GO hybrid exhibited a strong green fluorescence emission at 500 nm, but after incubation with HepG2 cells at 37 °C the CdQ2/GO hybrid was shown to adhere to the cell membrane, which resulted in a strong cellular green fluorescence observed using laser confocal fluorescence microscopy.

Recently, graphene sheet was used in an innovative fashion as a protective layer for a fluorescence imaging agent. For this, a novel hybrid of mesoporous silica nanoparticles (MSNPs) wrapped with GO was developed as an efficient dye-protecting vessel and was used to protect different types of organic dye [89]. Bis(2,4,6-trihydroxyphenyl) squaraine dye, a type of zwitterionic molecule, was used as a cargo model. The sealing mechanism was shown to rely on the charge–charge interaction between the GO surface and the charged surface of the MSNPs (Fig. 5). When the latter were wrapped with the GO sheets, the protected dye within the GO-MSNPs exhibited a remarkable stability, so that the loaded dye was very well protected against attacks by nucleophiles such as glutathione (GSH) and cysteine. Fluorescent images of HeLa cells recorded after incubation with dye-loaded GO-MSNPs revealed a clear accumulation of the hybrids in the cell cytoplasm, confirming the promising potential of the vessel for cellular bioimaging applications.

3.1.3 Multiphoton Excitation Imaging

The use of multiphoton active dyes could reduce the harmful effects of laser exposure and also enhance the penetration depth in the tissues being studied. The requisite excitation wavelength for multiphoton active dyes normally lies in the optical window of the biological tissue, which is perfect for tissue penetration due to a lack of absorption by hemoglobin, melanin, water molecules, and fat. Typically, two-photon and three-photon-induced PL from the PEG/GO hybrid under femtosecond pulsed laser excitation can be detected [90]. Application of the obtained hybrid as a contrast agent for *in-vivo* two-photon luminescence cellular imaging was achieved using two-photon scanning microscopy, and the hybrid was seen to be distributed in the mitochondria, the endoplasmic reticulum, Golgi body and lysosomes of the HeLa cells. Following injection into the tail vein of mice, delivery of the PEG/GO hybrid was tracked *in situ* by utilizing a

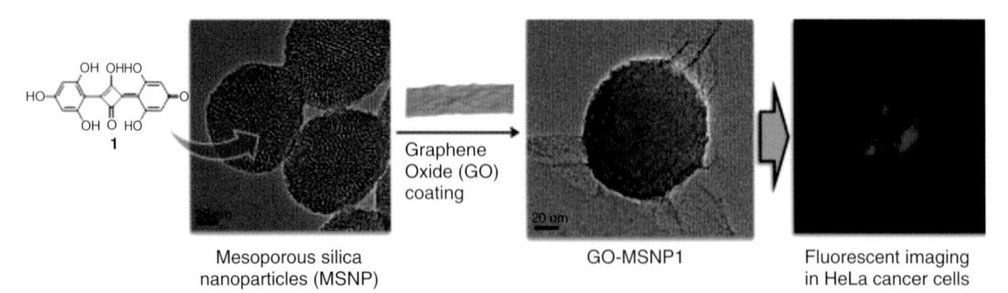

Fig. 5 Schematic representation showing the loading of squaraine dye inside MSNPs, followed by wrapping with ultrathin GO sheets, leading to the formation of a novel fluorescent hybrid for cellular imaging. Reprinted with permission from Ref. [89]; © 2012, American Chemical Society.

deep-penetrating, two-photon imaging technique.

3.2
Localized Surface Plasmon Resonance (LSPR)

MNPs such as silver nanoparticles (AgNPs) and AuNPs were employed as effective signal reporters for locating graphene in optical imaging [91]. The MNP/GO hybrids were used to directly illuminate graphene for optical imaging by employing a strong localized surface plasmon resonance (LSPR) light scattering of MNPs as an effective signal reporter, such that the profiles of graphene could be observed using dark-field microscopy.

3.3
Raman Imaging

Based on the intrinsic Raman-active property of graphene layers, graphene-based Raman imaging was demonstrated as an alternative approach to *in-vitro* imaging. The functionalization of GO with pillararenes through strong hydrogen-bonding interactions led to a homogeneous aqueous suspension, which could be kept for a long time. After confirming a good biocompatibility of the hybrid, the Raman signal generated from the GO layer was detected within HeLa cells, thus demonstrating the successful use of GO as a Raman imaging contrast agent [92].

Decorating the graphene sheet with some nanocrystals might enhance its surface Raman scattering capabilities, and the GO/AuNP hybrid did indeed show useful applications in Raman imaging. The presence of AuNPs on the graphene surface resulted in a remarkable boost of the Raman signal of GO by a surface-enhanced Raman scattering (SERS) effect, even in aqueous solution. Notably, HeLa 229 cells incubated with the GO/AuNP hybrid exhibited a much stronger Raman signal than those cells incubated with pristine GO [93].

3.4
Magnetic Resonance Imaging (MRI)

In order to develop new MRI contrast agents, a simple hybrid of superparamagnetic NPs with graphene sheets was investigated to determine if graphene could indeed enhance the MRI contrast. In the past, superparamagnetic Fe_3O_4 NPs have been widely used in biomedicine as MRI contrast agents, and the aggregate formation of these NPs has led to an enhancement in relaxation rate (r_2) and, in turn, to a new type of better-performing contrast agent [94–97]. A hybrid (Fe_3O_4-GO) with aminodextran-coated Fe_3O_4 NPs showed an increase in the T_2 relaxivity ($r_2 = 76$ Fe mM^{-1} s^{-1}), which was much higher than that of monodispersed Fe_3O_4 nanoparticles ($r_2 = 21$ Fe mM^{-1} s^{-1}) [98]. In this context, the GO served as a platform for the Fe_3O_4 NPs to assemble and form aggregates, which in turn increased the MRI sensitivity of the hybrid. The Fe_3O_4-GO hybrids with different iron concentrations of 10, 20, 40 and 80 g ml^{-1} were reported to show 100, 96, 92 and 91% cell viability, respectively, indicating their good biocompatibility [98]. In a similar approach, superparamagnetic $MnFe_2O_4$ nanocrystals were anchored onto an oleylamine-modified GO, yielding a hybrid with the r_2 relaxivity value as high as 256.2 Fe mM^{-1} s^{-1} [99]. This hybrid showed a negligible change in size after a four-day incubation in phosphate-buffered saline (PBS) 1× solution at 25 and 37 °C, thus proving its colloidal stability under physiological conditions [99].

The wrapping of several NPs within graphene sheets via an aerosol-phase synthesis has created a new identity, the

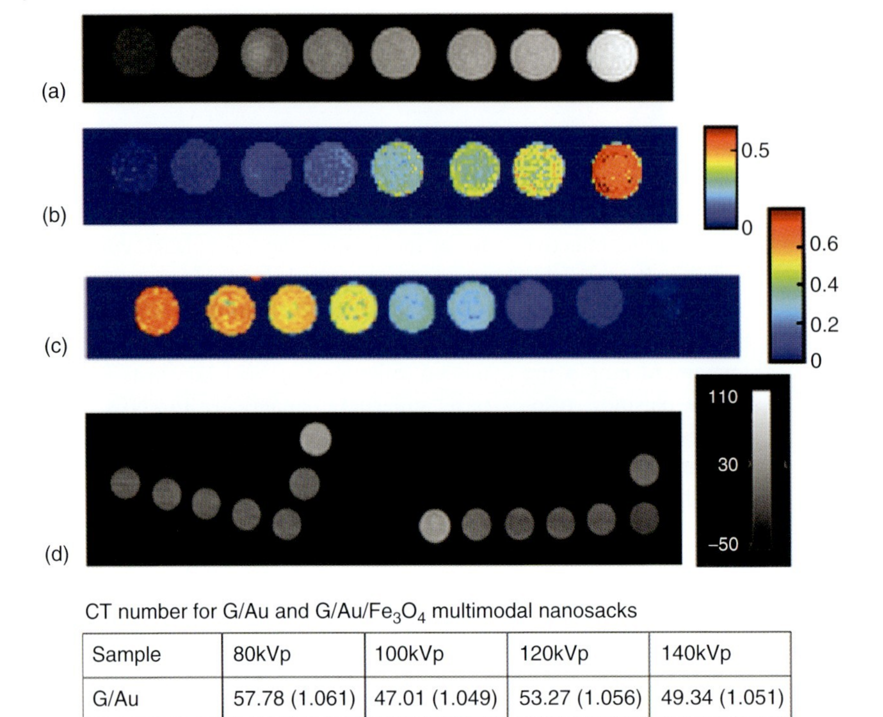

(e) CT number for G/Au and G/Au/Fe_3O_4 multimodal nanosacks

Sample	80kVp	100kVp	120kVp	140kVp
G/Au	57.78 (1.061)	47.01 (1.049)	53.27 (1.056)	49.34 (1.051)
G/Au/Fe_3O_4	25.61 (1.045)	22.26 (1.032)	23.65 (1.041)	16.67 (1.035)

Fig. 6 MRI and X-ray CT results of rGO/Fe_3O_4, rGO/AuNPs, and rGO/Au/Fe_3O_4 multifunctional probes. (a) T_2-weighted image of rGO/Fe_3O_4 with the concentration increase from right to left: 0, 0.5, 1, 2, 5, 10, 20, and 50 μg ml^{-1}; (b) Corresponding T_2 map computed from 24 echo image series. The color bar shows T_2 (in s); (c) MRI T_2 map of rGO/Au/Fe_3O_4 with the concentration increase from left to right: 0.05, 0.1, 0.5, 1, 5, 10, 50, 100, and 500 μg ml^{-1}; (d) X-ray CT image (80 kVp) of graphene/Au nanosack. Right-hand L-shaped sequence for the nanosack (concentrations in μg ml^{-1}) from left to right: 2000, 400, 200, 40, 20, gel control, and water control. Left-hand L-shaped sequence for free Au control (concentrations in μg ml^{-1}) from left to right: 400, 200, 40, 20, gel control, with 2000 μg ml^{-1} of empty nanosack at top, and with 200 μg ml^{-1} of empty nanosack beneath it; (e) Table of CT results showing the changes in CT number relative to the gel control samples shown in parentheses. Reprinted with permission from Ref. [100]; © 2013, American Chemical Society.

nanosack (Fig. 6). This technique demonstrated an efficient wrapping of various types of NPs, such as AuNPs, SiNPs, and Fe_3O_4 NPs, as well as different components such as Au–Fe_3O_4 and $BaTiO_3$–Fe_3O_4, by the graphene sheets [100]. A multifunctional capability was demonstrated by fabricating dual-purpose magnetically responsive contrast agents for diagnostic techniques, including MRI, X-ray computed tomography (CT). The imaging results indicated the practicality of using the aforementioned nanosack for dual-mode imaging of MRI and CT. Moreover, the wrapping architecture of the nanosack enabled both the loading and release of the cargo to be effected in a controlled manner. Compared to the GO/CsCl hybrid without carboxymethylcellulose (CMC) sealing, the rGO/CsCl/CMC nanosack demonstrated an ability to release loaded salt over a long period (days), compared to a rapid release of CsCl (maximum release within minutes). This controlled-release property of the nanosack, combined with an ability for dual-mode imaging for noninvasive MRI and CT, has opened up a new pathway for versatile applications.

3.5 Positron Emission Tomography (PET)

The possibility of a GO hybrid for tumor targeting in an animal tumor model was confirmed by quantitatively evaluating the pharmacokinetics and tumor-targeting efficacy through PET imaging using ^{66}Ga as the radiolabel [101]. GO linked covalently with PEG was conjugated to NOTA (1,4,7-triazacyclononane-1,4,7-triacetic acid) and TRC105 (an antibody that binds to CD105), making the hybrid specific towards CD105 (an endoglin-based marker for tumor angiogenesis). In 4T1 murine breast cancer tumor-bearing mice, the ^{66}Ga/NOTA/GO/TRC105 and ^{66}Ga/NOTA/GO hybrids were primarily cleared through the hepatobiliary pathway. However, ^{66}Ga/NOTA/GO/TRC105 was accumulated quickly in the 4T1 tumor, with uptake remaining stable over time (3.8 ± 0.4, 4.5 ± 0.4, 5.8 ± 0.3 and $4.5 \pm 0.4\%$ of the injected dose per gram at 0.5, 3, 7, and 24 h post-injection, respectively). Histological examination revealed that NOTA/GO/TRC105 could target tumor vasculature CD105 with minimal extravasation.

Clearly, by combining the advantages of graphene with biodegradable and functional materials, graphene-based hybrids are highly suited to bioimaging. In particular, new imaging technologies could be investigated using graphene-based hybrids in the clinical setting, for example in real-time cancer cell imaging. However, the above-mentioned properties of the graphene/nanoparticle hybrids, the size of graphene and the toxicity of the metal/metal oxide NPs remain the main challenges for their application *in vivo*. The size of graphene, which determines the overall size of the hybrids, is one of the main factors controlling the enhanced permeability and retention (EPR) effect. Oversizing of the graphene can be partially reduced by using a centrifugal separator or vacuum membrane filtration [102]. Whilst serious toxicity of these metal/metal oxide NPs has not yet been reported, the chronic effects that they may cause is of major concern for their clinical application.

4 Therapeutics

Biocompatible nanomaterials offer excellent platforms to interact with biological systems in desired pathways. By rationally controlling the interaction and functionalization on the surface of graphene derivatives, tasks that were considered ridiculous some 20 years ago could be achieved nowadays. Numerous reports have illustrated the potential applications of graphene-based nanomaterials for molecular therapeutics. In particular, by integrating the graphene sheet with drug molecules, targeting agents or NPs, novel types of multifunctional nanomaterial can be obtained for theranostics applications. Some recent achievements using different types of graphene-based hybrid, and their respective applications in the field of therapeutics, are highlighted in the following section.

Nanomaterials with functional units on the surface that are capable of drug loading and release are highly desirable in drug-delivery research. The fact that both sides of the graphene derivatives are accessible for drug binding contributes to an enhanced level of drug loading, while graphene derivatives can also integrate with aromatic drugs through noncovalent binding interactions such as $\pi-\pi$ stacking and van der Waals interactions. Other types of interaction, such as cleavable covalent bond linkage may also be applied in the targeting and treatment processes. The development of graphene-based targeted and controlled delivery systems is clearly an exciting area in biomedical research, and the various drug-loading mechanisms available, together with their targeting properties, controlled drug release approaches and different therapeutic methods by using graphene-based hybrids, are now discussed.

4.1 Drug Loading by Noncovalent Interactions

The drug-delivery properties of graphene-based materials have been well investigated. Initially, a successful graphene-based drug-delivery system was constructed by functionalizing GO sheets (lateral dimension <50 nm) with branched PEG to render a high biocompatibility to the GO hybrid under physiological conditions [103]. This PEGylated GO can bind with the water-insoluble aromatic molecule SN38, an anticancer drug camptothecin (CPT) analog, via van der Waals interaction, such that the GO/PEG/SN38 hybrid obtained has an excellent aqueous solubility and retains the same active property as free SN38 in organic solvents. Other drugs, such as various CPT analogs Iressa (gefitinib) [103] and doxorubicin (DOX) [83] can also be attached to the GO/PEG hybrid using a simple, noncovalent approach.

A biocompatible and aqueous soluble GO/chitosan hybrid was shown to possess a superior loading capacity for CPT. This hybrid, which consisted of 64 wt% chitosan [104], was developed in order to load water-insoluble CPT via $\pi-\pi$ stacking and hydrophobic interactions. The resultant GO/chitosan/CPT complex showed a remarkably high cytotoxicity in carcinoma HepG2 and cervical cancer HeLa cell lines compared to that of the individual drug. Gelatin/graphene, another biocompatible hybrid, exhibited an excellent stability in water and various physiological fluids [105], with cellular toxicity tests at a high concentrations (200 mg ml^{-1}) showing gelatin/graphene to be nontoxic in human breast carcinoma cells MCF-7. When DOX was loaded onto the hybrid at a high loading capacity (through physisorption), apoptosis of the MCF-7 cells was clearly observed after incubation with the gelatin/graphene/DOX hybrid, and gelatin-mediated sustained release *in vitro* was demonstrated.

Several types of biocompatible molecule, such as pluronic F38 (F38), maltodextrin (MD) and Tween 80 (T80) were covalently functionalized onto GO in order to increase the loading capacity of the water-insoluble antioxidant and anticancer drug, ellagic acid (EA) [106]. The loading capacities of EA onto the hybrids were 1, 1.14, and 1.22 g per gram of F38-, MD-, and T80-functionalized GO, respectively. The release of EA from these hybrids in solution at 37 °C indicated that F38-, MD-, and T80-functionalized GO could release 36–38% of the drug at pH 10 within 3 days. The cytotoxicity of loaded EA to MCF-7 cells and human colon adenocarcinoma cells HT29 was shown subsequently to be higher than that of free EA. Use of the di(phenyl)-(2,4,6-trinitrophenyl) iminoazanium (DPPH) assay to study antioxidant activity indicated that the loading of EA onto the functionalized GO did not obstruct its antioxidant activity.

It is well known, and a widely adopted clinical practice, that a combination of two or more drugs often displays a much better therapeutic efficacy than does a single drug, and that the controlled loading and targeted delivery of mixed drugs may further enhance their efficacy in molecular biomedicine. Thus, graphene-based delivery systems carrying mixed drugs were much sought after [107]. One such example was a folic acid (FA) -modified GO, which could be used for the loading of both DOX and CPT. In this case, GO was covalently functionalized by FA on the surface, and this allowed it to specifically target MCF-7 cells through the FA receptors. Two types of anticancer drug, DOX and CPT, were then loaded onto the FA/GO hybrid via $\pi-\pi$ stacking and hydrophobic interactions. The hybrid obtained with these mixed anticancer drugs showed specific targeting to MCF-7 cells and a remarkably higher cytotoxicity compared to that of GO loaded with either DOX or CPT.

4.2 Drug Loading by Host–Guest Interaction

Host–guest interaction takes advantage of the hydrophobic cavity of cyclodextrin (CD) and the hydrophobic anticancer drug molecules. CD, acting as a host, was covalently linked onto the graphene surface so as to provide the cavities for drug loading. Then, by co-reduction of GO and CD in the presence of ammonia, a portion of the primary and secondary hydroxyl groups of CD was converted to other groups such as aldehydes, carboxyl groups, and ketones (Fig. 7). These groups could then further react with hydroxyl groups on GO, resulting in the formation of hemiacetal or acetal and affording the graphene nanosheet–CD hybrid [108].

The hybrid of rGO-C_6H_4-CO-NH-PEI-NH-CO-CD-Biotin was developed by introducing the targeting property though the host–guest complexation, and the hybrid showed about 25% DOX loading [109]. The drug release was pH-dependent and salt-dependent at different DOX concentrations. Biotin conjugation on the hybrid improved the targeting effect towards HepG2 cancer cells that were over-expressed with biotin-specific receptors on the surface. In another example, a surface modification of GO with FA was carried out through a synthetic bifunctional molecule that consisted of a planar porphyrin moiety as the binding group and an adamantane moiety that could be encapsulated by the cavity of β-CD. By taking advantage of the mutual cooperation of FA-modified β-CD (FA-β-CD), adamantane-grafted porphyrin and GO, the structural integrity of GO was preserved. The resultant supramolecular assembly was able specifically to carry DOX into folate receptor-positive malignant cells, without imparting any enhanced toxicity [110].

4.3 Controlled Release System

Drug release is another important aspect in therapeutics, as the dosage of a drug must be controlled with respect to different timeframes during the treatment process. By preventing the random release of a drug during its circulation in the body but rather releasing it at a specific target – that is, the tumor site – a more effective therapy with fewer side effects would be expected. Controlled drug release systems have been shown feasible under different stimulus

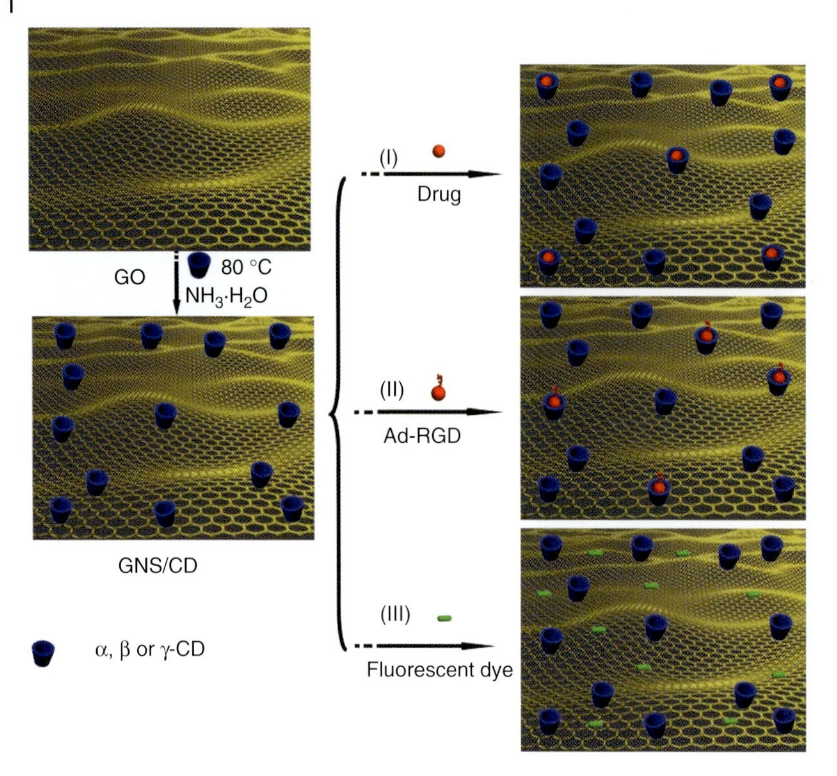

Fig. 7 Schematic illustration of graphene nanosheet (GNS)/CD formation and host–guest functionalization. The surface GNS/CD hybrid capped with a large quantity of CD cavities can be used to host various functional guests, including the antivirus drug amantadine or adamantane-containing RGD-targeting ligands, or fluorescent dyes. Reproduced with permission from Ref. [108]; © 2013, Wiley-VCH Verlag GmbH & Co. KGaA, Weinheim.

approaches, such as pH-triggered release [111, 112], thermal-triggered release [113, 114], NIR-triggered release [115], and redox-responsive release [116].

As some noncovalent interactions such as π–π stacking and hydrogen-bonding interactions can be influenced by pH, changing the pH of an investigated solution represents one of the most convenient stimulus methods used with drug-loaded graphene hybrids. A loading amount of DOX up to $2.35\,\mathrm{mg\,mg^{-1}}$ was obtained within the GO/DOX hybrid [117] through strong π–π stacking interactions and hydrogen-bonding interactions. The loading and release of DOX on GO showed a pH-mediated response on account of these interactions. Furthermore, the GO/chitosan hybrid can afford different loading amounts depending on its various affinities with drugs [118]; for example, loading of the hydrophobic and aromatic anti-inflammatory drug, ibuprofen, was $0.097\,\mathrm{mg\,mg^{-1}}$, which was higher than for the hydrophilic anticancer drug, 5-fluorouracil ($0.0053\,\mathrm{mg\,mg^{-1}}$). The release behavior of ibuprofen and 5-fluorouracil from the graphene-based hybrids was controlled by tuning the solution pH in order to change the affinity between the drug and the GO/chitosan hybrid. The pluronic F127/graphene hybrid was also found to

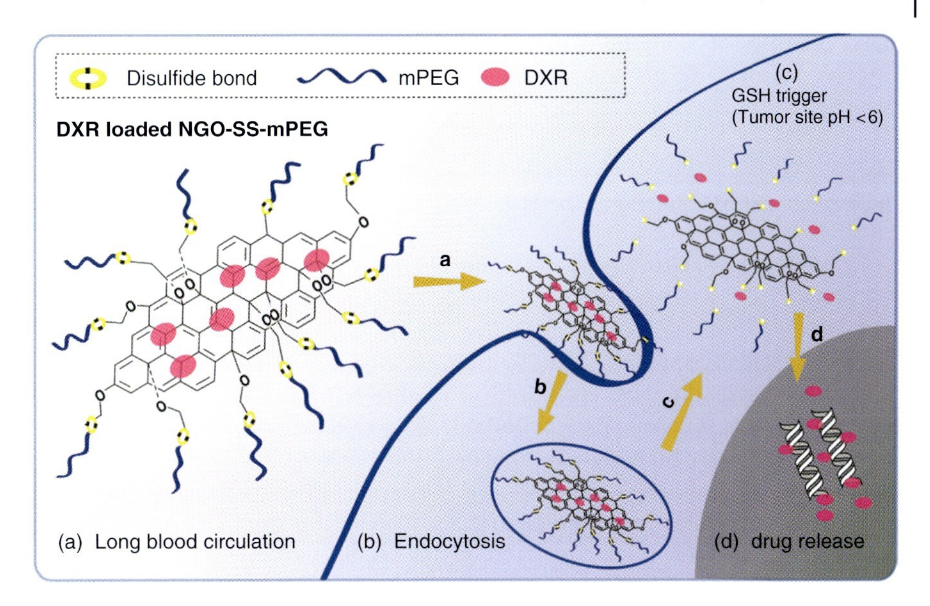

Fig. 8 Antitumor activity of redox-sensitive DOX/GO/PEG hybrid was accomplished through the following steps: (a) The DOX/GO/PEG hybrid was constructed with a disulfide bond linkage for prolonged blood circulation; (b) Endocytosis of the DOX/GO/PEG hybrid in tumor cells; (c) GSH-triggered PEG dissociation from DOX/GO/PEG; (d) Rapid DOX release in targeted tumor sites. Reproduced with permission from Ref. [120]; © 2012, Wiley-VCH Verlag GmbH & Co. KGaA, Weinheim.

be capable of effectively encapsulating DOX, with a drug-loading efficiency up to 289 w/w% at an initial DOX concentration of $0.9\,mg\,ml^{-1}$. The hybrid exhibited a higher drug-release capability under acidic conditions than under neutral or basic conditions [119].

The ability of intracellular GSH to cleave disulfide bonds was employed to develop a redox-responsive drug delivery carrier (Fig. 8). For this, PEG was covalently linked to GO through a disulfide bond [120] after which, by cleaving the disulfide bond linkage, the PEG shell was detached from the graphene sheet. The redox-detachable PEG shell not only provides the hybrid with high physiological solubility and stability during the circulation, but also is able to rapidly release an encapsulated payload under GSH stimulation. DOX release in the presence of GSH was noted to be 1.55-fold faster than in the absence of a redox trigger.

In order to carry out targeted drug delivery and realize an intelligent controlled release, a dual system was established which contained a targeted drug delivery and pH-controlled release based on a multicomponent GO [121]. A superparamagnetic GO/Fe_3O_4 hybrid was first prepared by decorating Fe_3O_4 nanoparticles onto the GO surface. 3-Aminopropyl triethoxysilane was modified onto the NPs to provide the amino group as a binding site for FA conjugation. A high DOX loading of $0.387\,mg\,mg^{-1}$ at an initial DOX concentration of $0.238\,mg\,ml^{-1}$ was obtained through the $\pi-\pi$ stacking interactions. The drug-release profile was shown to depend heavily on the pH changes, while cell uptake studies suggested that this multifunctional GO hybrid showed much promise in applications of tumor-targeted delivery and pH-controlled drug release.

4.4 Photodynamic and Photothermal Therapy

Both, photodynamic therapy (PDT) and PTT employ electromagnetic radiation for the treatment of various diseases, including cancer. In PDT, an appropriate light source is used to excite the photosensitizer to produce reactive oxygen species (ROS), which generally are free radical oxygen (type I PDT) or singlet oxygen $^1O_2$ (type II PDT), and are capable of destroying cancer cells. Unlike PDT, PTT does not require oxygen to interact with the target cells or tissues, and in this case the photosensitizer is excited with a specific wavelength to its excited state, which then releases vibrational energy in the form of heat to kill the targeted cells. Normally, NIR light is used as the excitation source for both PDT and PTT, as this provides a deep penetration into the human body and is less harmful to any tissues in its path.

Biocompatible graphene-based hybrids containing photosensitizers have been explored for potential PDT applications. For example, zinc phthalocyanine (ZnPc), a common photosensitizer for PDT, can be loaded up to 14 wt% in the GO/methoxy–PEG hybrid through $\pi-\pi$ stacking interactions [122]. Hydrophobic ZnPc, when introduced into MCF-7 cells through the GO/methoxy–PEG hybrid, showed pronounced phototoxicity to the cells under Xe light irradiation. Another photosensitizer, hypocrellin A (hA), was incorporated with GO to yield a GO/hA hybrid, the stability of which was superior to that of free hA in aqueous solution [123]. It should be noted that GO/hA can be excited by light irradiation to generate $^1O_2$, resulting in significant cell death *in vitro* for potential clinical PDT.

Owing to the their strong optical absorbance in the NIR region and dose-dependent low toxicity, graphene-based hybrids have during recent years been studied intensively for their promising applications in cancer phototherapy [124, 125]. Graphene has been widely accepted as a good photothermal agent, along with other NIR photothermal agents including gold nanomaterials and carbon nanotubes. The graphene/PEG hybrid was a successful example of efficient PTT *in vivo* [122], whereby the strong optical absorbance of the graphene/PEG hybrid in the NIR region was utilized for PTT *in vivo*. Following intravenous administration of the hybrid in mice and subsequent low-power NIR laser irradiation of the tumor, the tumor size was effectively reduced [126]. Moreover, no obvious side effects of the graphene/PEG hybrid were observed in the treated mice, whether histologically or by complete blood panel analyses.

Another targeting peptide bearing the Arg-Gly-Asp (RGD) unit was attached onto rGO, showing selective cellular uptake in U87MG cancer cells and highly effective tumor ablation *in vitro*. In contrast, cells incubated with a higher concentration of rGO than was needed for the photothermal effect showed a higher cell viability in the absence of NIR light irradiation (Fig. 9), clearly indicating photo-triggered cell destruction *in vitro* [127].

Combined therapeutic methods of PTT and PDT were also explored by using graphene derivatives and proper photosensitizers. As an example, when the photosensitizer chlorin e6 (Ce6) was loaded onto GO/PEG via $\pi-\pi$ stacking interactions [128], the obtained hybrid showed excellent water solubility and was

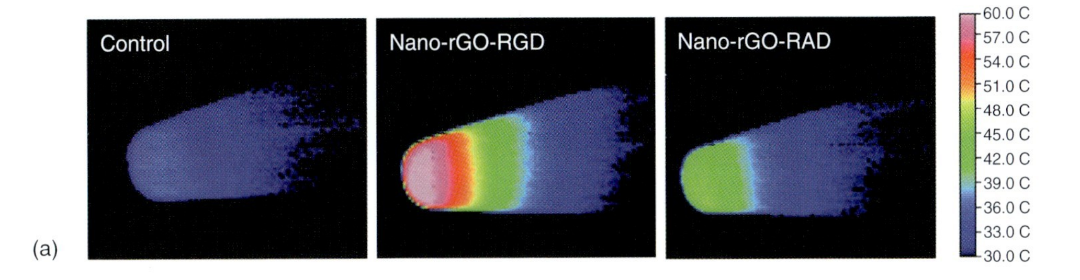

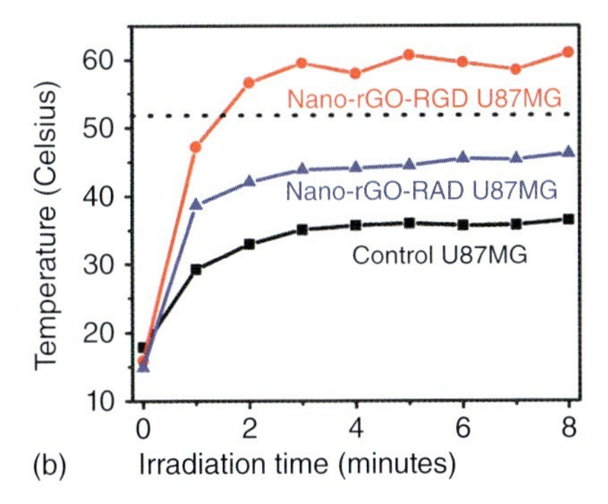

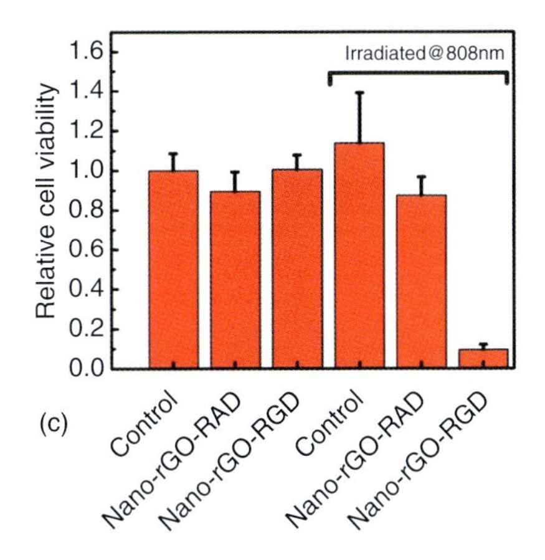

Fig. 9 (a) Thermal images of vials containing pellets of nontreated U87MG cells as control, the cells treated by rGO/RGD, and the cells treated by rGO/RAD (Arg-Ala-Asp), respectively, after 8 min irradiation with 808 nm laser at a power of 15.3 W cm^{-2}; (b) Cell pellet temperature versus time during the irradiation; (c) Cell viability at 24 h post irradiation. The results indicated successful targeting toward U87MG cells and selective photoablation of the targeted cells. Reproduced with permission from Ref. [127]; © 2011, American Chemical Society.

able to generate cytotoxic $^1O_2$ under light excitation for PDT. This GO/PEG/Ce6 hybrid exhibited a remarkably improved photodynamic destruction of cancer cells compared to that of free Ce6. More importantly, the photothermal effect of graphene introduced local heating to promote the delivery of Ce6 when exposed to an NIR laser at a low power density, and this further enhanced the PDT efficacy against cancer cells. Thus, integrating the merits of PTT, PDT and graphene paves the way for applying carbon materials in clinical cancer phototherapy.

4.5 Combination of Diagnostics and Therapeutics: Graphene-Based Theranostics System

Multifunctional nanomaterials are highly sought after in biomedicine. The combination of diagnostics and therapeutics – that is, theranostics – offers the real-time monitoring of treatment processes, and the combination can be made highly feasible by multifunctionalizing graphene with desired components. An example of this strategy is the rational combination between NIR imaging and PTT/PDT. The integration

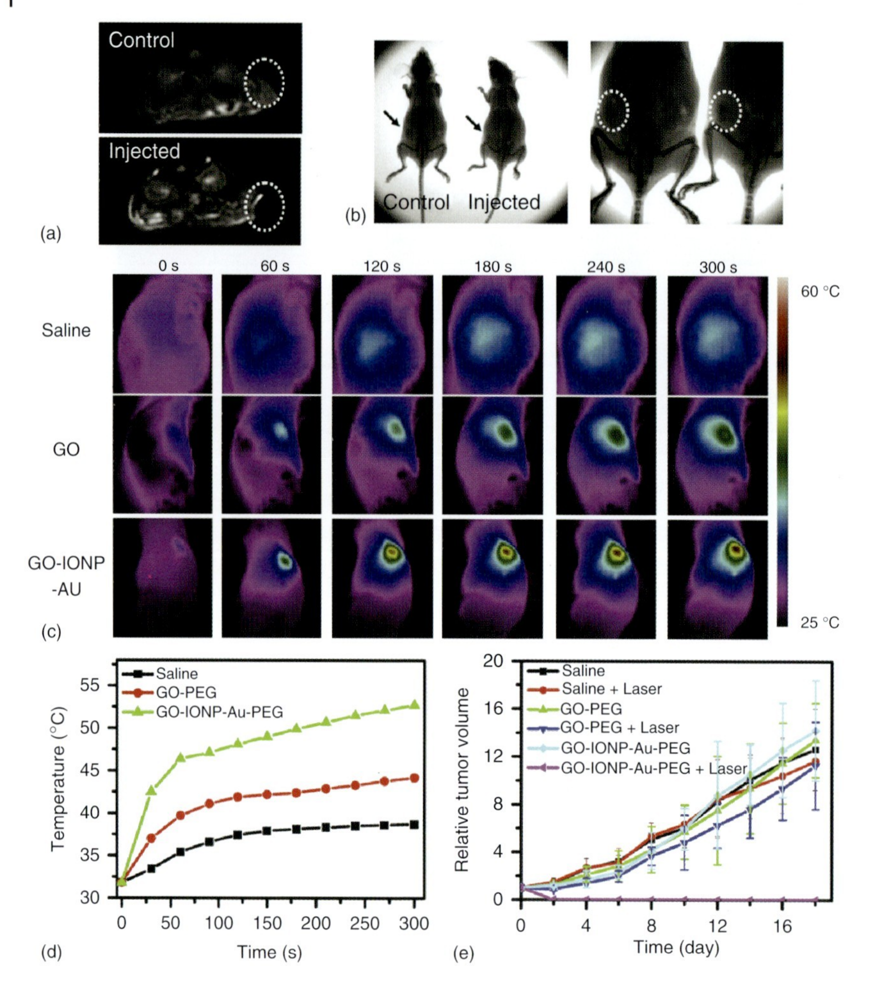

Fig. 10 (a) T_2-weighted MR images of 4T1 tumor-bearing mice before (top) and after (bottom) intratumoral injection of the GO-IONP-Au-PEG hybrid; (b) X-ray images of tumor-bearing mice before (left) and after (right) intratumoral injection of the hybrid; (c) Infrared thermal images of tumor-bearing mice injected with corresponding solutions under laser irradiation (808 nm, 0.75 W cm^{-2}); (d) Temperature changes on tumors under different treatments indicated; (e) Tumor growth curves under different treatments indicated [130]. Reprinted with permission from Ref. [130]; © 2013, Elsevier.

of upconversion nanoparticles (UCNPs) with GO led to promising hybrids, with $NaYF_4$:Yb^{3+},Er^{3+},Tm^{3+}/$NaYF_4$ UCNPs decorated onto PEGylated GO sheets being reported as effective NIR imaging agents whilst, at the same time, exhibiting PTD and PTT properties with enhanced anticancer efficacy [129]. *In-vitro* cancer therapy of the hybrid was demonstrated using HeLa cells, which showed a high therapeutic efficacy, while in another study a dual-mode bioimaging and photothermal tumor destruction was demonstrated *in vivo* using iron oxide/AuNP co-decorated GO (GO-IONP-Au-PEG) [130]. The results of this study validated the hypothesis that the

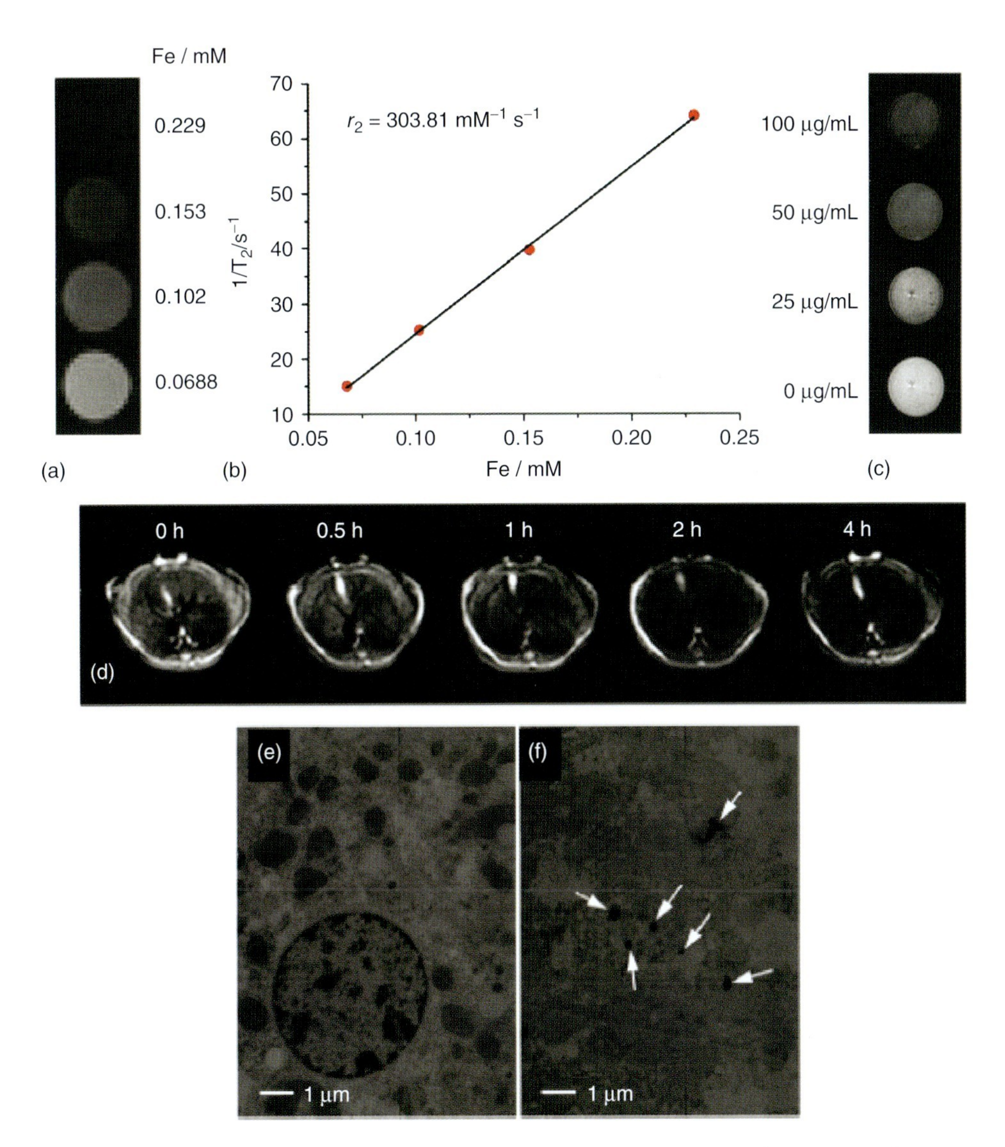

Fig. 11 (a) T_2-weighted MR images of the GO-PEG-β-FeOOH hybrid; (b) Plot of $1/T_2$ versus Fe concentration in GO-PEG-β-FeOOH, where the slope indicates the specific relaxivity (r_2); (c) T_2-weighted MR images of HeLa cells after 4 h incubation with different concentrations of GO-PEG-β-FeOOH; (d) *In-vivo* T_2-weighted MR images of liver before and after intravenous administration of GO-PEG-β-FeOOH corresponding to 1.0 mg Fe per kg body weight of mice (0.5, 1, 2, and 4 h post-treatment); (e,f) Transmission electron microscopy images of liver cells (e) before and (f) after intravenous injection of the hybrid. Reproduced with permission from Ref. [131]; © The Royal Society of Chemistry.

greater the number of components loaded onto the graphene sheets, the greater the number of synergistic properties gained by the hybrid (Fig. 10).

In another study, the *in-situ* growth of β-FeOOH nanorods on PEGylated GO sheets produced a novel hybrid with an ultra-high transverse r_2 relaxivity of 303.81 Fe $mM^{-1}\,s^{-1}$, which was more than 60-fold higher than that of hitherto reported β-FeOOH-based MRI contrast agents [131]. The GO-PEG-β-FeOOH hybrid also showed a DOX loading capability of 1.35 mg mg^{-1} at pH 7.4. Moreover, when the pH value of the buffer medium was adjusted to 5.5 (i.e., normal pH in the endosome/lysosome environment), approximately 57% of the DOX loaded onto the GO-PEG-β-FeOOH hybrid was released [131]. In addition, the hybrid obtained not only performed as an excellent MRI contrast agent on HeLa cells *in vitro*, but also demonstrated its practical use *in vivo* (Fig. 11).

5
Genetic Engineering

Genetic engineering, also known as *genetic modification*, presents much promise and opportunities in biomedicine. The technique is used to directly manipulate the genome of an organism by introducing new DNA sequences to that genome. Typically, a DNA sequence can be developed by isolating and copying suitable genetic materials through molecular cloning methods, after which the new DNA sequence is transported to the host organism to replicate development of the desired properties. Examples of this approach include RNA interference (RNAi) and antisense (AS) therapies, both of which are (potentially) powerful methods for the clinical treatment of various diseases, including cancer and acquired immunodeficiency syndrome (AIDS). Unfortunately, the therapeutic oligonucleotides can be easily degraded by cellular enzymes, or digested by cellular nucleases, before reaching their targets. In contrast to other carbon materials, the planar structure and large surface area of graphene enables it (and its derivatives) to serve as excellent delivery carriers. Moreover, pristine properties such as robust and inert surfaces endow graphene-based materials with steric hindrance effects that can protect any biomacromolecules that may be attached; an example of this is the protection of therapeutic oligonucleotides against cleavage in the intracellular environment. Thus, in genetic engineering systems graphene-based materials can serve not only as the carriers but also as protectors.

As noted above, strong interactions between oligonucleotides and graphene can protect the former from being attacked in the intracellular environment. In this respect, a modified GO was shown as a good vehicle for delivering hairpin-shaped DNA (a molecular beacon) into cells, and also to protect the loaded biospecies against enzymatic cleavage [132]. Thus, when the aptamer-carboxyfluorescein (FAM)/GO hybrid was synthesized based on particular interactions between FAM and GO [133], the GO served as a transporting vehicle to deliver the aptamer into cells while simultaneously protecting it against enzyme-mediated metabolism during the delivery process.

Modification of the surface of graphene with biocompatible units can lead to a dramatic enhancement of delivery efficiency. In this regard, nontoxic vehicles for efficient gene transfection were constructed by conjugating GO with low-molecular-weight branched polyethylenimine (BPEI) in order

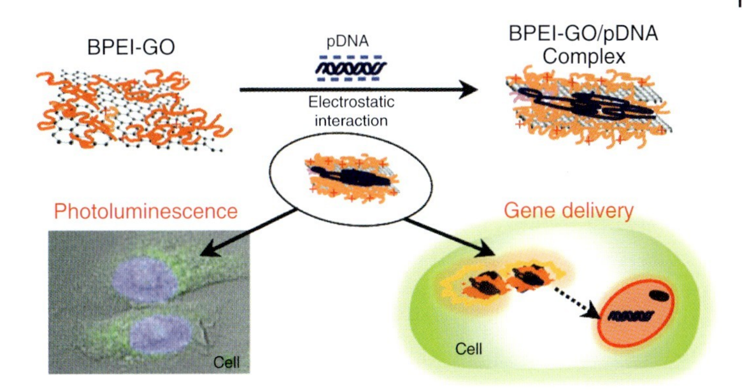

Fig. 12 Gene carrier, fabricated by the conjugation of BPEI with GO, increases the effective molecular weight of BPEI and consequently improves DNA binding, as well as condensation and transfection efficiency. Considering its high transfection efficiency, high cell viability and promising application as a bioimaging agent, the BPEI/GO hybrid could be extended for siRNA delivery and photothermal therapy. Reproduced with permission from Ref. [135]; © 2011, American Chemical Society.

to improve DNA binding and transfection efficiency. The PL properties of GO were enhanced through the conjugation of BPEI to GO, which led to the BPEI-GO hybrid being investigated as a fluorescence reagent for bioimaging [134]. In another study, positively charged GO/PEI hybrids were bound with plasmid DNA (pDNA) for intracellular transfection of the enhanced green fluorescence protein (EGFP) gene in HeLa cells. A high EGFP expression was observed when using GO/PEI-1.2 kDa as the transfection agent compared to PEI-1.2 kDa alone (Fig. 12). In contrast, GO/PEI-10 kDa showed a similar EGFP transfection efficiency but a lower toxicity than PEI-10 kDa [135].

Both, genes and drugs can be co-delivered by using graphene-based hybrids. For example, a sequential delivery of the Bcl-2 protein-targeted siRNA and anticancer drug DOX into cancer cells was accomplished by using the PEI/GO hybrid, exhibiting a synergistic effect and a significant enhancement in chemotherapeutic efficacy [104]. Likewise, a chitosan-functionalized magnetic graphene hybrid was used as a platform for simultaneous gene/drug delivery to tumors [136]. A pDNA encoding green fluorescent protein (GFP) and DOX was successfully loaded onto the GO sheet of the hybrid through the $\pi-\pi$ stacking interaction, and efficiently delivered into A549 lung cancer cells and C42b prostate cancer cells. Subsequent *in-vivo* investigations with tumor-bearing mice showed both GFP expression and DOX accumulation at the tumor sites at 24 and 48 h after administration. The hybrid presented integrated functions of chemo- and gene-therapeutics, as well as real-time diagnostics using MRI techniques. The results of each of these studies have suggested that graphene is a suitable gene delivery vector with a low cytotoxicity and a high transfection efficiency, presenting promising applications in non-viral-based gene therapy.

6 Conclusions

Recent achievements with different types of graphene (and derivatives) have been discussed in this chapter, based on the versatility of combinations of graphene and functional species. As well as an intrinsic biocompatibility, graphene can provide an excellent platform for various modifications suited to biomedical applications.

The biosensing applications of graphene (and derivatives) continue to influence current biotechnology, with the introduction of bioreceptors on graphene being accomplished via both covalent and noncovalent interactions, thereby presenting numerous methods for the functionalization of electronic interfaces. The modification of NPs on graphene may lead to the formation of electrochemical biosensors with enhanced sensitivities, better selectivities, and wide detection ranges. Indeed, multicomponent detection has also been achieved with a low detection limit and a high sensitivity. The biocompatibility of graphene-based hybrids also enables the *in-situ* detection of living cells, and rapid, accurate, multipurpose, cost-effective biosensors based on graphene hybrids are expected to be mass-produced in the near future.

The application of graphene (and derivatives) to bioimaging has been well recognized, and optical imaging, localized surface plasmon resonance, Raman imaging, MRI, and PET have been discussed to illustrate the many bioimaging applications of graphene hybrids. Flexible fabrication methods facilitate the fabrication of multimodal imaging techniques, which will be highly advantageous for future diagnostic applications.

Graphene (and derivatives) continues to bring novel applications to the field of therapeutics, while different approaches for drug loading such as noncovalent interactions, cleavable covalent linkage, and host–guest interactions continue to pave the way to versatile drug delivery where the release of a loaded drug can be controlled, permitting precise delivery during treatment processes. Moreover, graphene, with its intrinsic NIR absorption properties, provides another approach to tumor ablation, with both PTD and PTT having been shown effective for tumor treatment. The combination of imaging and therapeutics based on graphene hybrids will surely lead to the development of novel theranostic systems for "personalized" medicine in the future.

Unlike other carbon materials, the planar structure and large surface area of graphene (and derivatives) permits its use as an excellent delivery carrier in genetic engineering. Moreover, the robust and inert surface of graphene hybrids provides steric hindrance effects that can protect attached biomacromolecules from being degraded in the intracellular environment. Thus, in genetic engineering, graphene-based materials serve as carriers as well as protectors.

Clearly, graphene (and derivatives) has major potential in fields of biosensing, bioimaging, therapeutics, and genetic engineering, and in this respect some representative hybrids (as well as details of their performance in different areas) have been highlighted in this chapter. Current research investigations have identified the promise of graphene and its derivatives in real-life diagnostics and therapeutics, although in-depth investigations into the biocompatibility and toxicity of such hybrids in biological systems must be completed before their real-life application.

Acknowledgments

This research was supported by the National Research Foundation (NRF), Prime Minister's Office, Singapore under its NRF Fellowship (NRF2009NRF-RF001-015) and Campus for Research Excellence and Technological Enterprise (CREATE) Programme–Singapore Peking University Research Centre for a Sustainable Low-Carbon Future, and the NTU-A*Star Centre of Excellence for Silicon Technologies (A*Star SERC No.: 112 351 0003).

References

1. Geim, A.K. (2009) Graphene: status and prospects. *Science*, **324** (5934), 1530–1534.
2. Lee, C., Wei, X., Kysar, J.W., Hone, J. (2008) Measurement of the elastic properties and intrinsic strength of monolayer graphene. *Science*, **321** (5887), 385–388.
3. Wallace, P.R. (1947) The band theory of graphite. *Phys. Rev.*, **71** (9), 622–634.
4. Novoselov, K.S., Geim, A.K., Morozov, S.V., Jiang, D., Zhang, Y., Dubonos, S.V., Grigorieva, I.V., Firsov, A.A. (2004) Electric field in atomically thin carbon films. *Science*, **306** (5696), 666–669.
5. Dreyer, D.R., Park, S., Bielawski, C.W., Ruoff, R.S. (2010) The chemistry of graphene oxide. *Chem. Soc. Rev.*, **39** (1), 228–240.
6. Schwierz, F. (2010) Graphene transistors. *Nat. Nanotechnol.*, **5** (7), 487–496.
7. Eda, G., Chhowalla, M. (2010) Chemically derived graphene oxide: towards large-area thin-film electronics and optoelectronics. *Adv. Mater.*, **22** (22), 2392–2415.
8. Rao, C.N.R., Sood, A.K., Subrahmanyam, K.S., Govindaraj, A. (2009) Graphene: the new two-dimensional nanomaterial. *Angew. Chem. Int. Ed.*, **48** (42), 7752–7777.
9. Yan, H., Zhu, L., Li, X., Kwok, A., Pan, X., Zhao, Y. (2012) A photoswitchable [2]rotaxane array on graphene oxide. *Asian J. Org. Chem.*, **1** (4), 314–318.
10. Pan, X., Li, H., Nguyen, K.T., Grüner, G., Zhao, Y. (2012) Phonon energy transfer in graphene–photoacid hybrids. *J. Phys. Chem. C*, **116** (6), 4175–4181.
11. Dikin, D.A., Stankovich, S., Zimney, E.J., Piner, R.D., Dommett, G.H.B., Evmenenko, G., Nguyen, S.T., Ruoff, R.S. (2007) Preparation and characterization of graphene oxide paper. *Nature*, **448** (7152), 457–460.
12. Robinson, J.T., Perkins, F.K., Snow, E.S., Wei, Z., Sheehan, P.E. (2008) Reduced graphene oxide molecular sensors. *Nano Lett.*, **8** (10), 3137–3140.
13. Sanchez, V.C., Jachak, A., Hurt, R.H., Kane, A.B. (2012) Biological interactions of graphene-family nanomaterials: an interdisciplinary review. *Chem. Res. Toxicol.*, **25** (1), 15–34.
14. Zhang, H., Grüner, G., Zhao, Y. (2013) Recent advancements of graphene in biomedicine. *J. Mater. Chem. B*, **1** (20), 2542–2567.
15. Liao, K.H., Lin, Y.S., MacOsko, C.W., Haynes, C.L. (2011) Cytotoxicity of graphene oxide and graphene in human erythrocytes and skin fibroblasts. *ACS Appl. Mater. Interfaces*, **3** (7), 2607–2615.
16. Zhang, Y., Ali, S.F., Dervishi, E., Xu, Y., Li, Z., Casciano, D., Biris, A.S. (2010) Cytotoxicity effects of graphene and single-wall carbon nanotubes in neural phaeochromocytoma-derived PC12 cells. *ACS Nano*, **4** (6), 3181–3186.
17. Chang, Y., Yang, S.T., Liu, J.H., Dong, E., Wang, Y., Cao, A., Liu, Y., Wang, H. (2011) In vitro toxicity evaluation of graphene oxide on A549 cells. *Toxicol. Lett.*, **200** (3), 201–210.
18. Yang, K., Wan, J., Zhang, S., Zhang, Y., Lee, S.T., Liu, Z. (2011) In vivo pharmacokinetics, long-term biodistribution, and toxicology of pegylated graphene in mice. *ACS Nano*, **5** (1), 516–522.
19. Yan, L., Zhao, F., Li, S., Hu, Z., Zhao, Y. (2011) Low-toxic and safe nanomaterials by surface-chemical design, carbon nanotubes, fullerenes, metallofullerenes, and graphenes. *Nanoscale*, **3** (2), 362–382.
20. Liu, Y., Zhao, Y., Sun, B., Chen, C. (2013) Understanding the toxicity of carbon nanotubes. *Acc. Chem. Res.*, **46** (3), 702–713.
21. Johns, J.E., Hersam, M.C. (2013) Atomic covalent functionalization of graphene. *Acc. Chem. Res.*, **46** (1), 77–86.
22. Georgakilas, V., Otyepka, M., Bourlinos, A.B., Chandra, V., Kim, N., Kemp, K.C.,

Hobza, P., Zboril, R., Kim, K.S. (2012) Functionalization of graphene: covalent and non-covalent approaches, derivatives and applications. *Chem. Rev.*, **112** (11), 6156–6214.
23. Pumera, M. (2011) Nanotoxicology: the molecular science point of view. *Chem. Asian J.*, **6** (2), 340–348.
24. Kuila, T., Bose, S., Mishra, A.K., Khanra, P., Kim, N.H., Lee, J.H. (2012) Chemical functionalization of graphene and its applications. *Prog. Mater. Sci.*, **57** (7), 1061–1105.
25. Nguyen, P., Berry, V. (2012) Graphene interfaced with biological cells: opportunities and challenges. *J. Phys. Chem. Lett.*, **3** (8), 1024–1029.
26. Premkumar, T., Geckeler, K.E. (2012) Graphene-DNA hybrid materials: assembly, applications, and prospects. *Prog. Polym. Sci.*, **37** (4), 515–529.
27. Yang, K., Li, Y., Tan, X., Peng, R., Liu, Z. (2013) Behavior and toxicity of graphene and its functionalized derivatives in biological systems. *Small*, **9** (9-10), 1492–1503.
28. Yang, M., Yao, J., Duan, Y. (2013) Graphene and its derivatives for cell biotechnology. *Analyst*, **138** (1), 72–86.
29. Novoselov, K.S., Geim, A.K., Morozov, S.V., Jiang, D., Katsnelson, M.I., Grigorieva, I.V., Dubonos, S.V., Firsov, A.A. (2005) Two-dimensional gas of massless Dirac fermions in graphene. *Nature*, **438** (7065), 197–200.
30. Zhang, Y., Tan, Y.-W., Stormer, H.L., Kim, P. (2005) Experimental observation of the quantum Hall effect and Berry's phase in graphene. *Nature*, **438** (7065), 201–204.
31. Yang, W., Ratinac, K.R., Ringer, S.R., Thordarson, P., Gooding, J.J., Braet, F. (2010) Carbon nanomaterials in biosensors: should you use nanotubes or graphene. *Angew. Chem. Int. Ed.*, **49** (12), 2114–2138.
32. Song, Y., Wei, W., Qu, X. (2011) Colorimetric biosensing using smart materials. *Adv. Mater.*, **23** (37), 4215–4236.
33. Liu, Y., Dong, X., Chen, P. (2012) Biological and chemical sensors based on graphene materials. *Chem. Soc. Rev.*, **41** (6), 2283–2307.
34. Artiles, M.S., Rout, C.S., Fisher, T.S. (2011) Graphene-based hybrid materials and devices for biosensing. *Adv. Drug Delivery Rev.*, **63** (14-15), 1352–1360.
35. Iost, R.M., Crespilho, F.N. (2012) Layer-by-layer self-assembly and electrochemistry: applications in biosensing and bioelectronics. *Biosens. Bioelectron.*, **31** (1), 1–10.
36. Pumera, M. (2011) Graphene in biosensing. *Mater. Today*, **14** (7-8), 308–315.
37. Morales-Narváez, E., Merkoçi, A. (2012) Graphene oxide as an optical biosensing platform. *Adv. Mater.*, **24** (25), 3298–3308.
38. Marín, S., Merkoçi, A. (2012) Nanomaterials based electrochemical sensing applications for safety and security. *Electroanalysis*, **24** (3), 459–469.
39. Lee, W.C., Lim, C.H.Y.X., Shi, H., Tang, L.A.L., Wang, Y., Lim, C.T., Loh, K.P. (2011) Origin of enhanced stem cell growth and differentiation on graphene and graphene oxide. *ACS Nano*, **5** (9), 7334–7341.
40. Lim, H.N., Huang, N.M., Lim, S.S., Harrison, I., Chia, C.H. (2011) Fabrication and characterization of graphene hydrogel via hydrothermal approach as a scaffold for preliminary study of cell growth. *Int. J. Nanomed.*, **6**, 1817–1823.
41. Yang, M., Gong, S. (2010) Immunosensor for the detection of cancer biomarker based on percolated graphene thin film. *Chem. Commun.*, **46** (31), 5796–5798.
42. Yang, M., Javadi, A., Li, H., Gong, S. (2010) Ultrasensitive immunosensor for the detection of cancer biomarker based on graphene sheet. *Biosens. Bioelectron.*, **26** (2), 560–565.
43. Zhang, B., Li, Q., Cui, T. (2012) Ultra-sensitive suspended graphene nanocomposite cancer sensors with strong suppression of electrical noise. *Biosens. Bioelectron.*, **31** (1), 105–109.
44. Wan, Y., Lin, Z., Zhang, D., Wang, Y., Hou, B. (2011) Impedimetric immunosensor doped with reduced graphene sheets fabricated by controllable electrodeposition for the non-labelled detection of bacteria. *Biosens. Bioelectron.*, **26** (5), 1959–1964.
45. Feng, L., Chen, Y., Ren, J., Qu, X. (2011) A graphene functionalized electrochemical aptasensor for selective label-free detection of cancer cells. *Biomaterials*, **32** (11), 2930–2937.
46. Qu, F., Li, T., Yang, M. (2011) Colorimetric platform for visual detection of cancer biomarker based on intrinsic peroxidase activity of graphene oxide. *Biosens. Bioelectron.*, **26** (9), 3927–3931.

47. He, Y., Lin, Y., Tang, H., Pang, D. (2012) A graphene oxide-based fluorescent aptasensor for the turn-on detection of epithelial tumor marker mucin 1. *Nanoscale*, **4** (6), 2054–2059.
48. Luo, X., Morrin, A., Killard, A.J., Smyth, M.R. (2006) Application of nanoparticles in electrochemical sensors and biosensors. *Electroanalysis*, **18** (4), 319–326.
49. Lu, L.M., Li, H.B., Qu, F., Zhang, X.B., Shen, G.L., Yu, R.Q. (2011) In situ synthesis of palladium nanoparticle-graphene nanohybrids and their application in nonenzymatic glucose biosensors. *Biosens. Bioelectron.*, **26** (8), 3500–3504.
50. Chen, X., Cai, Z., Huang, Z., Oyama, M., Jiang, Y., Chan, X. (2013) Non-enzymatic oxalic acid sensor using platinum nanoparticles modified on graphene nanosheets. *Nanoscale*, **5** (13), 5779–5783.
51. Sun, C.L., Lee, H.H., Yang, J.M., Wu, C.C. (2011) The simultaneous electrochemical detection of ascorbic acid, dopamine, and uric acid using graphene/size-selected Pt nanocomposites. *Biosens. Bioelectron.*, **26** (8), 3450–3455.
52. Li, J., Kuang, D., Feng, Y., Zhang, F., Xu, Z., Liu, M., Wang, D. (2013) Green synthesis of silver nanoparticles-graphene oxide nanocomposite and its application in electrochemical sensing of tryptophan. *Biosens. Bioelectron.*, **42** (1), 198–206.
53. Gutés, A., Carraro, C., Maboudian, R. (2012) Single-layer CVD-grown graphene decorated with metal nanoparticles as a promising biosensing platform. *Biosens. Bioelectron.*, **33** (1), 56–59.
54. Parlak, O., Tiwari, A., Turner, A.P.F. (2013) Template-directed hierarchical self-assembly of graphene based hybrid structure for electrochemical biosensing. *Biosens. Bioelectron.*, **49**, 53–62.
55. Yang, J., Deng, S., Lei, J., Ju, H., Gunasekaran, S. (2011) Electrochemical synthesis of reduced graphene sheet-AuPd alloy nanoparticle composites for enzymatic biosensing. *Biosens. Bioelectron.*, **29** (1), 159–166.
56. Wang, H., Zhang, Y., Li, H., Du, B., Ma, H., Wu, D., Wei, Q. (2013) A silver-palladium alloy nanoparticle-based electrochemical biosensor for simultaneous detection of ractopamine, clenbuterol and salbutamol. *Biosens. Bioelectron.*, **49**, 14–19.
57. Wei, Q., Xiang, Z., He, J., Wang, G., Li, H., Qian, Z., Yang, M. (2010) Dumbbell-like Au-Fe_3O_4 nanoparticles as label for the preparation of electrochemical immunosensors. *Biosens. Bioelectron.*, **26** (2), 627–631.
58. Zhuo, Y., Chai, Y.Q., Yuan, R., Mao, L., Yuan, Y.L., Han, J. (2011) Glucose oxidase and ferrocene labels immobilized at Au/TiO_2 nanocomposites with high load amount and activity for sensitive immunoelectrochemical measurement of ProGRP biomarker. *Biosens. Bioelectron.*, **26** (9), 3838–3844.
59. Liu, M., Chen, Q., Lai, C., Zhang, Y., Deng, J., Li, H., Yao, S. (2013) A double signal amplification platform for ultrasensitive and simultaneous detection of ascorbic acid, dopamine, uric acid and acetaminophen based on a nanocomposite of ferrocene thiolate stabilized Fe_3O_4@Au nanoparticles with graphene sheet. *Biosens. Bioelectron.*, **48**, 75–81.
60. Yu, Y., Chen, Z., He, S., Zhang, B., Li, X., Yao, M. (2014) Direct electron transfer of glucose oxidase and biosensing for glucose based on PDDA-capped gold nanoparticle modified graphene/multi-walled carbon nanotubes electrode. *Biosens. Bioelectron.*, **52**, 147–152.
61. Gao, H., Xiao, F., Ching, C.B., Duan, H. (2011) One-step electrochemical synthesis of PtNi nanoparticle-graphene nanocomposites for nonenzymatic amperometric glucose detection. *ACS Appl. Mater. Interfaces*, **3** (8), 3049–3057.
62. Lu, W., Luo, Y., Chang, G., Sun, X. (2011) Synthesis of functional SiO_2-coated graphene oxide nanosheets decorated with Ag nanoparticles for H_2O_2 and glucose detection. *Biosens. Bioelectron.*, **26** (12), 4791–4797.
63. Zeng, Q., Cheng, J.S., Liu, X.F., Bai, H.T., Jiang, J.H. (2011) Palladium nanoparticle/chitosan-grafted graphene nanocomposites for construction of a glucose biosensor. *Biosens. Bioelectron.*, **26** (8), 3456–3463.
64. Xiao, F., Li, Y., Gao, H., Ge, S., Duan, H. (2013) Growth of coral-like PtAu-MnO_2 binary nanocomposites on free-standing graphene paper for flexible nonenzymatic glucose sensors. *Biosens. Bioelectron.*, **41**, 417–423.

65. Liu, M., Liu, R., Chen, W. (2013) Graphene wrapped Cu_2O nanocubes: non-enzymatic electrochemical sensors for the detection of glucose and hydrogen peroxide with enhanced stability. *Biosens. Bioelectron.*, **45**, 206–212.
66. Shan, C., Yang, H., Han, D., Zhang, Q., Ivaska, A., Niu, L. (2010) Graphene/AuNPs/chitosan nanocomposites film for glucose biosensing. *Biosens. Bioelectron.*, **25** (5), 1070–1074.
67. Yang, G., Cao, J., Li, L., Rana, R.K., Zhu, J.J. (2013) Carboxymethyl chitosan-functionalized graphene for label-free electrochemical cytosensing. *Carbon*, **51** (1), 124–133.
68. Zhong, Z., Wu, W., Wang, D., Shan, J., Qing, Y., Zhang, Z. (2010) Nanogold-enwrapped graphene nanocomposites as trace labels for sensitivity enhancement of electrochemical immunosensors in clinical immunoassays: carcinoembryonic antigen as a model. *Biosens. Bioelectron.*, **25** (10), 2379–2383.
69. Song, W., Li, D.W., Li, Y.T., Li, Y., Long, Y.T. (2011) Disposable biosensor based on graphene oxide conjugated with tyrosinase assembled gold nanoparticles. *Biosens. Bioelectron.*, **26** (7), 3181–3186.
70. Li, H., Wei, Q., He, J., Li, T., Zhao, Y., Cai, Y., Du, B., Qian, Z., Yang, M. (2011) Electrochemical immunosensors for cancer biomarker with signal amplification based on ferrocene functionalized iron oxide nanoparticles. *Biosens. Bioelectron.*, **26** (8), 3590–3595.
71. Xue, Y., Zhao, H., Wu, Z., Li, X., He, Y., Yuan, Z. (2011) The comparison of different gold nanoparticles/graphene nanosheets hybrid nanocomposites in electrochemical performance and the construction of a sensitive uric acid electrochemical sensor with novel hybrid nanocomposites. *Biosens. Bioelectron.*, **29** (1), 102–108.
72. Wang, Y., Yuan, R., Chai, Y., Yuan, Y., Bai, L., Liao, Y. (2011) A multi-amplification aptasensor for highly sensitive detection of thrombin based on high-quality hollow CoPt nanoparticles decorated graphene. *Biosens. Bioelectron.*, **30** (1), 61–66.
73. Lian, W., Liu, S., Yu, J., Xing, X., Li, J., Cui, M., Huang, J. (2012) Electrochemical sensor based on gold nanoparticles fabricated molecularly imprinted polymer film at chitosan-platinum nanoparticles/graphene-gold nanoparticles double nanocomposites modified electrode for detection of erythromycin. *Biosens. Bioelectron.*, **38** (1), 163–169.
74. Cao, S., Zhang, L., Chai, Y., Yuan, R. (2013) An integrated sensing system for detection of cholesterol based on TiO_2-graphene-Pt-Pd hybrid nanocomposites. *Biosens. Bioelectron.*, **42**, 532–538.
75. Zhang, J., Sun, Y., Xu, B., Zhang, H., Gao, Y., Song, D. (2013) A novel surface plasmon resonance biosensor based on graphene oxide decorated with gold nanorod-antibody conjugates for determination of transferrin. *Biosens. Bioelectron.*, **45**, 230–236.
76. Liu, M., Deng, J., Chen, Q., Huang, Y., Wang, L., Zhao, Y., Zhang, Y., Li, H., Yao, S. (2013) Sensitive detection of rutin with novel ferrocene benzyne derivative modified electrodes. *Biosens. Bioelectron.*, **41**, 275–281.
77. Hu, Y., Xue, Z., He, H., Ai, R., Liu, X., Lu, X. (2013) Photoelectrochemical sensing for hydroquinone based on porphyrin-functionalized Au nanoparticles on graphene. *Biosens. Bioelectron.*, **47**, 45–49.
78. Wang, L., Hua, E., Liang, M., Ma, C., Liu, Z., Sheng, S., Liu, M., Xie, G., Feng, W. (2014) Graphene sheets, polyaniline and AuNPs based DNA sensor for electrochemical determination of BCR/ABL fusion gene with functional hairpin probe. *Biosens. Bioelectron.*, **51**, 201–207.
79. Wang, Z., Zhang, J., Yin, Z., Wu, S., Mandler, D., Zhang, H. (2012) Fabrication of nanoelectrode ensembles by electrodepositon of Au nanoparticles on single-layer graphene oxide sheets. *Nanoscale*, **4** (8), 2728–2733.
80. Yu, S., Wei, Q., Du, B., Wu, D., Li, H., Yan, L., Ma, H., Zhang, Y. (2013) Label-free immunosensor for the detection of kanamycin using Ag@Fe_3O_4 nanoparticles and thionine mixed graphene sheet. *Biosens. Bioelectron.*, **48**, 224–229.
81. Yang, J., Strickler, J.R., Gunasekaran, S. (2012) Indium tin oxide-coated glass modified with reduced graphene oxide sheets and gold nanoparticles as disposable working electrodes for dopamine sensing in meat samples. *Nanoscale*, **4** (15), 4594–4602.
82. Sun, X., Ji, J., Jiang, D., Li, X., Zhang, Y., Li, Z., Wu, Y. (2013) Development of a novel

electrochemical sensor using pheochromocytoma cells and its assessment of acrylamide cytotoxicity. *Biosens. Bioelectron.*, **44**, 122–126.

83. Sun, X., Liu, Z., Welsher, K., Robinson, J.T., Goodwin, A., Zaric, S., Dai, H. (2008) Nano-graphene oxide for cellular imaging and drug delivery. *Nano Res.*, **1** (3), 203–212.
84. Zhu, S., Zhang, J., Qiao, C., Tang, S., Li, Y., Yuan, W., Li, B., Tian, L., Liu, F., Hu, R., Gao, H., Wei, H., Zhang, H., Sun, H., Yang, B. (2011) Strongly green-photoluminescent graphene quantum dots for bioimaging applications. *Chem. Commun.*, **47** (24), 6858–6860.
85. Peng, J., Gao, W., Gupta, B.K., Liu, Z., Romero-Aburto, R., Ge, L., Song, L., Alemany, L.B., Zhan, X., Gao, G., Vithayathil, S.A., Kaipparettu, B.A., Marti, A.A., Hayashi, T., Zhu, J.J., Ajayan, P.M. (2012) Graphene quantum dots derived from carbon fibers. *Nano Lett.*, **12** (2), 844–849.
86. Zhang, M., Bai, L., Shang, W., Xie, W., Ma, H., Fu, Y., Fang, D., Sun, H., Fan, L., Han, M., Liu, C., Yang, S. (2012) Facile synthesis of water-soluble, highly fluorescent graphene quantum dots as a robust biological label for stem cells. *J. Mater. Chem.*, **22** (15), 7461–7467.
87. Sheng, Y., Tang, X., Peng, E., Xue, J. (2013) Graphene oxide based fluorescent nanocomposites for cellular imaging. *J. Mater. Chem. B*, **1** (4), 512–521.
88. Xie, G., Cheng, J., Li, Y., Xi, P., Chen, F., Liu, H., Hou, F., Shi, Y., Huang, L., Xu, Z., Bai, D., Zeng, Z. (2012) Fluorescent graphene oxide composites synthesis and its biocompatibility study. *J. Mater. Chem.*, **22** (18), 9308–9314.
89. Sreejith, S., Ma, X., Zhao, Y. (2012) Graphene oxide wrapping on squaraine-loaded mesoporous silica nanoparticles for bioimaging. *J. Am. Chem. Soc.*, **134** (42), 17346–17349.
90. Qian, J., Wang, D., Cai, F.H., Xi, W., Peng, L., Zhu, Z.F., He, H., Hu, M.L., He, S. (2012) Observation of multiphoton-induced fluorescence from graphene oxide nanoparticles and applications in in vivo functional bioimaging. *Angew. Chem. Int. Ed.*, **51** (42), 10570–10575.
91. Wang, Y., Zhen, S.J., Zhang, Y., Li, Y.F., Huang, C.Z. (2011) Facile fabrication of metal nanoparticle/graphene oxide hybrids: a new strategy to directly illuminate graphene for optical imaging. *J. Phys. Chem. C*, **115** (26), 12815–12821.
92. Zhang, H., Ma, X., Nguyen, K.T., Zeng, Y., Tai, S., Zhao, Y. (2014) Water-soluble pillararene-functionalized graphene oxide for in vitro Raman and fluorescence dual-mode imaging. *ChemPlusChem*, **79** (3), 462–469.
93. Liu, Q., Wei, L., Wang, J., Peng, F., Luo, D., Cui, R., Niu, Y., Qin, X., Liu, Y., Sun, H., Yang, J., Li, Y. (2012) Cell imaging by graphene oxide based on surface enhanced Raman scattering. *Nanoscale*, **4** (22), 7084–7089.
94. Jaganathan, H., Hugar, D.L., Ivanisevic, A. (2011) Examining MRI contrast in three-dimensional cell culture phantoms with DNA-templated nanoparticle chains. *ACS Appl. Mater. Interfaces*, **3** (4), 1282–1288.
95. Liu, G., Wang, Z., Lu, J., Xia, C., Gao, F., Gong, Q., Song, B., Zhao, X., Shuai, X., Chen, X., Ai, H., Gu, Z. (2011) Low molecular weight alkyl-polycation wrapped magnetite nanoparticle clusters as MRI probes for stem cell labeling and in vivo imaging. *Biomaterials*, **32** (2), 528–537.
96. Ai, H., Flask, C., Weinberg, B., Shuai, X.T., Pagel, M.D., Farrell, D., Duerk, J., Gao, J. (2005) Magnetite-loaded polymeric micelles as ultrasensitive magnetic-resonance probes. *Adv. Mater.*, **17** (16), 1949–1952.
97. Lu, J., Ma, S., Sun, J., Xia, C., Liu, C., Wang, Z., Zhao, X., Gao, F., Gong, Q., Song, B., Shuai, X., Ai, H., Gu, Z. (2009) Manganese ferrite nanoparticle micellar nanocomposites as MRI contrast agent for liver imaging. *Biomaterials*, **30** (15), 2919–2928.
98. Chen, W., Yi, P., Zhang, Y., Zhang, L., Deng, Z., Zhang, Z. (2011) Composites of aminodextran-coated Fe_3O_4 nanoparticles and graphene oxide for cellular magnetic resonance imaging. *ACS Appl. Mater. Interfaces*, **3** (10), 4085–4091.
99. Peng, E., Choo, E.S.G., Chandrasekharan, P., Yang, C. T., Ding, J., Chuang, K.H., Xue, J.M. (2012) Synthesis of manganese ferrite/graphene oxide nanocomposites for biomedical applications. *Small*, **8** (23), 3620–3630.

100. Chen, Y., Guo, F., Qiu, Y., Hu, H., Kulaots, I., Walsh, E., Hurt, R. H. (2013) Encapsulation of particle ensembles in graphene nanosacks as a new route to multifunctional materials. *ACS Nano*, **7** (5), 3744–3753.
101. Hong, H., Zhang, Y., Engle, J.W., Nayak, T.R., Theuer, C.P., Nickles, R.J., Barnhart, T.E., Cai, W. (2012) In vivo targeting and positron emission tomography imaging of tumor vasculature with ^{66}Ga-labeled nano-graphene. *Biomaterials*, **33** (16), 4147–4156.
102. Yang, K., Feng, L., Hong, H., Cai, W., Liu, Z. (2013) Preparation and functionalization of graphene nanocomposites for biomedical applications. *Nat. Protoc.*, **8** (12), 2392–2403.
103. Liu, Z., Robinson, J. T., Sun, X., Dai, H. (2008) PEGylated nanographene oxide for delivery of water-insoluble cancer drugs. *J. Am. Chem. Soc.*, **130** (33), 10876–10877.
104. Zhang, L., Lu, Z., Zhao, Q., Huang, J., Shen, H., Zhang, Z. (2011) Enhanced chemotherapy efficacy by sequential delivery of siRNA and anticancer drugs using PEI-grafted graphene oxide. *Small*, **7** (4), 460–464.
105. Liu, K., Zhang, J.J., Cheng, F.F., Zheng, T.T., Wang, C., Zhu, J.J. (2011) Green and facile synthesis of highly biocompatible graphene nanosheets and its application for cellular imaging and drug delivery. *J. Mater. Chem.*, **21** (32), 12034–12040.
106. Kakran, M., Sahoo, N.G., Bao, H., Pan, Y., Li, L. (2011) Functionalized graphene oxide as nanocarrier for loading and delivery of ellagic acid. *Curr. Med. Chem.*, **18** (29), 4503–4512.
107. Zhang, L., Xia, J., Zhao, Q., Liu, L., Zhang, Z. (2010) Functional graphene oxide as a nanocarrier for controlled loading and targeted delivery of mixed anticancer drugs. *Small*, **6** (4), 537–544.
108. Dong, H., Li, Y., Yu, J., Song, Y., Cai, X., Liu, J., Zhang, J., Ewing, R. C., Shi, D. (2013) A versatile multicomponent assembly via β-cyclodextrin host–guest chemistry on graphene for biomedical applications. *Small*, **9** (3), 446–456.
109. Wei, G., Dong, R., Wang, D., Feng, L., Dong, S., Song, A., Hao, J. (2014) Functional materials from the covalent modification of reduced graphene oxide and β-cyclodextrin as a drug delivery carrier. *New J. Chem.*, **38** (1), 140–145.
110. Yang, Y., Zhang, Y. M., Chen, Y., Zhao, D., Chen, J. T., Liu, Y. (2012) Construction of a graphene oxide based noncovalent multiple nanosupramolecular assembly as a scaffold for drug delivery. *Chem. Eur. J.*, **18** (14), 4208–4215.
111. Liang, K., Such, G.K., Johnston, A.P.R., Zhu, Z., Ejima, H., Richardson, J.J., Cui, J., Caruso, F. (2014) Endocytic pH-triggered degradation of nanoengineered multilayer capsules. *Adv. Mater.*, **26** (12), 1901–1905.
112. Tabatabaei Rezaei, S.J., Abandansari, H.S., Nabid, M.R., Niknejad, H. (2014) pH-responsive unimolecular micelles self-assembled from amphiphilic hyperbranched block copolymer for efficient intracellular release of poorly water-soluble anticancer drugs. *J. Colloid Interface Sci.*, **425**, 27–35.
113. Yan, H., Teh, C., Sreejith, S., Zhu, L., Kwok, A., Fang, W., Ma, X., Nguyen, K.T., Korzh, V., Zhao, Y. (2012) Functional mesoporous silica nanoparticles for photothermal-controlled drug delivery in vivo. *Angew. Chem. Int. Ed.*, **51** (33), 8373–8377.
114. Paasonen, L., Sipilä, T., Subrizi, A., Laurinmäki, P., Butcher, S. J., Rappolt, M., Yaghmur, A., Urtti, A., Yliperttula, M. (2010) Gold-embedded photosensitive liposomes for drug delivery: triggering mechanism and intracellular release. *J. Controlled Release*, **147** (1), 136–143.
115. Timko, B.P., Dvir, T., Kohane, D.S. (2010) Remotely triggerable drug delivery systems. *Adv. Mater.*, **22** (44), 4925–4943.
116. Ma, X., Ong, O.S., Zhao, Y. (2013) Dual-responsive drug release from oligonucleotide-capped mesoporous silica nanoparticles. *Biomater. Sci.*, **1** (9), 912–917.
117. Yang, X., Zhang, X., Liu, Z., Ma, Y., Huang, Y., Chen, Y. (2008) High-efficiency loading and controlled release of doxorubicin hydrochloride on graphene oxide. *J. Phys. Chem. C*, **112** (45), 17554–17558.
118. Rana, V.K., Choi, M.C., Kong, J.Y., Kim, G.Y., Kim, M.J., Kim, S.H., Mishra, S., Singh, R.P., Ha, C.S. (2011) Synthesis and drug-delivery behavior of chitosan-functionalized graphene oxide hybrid nanosheets. *Macromol. Mater. Eng.*, **296** (2), 131–140.
119. Hu, H., Yu, J., Li, Y., Zhao, J., Dong, H. (2012) Engineering of a novel pluronic

F127/graphene nanohybrid for pH responsive drug delivery. *J. Biomed. Mater. Res. Part A*, **100** (1), 141–148.
120. Wen, H., Dong, C., Dong, H., Shen, A., Xia, W., Cai, X., Song, Y., Li, X., Li, Y., Shi, D. (2012) Engineered redox-responsive PEG detachment mechanism in PEGylated nano-graphene oxide for intracellular drug delivery. *Small*, **8** (5), 760–769.
121. Yang, X., Wang, Y., Huang, X., Ma, Y., Huang, Y., Yang, R., Duan, H., Chen, Y. (2011) Multi-functionalized graphene oxide based anticancer drug-carrier with dual-targeting function and pH-sensitivity. *J. Mater. Chem.*, **21** (10), 3448–3454.
122. Dong, H., Zhao, Z., Wen, H., Li, Y., Guo, F., Shen, A., Pilger, F., Lin, C., Shi, D. (2010) Poly(ethylene glycol) conjugated nano-graphene oxide for photodynamic therapy. *Sci. China: Chem.*, **53** (11), 2265–2271.
123. Zhou, L., Wang, W., Tang, J., Zhou, J.H., Jiang, H.J., Shen, J. (2011) Graphene oxide noncovalent photosensitizer and its anticancer activity in vitro. *Chem. Eur. J.*, **17** (43), 12084–12091.
124. Liu, Z., Robinson, J.T., Tabakman, S.M., Yang, K., Dai, H. (2011) Carbon materials for drug delivery and cancer therapy. *Mater. Today*, **14** (7-8), 316–323.
125. Hönigsmann, H. (2013) History of phototherapy in dermatology. *Photochem. Photobiol. Sci.*, **12** (1), 16–21.
126. Yang, K., Zhang, S., Zhang, G., Sun, X., Lee, S.T., Liu, Z. (2010) Graphene in mice: ultrahigh in vivo tumor uptake and efficient photothermal therapy. *Nano Lett.*, **10** (9), 3318–3323.
127. Robinson, J.T., Tabakman, S.M., Liang, Y., Wang, H., Sanchez Casalongue, H., Vinh, D., Dai, H. (2011) Ultrasmall reduced graphene oxide with high near-infrared absorbance for photothermal therapy. *J. Am. Chem. Soc.*, **133** (17), 6825–6831.
128. Tian, B., Wang, C., Zhang, S., Feng, L., Liu, Z. (2011) Photothermally enhanced photodynamic therapy delivered by nano-graphene oxide. *ACS Nano*, **5** (9), 7000–7009.
129. Wang, Y., Wang, H., Liu, D., Song, S., Wang, X., Zhang, H. (2013) Graphene oxide covalently grafted upconversion nanoparticles for combined NIR mediated imaging and photothermal/photodynamic cancer therapy. *Biomaterials*, **34** (31), 7715–7724.
130. Shi, X., Gong, H., Li, Y., Wang, C., Cheng, L., Liu, Z. (2013) Graphene-based magnetic plasmonic nanocomposite for dual bioimaging and photothermal therapy. *Biomaterials*, **34** (20), 4786–4793.
131. Chen, M.-L., Shen, L.-M., Chen, S., Wang, H., Chen, X.-W., Wang, J.-H. (2013) In situ growth of β-FeOOH nanorods on graphene oxide with ultra-high relaxivity for in vivo magnetic resonance imaging and cancer therapy. *J. Mater. Chem. B*, **1** (20), 2582–2589.
132. Lu, C.H., Zhu, C.L., Li, J., Liu, J.J., Chen, X., Yang, H.H. (2010) Using graphene to protect DNA from cleavage during cellular delivery. *Chem. Commun.*, **46** (18), 3116–3118.
133. Wang, Y., Li, Z., Hu, D., Lin, C.T., Li, J., Lin, Y. (2010) Aptamer/graphene oxide nanocomplex for in situ molecular probing in living cells. *J. Am. Chem. Soc.*, **132** (27), 9274–9276.
134. Feng, L., Zhang, S., Liu, Z. (2011) Graphene-based gene transfection. *Nanoscale*, **3** (3), 1252–1257.
135. Kim, H., Namgung, R., Singha, K., Oh, I.K., Kim, W.J. (2011) Graphene oxide-polyethylenimine nanoconstruct as a gene delivery vector and bioimaging tool. *Bioconjugate Chem.*, **22** (12), 2558–2567.
136. Wang, C., Ravi, S., Garapati, U.S., Das, M., Howell, M., Mallela, J., Alwarappan, S., Mohapatra, S.S., Mohapatra, S. (2013) Multifunctional chitosan magnetic-graphene (CMG) nanoparticles: a theranostic platform for tumor-targeted co-delivery of drugs, genes and MRI contrast agents. *J. Mater. Chem. B*, **1** (35), 4396–4405.

5
Synthetic Gene Circuits

Barbara Jusiak[1,2], *Ramiz Daniel*[1,3], *Fahim Farzadfard*[2–5], *Lior Nissim*[1,3], *Oliver Purcell*[2,3], *Jacob Rubens*[1,3,5], *and Timothy K. Lu*[1–6]

[1] *Massachusetts Institute of Technology, Research Laboratory of Electronics, Cambridge, MA 02139, USA*

[2] *Massachusetts Institute of Technology, Department of Biological Engineering, Cambridge, MA 02139, USA*

[3] *Massachusetts Institute of Technology, Synthetic Biology Center, Cambridge, MA 02139, USA*

[4] *Massachusetts Institute of Technology, Department of Electrical Engineering and Computer Science, Cambridge, MA 02139, USA*

[5] *Massachusetts Institute of Technology, MIT Microbiology Program, Cambridge, MA 02139, USA*

[6] *Massachusetts Institute of Technology, MIT Computational and Systems Biology Program, Cambridge, MA 02139, USA*

Translational Nanomedicine. First Edition. Edited by Robert A. Meyers.

Keywords

Analog circuit
A circuit that produces a graded response based on the levels of inputs it receives. Analog circuits can have a graded transfer function over a wide range of input concentrations.

Digital circuit
A circuit that produces an all-or-none response, depending on whether its inputs are above or below a defined threshold level. Digital circuits include digital logic gates. Complex circuits may include layered logic gates, so that the output of one gate provides the input for another.

Logic gate
A device that accepts inputs in the form of TRUE and FALSE (also represented as 1 and 0, respectively) and returns a single TRUE or FALSE output based on Boolean logic operations such as AND NOT OR, and others. For example, an AND gate returns the TRUE output if, and only if, all inputs are TRUE.

Orthogonality
The ability of circuit components to function in the same cell without crosstalk; for example, two transcription factors that bind distinct DNA motifs, or RNA molecules that regulate distinct transcripts. Orthogonal components are vital for building complex synthetic circuits.

Oscillator
A circuit that cycles repeatedly between states, such as high and low levels of expression of a particular protein; one of the prototypical examples of a synthetic gene circuit.

Synthetic transcription factor (sTF)
A human-made transcription factor designed to regulate transcription from a specific promoter; often designed to be responsive to an input, such as a small molecule. May include domains from naturally occurring transcription factors, such as a transactivation or ligand-binding domain, and rationally designed motifs, such as sequence-specific DNA-binding zinc fingers.

Toggle switch
A circuit that can exist in one of two stable states and may be switched (toggled) between the two states by a defined input; along with the oscillator, the toggle switch is a classic example of a proof-of-principle synthetic gene circuit.

Transfer function
The output level of a gene circuit as a function of the input level(s) (e.g., the activity of the fluorescent reporter output as a function of the concentration of a small molecule input).

Tunability
The ability to adjust the activity level of a synthetic circuit component, such as the strength of gene expression from a synthetic promoter.

The past decade has witnessed tremendous advances in the design and implementation of synthetic gene circuits that program living cells to perform specific user-defined tasks. Synthetic circuits have been implemented in bacteria, yeast, and mammalian cells, using a variety of transcriptional and post-transcriptional regulatory mechanisms. These devices, which lie at the intersection of biology and engineering, have provided insights into the function of naturally occurring gene regulatory networks. Furthermore, they hold the potential for transformative future applications in medicine, bioremediation, manufacturing, and more. In this review, some examples are presented of commonly used synthetic circuits, including oscillators, switches, memory devices, and circuits that perform digital and analog computation. The building blocks of synthetic gene circuits, as well as the challenges and considerations of circuit design, are also discussed. Finally, an overview is provided of the potential practical applications of this dynamic field of research.

1 Introduction to Synthetic Gene Circuits

Living cells monitor their environment and respond to a variety of inputs with sophisticated behaviors, including changes in gene expression, cell morphology and motility, regulation of the cell cycle and growth, and protein secretion. The genetic circuits that control these behaviors have been selected over evolutionary time to enhance the fitness of the cell (or multicellular organism). The emerging and rapidly expanding discipline of synthetic biology aims to engineer synthetic genetic circuits that perform user-defined functions in a predictable and reliable manner.

The synthetic biology approach is modeled explicitly after other engineering disciplines, such as electrical and mechanical engineering. Like other forms of engineering, synthetic biology aims to adopt different levels of abstraction for design, from a very general, high-level description of desired circuit behavior (e.g., "design a circuit that recognizes and kills cancer cells"), to a detailed description of the molecular mechanisms used to implement the desired behavior, including the sequences of all the DNA-based components.

The motivations for building synthetic gene circuits include both discovery and applications. Early examples of synthetic circuit design include relatively small and simple genetic devices that were used to study the principles underlying the behavior of gene regulatory networks (for reviews, see Refs [1–4]). More recent studies have rewired the native regulatory pathways of living cells in order to study the design principles of gene networks in a "learn by design" approach [5], as well as to gain insights into complicated processes such as malignant transformation [6]. Meanwhile, many synthetic biologists are aiming to build circuits with possible practical applications in areas such as medicine, environmental bioremediation, the manufacture of biofuels and valuable chemicals, and biological computation [1, 7].

The review begins with a summary of the building blocks of synthetic gene circuits (Sect. 2). In Sections 1–7 are presented a number of commonly used and studied types of synthetic gene circuit, including oscillators, toggles and cascades (Sect. 3); memory devices (Sect. 4); digital and analog circuits (Sects. 5 and 6); and multicellular systems (Sect. 7). While synthetic biology has progressed greatly during the past few years, significant challenges remain, and these are discussed briefly in Sect. 8. The review concludes with a survey of applications of synthetic circuits (Sect. 9) that provide an overview of the potential for the tremendous advances in technology and medicine offered by this burgeoning field of research.

2 Building Blocks of Synthetic Gene Circuits

A genetic circuit often consists of three parts: (i) a sensor, which accepts an input or inputs; (ii) a processor, which computes the desired response to the input(s); and (iii) an actuator, which produces the corresponding output. The function of a synthetic gene circuit depends on its building blocks (synthetic DNA, RNA, and proteins), as well as the way these components are wired together into sensor, processor, and actuator modules. The building blocks of synthetic gene circuits may be rationally designed or harvested from Nature, sometimes accompanied by directed evolution to alter their performance in a desired way.

In this section, the choice of host cells for implementing synthetic circuits (the chassis), circuit inputs and outputs is discussed, as well as the molecular implementation of the circuits themselves. Attention is focused mainly on transcriptional control, which has been used extensively in synthetic gene circuits; RNA- and protein-based approaches are also briefly discussed. Considerations of circuit topology are provided in Sect. 8.

2.1 The Chassis: Choice of the Host Cell

Almost all synthetic gene circuits developed to date have been implemented in the bacterium *Escherichia coli*, the budding yeast *Saccharomyces cerevisiae*, or mammalian cells. Each of these hosts presents a unique set of advantages and challenges, as well as distinct host–circuit interactions that must be considered when designing circuits for operation in living cells.

Many of the earliest synthetic circuits were implemented in the model bacterium *E. coli* [8, 9] (see Sect. 3). The advantages of *E. coli* include its relatively small genome, extensive toolbox for genetic manipulation, rapid and easy growth characteristics, and simple and well-understood manner of transcriptional regulation. In addition to providing a fertile testing ground for proof-of-principle genetic circuits, *E. coli* is also an interesting organism for practical applications in bioremediation; for the manufacture of biofuels, pharmaceuticals, and other valuable chemicals; and for human health (e.g., identifying mechanisms to combat antibiotic-resistant pathogenic bacteria, or engineering bacteria to find and destroy cancer cells).

The budding yeast *S. cerevisiae* presents an excellent model system for designing synthetic circuits in eukaryotes (for a review, see Ref. [10]). Like *E. coli*, *S. cerevisiae* is quick and easy to grow in the laboratory, and it offers a well-developed suite of genetic tools, including the ability to maintain foreign genetic elements stably on plasmids and to achieve an efficient homologous recombination of synthetic constructs into the genome. At the same time, *S. cerevisiae* offers a variety of useful characteristics for designing logic circuits that bacteria lack, including intracellular compartmentalization and a rich regulatory repertoire with complex transcriptional regulation and protein signaling cascades. Therefore, yeast can be thought of as a "testing platform" for synthetic biology approaches that can then be adapted to more complicated mammalian cells [11]. *S. cerevisiae* is also a workhorse of industrial synthetic biology, and there is great interest in programming yeast strains for improved production of biofuels and commodity chemicals [12].

Mammalian cells are highly desirable targets for synthetic circuit engineering, due to a myriad possible applications such as therapeutics and tissue engineering [13]. However, these cells have highly complex functions, gene regulation, and intercellular interactions. Furthermore, it is more difficult to work with mammalian cells compared to microorganisms in many technical aspects, such as culture maintenance, DNA delivery, and experimental turnover time. Thus, compared to yeast and bacteria, mammalian cells pose a significantly greater challenge for synthetic biologists. Nonetheless, the past few years have seen many advances in programming circuits in mammalian cells, as described below. Popular mammalian cell types that are well adapted to laboratory conditions include HeLa (human cancer), HEK293 (human embryonic kidney), and CHO (Chinese hamster ovary) cells.

2.2 Inputs and Outputs of Synthetic Circuits

Many synthetic gene circuits are designed to perform a specific action in response to a defined input. Commonly used inputs during circuit design are convenient small molecules that cells can be engineered to respond to, including antibiotics such as anhydrotetracycline (aTc); metabolites or their analogs, such as arabinose or isopropyl β-D-thiogalactopyranoside (IPTG), which mimics a lactose metabolism intermediate; and acyl homoserine lactones (AHLs), diffusible molecules that bacteria use to communicate with each other (see Sect. 7.1) [14–16]. In Nature, bacteria respond to such inputs by activating or repressing target gene expression; for example, arabinose triggers the transcription of genes that encode enzymes and transporters needed to utilize this sugar [17]. Synthetic biology rewires the cell's response to these small-molecule inputs. Cells can also trigger gene expression in response to an external stressor, such as a pulse of heat or DNA-damaging ultraviolet radiation, and these stimuli have been used as inputs to synthetic gene circuits [9, 18]. Their disadvantage, however, is that they can stimulate wide-ranging responses in the host cell that may interfere with desired function of the circuit, and prolonged or repeated exposure may kill the host cell.

More recently, various research groups have harnessed specific wavelengths of light as inputs into synthetic gene circuits, for example, by using light-sensitive proteins from photosynthetic or light-sensitive organisms, including cyanobacteria and plants [19, 20]. One of the advantages of light over a diffusible small molecule is the exquisite level of spatiotemporal control that it offers; it is possible to shine a light specifically onto one part of a plate covered with engineered cells. Moreover, light can be switched on and off instantly, unlike a small molecule which, once applied, can only be removed through washing or gradual dilution. The field of controlling cell behavior with light, named optogenetics, has undergone explosive growth during the past few years (for a review, see Ref. [21]).

While many proof-of-principle synthetic circuits use exogenous small molecules or other inputs that are easy to measure and apply, synthetic biology also aims to create circuits that respond to endogenous and functionally relevant inputs, such as disease markers [22–24]. Designing a circuit to respond to such input is challenging, particularly because of the challenge in finding or designing relevant sensors [13, 22–24]. However, a number of circuits responsive to endogenous inputs have been constructed (see Sect. 9).

The outputs of synthetic gene circuits are frequently fluorescent reporter proteins, because such outputs are easy to detect and quantify, enabling characterization of circuit performance and dynamics. Flow cytometry allows several different fluorescent species to be quantified per cell in a high-throughput manner, and time-lapse microscopy allows the observation of changes in output levels in response to defined inputs in single cells over time. Hence, fluorescent proteins are frequently used in proof-of-principle demonstrations as well as for troubleshooting circuits; subsequently, once the circuit displays a desired performance based on fluorescence assays, the reporter genes may be replaced by output genes that perform a desired function.

Future circuits will increasingly aim to produce outputs relevant for industrial or therapeutic applications, such as control over the cell's proliferation, survival, differentiation, morphogenesis, migration, or synthesis and secretion of a therapeutic protein or commodity chemical. Examples of circuits with functionally relevant outputs are described in Sect. 9.

2.3 Properties of Synthetic Building Blocks

The function of a synthetic circuit depends upon the function of its components (for reviews, see Refs [1, 2]). Synthetic building blocks, such as transcription factors (TFs) or RNA regulatory elements, should ideally be:

- *Modular*: The component should have a defined function that persists regardless of context; for example, a promoter drives the expression of a downstream gene at the same level, regardless of the identity of the downstream gene. Modularity is the property that allows basic building blocks to be assembled into complex devices in a predictable manner.
- *Composable*: The components should be arranged together into a functional circuit. The output of one part of the circuit can serve as the input for a downstream part.
- *Orthogonal*: The components should avoid unwanted crosstalk with each other, or with the host cell's molecular machinery. Unlike an electronic device, in which components are connected by physical wires, a gene circuit consists largely of building blocks that freely diffuse and mix inside the cell, creating the potential for unwanted interactions. For example, if a circuit includes the TFs A and B that regulate promoters P_A and P_B, respectively, then the unwanted regulation of P_A by B may lead to circuit failure. Likewise, unplanned interactions between a synthetic factor and the cell's native genes may disrupt circuit function or harm the host cell.
- *Tunable*: Different circuit functions require components with different levels of activity, such as promoters that drive expression of downstream genes at different levels. Moreover, successful circuit design requires level matching between upstream and downstream parts of the circuit (see Sect. 8.1). Hence, synthetic biology requires components that display tunability – the ability to achieve different output levels through small changes, such as point mutagenesis of a promoter or a RNA regulatory device.

Orthogonality and tunability require the construction of large libraries of each type of circuit component, such as promoters, TFs, ribosome-binding sites (RBSs), regulatory RNA molecules, and others. Efforts have been made to construct libraries of orthogonal parts with reduced unwanted crosstalk [25–27]. Libraries of building blocks may be obtained through a combination of three approaches [28]: (i) parts mining, which involves searching the annotated genomes of different species for components that are predicted to have the desired function while avoiding crosstalk with the host cell; (ii) directed evolution, whereby a library of mutagenized components is constructed and subjected to selection *in vitro* or *in vivo* in order to identify components with desired activity; and (iii) rational design, which is applicable to the types of device that operate according to well-understood rules. For example, synthetic transcription factors (sTFs) based on transcription activator-like effectors (TALEs) offer a way to obtain orthogonal TFs by rational design [29] (see below).

It is possible to use a combination of rational design and directed evolution to achieve tunability and to produce libraries of components with the same function, but with different activity levels. For example, the expression level of a protein depends partly on the sequence of the RBS on its corresponding mRNA; hence, protein expression levels can be adjusted through point mutations in the RBS. The RBS calculator [30] is a program for rationally designing an RBS sequence that results in a user-defined protein expression level. Similarly, promoters can be tuned via point mutations [31]. In addition, some circuits are designed to be tunable *in vivo* through the addition of small-molecule inputs. For example, a bandpass filter has been designed that allows its host cells to survive only at a specific concentration of two antibiotics, ampicillin and tetracycline [32]. Varying the levels of the inducer IPTG changes the target concentration of ampicillin and tetracycline at which cells can survive by adjusting the expression of the bandpass filter's components [32] (see Sect. 3.4).

Below are described several of the most commonly used families of synthetic circuit components (also see Table 1 in Ref. [1]). Some examples of commonly used synthetic circuit building blocks are shown in Fig. 1.

2.4
Building Blocks of Synthetic Transcriptional Regulation

Transcription is a key mechanism for gene regulation in living cells. In its most basic form, a transcriptional unit consists of a TF and its cognate promoter/regulator. The TF binds its regulator region in a sequence-specific manner and activates or represses its transcription, in some cases in response to an input such as a small molecule. The modular nature of eukaryotic TFs enables the construction of sTFs by combining different DNA-binding, ligand-binding, and regulatory domains [26, 31, 33–36]. Some transcriptional regulatory domains, such as the commonly used viral VP16 activation domain or its derivative, the VP64 domain [37], activate gene expression from their target promoter. In this case, the cognate synthetic promoter is either a very weak or a minimal promoter located downstream of the TF recognition site, and transcription is initiated from this promoter in the presence of the TF. Other transcriptional regulatory domains inhibit gene expression, such as the Krüppel-associated box (KRAB) domain found in vertebrates [38], which inhibits transcription by recruiting chromatin-modifying proteins that cause formation of repressive heterochromatin. In this case, the cognate synthetic promoter is usually a strong constitutive promoter with associated TF recognition sites that is active by default until silenced by the TF. Depending on its mechanism of action, the same sTF may activate or repress transcription of a target gene, depending on where in the gene it binds [43].

In bacteria, transcriptional repressors downregulate gene expression at close range, via directly interfering with RNA polymerase (RNAP) or activator binding [44]. Transcriptional repression in *S. cerevisiae* also tends to occur over short distances due to the compact nature of the genome, and may involve interference with the recruitment of basal transcriptional machinery as well as chromatin-mediated silencing [45, 46]. In contrast, mammalian repressors often involve chromatin remodeling and can have long-range effects of up to tens of kilobases both upstream and downstream of their binding sites [47]. Thus, transgenes inserted into the mammalian genome may become silenced by repressive chromatin spreading from adjacent loci, a phenomenon that may be countered by flanking the transgene with insulator sequences [48]. Furthermore,

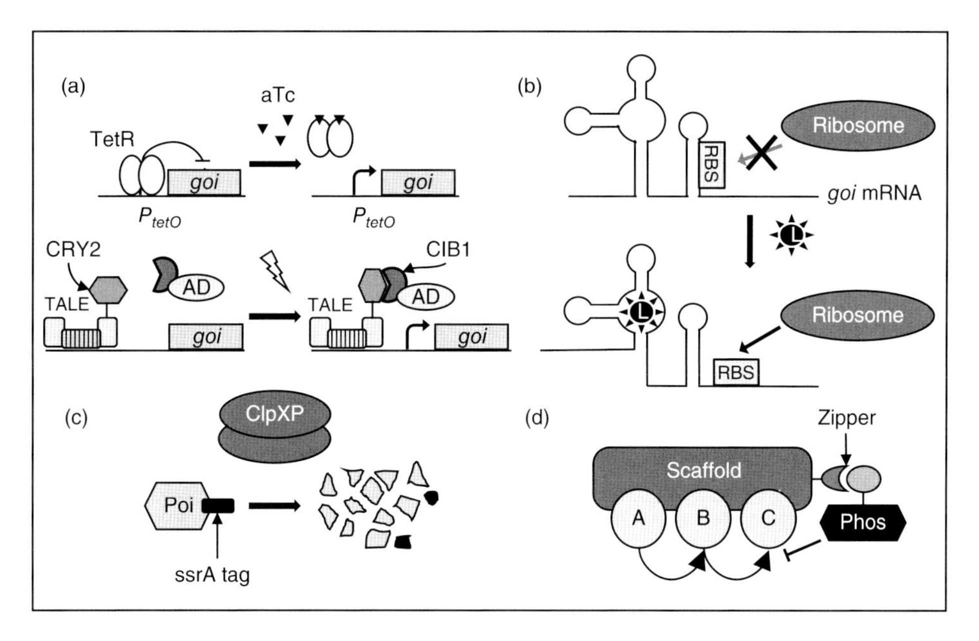

Fig. 1 Examples of commonly used synthetic building blocks. (a) Transcriptional regulation is the backbone of many synthetic circuits (see Sect. 2.4). Top: The bacterial TetR transcriptional repressor forms homodimers that bind tightly to their DNA target motifs, sterically blocking RNA polymerase recruitment to the PtetO promoter. Binding of the small molecule anhydrotetracycline (aTc) causes TetR to dissociate from DNA, allowing expression of the downstream gene of interest (goi). TetR-based transgene regulation is commonly used in *E. coli* and has also been adapted for use in eukaryotes. Bottom: Transcription Activator-Like Effectors (TALEs) bind to user-defined DNA sequences. Here, a synthetic TALE-TF is used to activate a goi in mammalian cells in response to blue light [39]. Upon illumination, the light-sensitive CRY2 domain fused to the TALE-TF undergoes a conformational change, resulting in recruitment of its partner CIB1 domain along with a transcriptional activator domain (AD) that stimulates target gene expression. Lower panel adapted from Ref. [39]; (b) RNA-based devices can regulate target gene expression at different levels, including transcriptional elongation, translation initiation, and mRNA stability (see Sect. 2.5). In this example, an aptamer (ligand-binding RNA molecule) is fused to an mRNA encoding a transgene of interest. In the absence of ligand (top), the aptamer folds into a stem-loop structure that blocks the ribosome binding site (RBS), preventing mRNA translation. The ligand (L) causes a conformational change in the aptamer, exposing the RBS and allowing translation (bottom). Adapted from Ref. [40]; (c) Regulated protein degradation can be used to control synthetic circuit activity at a post-translational level (see Sect. 2.7). Here, an ssrA tag fused to the protein of interest (Poi) targets the protein for degradation by the ClpXP protease. Adapted from Ref. [41]; (d) Synthetic fusion proteins can "rewire" intracellular signaling cascades. Here, a scaffold protein brings together three protein kinases (A, B, and C), resulting in efficient signal propagation in an *S. cerevisiae* Mitogen Activated Protein Kinase (MAPK) pathway. This naturally occurring system is "rewired" by fusing the scaffold to a leucine zipper domain, which recruits a protein phosphatase (Phos) to the signaling complex [42]. The phosphatase attenuates MAPK signaling by dephosphorylating and hence inactivating the MAP kinase. In contrast, the recruitment of an activating protein to the scaffold potentiates MAPK signaling (not shown) [42]. Adapted from Ref. [42].

chromatin remodeling is time-consuming, resulting in slow kinetics of transcriptional regulation in mammalian cells [49].

As mentioned in Sect. 2.3, one of the key requirements of good synthetic building blocks is orthogonality: the ability of multiple elements to exist in a single cell without crosstalk. The regulatory domain can be the same among multiple sTFs operating in parallel without impairing orthogonality. However, off-target binding of the sTF to DNA might result in an unwanted deregulation of native genes and shunting of the sTF away from its target promoter. Thus, the DNA-binding domain of an sTF must be highly specific and target DNA sequences that are very rare or absent in the host genome. This is a greater challenge for mammalian cells than for bacteria and yeast, due to the difference in genome sizes (3×10^9 bp in humans [50] compared to $\sim 5 \times 10^6$ in *E. coli* [51] and $\sim 12.5 \times 10^6$ in *S. cerevisiae* [52]).

The development of sTFs has occurred in stages, with each new generation of sTFs providing new ways to improve orthogonality and tunability. The "first-generation" synthetic transcriptional circuits repurposed TFs that occur naturally in bacteria or in phages (viruses that infect bacteria). Examples include the LacI, TetR, and λ CI TF-promoter pairs [8, 9], which were initially used to build synthetic circuits in bacteria and later optimized for use in eukaryotic cells (Fig. 1a). Transcriptional repression by LacI and TetR is relieved by binding to the small molecules IPTG and aTc, respectively; hence, these TFs can be used to build synthetic gene circuits that respond to small-molecule inputs [8, 9]. Early experiments in mammalian cells utilized the DNA-binding domain of the *S. cerevisiae* GAL4 TF, which recognizes an upstream activating sequence (UAS) [53]. The GAL4 DNA-binding domain fused to the VP16 transcription activation domain (VP16AD) can activate gene expression from a UAS-containing cognate target promoter [37]. This useful system exhibits tight regulation and a wide dynamic range.

In order to control the level of transcription output in response to a small-molecule input, ligand-inducible systems were further developed in mammalian cells. The most-characterized system is based on the bacterial tetracycline-dependent transactivator (tTA) that, in its native form, binds its target DNA motif in the absence of tetracycline. Thus, a tTA–VP16AD fusion was utilized for the construction of a tetracycline-off system, which activates its target gene in absence of tetracycline [38]. Later, tTA was mutated to generate a reverse tetracycline-dependent transactivator version (rtTA), which binds its target site only in the presence of the tetracycline analog doxycycline (Dox), constituting a tetracycline-on system when fused to VP16AD [54].

An alternative ligand-responsive transcriptional regulatory system was based on mammalian nuclear hormone receptors, TFs that regulate target genes in response to steroid hormones such as estrogen and progesterone, which diffuse inside the target cell and bind to the hormone receptor's ligand-binding domain (LBD), enabling regulation of the target gene [55]. The simplicity of such systems, which do not require an elaborate signal-transduction pathway inside the host cell, make LBDs attractive for engineering ligand-responsive sTFs by fusing LBDs to DNA-binding and regulatory domains of choice for use in mammalian cells [56–58] and in *S. cerevisiae* [59]. To avoid crosstalk with the mammalian host cell's native hormone signaling, LBDs have been engineered to respond specifically to synthetic compounds, such as the progesterone antagonist RU486 [57]. Alternatively, LBDs responsive to insect hormones such as ecdysone may be used [56]. Additional ligand-dependent mammalian sTFs have been developed to respond to a variety of ligands such as macrolides [60] and streptogramins [61], L-arginine [62], biotin [63], urate [24], and others [64].

Although well-characterized TFs based on naturally occurring proteins such as bacterial TetR or yeast GAL4 have many advantages, generating a library of sTFs sufficient for building complex gene networks requires another approach. The "second-generation" sTFs have custom-designed DNA-binding domains that can bind any user-defined DNA sequence with high specificity. The first attempt to develop such sTFs focused on engineering the DNA-binding domain of zinc finger transcription factors (ZF-TFs) [33, 65, 66]. A zinc finger is a small protein domain of about 30 amino acids that recognizes a specific 3 bp DNA sequence [67]. A synthetic protein that includes a tandem array of zinc fingers will specifically bind to a user-defined sequence in the genome [68], circumventing the need to rely on the small number of DNA-binding domains found in naturally occurring TFs. Pioneering studies demonstrated that synthetic ZF-TFs can regulate endogenous human genes in a sequence-specific manner [69]. More recently, suites of orthogonal ZF-TFs have been constructed for use in synthetic gene circuits in mammalian cells [70] and in *S. cerevisiae* [26]. The design of custom ZF-TFs was facilitated by publicly available online tools such as the Zinc Finger Targeter (ZiFiT) that identifies potential ZF-binding sites in user-supplied DNA sequences [71], as well as the publicly available Zinc Finger Database (ZiFDB) that provides information on functional zinc fingers [72]. Fusing ZF DNA-binding arrays to the LBD of a nuclear hormone receptor produced ZF-TFs that regulate their targets in response to hormone inputs [73, 74].

Although they represented a great technical advance, ZF-TFs are not trivial to design, and in some cases can lack specificity. For example, adjacent zinc fingers can influence each other's binding specificity, thus complicating the design of tandem arrays of these domains [75]. An alternative approach utilizes the TALE proteins from *Xanthomonas* spp. bacteria, which are plant pathogens that use TALEs to modulate their host cell's gene expression [29, 76]. TALEs feature arrays of short amino acid repeats, with each repeat specifically binding a single DNA base pair. Hence, similar to ZF-TFs, synthetic TALEs can be programmed to bind a specific DNA sequence by combining specific base-pair-binding amino acid sequences in the correct order [29, 77]. Furthermore, the individual TALE domains are relatively independent of one another, enabling modular assembly into higher-order arrays to target longer DNA sequences. Custom-designed TALE-TFs have been used to activate reporter genes in mammalian cells [34, 76]. Although TALEs naturally act as transcriptional activators, targeting TALE-TFs to the core promoter region of target genes in order to block RNAP recruitment can turn the TALE-TFs into transcriptional repressors, as shown in *S. cerevisiae* [31]. Like ZF-TFs, TALE-TFs can also respond to hormone-mediated induction when fused to LBDs of nuclear hormone receptors [78]. Another recent study presented TALE-TFs that regulate genes of interest in response to blue light (Fig. 1a) [39]. The system consisted of two parts: a TALE DNA-binding domain fused to the *Arabidopsis thaliana* light-sensitive cryptochrome 2 (CRY2) protein, and the CRY2 binding partner CIB1 fused to a transcriptional effector domain, such as VP64. When illuminated by blue light, CRY2 recruits CIB1, leading to the formation of a functional TF that can regulate the gene of interest. The system displays high specificity and signal-to-noise ratio, as well as faster response time compared with small-molecule-inducible TALE-TFs [39].

In addition to using TALE-TFs to regulate target genes via regulatory domains such as VP16, recent studies have harnessed TALEs for targeting chromatin modifications to endogenous loci of interest. For example, TALEs fused to the TET1 DNA demethylase or the LSD1 histone demethylase allow site-specific DNA or histone demethylation in mammalian cell culture [79, 80]. Zhang and colleagues used the light-sensitive TALE system described above [39] to target chromatin modifications to specific genes in response to light by fusing the CIB1 protein to various histone-modifying enzymes. This ability to modify the host cell's chromatin at selected loci has exciting implications for future research in areas such as cell fate specification and maintenance, as well as cancer.

A recent breakthrough in TF engineering came from the bacterial CRISPR (Clustered Regularly Interspaced Short Palindromic Repeats) system, in which sequence specificity can be easily determined by the guide RNA (gRNA) sequence rather than protein engineering [81]. The CRISPR system serves as a form of immune memory in many Bacteria and Archaea, as it causes the cell to cleave the DNA of viruses it has previously encountered and "remembered" by incorporating viral sequences into its CRISPR locus [82]. Short RNAs guide a Cas (CRISPR-associated) endonuclease to cleave DNA that complements the RNA sequence. The results of recent studies have shown that the Cas9 protein from *Streptococcus pyogenes* and synthetic gRNAs can cleave specific DNA sequences when expressed in heterologous hosts such as *E. coli* [83] and mammalian cells [84, 85]. The CRISPR system presents an attractive, efficient method for reprogramming cells by genome engineering (e.g., knocking out a specific gene of interest) [84, 85]. In addition, it was shown that a deactivated Cas9 protein that lacks endonuclease activity (abbreviated dCas9) can be recruited by its associated gRNAs to specific DNA loci, where it can act as a transcriptional regulator without cleaving the target DNA [43, 81]. Once recruited to the locus of interest, dCas9 can repress gene expression by sterically inhibiting RNAP binding to the promoter; alternatively, dCas9 fused to a transcriptional coactivator domain such as VP16 can activate target gene transcription [43, 86]. The potential advantages of the CRISPR system over ZF-TFs and TALEs include a greater ease of design, as sequence-specific gRNAs are easier to design than are zinc fingers or TALE motifs, as well as the possibility of building a circuit with many orthogonal parts. In the ZF-TF and TALE systems, each target promoter requires a separate regulatory protein, while in the CRISPR system, a single dCas9 protein may regulate multiple genes by associating with distinct gRNAs, although this configuration may present problems concerning the allocation of a fixed resource (the dCas9 protein) among multiple gRNAs targeting different promoters.

In order to facilitate the design of sTFs based on zinc fingers, TALEs, and CRISPR, a recent study presented a set of formal rules for building functional sTFs based on 11 possible architectures [35]. Besides considering the sTF architecture, the sTF must also be designed to target a DNA motif that is not expected to cause unwanted crosstalk with the host cell's native genes. To this end, Lu and colleagues [36] identified

9-, 12-, and 15-bp DNA motifs that are underrepresented in or absent from the genomes of six model microorganisms, including *E. coli* and *S. cerevisiae*. This data set can aid in the design of orthogonal promoters for synthetic gene circuits. Another study identified over 180 DNA sequences, each of 20 bp, that differed by at least 3 bp from all possible 20-mers in annotated human promoter regions [34]. Altogether, recent advances in orthogonal sTFs present a resource of great value for building synthetic regulatory circuits.

2.5 Post-Transcriptional Regulation: RNA-Based Circuit Engineering

While most synthetic circuits rely on transcriptional regulation, RNA-based regulatory devices are also common, due to the many attractive features of RNA regulatory molecules (for reviews, see Refs [87, 88]). First, the rules for rational construction of RNA devices with desired function are relatively well understood, enabling the construction of libraries of orthogonal RNA components. Rules for the rational development of synthetic RNA devices with desired properties have been described, such as the theoretical framework for the development of aptazymes [89] (see below). RNA devices are also tunable, as point mutations in the RNA molecule affect its folding and/or strength of interaction with its partner RNA or DNA molecule, and hence its activity. Second, unlike TFs, regulatory RNA molecules do not require translation in order to function. Because of these advantages, synthetic RNA-based regulation is an expanding and exciting field. Some examples of synthetic RNA-based regulation are presented below.

Gene expression in mammalian cells can be knocked down by RNA interference (RNAi), complementary RNA sequences that target mRNA of interest for degradation in a sequence-specific manner [90]. RNAi molecules, such as short interfering RNAs (siRNAs) or microRNAs (miRNAs), can be chemically transfected into the cells or expressed from designated vectors to knock down a specific gene. RNAi may be used to build complex regulatory devices through the combinatorial regulation of a common output transcript by multiple miRNAs. Benenson and colleagues used this property of miRNAs to build complex logic circuits in mammalian cells [23, 91, 92] (see Sect. 5 for details). In addition, in mammalian systems, specific 5′ intronic sequences and 3′ polyadenylation sites can be added to increase RNA stability and transgene translation [93]. Conversely, an mRNA can be destabilized with 3′ degradation tags, such as AU-rich elements (AREs) [94] to reduce transgene levels and decrease promoter leakiness.

Synthetic RNA-based devices also include small, *trans*-acting RNA molecules that base-pair with their target transcripts to either allow or inhibit transcriptional elongation or translation. Examples of such systems in *E. coli* include the riboregulator [95], the pT181 attenuator [96], and the RNA-IN-RNA-OUT system [97].

The riboregulator consists of two components: a *cis*-repressed RNA (crRNA) encoding the protein of interest, and a cognate small *trans*-activating RNA (taRNA) [95]. In absence of the taRNA, the 5′ untranslated region (UTR) of the crRNA forms a stem loop that blocks the RBS and prevents translation. When the taRNA base-pairs with its partner crRNA, the stem loop unfolds, allowing protein translation [95]. Conversely, the pT181 attenuator allows target gene expression only in the absence of the *trans*-acting RNA: the 5′ UTR of the target gene is engineered with an attenuator loop. In the presence of its partner antisense RNA, the attenuator loop folds into a hairpin that exposes a terminator site, preventing transcription of the downstream target gene [96]. In the absence of the antisense RNA, the target gene is transcribed [96]. While the riboregulator controls translation, and the attenuator regulates transcription, the RNA-IN-RNA-OUT system couples translational control to transcriptional elongation of a target gene [97]. The system consists of pairs of RNA molecules: an RNA-IN molecule and its cognate RNA-OUT molecule. The RNA-IN element is placed upstream of a sequence encoding a small regulatory peptide, *tnaC*, followed by the coding sequence of the target gene. RNA-OUT forms a complex with RNA-IN that blocks the translation of *tnaC*. In this system, the translation of *tnaC* is necessary to enable transcriptional elongation of the downstream target gene; hence, the presence of RNA-OUT blocks transcription of the gene coupled to RNA-IN [97].

Another widely used class of regulatory RNA molecules is the aptamer, an RNA molecule that specifically binds a ligand such as a small molecule or protein [40]. Aptamers may be used to build riboswitches, which regulate target gene expression in response to ligand binding. Naturally occurring riboswitches are found in bacterial RNAs encoding metabolic enzymes, where they regulate the enzyme's expression in response to the levels of the corresponding metabolite [98]. In synthetic circuits, riboswitches inserted into the 5′ UTR of a target gene modulate target gene expression by either permitting or blocking transcriptional elongation or translation initiation of the riboswitch-coupled transcript in response to a specific ligand [99, 100] (Fig. 1b; for a review, see Ref. [40]). Because of their usefulness, many studies have focused on screening for aptamers and riboswitches with desired activity. Aptamers that bind a ligand of interest may be selected *in vitro* from large pools of random nucleotides [101, 102]. Gallivan and colleagues screened large libraries of riboswitches by coupling riboswitch activity to an easily observable readout, such as bacterial cell motility or fluorescence [103, 104]. Examples of ligands recognized by synthetic riboswitches include the small molecule theophylline [105], antibiotics [106], and dyes [107]. As an example of how aptamers can be used to control cell behavior, one study coupled the theophylline-responsive aptamer to the expression of CheZ, a protein necessary for chemotaxis in *E. coli*, resulting in *E. coli* cells that migrated toward a source of theophylline [108].

Aptamers may also be linked with catalytically active RNA molecules called *ribozymes* that can cleave or splice RNA. Such molecules, known as *aptazymes*, carry out RNA-modifying reactions in a ligand-responsive manner [109, 110]. In one example, the small molecule theophylline binds to its target aptamer and activates a ribozyme that cleaves the target mRNA, resulting in a downregulation of target gene expression in response to theophylline in *S. cerevisiae* cells [110]. A similar system was later implemented in mammalian cells to achieve theophylline-mediated mRNA degradation [111, 112]. Together, riboswitches, aptazymes, and *trans*-regulatory noncoding RNAs present an expansive toolbox for engineering fine-tuned gene expression in synthetic circuits.

2.6 Insulator Elements

In order for a synthetic circuit to function correctly, its component elements must be protected from unwanted interactions with other circuit components and with the host cell. Therefore, synthetic circuits may include insulators – DNA or RNA elements that insulate one part of a circuit from interference by an adjacent synthetic part or by the host's genome. For example, insulator sequences flanking a genetic circuit inserted into a mammalian chromosome help to prevent the circuit from becoming silenced due to local chromatin effects [113].

In another example, two RNA regulatory elements placed on one mRNA molecule may base-pair with each other and disrupt each other's function. To avoid this interference, the RNA regulatory elements may be separated physically through ribozyme-mediated cleavage [87]. A recently described alternative to ribozymes is the Csy4 endonuclease from the CRISPR system [114] (see Sect. 2.4), which cleaves RNAs at a 28-nucleotide recognition sequence that does not occur naturally in *E. coli*. Csy4 can efficiently cleave mRNAs engineered with the recognition sequence in *E. coli* and *S. cerevisiae*, and Csy4-mediated separation of transcript components such as the 5′ UTR and the coding region allowed expression levels to be protected from context effects in *E. coli* [114].

2.7 Post-Translational Regulation and Protein-Based Circuits

The behavior of transcriptional regulatory circuits depends in part on the stability of the proteins encoded by the circuit genes. Some applications, such as the oscillator (Sect. 3.1), require rapid degradation of the component TFs. In eukaryotes, protein lifetime can be tuned by fusing the protein to a tag that targets it for ubiquitin-dependent degradation [115]. In *E. coli*, the 11-amino acid *ssrA* tag targets proteins for rapid degradation by the ClpXP protease [116]. To achieve inducible, fast degradation of synthetic proteins, Hasty and colleagues engineered an *S. cerevisiae* strain for expression of ClpXP from an IPTG-inducible promoter [41] (Fig. 1c). The addition of IPTG stimulates the production of ClpXP, which leads to a specific degradation of *ssrA*-tagged target proteins. The amount of ClpXP, and hence the degradation rate, can be tuned by the amount of IPTG added [41]. Moreover, specific tags have been developed that enable inducible protein degradation, in which protein degradation is inhibited by small molecules [117].

In addition, protein-based regulatory cascades may be designed and modeled after intracellular signal transduction pathways found in eukaryotic cells [118]. Protein-based signaling pathways have several advantages over transcription-based circuits. Protein-based signaling occurs at much faster timescales than transcription; moreover, protein-based signal transduction relies on processes (such as protein phosphorylation or conformational change) that require little of the cell's energy and resources. However, to design protein-based synthetic devices presents unique challenges, mainly because it is currently extremely difficult to predict a protein's function and activity from its amino acid sequence. Instead of rationally designing proteins *de novo*, the construction of synthetic post-translational regulation relies on repurposing catalytic and regulatory domains from naturally occurring signaling proteins, and wiring them together in novel ways. Key examples include rewiring the pheromone signaling response in *S. cerevisiae* [42] (Fig. 1d), as well as using bacterial proteins to attenuate signaling pathways in yeast and in human T cells, with possible applications for T-cell-based therapy [119].

3
Dynamical Circuits

The building blocks described above have been used to construct synthetic gene circuits of varying complexities and purposes. The first synthetic gene circuits that were constructed were not intended for practical applications, but were simple dynamical networks intended to demonstrate the proof-of-principle ability to engineer gene circuits with desired behaviors. Early examples of dynamical circuits were described in 2000 [8, 9], and have formed the basis (either conceptually or practically) for much work on synthetic circuits since then, whether applied or not. These two circuits were the oscillator and the switch.

3.1
Oscillators

Oscillators are common components in electronic devices [120]. An oscillator stably interchanges between two states and can be characterized by its amplitude and period. The first synthetic gene oscillator circuit was the *E. coli* "repressilator" [8], a ring of three transcriptional regulators, LacI, TetR, and CI, each repressing the next one in the ring (Fig. 2a). As LacI represses the transcription of *tetR*, it relieves the repression by TetR on the *cI* gene, allowing CI levels to rise and repress *lacI*; the now de-repressed *tetR* gene product begins to accumulate until it can repress *cI*, and the cycle continues. Studies on the repressilator sparked almost a decade of focus on synthetic gene oscillators (see Ref. [121] for an extensive analysis). Studies on oscillators illustrate the importance of circuit topology and parameters on overall circuit function, and provide an example of the "design–build–test" cycle of synthetic circuit engineering. Following the initial repressilator, a number of improved versions were designed and implemented.

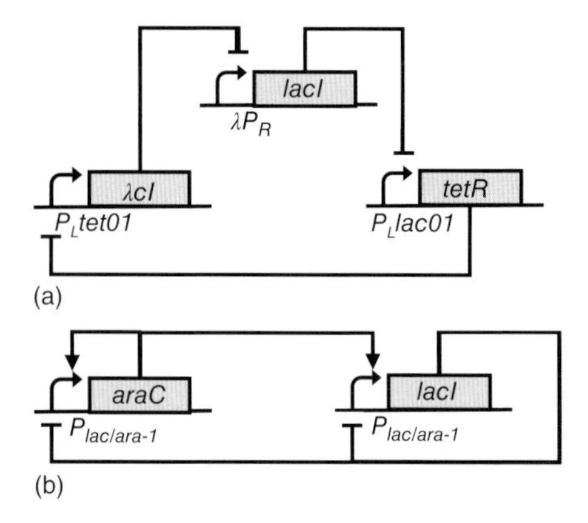

Fig. 2 Synthetic gene oscillators. (a) The repressilator consists of three genes in a ring, each repressing transcription from the next one in the chain [8]. Gene and promoter names are given; (b) The robust oscillator constructed by Stricker *et al.* uses a combination of positive and negative feedback [122]. In both circuits the arrows represent transcriptional activation, while flat-ended arrows represent transcriptional repression. Modified from Ref. [122].

Despite being a significant breakthrough, the repressilator was not a robust circuit as it only functioned in about 40% of cells. It was not until 2008 that a highly robust oscillator circuit was constructed [122]; this oscillator had a different topology from the repressilator, comprising one gene that activates itself and another gene, this second gene feeding back to and repressing the first gene (Fig. 2b). This design is known as an *amplified negative feedback oscillator*, where "amplified" refers to the first gene's positive feedback on itself. A number of other oscillators have been constructed, the majority within the bacterium *E. coli* [123, 124], and one in *Salmonella typhimurium*

[125]. For example, one study implemented a metabolic oscillator using fluctuations in metabolite pools [123].

Oscillators have also been successfully introduced into mammalian cells. In a negative feedback loop-based oscillator, an intron placed upstream of a repressor protein coding region increases oscillatory period length with the size of the intron, due to increased transcript length [126]. In another example, a tunable oscillator was constructed by combining a feedforward loop and a time-delayed negative feedback loop based on an autoregulated sense–antisense transcription control [127]. A subsequent study produced a low-frequency oscillator, in which a time-delayed negative feedback loop was based on a short hairpin RNA (shRNA) encoded in the intron of a self-regulated transactivator [128]. Finally, synthetic–natural hybrid oscillators can be constructed in human cells based on the structure of natural networks, such as the p53 pathway [129].

Now that oscillators can be constructed routinely, the challenge is to integrate them into larger synthetic circuits. A potential application is to use oscillators as timer circuits, keeping time for the rest of a synthetic circuit. Oscillators may also be used in frequency multiplier circuits, allowing different parts of a larger circuit to be kept at different timings [130]. Attempts to connect oscillators to downstream circuits have highlighted a common issue that is known to electrical engineering when connecting up different circuits: *retroactivity*, which occurs when a downstream component of a circuit affects its upstream components [131, 132]. In genetic oscillators, retroactivity occurs because some of the protein used in the oscillator itself has to be used to drive the downstream circuit by binding to the downstream gene promoters. This binding sequesters the oscillator protein, affecting the oscillator dynamics. One approach to minimizing retroactivity is to use an amplifier circuit as an intermediate between the oscillator and the downstream circuit. The amplifier requires only a small amount of protein from the oscillator, minimizing the effect on the oscillator, but is strongly expressed, giving a large output sufficient to drive the downstream circuit. This approach has been successfully demonstrated in an *in vitro* synthetic oscillator based on hybridization of synthetic DNA and RNA oligonucleotides (see Ref. [133] for details).

3.2 Toggle Switches

Along with the oscillator, another example of a simple functioning device is a toggle switch: a circuit that can switch between two stable states, ON and OFF, similar to the light switch of a lamp. As a result, a switch possesses a form of memory, which is vital for cellular computation (see Sect. 4). The genetic switch, commonly called a *toggle switch*, consists of two transcriptional repressors, each repressing the other (Fig. 3). Collins and colleagues described a functional toggle switch in a living *E. coli* cell [9]; the switch uses two small molecule-responsive transcriptional repressors – TetR, which binds aTc, and LacI, which binds IPTG (see Sect. 2.2). The addition of aTc inactivates TetR, switching the system into a state with high expression of LacI, which then maintains a low concentration of TetR. Conversely, addition of IPTG inactivates LacI, switching the system into a high-TetR state, which then maintains a low concentration of LacI. Another version of the toggle switch uses a heat pulse as a means of switching off a temperature-sensitive transcription repressor [9].

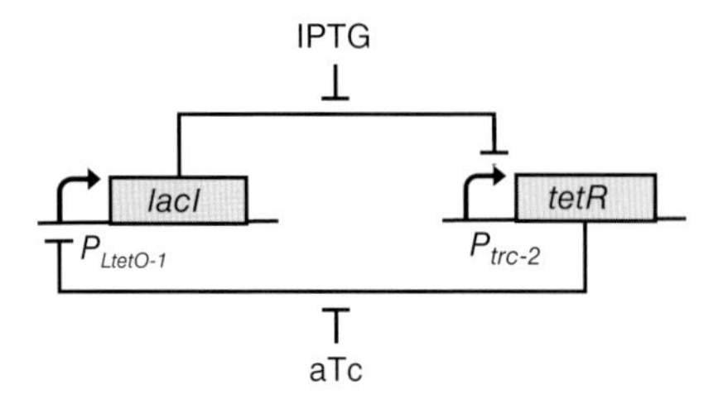

Fig. 3 The toggle switch. The original synthetic toggle switch in *E. coli* consists of two repressors, *lacI* and *tetR*, which repress each other's transcription [9]. The small molecules aTc and IPTG bind to TetR and LacI proteins, respectively, preventing them from repressing transcription and therefore allowing expression of *lacI* and *tetR*, respectively. Flat-headed arrows depict transcriptional repression. Adapted from Ref. [9].

A mammalian toggle switch was also developed based on two transcriptional repressors, enabling it to switch between states according to two antibiotic inputs [134]. Subsequently, an RNAi-based toggle switch was designed for the tight regulation of gene expression in mammalian cells that switch between states according to small-molecule inputs [135]. The switch could be used to tightly control the expression of potentially useful proteins, such as the toxic alpha chain of diphtheria toxin and the pro-apoptotic protein BAX (BCL-2-associated X protein) that can kill cancer cells. Recently, a light-responsive toggle switch was developed, in which gene expression is induced by red light and silenced by far-red light [136].

As is the case with the oscillator, the function of the toggle switch depends on the relative expression levels of its component transcription repressors – the mutual repression between the two genes must be approximately balanced. By unbalancing the relative levels of the two repressors, Ellis *et al.* were able to convert the *E. coli* toggle switch into a genetic timer [137]. In this way, the authors changed the dynamics of the switch from bistable to monostable, so that the circuit would return to its initial state after perturbation. In this implementation, the toggle switch was set up to return to a stable state with low LacI expression. Induction with aTc moved the system into a high-LacI, low-TetR state. Upon the removal of aTc, the system eventually returned to low LacI and high TetR conditions. The time needed to reset the system confers the circuit's timer action, which can be controlled by changing the strengths of the various promoters [137].

A further advance in switches was made with the construction of a push-on-off switch [138]. Whereas, the toggle switch requires two inputs, the push-on-off switch switches repeatedly back and forth between two states using just one input. This circuit was larger and more complex than the original toggle switch (see Ref. [138] for details), and was not particularly robust; the fraction of cells switching decreased quickly as the number of rounds of switching increased, in part because the input used – UV light – caused cell lethality. The authors proposed ways to improve future implementations of the circuit, including using an input that would be less harmful to the host cell [138].

Like oscillators, toggle switches can also be coupled to cell behaviors of interest. For example, Kobayashi *et al.* built a toggle switch circuit in *E. coli* to induce biofilm formation in response to DNA damage [139]. The toggle switch consisted of the CI and LacI TFs mutually repressing each other. This circuit starts out in a low-LacI, high-CI state. A pulse of DNA-damaging radiation then triggers expression of the native RecA protease, which cleaves CI and flips the toggle switch into low-CI, high-LacI state, which in turn promotes expression of a gene required for biofilm formation [139].

In a recent study, a synthetic toggle switch was used to address a fundamental question in cell biology: How can cells in an isogenic population achieve distinct cell fates when starting from the same state [140]? The authors implemented a toggle switch in *S. cerevisiae* that can exist in two states: the high-LacI/low-TetR state, which expresses green fluorescent protein (GFP) but not mCherry; and the low-LacI/high-TetR state, which expresses mCherry but not GFP. Addition of the TetR inhibitor aTc switches the cells into the high-LacI, GFP-positive state. The system was "initialized" by growing the cells in glucose, which blocked expression of all synthetic proteins. The cells were then transferred to galactose media to permit protein expression, and their fate (high versus low GFP) was observed in the presence of varying amounts of aTc. In the presence of an intermediate aTc concentration, the isogenic cell population eventually reached a bimodal distribution, with approximately equal numbers of cells becoming GFP-positive and GFP-negative [140]. In contrast, when given a high or low dose of aTc, all cells reached a high-GFP or low-GFP state, respectively. Moreover, the authors showed that the intermediate aTc concentration that leads to a bimodal distribution of cell fates depends on the expression levels of LacI and TetR: for example, a cell line engineered to express a lower amount of TetR relative to LacI will need less aTc to reach a bimodal distribution. The experimental evidence and accompanying mathematical model indicate that isogenic cells can stochastically reach distinct cell fates when exposed to a signal whose concentration is in a critical intermediate range, which depends in turn on the expression levels of the cell's regulators that respond to the signal [140]. In summary, the authors used a simple synthetic circuit in order to gain insight into a complex biological phenomenon.

3.3 Gene Cascades

In addition to the repressilator and toggle switch, synthetic biologists have built genetic cascades: systems in which an upstream gene product regulates a downstream gene, which in turn regulates a gene further downstream, and so on. Weiss and colleagues [141] constructed synthetic cascades of various lengths in *E. coli* (Fig. 4) in order to understand how the characteristics of the cascade relate to its length (i.e., the number of genes in the cascade). Longer cascades were shown to have sharper switching between ON and OFF states, and to display noisier dynamics. As predicted, the response time increased with the length of the cascade. This feature allowed the longer cascade to act as a low-pass filter, with the output less affected by rapid fluctuations in the input than fluctuations persisting for a longer time. However, this filtering ability came at the expense of synchrony: the longer the cascade, the more variability was observed in response time across the cell population. The authors pointed out that genetic cascades are common in Nature, but that if the cascade regulates a process that requires the cells to act in concert (such as development of a multicellular organism), then additional regulatory mechanisms must exist to ensure synchrony [141].

3.4 Bandpass Filters

In electronics, bandpass filters are made to block the transmission of any wavelength under and above a specific range, thus allowing only a specific range of wavelength

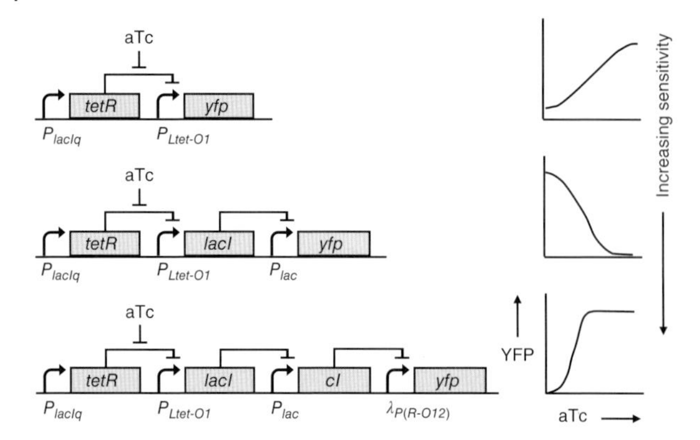

Fig. 4 The properties of synthetic gene cascades vary with cascade length. Cascades of two, three, and four genes were constructed [141]. aTc binds to TetR protein, preventing it from repressing transcription from a downstream promoter. The graphs on the right show the steady-state relationship between the concentration of aTc (input) and yellow fluorescent protein (YFP) fluorescence (output). As the length of the cascade increases, this relationship becomes more sensitive, with sharper switching between ON and OFF states [141]. Modified from Ref. [141].

with a defined minimum and maximum to pass [142]. In biology, a bandpass filter can be used to limit the response of a gene network to predefined biological input levels, ignoring inputs that are too weak or too strong.

A tunable bandpass filter was implemented in *E. coli* using a feedforward loop in which a β-lactamase gene (*bla*) and a tetracycline resistance gene (*tetC*) are linked by mutually repressive interactions [32]. The Bla enzyme allows the bacteria to survive in ampicillin (Amp) by hydrolyzing Amp. A minimum threshold level of Bla is required for the cell to survive in the presence of Amp, but if Bla accumulates above the maximum threshold level, this leads to a repression of the *tetC* gene and hence makes the cell sensitive to tetracycline (Tet). As a result, the circuit acts as a bandpass filter when the cells are grown in the presence of both Amp and Tet: only cells with a specific range of circuit activity, assayed by a GFP reporter, will survive. An interesting feature of the circuit is the control of the *bla* gene by an IPTG-inducible promoter, which makes the circuit tunable: varying the IPTG concentration in the media changes the properties of the bandpass filter (the concentration of Amp and Tet that allows the cell to survive) [32]. A bandpass filter was also built in mammalian cells. Through a combination of interconnected transcriptional activators and repressors, the cells were programmed to express a reporter gene only in the presence of intermediate input concentrations [143].

4 Memory Devices

Many synthetic circuits respond to specific inputs quickly and reversibly: the cell performs a desired action in the presence of input such as a small molecule or a specific wavelength of light, and then, once the input is withdrawn, the cell returns to its "ground" state. The kinetics of the circuit's response to the input are important for its function, and many applications require the circuit to return to the pre-input state promptly.

In contrast, some applications require the cell to maintain memory of a transient stimulus, even over long time periods and multiple generations. One potential use of a memory circuit is to record the cell's history of intrinsic or extrinsic events, such as exposure to toxic chemicals or DNA damage, that are relevant in environmental remediation and medical diagnostics. A counter enables the cell to remember not only prior exposure to a stimulus, but also the number of times it was exposed. Memory circuits may enable engineered microbes to respond to complex and uncertain environments, based not only on current input but also on memory of past events; this ability would allow more complex cell responses than those permitted by simple digital logic (see Sect. 5). Potential uses include bioremediation, support for crops growing in marginal soil, or disease treatment [144]. Moreover, during the development of a complex multicellular organism, each cell must at some point commit to a specific cell fate, and its progeny must maintain memory of the cell fate choice. Synthetic memory devices may mimic the complex cellular logic implemented during differentiation; for example, a memory device may produce distinct outputs depending on the sequence of inputs (A then B versus B then A). These memory circuits may be useful in studying or programming cellular differentiation for applications such as tissue engineering.

Synthetic memory can be divided into volatile and nonvolatile [145]. Volatile memory requires the cell to maintain the memory of the past event actively, whereas nonvolatile memory is passively maintained by the cell following the initial stimulus. These memory systems can be analogized to dynamic and static memory in electronic systems, respectively. Many of the synthetic memory devices, such as the autoregulatory feedback loop described below, resemble those that occur naturally in cells and function extensively in processes such as cell development [146, 147]. In mammalian cells, naturally occurring long-term memory also relies heavily on the chromatin state, which may present another opportunity for implementing synthetic memory devices [148]. Currently, an incomplete knowledge of the mechanisms and outcomes of specific chromatin modifications has limited the ability to design chromatin-based synthetic memory devices. However, synthetic chromatin-modifying factors have been described [149], and synthetic zinc finger- and TALE-based DNA-binding domains have allowed chromatin-modifying proteins to be targeted to specific loci in living cells [39, 80] (Sect. 2.4). A greater understanding of chromatin modifications may enable a more sophisticated and reliable use of chromatin-based synthetic circuits in the future.

4.1 Volatile Memory

A volatile memory device requires active maintenance of the memory state by the cell (e.g., through transcriptional regulation). The toggle switch (see Sect. 3.2), in which two genes, each responsive to a distinct input, repress each other's expression, is an example of volatile memory: the cell remembers the last input it received, even after the input is withdrawn, but maintenance of this memory requires active gene expression [9, 134].

Volatile memory has also been implemented through autoregulatory feedback loops, which consist of two sTFs: a sensor TF and a self-activating "loop" TF. A transient stimulus activates the sensor TF, which in turn initiates expression of the loop TF. Once the loop TF accumulates above a threshold level, it maintains its own expression in a positive autoregulatory feedback loop even after the sensor TF becomes inactivated [18, 150]. In an early example, *S. cerevisiae* cells were programmed to turn on long-term yellow fluorescent protein (YFP) expression in response to a transient galactose input [150]. In a subsequent study, mammalian cells were programmed to remember past exposure to hypoxia and DNA damage, both of which are linked with cancer [18].

Another circuit design exhibits a different form of memory: the ability to count three pulses of a defined small-molecule input [151]. In this circuit, expression of the final output (GFP) depends on a series of orthogonal phage-derived RNAPs that are under the control of a riboregulator (see Sect. 2.5): the RNAP-encoding transcripts cannot be translated, due to *cis*-repressive RNA structures present in the transcripts. Each pulse of the inducer triggers the expression of a small *trans*-acting RNA, which relieves the RNA-based repression and allows each RNAP transcript to be translated. After three pulses of the inducer, the cascade is complete and the GFP reporter is highly expressed [151]. The device requires the pulses to be spaced closely together, to prevent each RNAP from being degraded or diluted by cell division before the next RNAP in the cascade can be expressed.

4.2 Nonvolatile Memory

In contrast to volatile memory, nonvolatile memory does not require continuous action by the host cell; rather, the memory device retains its state passively after the initial switching event. Nonvolatile memory devices have been implemented using site-specific DNA recombinases that can insert, excise, and invert DNA fragments at specific DNA sequence motifs. Advantages of nonvolatile memory devices include a lower metabolic burden on the cell, because the cell does not expend energy to maintain the memory state, as well as the ability to construct a multistate circuit from relatively few parts by arranging the recognition sites for two or more orthogonal DNA recombinases. The number of possible states of the system grows exponentially with the number of orthogonal recombinases available [144].

Notably, unlike volatile memory, nonvolatile memory based on DNA recombinases persists after the removal of the memory device (the rearranged DNA fragment) from the host cell. This has led to the intriguing idea that such memory devices could be shared among cells, via the transformation of a "naïve" cell with DNA from a lysed memory cell [144], leading in turn to the possibility of complex multicellular computation.

Examples of synthetic nonvolatile memory include the use of bacterial invertases that catalyze inversion of a fragment of DNA flanked by specific inverted repeat sequences [144, 152]. In one example, induction of the *E. coli* FimE invertase leads to an irreversible flipping of the promoter of the *gfp* reporter gene from the OFF state to the ON state [152]. In a more complex follow-up study, the use of two orthogonal invertases (Fim from *E. coli* and Hin from *Salmonella*) allows for the programming of devices with many possible states, including devices that can produce different output depending on the history of the system: Fim followed by Hin, versus Hin then Fim (Fig. 5a) [144]. While these studies presented an exciting proof-of-principle, the resulting circuits did not behave entirely as expected, due partly to the very low rate of inversion by Fim [144].

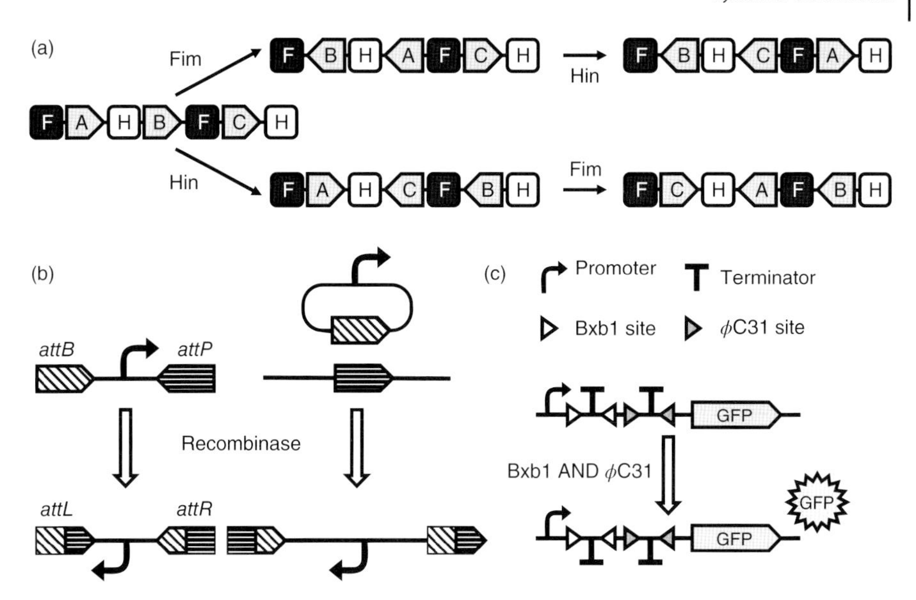

Fig. 5 Synthetic nonvolatile memory devices. (a) The bacterial invertase-based memory device uses two orthogonal enzymes, Fim and Hin, which mediate DNA inversion between sites marked "F" and "H," respectively. Note that the final state of the system depends on the order of inputs: Fim followed by Hin versus Hin followed by Fim [144]; (b) Phage-derived serine recombinases catalyze DNA inversion (left) or integration/excision (right) between specific recognition sites, termed *attB* and *attP*. The reaction produces *attL* and *attR* sites, which are not recognized by the recombinase, making the reaction irreversible unless a cognate excisionase is added; (c) An example of a two-input digital logic gate implemented with two orthogonal serine recombinases, Bxb1 and ϕC31 [153]. In this AND gate, two transcriptional terminators prevent expression of the GFP reporter from the upstream promoter. When both recombinases are present, the two terminators are switched to the "off" position, and GFP expression can proceed. Panel (a) modified from Ref. [144]; panel (c) modified from Ref. [153].

Another early example of a recombinase-based synthetic memory device is the DNA invertase cascade in *E. coli* that can count to three: reporter GFP expression becomes active only after three pulses of an inducer molecule [151]. In this system, each pulse of the small molecule input drives expression of a recombinase, which in turn catalyzes a DNA inversion that prepares the circuit to respond to the next inducer pulse. One version of the circuit counts three pulses of the same small molecule, arabinose. Another version respond to pulses of three different inducers given in a specified order: for example, the circuit expresses GFP if aTc, arabinose, and IPTG are supplied in that order, but not if the order of the stimuli is rearranged [151]. This system is useful for modeling processes such as development, where the timing of the stimuli matters in addition to the identities of the stimuli.

Other studies in this area have utilized phage-derived serine recombinases. Originally used in bacteria, codon-optimized recombinases also function in eukaryotic cells. These enzymes catalyze recombination between two recognition sites, termed *attP* and *attB* (attachment phage and attachment bacterium, respectively), resulting in the formation of *attL* and *attR* sites (Fig. 5b). Because the recombinase does not recognize *attL* and *attR* sites, the reaction is irreversible unless an excisionase is added (see below). Depending on the location and orientation of the starting *attB* and *attP* sites with respect to each other, the reaction results in excision, integration, or inversion ("flipping") of a DNA fragment (Fig. 5b). Site-specific recombinases from multiple phages have been described, including TP901, Bxb1, and ϕC31 [154]. Notably, these recombinases are orthogonal, with each recognizing a specific and unique pair of *attP* and *attB* sites, a property that allows the use of multiple recombinases in a single cell.

A recent study demonstrated the construction of all 16 two-input Boolean logic gates (see Sect. 5) in *E. coli* using two orthogonal site-specific recombinases, Bxb1 and ϕC31 [153]. In this study various combinations of promoters, transcriptional terminators and reporter coding regions flanked by recombinase target sites were used to achieve the desired logic function. For example, AND logic was implemented by inserting two transcriptional terminators, one flanked by Bxb1 sites and the other by ϕC31 sites, between a promoter and a GFP-coding region, such that the two terminators were flipped to the OFF orientation and GFP was expressed only in the presence of both recombinases (Fig. 5c). The expression of the two recombinases was coupled to two different small-molecule inputs, so that Boolean logic computations were performed upon addition of the specified input molecules. As with the previous examples of recombinase-based circuits, the circuits passively maintained their state upon removal of the stimulus.

Although a serine integrase alone catalyzes an irreversible reaction (*attB* +*attP* → *attL* + *attR*), the combination of the integrase with a cognate excisionase can catalyze the reverse reaction and restore the *attB* and *attP* sites. A recent study harnessed this phenomenon to build a reversible memory module in *E. coli* [155]. In this system, expression of the Bxb1 integrase alone leads to red fluorescent protein (RFP) reporter expression; coexpressing the Bxb1 integrase with its partner excisionase inverts the orientation of the reporter gene promoter, resulting in GFP expression [155]. In theory, the system can "flip" repeatedly between the GFP-on state and the RFP-on state through repeated induction of integrase alone or integrase with excisionase. However, functioning of the system requires careful adjustment of the integrase:excisionase ratio [155].

5
Boolean Logic and Digital Circuits

Most of the synthetic gene circuits that have been implemented so far have benefited from digital logic design, in which cellular networks are thought of as assemblies of digital logic gates. A digital logic gate is a device that carries out a Boolean operation and produces a Boolean output based on the different combinations of inputs it receives. Both the inputs and the output of a logic gate are in the form of TRUE and FALSE, often represented as 1 and 0, respectively. Examples of basic Boolean operations include AND (output is TRUE only if all inputs are TRUE), NAND (output is TRUE unless all inputs are TRUE – the negation of AND), OR (output is TRUE if at least one input is TRUE), and NOR (output is true if none of the inputs is TRUE – the negation of OR) (Fig. 6). Most digital logic gates used in synthetic biology have been two-input gates, of which there are 16 types. Interconnecting such two-input logic gates so that the output of some gates acts as input for others can be used to design more complex logic circuits.

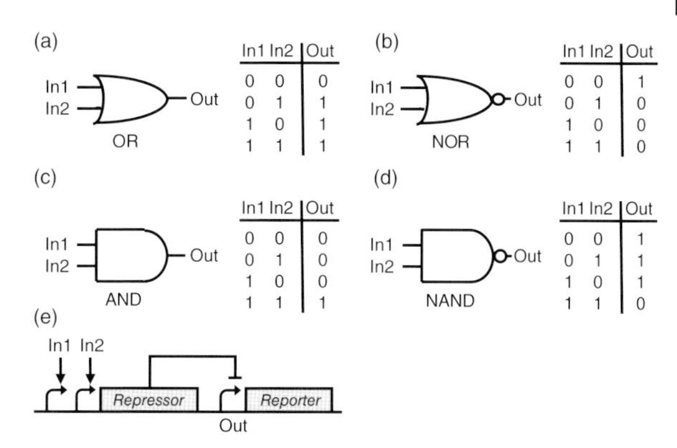

Fig. 6 Digital logic gates. A few examples of two-input logic gates as well as the truth table that each one encodes are shown. (a) OR gate; (b) NOR gate; (c) AND gate; (d) NAND gate. A truth table summarizes the outputs to each possible combination of inputs of a logic gate. The logic gates are drawn according to the conventions of electrical engineering. The bubble in (b) and (d) symbolizes negation; (e) Schematic representation of a genetic NOR gate: the presence of either or both of the two inducers activates the expression of the repressor protein, which in turn represses the reporter output. Hence, the output is expressed only when neither of the inducers is present, as per the truth table shown in panel (b). In1, input 1; In2, input 2; Out, output.

The domain of digital logic computation in living cells is modeled after engineering electronic circuits, with the goal of programming cells similarly to programming a digital computer, only with biological inputs and outputs rather than electronic ones. Instead of physical wires, molecules such as RNA and protein connect logic gates inside the cells. Such circuits could be implemented for many purposes, including diagnostics and therapeutics, in which the cell is designed to perform a specific behavior in response to specific stimuli, such as "if the inputs match the profile of a cancer cell, kill the cell" [22, 23, 156–158]. Moreover, since biological inputs can be converted to electric ones and vice versa [136, 159 161], cells could potentially be integrated as biological computational units for electronic computers.

Compared to the assembly of their silicon-based counterparts, the construction of complex biological logic circuits by layering elementary logic gates has proven to be extremely challenging. Over the past decade, many different designs with various functions have been implemented in both prokaryotic and eukaryotic cells, but the level of complexity of these designs has not improved significantly [162]. This is in part due to unwanted crosstalk between components of synthetic circuits (Sect. 2.3), the propagation and amplification of noise through the circuit (Sect. 8.3), and the metabolic burden that accompanies the expression of exogenous genes in host cells (Sect. 8.2). In this section, a few hallmark examples of synthetic digital logic circuits that have been implemented are briefly provided.

An early example of a digital logic circuit used chimeric promoters that bind small-molecule-responsive transcriptional activators and repressors. Depending on the combination of activators and repressors used, the system gave rise to different logic gates, including AND, NAND, and NIMPLY (output is TRUE only when one specific input is TRUE) gates [163]. Subsequent studies employed split inteins which, when fused to separate proteins, would mediate fusion of the two proteins into one. Hence, split inteins would allow the reconstitution of a functional TF that can activate or repress an output promoter, forming an AND or NAND logic gate, respectively. Split inteins fused to sequence-specific ZF-TFs (see Sect. 2.4) were used to build a variety of logic gates in mammalian cells [70]. Recently, the split intein strategy was used to build AND gates in mouse stem cells [164], suggesting the possibility of future applications in programming stem cell fate for research and therapy.

Other digital logic circuits combined transcriptional and post-transcriptional regulation. For example, Arkin and colleagues [165] constructed a two-input AND gate in *E. coli*. The first input activates transcription of a gene encoding T7 RNAP, but the protein cannot be translated due to the presence of two premature amber stop codons. The second input turns on expression of an amber suppressor tRNA, so that the presence of both inputs allows expression of T7 RNAP protein and subsequent transcription of the reporter gene from a T7 promoter [165]. The authors linked the AND gate to the expression of Invasin in bacteria, such that the bacteria invaded human cancer cells *in vitro* only in the presence of two user-defined inputs [165, 166]. This circuit represents a proof-of-principle strategy that may be used for cancer therapy in the future.

Digital logic circuits may also be implemented in mammalian cells using miRNAs, as shown by Benenson and colleagues, who built computational devices in mammalian cells by targeting multiple miRNAs to the same output transcript [91, 92]. Expression of the miRNAs may be coupled to the presence or absence of user-defined transcriptional activators or repressors, which act as inputs to the circuit [92]. For example, if a transcript that encodes a fluorescent protein output has recognition sequences for two miRNAs – miR-a, which is repressed by TF-A, and miR-b, which is activated by TF-B – then the resulting circuit will produce fluorescent output only if A is present and B is absent, implementing the digital logic function A AND NOT B. More complex circuits may be built by combining multiple miRNA-responsive output transcripts in one cell [91, 92].

The potential for practical applications of miRNA-based logic was demonstrated by a circuit that induces cell death specifically in the HeLa cancer cell line [23] (Fig. 7). The circuit accepts as inputs the levels of two user-defined sets of miRNAs: those expected to be highly expressed and those expected to be absent from HeLa cells ("HeLa-high" and "HeLa-low" miRNAs, respectively). The output of the circuit is the pro-apoptotic protein BAX, which triggers cell death. The "HeLa-high" miRNAs prevent expression of the LacI transcriptional repressor, which in turn represses the BAX-encoding gene. Hence, BAX protein expression is possible only when the "HeLa-high" miRNAs are present above a threshold level. In parallel, "HeLa-low" miRNAs downregulate the BAX transcript via their recognition sites in the BAX transcript 3′ UTR. Consequently, the circuit leads to BAX expression if, and only if, the host cell matches the HeLa cell profile – that is, all "HeLa-high" miRNAs are above a threshold level and all "HeLa-low" miRNAs are below a threshold level [23]. The circuit triggers significantly higher levels of apoptosis in HeLa cells compared to control human embryonic kidney (HEK) cells, indicating its potential for future cancer therapy [23]. Depending on the miRNA inputs used, the circuit could be designed to target any cell type that has a unique miRNA profile.

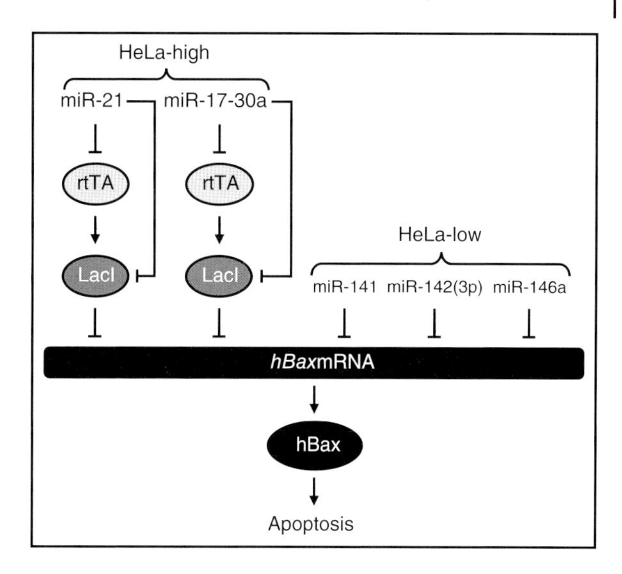

Fig. 7 Logic computation in mammalian cells: the miRNA classifier circuit. The HeLa cancer cell classifier causes production of the pro-apoptotic protein hBax and hence the death of the host cell if, and only if, all "HeLa-high" miRs are highly expressed and none of the "HeLa-low" miRs is expressed above a threshold level. The "HeLa-high" miRs regulate an inverter module: the miRs prevent expression of the rtTA transcription activator, which in turn activates expression of the LacI transcription repressor (the miRs also target the lacI transcript to improve the signal-to-noise ratio). Hence, in HeLa cells, the "HeLa-high" miRs prevent expression of the LacI protein, permitting transcription of hBax mRNA, which has sequences complementary to "HeLa-low" miRs in its 3′ UTR. In a healthy cell, high levels of one or more of the "HeLa-low" miRs block hBax expression, whereas in HeLa cells hBax translation can proceed. As a result, only HeLa cancer cells undergo apoptosis in the presence of the classifier circuit. Adapted from Ref. [23].

A subsequent study in mammalian cells used two small-molecule inputs – erythromycin and phloretin – to drive the expression of two transgenes: one that encodes the RNA-binding proteins MS2 or L7Ae; and one that contains the fluorescent reporter coding region fused to a RNA box sequence that specifically binds MS2 or L7Ae [167]. When bound to their cognate RNA boxes, MS2 and L7Ae block translation of the reporter transcript. This enables construction of NIMPLY gates, in which the fluorescent protein is expressed only if the input that activates its promoter is on, and the input that activates expression of the RNA-binding protein is off. By combining two NIMPLY gates in one cell, the technically challenging task was accomplished of constructing a two-input XOR gate, which outputs TRUE only if exactly one input is TRUE [167].

While most digital logic gates involve transcriptional regulation, RNA-based, post-transcriptional mechanisms have also been used [87] (see Sect. 2.5). In a recent study, digital logic gates were demonstrated in mammalian cells based on DNAzymes, which are synthetic DNA molecules that bind to and cleave a transcript in a sequence-specific manner [168]. The DNAzymes include inhibitory stem-loop structures, which block their catalytic activity unless relieved by binding to a specific miRNA. The use of DNAzymes allowed the construction of logic gates with specific miRNAs as inputs, and translation of the target mRNA as output. This method could potentially be used to detect miRNA profiles associated with cancer [168].

The above-described studies demonstrated the potential of using digital logic circuits to customize cellular signaling and regulatory networks to achieve various useful applications. However, these logic circuits were relatively simple, and most of them had single-layer gates. In an effort to build a complex layered logic circuit in *E. coli*, Moon *et al.* layered individual AND gates to integrate signals from four different inputs [169]. In their design, AND gates accept two promoter inputs and control one promoter output. Each gate is composed of a TF that needs a second chaperone protein in order to activate the output promoter. The authors applied directed evolution and part mining (see Sect. 2.3) to minimize crosstalk between the gate components, allowing the gates to be layered in a single cell. The result was a four-input AND gate, the most complicated circuit implemented in single cells to date [169].

TF-based logic circuits are limited by the shortage of truly orthogonal parts, as well as the metabolic burden they impose on the host cell; for example, the four-input AND gate requires 11 regulatory proteins [169]. One alternative approach would be to use recombinase-based circuits, which combine digital logic and memory [153, 170] (see Sect. 4.2). The memory module can be used to design robust and multilayered synthetic circuits, and the one-time inversion of a piece of DNA by a recombinase places less metabolic load on the cell than does the continued expression of a TF. Alternatively, RNA-based regulatory devices (see Sect. 2.5) can be composed into logic gates and cascades that place a lower metabolic burden on the cell than does the expression of heterologous regulatory proteins. For example, a four-input NOR gate based on the RNA-IN-RNA-OUT system in *E. coli* [97] achieved robust digital behavior while requiring a fraction of the resources of the protein-based four-input AND circuit [169] (see Sect. 2.5 for details).

6 Analog Circuits

As described above (Sect. 5), the majority of synthetic gene circuits built thus far were designed to perform in the digital domain. However, digital computation requires the composition of a large number of orthogonal devices together, each performing a simple binary computation, to achieve more complex functionalities. Although this is possible in the world of semiconductors due to the ability to assemble billions of transistors on a single substrate and to wire them together in a precise fashion, it is challenging to do in biological systems due to the paucity of available parts as well as resource limitations such as energy and space.

An alternative strategy would be to use analog computation, which leverages the inherent physics of a system to calculate mathematical functions and can thus achieve a greater computational complexity with fewer resources. Synthetic biologists have designed and fabricated circuits that function in the analog domain, where a graded input is converted into graded levels of output. The digital paradigm can be considered as a special case of graded analog functions where values below a given threshold are defined as "0" and values above that threshold are classified as "1" (Fig. 8a). A key difference between digital and analog circuits is the way in which they can be composed together in order

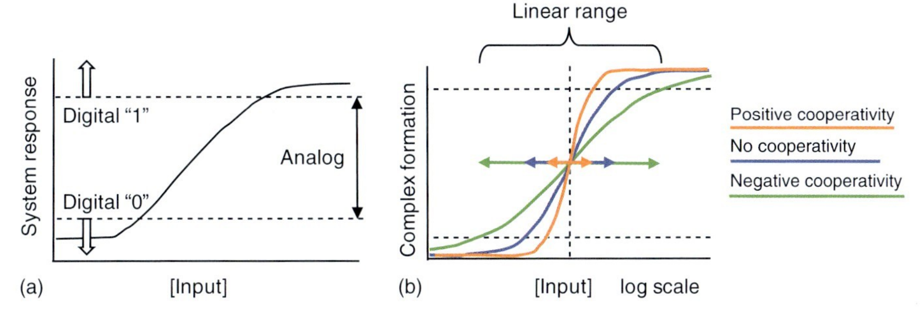

Fig. 8 Schematic representation of the difference between analog and digital circuits. (a) Abstraction of analog and digital circuits. A digital circuit recognizes only the difference between values above (1) and below (0) specific threshold levels. An analog circuit recognizes differences between graded levels of input in the linear portion of the input–output function; (b) Implementation of Weber's law and linear range in basic biochemical reactions. Negative cooperativity expands the linear range of the reaction. Modified from Ref. [171].

to construct higher-order functions. For example, an analog adder can be achieved by simply combining two parallel circuits that each have different input molecules but generate the same output molecule [171]. However, a digital adder cannot function correctly using the same principle; for example, adding two "1" binary numbers together requires another stage to hold the new bit "Carry out" ("10") (see Ref. [172] for an example of an adder implemented with digital logic in yeast).

The main advantage of analog computation in synthetic biology is the ability to implement complex mathematical functions using a small number of synthetic components with high efficiency. For example, in theory a circuit can be built that computes an integral function by measuring the accumulation of a protein over time, or a differential function by measuring the consumption of a protein over time. Moreover, analog circuits are of interest to synthetic biology because many cellular processes do not rely on the all-or-none responses found in digital circuits; rather, the cell responds in a graded fashion to changes in input concentration [173]. For example, increasing the levels of arabinose in the media activates progressively higher levels of expression of arabinose transporters and metabolic enzymes in *E. coli* [174]. Hence, analog circuits are suitable for programming cell behaviors for possible practical applications.

Living cells sense their environment using sensory systems, many of which follow Weber's law, to integrate environmental information such as light, sound, and chemotaxis [175]. Weber's law represents an analog rather than digital behavior and defines situations where the fold-change between signal levels and its background is a constant, thus enabling fold-change detection as opposed to absolute level detection [173]. Weber's law also applies to molecular signaling networks and biochemical reactions. Figure 8b shows the activity of forming a complex in steady state as a result of binding two molecules, such as a substrate and an enzyme or a TF and its target promoter.

Weber's law holds in the linear input dynamic range of a transcriptional circuit, which can be either increased or decreased depending on the binding cooperativity of the reaction. Cooperativity refers to the effect that the first binding interaction has on the probability of the second binding event. Positive cooperativity means that the binding of one molecule to its target site increases the binding affinity for the second molecule. Many commonly used transcriptional repressors, such as LacI, display a positive cooperativity in binding their target DNA [14]. Higher cooperativity leads to a steeper input-output transfer function [176], which in turn narrows the dynamic range of the system and produces an all-or-none response suitable to a digital logic circuit. In contrast, negative cooperativity, whereby the binding of the first molecule lowers the affinity for the second binding, is commonplace in cell signaling networks [177] and it allows a system to respond in analog fashion over a wide input dynamic range.

Building an analog circuit requires a wide dynamic range (the range of input concentrations over which the system displays a linear input–output response). One challenge in constructing analog synthetic biology circuits lies in the switch-like behavior of the synthetic biology parts that results from the narrow dynamic range of many biological components. In order to build an analog circuit with these components, the circuit topology must be constructed in a way that promotes a wide dynamic range.

Negative feedback loops are commonly used to linearize the input–output transfer functions of electronic and biological systems – that is, to make the input–output transfer function linear over a wider range of input concentrations. For example, one study implemented an autonegative feedback loop to linearize the response of a simple circuit to the small molecule aTc in *S. cerevisiae*[176] (Fig. 9a). The authors compared two circuits. In the first circuit, the TetR repressor is expressed from a constitutive promoter, and it represses transcription of a GFP reporter gene. The addition of aTc relieves transcriptional repression by TetR, allowing GFP expression. The circuit shows a digital response, with a narrow transition region between the GFP-off state (low aTc) and GFP-on state (high aTc). The second circuit is identical to the first, except that TetR is expressed from a TetR-repressed promoter – that is, autonegative feedback is introduced into the circuit. In contrast to the first circuit, the negative feedback loop allows linear GFP expression. A possible explanation is that the autonegative feedback reduces the basal level of the TetR repressor, because at low aTc concentrations the TetR protein represses its own production; this leads to higher basal GFP levels and, consequently, a linear input–output function at low levels of aTc [176]. The autonegative feedback circuit has also been constructed in mammalian cells and has achieved a linear response [11].

Madar *et al.* [178] studied the naturally occurring autonegative feedback loop in the arabinose utilization system of *E. coli* (Fig. 9b). These authors showed that removing the autonegative feedback from the arabinose regulatory network decreased the input dynamic range by 10-fold. The arabinose system is regulated by cAMP receptor protein (CRP), which is activated by cAMP, and AraC TF, which is activated by L-arabinose. AraC also represses its own promoter, creating an autonegative feedback loop (see [179–183] for details). To understand the role of the autonegative feedback loop, the authors put AraC under control of a constitutive promoter (Fig. 9b), and found that a loss of the autonegative feedback decreased the L-arabinose dynamic range by an order of magnitude [178].

Recently, Daniel *et al.* [171] implemented a strong negative feedback loop in a genetic

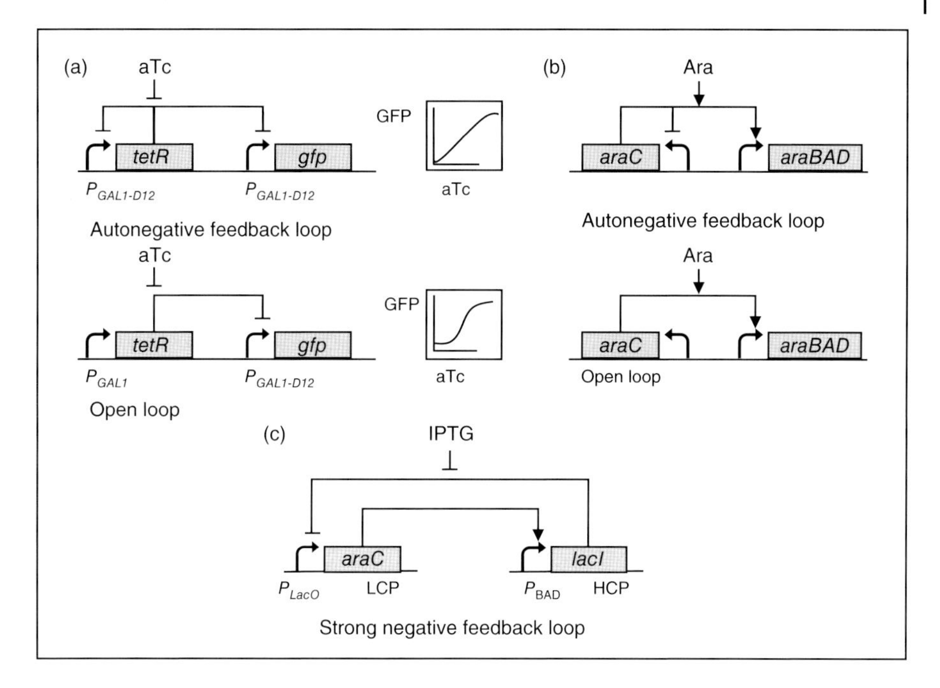

Fig. 9 Negative feedback loops in synthetic biology. (a) Autonegative feedback linearizes the dose response in a simple synthetic circuit [176]. The autonegative feedback loop (top), in which TetR represses its own expression as well as that of the GFP-expressing promoter, shows a graded, linear input–output function, in contrast to the steep input–output function of the open loop (bottom); (b) Autonegative feedback loop in the arabinose utilization system of *E. coli* creates a linear response to increasing concentrations of arabinose. In the open loop system, which lacks negative autoregulation by AraC, the dynamic range is narrower by an order of magnitude compared to the natural system [178]; (c) Strong negative feedback achieves a power law function [171]. HCP, high-copy plasmid; LCP, low-copy plasmid. Panel (a) modified from Ref. [176]; panel (c) modified from Ref. [171].

circuit, yielding a power law function relation between the input and the output. The input to this circuit is IPTG, which inhibits the binding of LacI to the P_{LacO} promoter that drives the expression of AraC from a low-copy plasmid. AraC in turn binds to the P_{BAD} promoter located on a high-copy plasmid and activates expression of the LacI repressor when the arabinose concentration is high. A strong and tunable negative feedback loop was achieved by adjusting the ratio between the low-copy and high-copy plasmid (Fig. 9c). The LacI-IPTG transfer function exhibited a power law function ($y = x^{0.7}$) over two orders of magnitude (see Ref. [171] for details), thus enabling power law computations in living cells with few synthetic parts.

A properly tuned positive-feedback loop can also linearize the dose response and extend the input dynamic range of a circuit. Daniel *et al.* [171] implemented graded positive feedback loops in *E. coli* for two different types of TF, LuxR and AraC, and their small-molecule inputs (AHL and arabinose, respectively). The input–output transfer functions exhibited a wide region of linearity (over three orders of magnitude) when plotted on a semi-log plot. The circuit consists of two parts: the positive-feedback

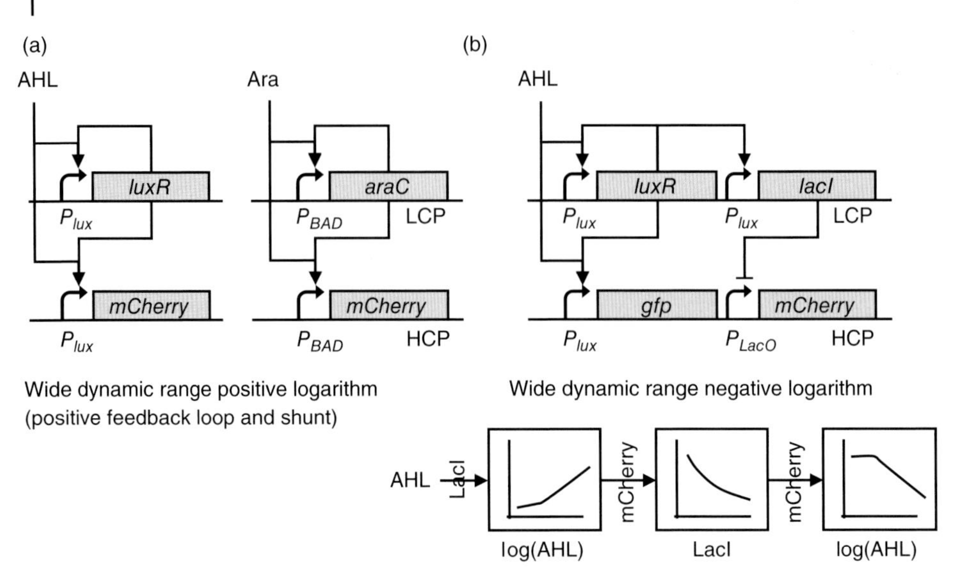

Fig. 10 Positive-feedback loops in analog synthetic biology circuits [171]. (a) Wide-dynamic-range positive-logarithm circuits for AHL and arabinose (Ara). The circuit achieves a wide dynamic range by using a positive-feedback loop and a high-copy plasmid shunt. The positive-feedback loop prevents saturation of the transcription factor at intermediate concentrations of the inducer, and the high-copy plasmid prevents saturation of TF-binding sites. This results in an input-inducer-to-output-protein transfer function that exhibits logarithmically linear behavior with a positive slope; (b) Wide-dynamic-range negative logarithm circuit for AHL. HCP, high-copy plasmid; LCP, low-copy plasmid. Modified from Ref. [171].

loop placed on a low-copy plasmid; and decoy binding sites encoded on a high-copy "shunt" plasmid, which reduces the positive-feedback loop strength by shunting away a proportion of the TFs that are produced by the positive feedback circuit (Fig. 10a). The shunt prevents the system from saturating at intermediate concentrations of the input, thus extending the dynamic range. When the "shunt" circuit was removed, the input–output dynamic range decreased by two to three orders of magnitude [171].

Like digital circuits, analog circuits can also be composed into cascades and more complex devices. The first successful two-stage analog cascade in living cells was used to linearize the P_{Laco} promoter transfer function, which is repressed by LacI [171] (Fig. 10b). The first stage consisted of the positive-feedback-and-shunt motif with AHL as the input and LacI as the output, while the second stage consisted of a LacI-repressed P_{LacO} promoter driving the expression of an mCherry fluorescent output from a high-copy number plasmid. The input–output transfer function of the cascade exhibited a wide region of linearity when plotted on a semi-log plot with a negative-slope function that spanned over four orders of magnitude (Fig. 10b).

Finally, combining analog circuits can enable the implementation of complex mathematical functions [171]. An analog adder was built by integrating, in parallel, two wide-dynamic-range positive-logarithm circuits (positive-feedback-and-shunt motifs) that each take in distinct input molecules (AHL and arabinose) and

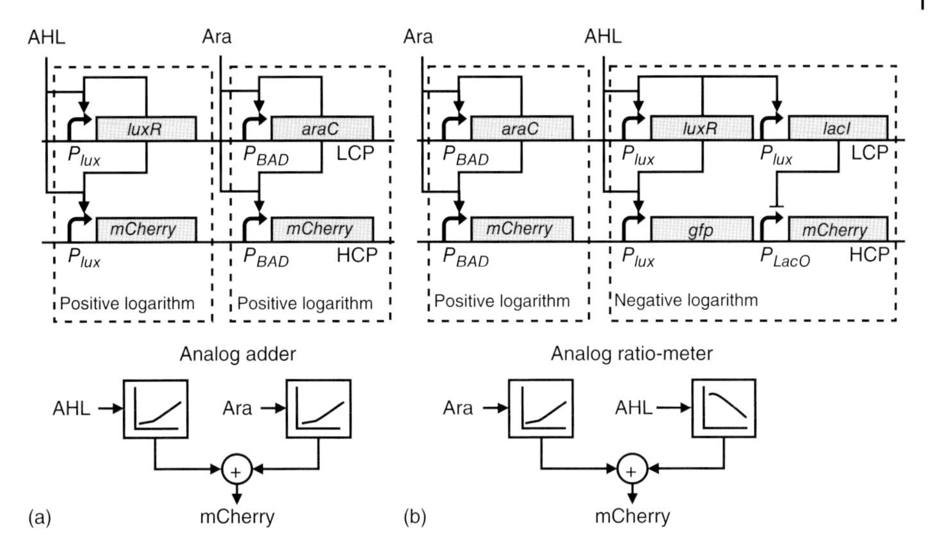

Fig. 11 Analog circuits perform mathematical functions in living cells. (a) An analog adder and (b) an analog ratio-meter were implemented through combinations of positive-logarithm and negative-logarithm circuits that accept distinct inputs (AHL and arabinose) and generate a shared output, mCherry. Modified from Ref. [171].

generate an output signal that is common between the two circuits, mCherry (Fig. 11a). An analog ratio-meter was also constructed using the same concept as the analog adder; however, the AHL-responsive positive slope circuit was replaced with an AHL-responsive negative slope (Fig. 11b). The output of the circuit is proportional to the ratio of the inputs (AHL and arabinose). The ratio-meter operates over four orders of magnitude [171]. Circuits capable of calculating the ratio between two inputs are potentially useful, as they enable synthetic biologists to mimic natural biological systems, many of which are balanced between two competing inputs, and to normalize inputs with respect to each other for biosensing and control applications.

The design and construction of synthetic analog circuits in living cells is a novel approach that poses new challenges. The accumulation of noise and the need for a high signal-to-noise ratio will be among the main challenges in scaling-up analog genetic circuits. However, these challenges have been faced and addressed by scientists in other fields and their solutions can surely be adapted to synthetic biology. For example, hybrid analog–digital designs can optimize energy efficiency and information precision in one system (see Ref. [171]).

7 Intercellular Communication and Synthetic Multicellular Devices

While the above-described circuits function at the level of a single cell, there is growing interest in programming multicellular systems in which multiple cells communicate with each other to accomplish a specific task [184]. One reason for this is the limitation on the size and complexity of a circuit that can be built in a single cell, due to the crosstalk among circuit components and the metabolic load imposed on the cell (see Sect. 8). Yet, the designers of multicellular systems may overcome this limitation by dividing a large, complex circuit into smaller subsystems that are implemented in separate cell strains programmed to communicate with each other. In addition, populations of cells may carry out behaviors that cannot be implemented in a single cell, such as spatial patterning, thus opening up the possibility of practical applications such as tissue patterning for transplants. Currently, there is also much interest in developing consortia of multiple microbial strains or species that can cooperate in the synthesis of biofuels and other valuable chemicals [185].

7.1 Intercellular Communication Mechanisms

In a multicellular system, cells must be able to send and receive signals among each other. Multicellular synthetic circuits have harnessed the molecules that cells use naturally to communicate. One well-known example is quorum sensing (QS), a process by which bacteria communicate with each other via the diffusion of small molecules that they synthesize. A QS module includes an enzyme that synthesizes the diffusible signaling molecule and a TF that regulates its target genes when bound to the signaling molecule. In *E. coli*, commonly used orthogonal QS systems are LuxR/LuxI from *Vibrio fischeri* and LasR/LasI or RhlR/RhlI from *Pseudomonas aeruginosa* [15, 16]. The LuxI enzyme synthesizes the diffusible signaling molecule, an AHL which, when bound to the LuxR TF, causes gene expression from the P_{lux} promoter. Other forms of intercellular communication exist among eukaryotes: *S. cerevisiae* cells communicate using mating pheromones [186], while mammalian cells possess a suite of signaling proteins that trigger complex signaling cascades and elaborate behaviors in recipient cells [187–192]. Alternatively, a cell may be engineered to synthesize a molecule that does not normally function in signaling, and that molecule may then diffuse out of the sender cell and into a recipient cell engineered to perform a specific action in response to the input.

One disadvantage of small-molecule-based intercellular communication is that only one type of information can be transmitted through a given channel (namely, high versus low concentrations of the signaling molecule) [193]. More advanced multicellular devices could be programmed if the communication channel could transmit diverse user-defined messages. An obvious candidate for such communication channel is intercellularly transmitted DNA. In a recent study [193], communication was engineered between *E. coli* cells via M13 bacteriophages, which exit the host cell without killing it and can transmit multi-kilobase single-stranded DNA molecules between cells. In this proof-of-concept study, the authors programmed sender cells with M13 phages encoding T7 RNAP. Upon receiving the phage message, the receiver cells activate GFP reporter expression from a T7 RNAP-responsive promoter [193]. Another possibility for DNA-based intercellular communication would be to use bacterial conjugation, the process by which bacteria exchange plasmids. While this has not yet been implemented, a recent computational simulation suggests that digital logic gates may be constructed by coculturing bacterial strains that can exchange plasmid-based messages [194].

7.2 Examples of Synthetic Multicellular Systems

Some multicellular systems are isogenic: all cells in the system are programmed with the same genetic circuit, and they communicate

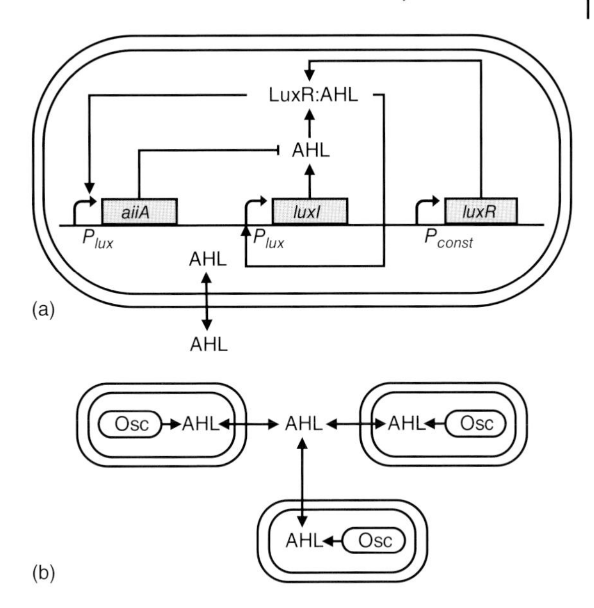

Fig. 12 Using quorum sensing (QS) to couple and synchronize oscillators in an *E. coli* population. (a) Circuit diagram of the synchronized oscillator [195]. AHL is a diffusible QS molecule produced by the LuxI enzyme. The LuxR TF is expressed from a constitutive promoter (P_{const}) and complexes with AHL to make LuxR:AHL, which activates the P_{lux} promoter. AiiA catalyzes the degradation of AHL. As AHL accumulates, more AiiA is produced from the P_{lux} promoter, leading to degradation of AHL; subsequent loss of AiiA expression allows AHL levels to rise again after a delay, leading to oscillations; (b) AHL is used to couple together the dynamics of oscillators in different cells across the population. Osc, oscillators. Panel (a) modified from Ref. [195].

with each other to perform a specific task [184]. In an isogenic system, intercellular communication provides a way for cells to synchronize their behavior and to form spatial patterns. In one example, bacterial QS was used to couple and synchronize oscillators (see Sect. 3.1) across a bacterial population [195] (Fig. 12). It should be noted that, at low cellular concentrations, oscillations were not observed as the concentration of the QS molecule AHL was too low to sufficiently activate gene expression. However, concentrating the bacterial cells in a microfluidic chamber allowed AHL to reach sufficient levels to trigger oscillations [195]. In a related study [196], a collection of contained cellular populations arranged in a grid was synchronized using hydrogen peroxide that had been synthesized by the cells and diffused among the cell populations.

Multicellular systems allow more sophisticated computation than is possible in a single cell, such as spatial patterning in a cell population. An outstanding example is the bacterial edge detection circuit [197]. By coupling a light sensor (Cph8), logic (NOT and AND) gates, and QS modules, the authors programmed a population of *E. coli* cells to detect and outline the (dark-light) edges of a projected image on a lawn of bacteria (Fig. 13). In this example, a cell produces an output (pigment) only if the cell itself is exposed to light whilst its neighbors, which communicate with it via QS, are in the dark [197]. In another study, *E. coli* cells were programmed to form a pattern of alternating stripes of high cell

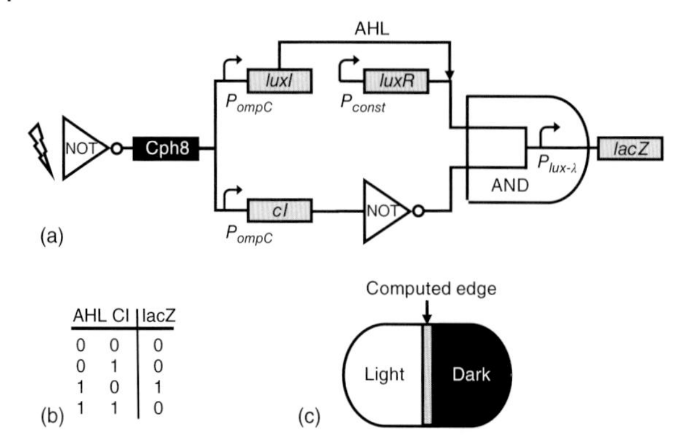

Fig. 13 The bacterial dark/light edge detection circuit in *E. coli* [197]. (a) Cph8 is a light-sensing protein, which can activate the P_{ompc} promoter only in the absence of red light. As a result, only cells in the dark can express LuxI and synthesize AHL molecules, which then diffuse to neighboring cells and allow LuxR to activate expression of the *lacZ* reporter. Cells in the dark also express the λ CI transcription repressor from a P_{ompc} promoter. Only cells that do not express CI and that receive AHL from nearby cells can express *lacZ* from the $P_{luxR} - \lambda$ hybrid promoter. The expression of *lacZ*, which encodes a blue pigment-producing enzyme, occurs only at the edge of light and dark, in cells that receive light (no CI expressed) and are near cells in the dark (receive AHL); (b) Truth table for the circuit shown in panel (a); (c) A schematic representation of the computed dark/light edge (gray) on a lawn of bacteria. Modified from Ref. [197].

density and low cell density on a culture plate by coupling QS to cell motility [198].

Unlike isogenic cell systems, some multicellular systems consist of two different cell strains, each engineered with a different circuit. Such multistrain consortia can be used for constructing digital logic gates (see Sect. 5). The construction of multilayered logic gates in single cells is hampered by crosstalk among different parts of the circuit, as well as by the large metabolic load that a large genetic circuit places on the host cell (Sect. 8). Both problems may be circumvented by distributed multicellular computation, whereby each subpopulation processes a specific logic gate and produces output in the form of a diffusible small molecule, which in turn can diffuse through the cell population and act as an input for a logic gate in another cell. Using this distributed computing strategy, Tamsir *et al.* constructed all 16 possible two-input logic gates in *E. coli* [199]. In another example, populations of yeast cells, each encoding an individual logic gate, were connected via diffusible pheromone "wires" to make higher-order digital logic circuits, including a multiplexer and 1-bit adder [172]. In both of these complex circuits layering of the gates was made possible through controlled interactions of subpopulations of cells.

The first synthetic intercellular communication device in mammalian cells used sender cells engineered to synthesize the volatile molecule acetaldehyde, which diffused into neighboring receiver cells and triggered reporter gene transcription from an acetaldehyde-responsive promoter [200]. Replacing the mammalian sender cells with acetaldehyde-producing *E. coli* or *S. cerevisiae* cells allowed interspecies signaling [200]. A subsequent study demonstrated two-way communication in mammalian cells, whereby one cell strain produced L-tryptophan and expressed a reporter gene in response to acetaldehyde, while the second cell strain synthesized acetaldehyde and responded to L-tryptophan [201]. Using this bidirectional system, the authors programmed the cell consortium for sequential production of angiopoietin-1 and vascular endothelial growth factor, two proteins required for the formation of mature blood vessels [201]. This study presents an example of how synthetic intercellular signaling may be used for tissue engineering, with potential therapeutic benefits.

8 Synthetic Circuit Construction: Challenges and Solutions

Synthetic biology aims to make the design of biological devices as predictable as in other disciplines, such as mechanical or electrical engineering. While great progress has been achieved, this goal remains elusive, due to the immense complexity and an incomplete understanding of living organisms. In order for a synthetic circuit to function correctly, the following conditions must be met [202]:

- Each component of the circuit (TF, promoter, regulatory RNA element, and others) should function as expected.
- The circuit design and the parameters of the circuit components (e.g., the rates of TF synthesis and degradation) are suitable to the specified task.
- The circuit avoids unwanted interactions with the host or the host's environment that disrupt circuit function or impair the host's viability.
- The circuit is robust, that is, it is capable of maintaining function in the presence of intrinsic and extrinsic noise.

Synthetic building blocks and their functions are described in Sect. 2. Below are discussed various aspects of circuit design, circuit–host interactions, and robustness to noise. Further discussions of synthetic circuit troubleshooting are available in recent reviews [202, 203].

8.1 Circuit Design: Topology and Parameters

Previous studies in synthetic biology have shown that circuit topology (the way in which the circuit components are wired together) and the parameters of the circuit components qualitatively affect circuit behavior. For example, the damping behavior of an oscillator depends on its topology: a repressilator that consists of two genes repressing each other shows damping after a limited number of cycles, but an oscillator with an amplifier loop shows sustained oscillations [121] (Sect. 3.1). In the case of a toggle switch, changing the relative expression levels of the two transcriptional repressors can change the behavior of the circuit from a switch into a timer [137]. Recently, a gene circuit in *E. coli* was converted from analog to digital function through adjusting plasmid copy numbers with a small-molecule inducer [171].

Computational modeling of different circuit topologies and sets of parameters has been used to identify network motifs that are necessary for, or enriched in, circuits that carry out desired functions. For example, the computational modeling of 1.6×10^8 three-node enzyme-based networks has been carried out to identify networks capable of adaptation (defined as an initial response to a stimulus, followed by return to the initial state even in presence of continued stimulus, a common property of sensory systems) [204]. The authors found that all 395 networks that display robust adaptation shared either one or two of the following motifs: a negative feedback loop or an incoherent feedforward loop (see

Sect. 8.3) [204]. In another study, Lim and colleagues used computational modeling to identify candidate network motifs that can induce cellular polarization, a vital property of living cells that is a prerequisite to behaviors such as directional migration [205]. These authors used the predictions to construct synthetic networks that trigger asymmetric accumulation of the membrane phospholipid PIP3 in the cell membrane of *S. cerevisiae* cells. Robust polarization was accomplished using networks that combined positive feedback with mutual inhibition between the synthetic proteins [205].

Another important consideration in circuit design includes level matching: in any circuit where the output of one part of the circuit serves as the input for another part, the concentration of the output must be in the correct range to trigger the desired response from the downstream part of the circuit. Level matching may be achieved in different ways, such as adjusting the copy number of a gene, or mutating its RBS to affect translation initiation rates and hence the level of protein expression [30].

Moreover, the time required for each layer of the circuit to process its input and produce an output must be considered [206]. The time delay due to biological processes such as transcription and translation affects the timing of its response, and may limit the complexity of circuits that can be implemented in living cells. For example, RNA-based cascades, which do not require translation, are faster than TF-based regulatory cascades [87]. Protein-based signal transduction pathways are faster still, but the ability to design them is still rudimentary despite advances made over the past few years [118].

The response time of a transcription regulator may be shortened by using a negative autoregulatory feedback loop, in which the TF represses its own promoter. This strategy relies on using a strong promoter, which drives TF expression at a high level when initially induced. When the TF has accumulated above a threshold it binds and represses its own promoter. This combination of a rapid initial build-up in TF concentration followed by negative autoregulation allows the TF to reach its steady-state level very quickly [207].

In order to design a circuit with the desired behavior, synthetic biologists take advantage of mathematical modeling. Commonly used mathematical modeling techniques include sensitivity analysis, which quantifies the effect of a parameter (or parameters) on overall circuit performance, and bifurcation analysis, which identifies the boundary in parameter space that separates circuits with qualitatively different behaviors, such as a stable steady state versus an oscillator [208]. Multiple software programs are available for modeling circuit behavior *in silico* (for reviews, see Refs [162, 208]). As circuits become more complex, however, reliance on mathematical models for their design will increase. For example, Purcell *et al.* recently described a platform for simulating synthetic circuits in the context of whole-cell models [209].

8.2
Circuit–Host and Circuit–Environment Interactions

Unwanted interactions between the circuit and its host cell may lead to circuit failure or cell death. For example, a component of the circuit may be toxic to the host cell and indeed, a recent study identified over 15 000 heterologous genes as toxic to *E. coli* [210]. Conversely, endogenous host genes or proteins may interfere with circuit function; for example, a native DNA sequence may bind a sTF and titrate it away from its target promoter. An increased knowledge of the host cell's regulatory and metabolic networks through the analysis of large-scale datasets will help synthetic biologists in designing circuits that will function as expected in a given host.

Even when none of the circuit components is toxic to the host, the designers must consider the metabolic load that the circuit imposes on the host cell in terms of requirement for ATP, RNAPs, ribosomes, and nucleotides [211, 212]. A metabolically expensive circuit may interfere with the cell's normal function and place selective pressure on the host to inactivate the synthetic device.

Certain circuit designs may help to alleviate metabolic load. Different ways of implementing the same type of function may have different energy requirements; for example, a memory device based on one-time recombinase-mediated DNA inversion is less metabolically demanding than a memory device that requires continued protein synthesis [153]. Moreover, regulatory RNA devices, which do not rely on the host's translation machinery, have a smaller metabolic footprint than TF-mediated regulation [87]. Another way to reduce the metabolic burden on any single cell is to distribute a large circuit among several strains of cells by placing a smaller subcircuit in each cell strain [172, 199] (Sect. 7). While powerful, distributed computing adds a level of complexity to the circuit design; the investigator must consider not only the effect of the circuit on each host strain but also the way in which two or more host strains interact to accomplish their task. For example, if cells of different strains communicate via a diffusible molecule, they must be in close physical proximity in order for the receiver cell to detect the signal [199]. If distributed computing is carried out by multiple engineered strains in a common culture medium [172], the computation may fail if one of the strains outcompetes the others.

8.3
Noise and Robustness

Living cells are noisy, due to differences among individual cells (or in a single cell over time) in parameters such as the number of TFs, mRNAs, and ribosomes; cell volume; state of cell cycle; and chromatin modifications [213]. Every component of a genetic circuit – whether synthetic or natural – experiences and propagates this noise to some extent, which in turn can be amplified by the noise from other components of the circuit. This will intensify the overall noise and hence may disrupt the performance of a genetic circuit. Moreover, the cell must cope with extrinsic noise, such as fluctuations in nutrient availability, temperature, pH, and other environmental variables. Notably, many practical applications (e.g., medicine or bioremediation) will require the engineered cell to function in a more unpredictable environment than a culture flask in the laboratory. Hence, a synthetic device must be able to minimize noise where possible and to maintain a desired function in the presence of unavoidable noise; this property is called *robustness* [214].

Over long time scales, evolution has selected for robustness in natural gene circuits [215]. However, in synthetic circuits, which lack an evolutionary tuning process, noise can blur the desired output and therefore is not usually considered a favorable factor. Without a mechanism for controlling noise, increasing the complexity of a circuit will likely increase the uncertainty of the output.

The noisy nature of biological systems is part of the reason why many synthetic circuits employ digital logic (see Sect. 5). Digital designs reduce the effect of noise by reducing the number of outputs of a given component of a circuit to two states: TRUE and FALSE. Ideally, the states are defined in such a way that passing from one state to another requires a significant change in concentration of input(s) of the component (e.g., an inducer or a TF), and therefore transition between states by cellular noise alone is very unlikely. Given that the number of distinguishable states that a circuit can possess can be considered a measure of complexity of the circuit, reduced complexity is a price that digital designs pay to overcome cellular noise. Digital gene circuits are generally more robust with respect to cellular noise than their corresponding analog circuits; however, they require more components than an analog circuit to achieve the same level of complexity, and hence they place a higher metabolic load on the cell [216].

Computational and experimental analysis of naturally occurring gene network motifs has revealed the robustness of certain motifs. For example, negative autoregulation, whereby a TF represses transcription of its own gene, helps to minimize noise in the TF expression levels: cells that initially have more TF will produce less of it, and cells that initially have less TF will produce more [217]. Another example of a motif that occurs commonly in natural gene regulatory networks is the feedforward loop (FFL), which consists of three nodes: node X regulates node Y, and X and Y regulate node Z. In coherent FFLs, the regulatory interaction (either activating or repressing) between X and Z is the same for both branches of the regulatory pathway; that is, if X activates Z directly, then X also activates Z via Y. In an incoherent FFL, the direction of the regulatory interaction is different for the two branches of the pathway so that, for example, X activates Z directly and inhibits Z via Y [218]. Mathematical modeling indicates that the coherent type 1 FFL, which consists entirely of positive interactions – X activates Y, and X and Y both activate Z – is the most robust to noise among the coherent FFLs [219]. The robustness of the coherent type 1 FFL may account for the large number of times this motif occurs in the gene regulatory networks of *E. coli* and *S. cerevisiae* [219]. Synthetic circuit design may benefit from using robust genetic motifs such as the coherent type 1 FFL that occur repeatedly in many natural systems, providing them with properties such as fast response time, robustness, or memory (see Ref. [218] for a detailed discussion of network motifs).

8.4 Evolution

In addition to making a system robust to short-term fluctuations in the levels of metabolites, circuit components, and so forth, synthetic biologists must also consider the impact of evolution on the function of the circuit over time. While evolution has been harnessed to improve circuit function and develop libraries of diverse components [220], it is also problematic in synthetic biology: circuits undergo point mutations, rearrangements, and deletions that disrupt their function, and may be lost from the host cell altogether after a number of generations. Repetitive sequences, high metabolic load, high plasmid copy number, and a quickly replicating host cell such as *E. coli* increase the probability of the circuit being lost or mutated [214].

The issues of circuit–host interactions, robustness and evolution are interrelated (e.g., a high metabolic load is likely to lead to evolutionary instability of the circuit), and may present trade-offs. For example, a reduction in protein level noise appears to come at the cost of higher energy requirements [221]. Synthetic biology is an iterative process: successes and failures in circuit construction teach about the mechanisms of gene network function, and these lessons can then be applied to the next round of circuit modeling and design. A deeper understanding of living systems, larger and better-characterized component libraries, and better models will help synthetic biology to advance the construction of predictable, robust circuits for practical applications.

9 Applications of Synthetic Circuits

Synthetic gene circuits used in applications must function reliably in adverse and often undetermined conditions. Consequently, applications push the field forward by challenging synthetic biologists to design robust, efficient circuits [7]. Examples of three types of practical applications – biosensors, therapeutics, and biomanufacturing – are presented below.

9.1 Biosensors

Biosensors translate the concentration of a specific analyte in the environment into a measurable signal by combining a living cell with a hardware platform that enables detection [222]. The output is usually a measurable signal, such as GFP or the blue pigment produced by the β-galactosidase enzyme. Biosensors may be judged by their selectivity for binding the target analyte, their sensitivity for low concentrations of the analyte, the input dynamic range of analyte concentrations, and the output signal-to-noise ratio. Many biosensors use transcription-factor-based circuits to measure the concentration of a toxic compound such as arsenite [196, 223, 224]. The goal is to use these biosensors in the developing world for testing water before its consumption. These field conditions make standardized measurements challenging, because the physiological state of the living cells in the biosensor is highly variable. To overcome this, Wackwitz *et al.* [224] developed a series of complementary cell strains that are tuned to respond to different arsenite concentrations by changing the strength of the RBS in front of the reporter gene. When calibrated and used in concert, these strains greatly improved arsenite detection. Rather than having the steady-state expression level of a reporter protein as their output, Hasty and colleagues [196] developed an oscillator whose frequency varies as a function of arsenite concentration; this decoupled the biosensor from the imaging conditions, such as beam power and exposure time. Further, the biosensor was an array of bacterial colonies in a microfluidic device that synchronized their oscillations at both the micro and macro-scale through diffusible and gaseous vapor molecules, respectively, enabling an accurate, high-strength signal that allows the biosensor to function as a handheld device.

In addition to cellular biosensors, synthetic biology has enabled the development of real-world deployable microbial sensors based on engineered phages. For example, Sample6 Technologies is commercializing a near-real-time microbial pathogen detection system based on engineered phages [225]. Bacteriophages ("bacteria eaters") are viruses that infect specific species and strains of bacteria [226]. By engineering phage libraries to express reporter genes during the infection of target bacteria, the presence or absence of pathogens can be detected. Reporter phages have been designed for the detection of clinically relevant pathogens such as *Mycobacterium tuberculosis*, *Staphylococcus aureus*, and food-borne pathogenic *E. coli* [227]. The advantages of engineered phage diagnostics include high sensitivity, high specificity, and short time-to-detection, which are shortcomings that are not addressable by conventional microbial detection approaches such as polymerase chain reaction (PCR), immunoassays, and culture. Moreover, phages specifically detect live bacteria that support phage replication, whereas some other rapid detection methods, such as PCR, do not distinguish between live and dead bacteria [227]. The tools of synthetic biology enable a rapid design–build–test cycle for new diagnostic systems, such as engineered phages, thus allowing assays to be built, tested and improved for real-world applications.

9.2 Therapeutic Applications

Synthetic biology holds tremendous promise in the area of drug development, diagnosis and treatment of disease [13]. Cell-based therapeutics constitute an emerging therapeutic class, and central to their efficacy are synthetic gene circuits that process environmental signals and actuate treatment, enabling appropriate selectivity, distribution, and dosage [13]. Microbial cells have been engineered to combat infectious diseases by either killing [228] or downregulating the growth [229] of pathogenic organisms. Both therapeutics rely on QS circuits (see Sect. 7) that trigger the release of active agents only in the presence of the target pathogen's quorum signal. Such targeted killing is a promising application for synthetic biology, as antibiotic resistance becomes more prevalent and broad-spectrum antibiotics lose favor. Microbes have also been engineered to invade cancer cells [166] using a synthetic gene circuit that triggers invasion only when the cells reach a critical density (detected via QS), or when they sense the hypoxic environment that is a hallmark of cancer cells with hyperactive metabolism. Furthermore, microbial-based therapies are an increasingly promising area of research as more information is gleaned about the microbiome and the role that symbiotic microorganisms play in human physiology.

Other studies have focused on developing mammalian cell-based therapeutics. Engineered cells may be enclosed in a semipermeable capsule, which allows them to be implanted in the body and to interface with human physiology, while at the same time being isolated from direct contact with the patient's tissues to prevent an immune response or metastasis. This microencapsulation technique was used in a system for regulating urate homeostasis to combat tumor lysis syndrome and gout, two diseases caused by abnormally high urate levels [24]. A synthetic gene circuit in the mammalian cells within the microcapsule utilizes a bacterial TF to sense urate concentrations and control the expression of an enzyme that degrades urate. This system restored urate homeostasis in a mouse model of acute hyperuricemia [24]. A similar study utilized a light-inducible gene circuit to control the production of glucagon-like peptide 1 and reduce glycemic excursions in type II diabetic mice [160].

A T-cell-based system represents another form of emerging mammalian cell-based therapeutic. In this case, the immune cells are removed from a patient, genetically engineered to target pathogens or cancer cells, and then transferred back into the patient. However, these therapies suffer from side effects such as hyperactivity and autoimmune off-target attacks, and would benefit from a synthetic control of proliferation. Two pioneering studies have demonstrated control over T-cell proliferation with synthetic RNA-mediated [230] or signaling protein-mediated [119] regulation of T-cell replication.

Synthetic gene circuits can also target therapeutics to different cell types via classifier circuits. Two examples demonstrate this approach to target cancer cells. In the first example, Nissim and Bar-Ziv [22] constructed a transcription-based AND gate that only expresses its therapeutic payload in cells in which both input promoters are on. In this design, two input promoters active in a specific cancer cell type are used to drive the expression of two fusion proteins, which form a transcription activator complex that activates the expression of a cytotoxic effector protein. Different input promoters may be chosen to target different cancer cell types. The use of two input promoters rather than one allows greater flexibility in the choice of input promoters, and produces a sharp activation threshold between premalignant and cancer cells, with the magnitude of the response increasing in more malignant cell lines. In order to avoid unwanted activation of the circuit in healthy cells, the level of effector gene expression may be tuned using point mutations in one of the fusion proteins to lower the efficiency of the synthetic TF complex formation. The digital logic circuit provides a precise and efficient way to target specific cell types while minimizing off-target effects on healthy cells [22]. In a second example, Xie *et al.* [23] built a classifier circuit that determines whether the levels of six different miRNAs match the reference profile of cancer cell miRNA expression, and based on this information controls expression of the BAX protein, which triggers apoptosis (see Sect. 5 and Fig. 7). Such deliverable cell-based classifiers utilizing synthetic gene circuits will play an important role in targeting therapeutics for difficult problems such as cancer or gene therapy.

In other examples of cancer therapy, oncolytic viruses were optimized for cancer targeting with improved specificity [156, 231]. Recently, synthetic constructs linking diphtheria-toxin gene expression under the control of the H19 promoter were tested in human patients [232].

Synthetic biology may also be applied to combating antibiotic-resistant bacteria, which present an emergent health threat worldwide [233]. The shortage of effective new antibiotics [234] necessitates the development of novel therapies. One possibility is to use synthetic biology to engineer phages (viruses that naturally infect bacteria) to combat antibiotic-resistant bacterial strains [235]. The engineered phages may be used to target biofilms, surface-associated bacterial communities encased in an extracellular matrix of polysaccharides and proteins that protects the bacteria from antibiotics and from the patient's immune system [236]. For example, T7 bacteriophage engineered to express Dispersin B, an enzyme that hydrolyzes a key biofilm component, efficiently disrupts *E. coli* biofilms [237].

Phages may also supplement antibiotic therapy. In one study, M13 phages were engineered to overexpress genes predicted to make host cells more vulnerable to antibiotics; these genes included *lexA3*, which inhibits the bacterial DNA damage response, leaving the cell vulnerable to DNA damage-inducing antibiotics; *csrA*, which inhibits biofilm formation; and *ompF*, which encodes a membrane protein through which antibiotics can enter the cell [238]. A combination of each engineered phage and antibiotic kills *E. coli* significantly more efficiently than either a combination of antibiotic and control (unmodified) phage or antibiotic alone [238]. Notably, M13 phage does not kill the host cell in absence of antibiotics, and hence it is less likely to give rise to bacterial resistance, which presents an important problem in antibacterial therapy [235]. In addition to resistance, many other challenges remain on the road to effective phage-based therapy, including potential side effects, the need for the phage to evade the patient's immune system, and a limited phage host range [235]. Nonetheless, the studies described above suggest that phage-based therapy is a promising strategy to pursue in combating antibiotic-resistant bacteria [235].

Finally, synthetic circuits can be utilized for the discovery and optimization of novel pharmaceuticals, such as new antituberculosis compounds [239] and novel classes of treatments for antibiotic-resistant-bacteria, such as lysins (phage-derived proteins that lyse bacteria) or bacteriocins (small peptides that kill bacteria by forming pores in their cell membranes) [233]. Synthetic biology also holds the potential for improving the yield of biopharmaceuticals and other valuable chemicals, as described below [233].

9.3 Synthetic Biology in Manufacturing

Synthetic biology has revolutionized industrial biotechnology by allowing engineers to optimize living cells rationally for the production of pharmaceuticals and other valuable chemicals [240]. Central to these efforts is repurposing biosynthetic genes from various organisms and tuning their level of gene expression. Often, gene expression is engineered to be static, so that it is constant throughout the course of production, or it is controlled with a simple gene circuit, in which an externally added small molecule induces a constitutively expressed TF to switch on the biosynthetic genes. However, these systems suffer from the metabolic burden that they exert on cells, slowing cellular growth and thus impairing productivity. Complex biosynthetic systems would benefit from the dynamic regulation enabled by synthetic gene circuits, wherein the expression of biosynthetic genes is adjusted based on a cell's physiological state, allowing adjustments based on the concentration of pathway intermediates or environmental conditions in the bioreactor such as nutrient availability, oxygen level, temperature, and cell-density [241]. Such dynamic regulation is akin to how cells naturally adjust their own physiology, and is expected to improve culture growth rates.

In an early example of dynamic regulation, the yield of lycopene in *E. coli* was improved by placing the expression of rate-limiting lycopene biosynthetic genes under the control of a TF that sensed excess glucose concentrations [242]. As a result, the pathway was only turned on when the cell had enough energy to continue growing, and this led to an 18-fold higher lycopene production. More recently, Zhang *et al.* engineered circuits to control biosynthetic pathways using an approach called *"dynamic sensor-regulator system"* (DSRS) [243]. These circuits are based on TFs that bind pathway intermediates, and affect synthetic promoters so that the expression of biosynthetic genes is activated only when needed, at the level needed. Such an approach to biosynthetic gene regulation is theoretically applicable to many different pathways, given the number of known metabolite-binding TFs, and could be further expanded with the use of synthetic aptamers evolved to bind different molecules of interest (see Sect. 2.5). An alternative method of improving the yield would be to tie the regulation of biosynthetic genes to the density of cells in the bioreactor in order to maximize cell growth before activating biosynthesis. This approach has been implemented with a QS circuit as an input to a toggle switch, and it has improved the yield of both recombinant proteins [139] and metabolic molecules [244, 245].

10 Conclusions

The past two decades of synthetic biology have led to a tremendous explosion in the design of ever more powerful and complex synthetic gene circuits. These circuits confer a greater degree of control over engineered biological systems than was ever possible before. However, significant challenges remain in the design and application of synthetic gene circuits due to incomplete biological knowledge, slow design–build–test cycles, nonpredictive *in-silico* models, challenges in designing circuits that can function reliably in different contexts from the laboratory environment in which they were engineered [246], and outstanding issues concerning orthogonality, modularity and noise (for reviews, see Refs [202, 203]). Yet, synthetic gene circuits clearly have much to offer society, and advances in fundamental circuit design and their implementation in biosensing, medicine, biomanufacturing and other application areas are expected to continue substantially during the next decade.

Note Added in Proof

While this chapter was under review, additional articles on synthetic gene circuits have been published. Chen *et al.* [247] programmed *E. coli* for inducible production of amyloid protein-based extracellular fibrils carrying affinity tags for inorganic nanoparticles. As proof of principle, the authors demonstrated inducible formation of a conductive biofilm composed of proteins decorated with gold nanoparticles. The system shows potential future use of synthetic biology for the formation of materials that combine the properties of living systems and inorganic matter to achieve novel functions [247]. Another recent study combined CRISPR gRNA and RNAi for tunable, multiplexed regulation of transcription in mammalian cells [248]. Notably, the study was the first to achieve inducible synthesis of gRNAs, opening up novel possibilities for inducible regulation of target genes by CRISPR [248]. In addition, a detailed protocol has been published for combining recombinase-based digital logic and memory in living cells [249], and two new reviews discuss features and potential applications of digital vs. analog synthetic gene circuits [250, 251].

References

1. Wang, Y.H., Wei, K.Y., Smolke, C.D. (2013) Synthetic biology: advancing the design of diverse genetic systems. *Annu. Rev. Chem. Biomol. Eng.*, **4**, 69–102.
2. Cheng, A.A., Lu, T.K. (2012) Synthetic biology: an emerging engineering discipline. *Annu. Rev. Biomed. Eng.*, **14**, 155–178.
3. Lu, T.K. (2010) Engineering scalable biological systems. *Bioeng. Bugs*, **1**, 378–384.
4. Lu, T.K., Khalil, A.S., Collins, J.J. (2009) Next-generation synthetic gene networks. *Nat. Biotechnol.*, **27**, 1139–1150.
5. Bashor, C.J., Horwitz, A.A., Peisajovich, S.G., Lim, W.A. (2010) Rewiring cells:

synthetic biology as a tool to interrogate the organizational principles of living systems. *Annu. Rev. Biophys.*, **39**, 515–537.

6. Pawson, T., Warner, N. (2007) Oncogenic re-wiring of cellular signaling pathways. *Oncogene*, **26**, 1268–1275.
7. Khalil, A.S., Collins, J.J. (2010) Synthetic biology: applications come of age. *Nat. Rev. Genet.*, **11**, 367–379.
8. Elowitz, M.B., Leibler, S. (2000) A synthetic oscillatory network of transcriptional regulators. *Nature*, **403**, 335–338.
9. Gardner, T.S., Cantor, C.R., Collins, J.J. (2000) Construction of a genetic toggle switch in *Escherichia coli*. *Nature*, **403**, 339–342.
10. Blount, B.A., Weenink, T., Ellis, T. (2012) Construction of synthetic regulatory networks in yeast. *FEBS Lett.*, **586**, 2112–2121.
11. Nevozhay, D., Zal, T., Balazsi, G. (2013) Transferring a synthetic gene circuit from yeast to mammalian cells. *Nat. Commun.*, **4**, 1451.
12. Nielsen, J., Larsson, C., van Maris, A., Pronk, J. (2013) Metabolic engineering of yeast for production of fuels and chemicals. *Curr. Opin. Biotechnol.*, **24**, 398–404.
13. Bacchus, W., Aubel, D., Fussenegger, M. (2013) Biomedically relevant circuit-design strategies in mammalian synthetic biology. *Mol. Syst. Biol.*, **9**, 691.
14. Lutz, R., Bujard, H. (1997) Independent and tight regulation of transcriptional units in *Escherichia coli* via the LacR/O, the TetR/O and AraC/I1-I2 regulatory elements. *Nucleic Acids Res.*, **25**, 1203–1210.
15. Pesci, E.C., Pearson, J.P., Seed, P.C., Iglewski, B.H. (1997) Regulation of las and rhl quorum sensing in *Pseudomonas aeruginosa*. *J. Bacteriol.*, **179**, 3127–3132.
16. Egland, K.A., Greenberg, E.P. (1999) Quorum sensing in *Vibrio fischeri*: elements of the luxl promoter. *Mol. Microbiol.*, **31**, 1197–1204.
17. Schleif, R. (2000) Regulation of the L-arabinose operon of *Escherichia coli*. *Trends Genet.*, **16**, 559–565.
18. Burrill, D.R., Inniss, M.C., Boyle, P.M., Silver, P.A. (2012) Synthetic memory circuits for tracking human cell fate. *Genes Dev.*, **26**, 1486–1497.
19. Shimizu-Sato, S., Huq, E., Tepperman, J.M., Quail, P.H. (2002) A light-switchable gene promoter system. *Nat. Biotechnol.*, **20**, 1041–1044.
20. Tabor, J.J., Levskaya, A., Voigt, C.A. (2011) Multichromatic control of gene expression in *Escherichia coli*. *J. Mol. Biol.*, **405**, 315–324.
21. Muller, K., Weber, W. (2013) Optogenetic tools for mammalian systems. *Mol. Biosyst.*, **9**, 596–608.
22. Nissim, L., Bar-Ziv, R.H. (2010) A tunable dual-promoter integrator for targeting of cancer cells. *Mol. Syst. Biol.*, **6**, 444.
23. Xie, Z., Wroblewska, L., Prochazka, L., Weiss, R. *et al.* (2011) Multi-input RNAi-based logic circuit for identification of specific cancer cells. *Science*, **333**, 1307–1311.
24. Kemmer, C., Gitzinger, M., Daoud-El Baba, M., Djonov, V. *et al.* (2010) Self-sufficient control of urate homeostasis in mice by a synthetic circuit. *Nat. Biotechnol.*, **28**, 355–360.
25. Cox, R.S. III, Surette, M.G., Elowitz, M.B. (2007) Programming gene expression with combinatorial promoters. *Mol. Syst. Biol.*, **3**, 145.
26. Khalil, A.S., Lu, T.K., Bashor, C.J., Ramirez, C.L. *et al.* (2012) A synthetic biology framework for programming eukaryotic transcription functions. *Cell*, **150**, 647–658.
27. Takahashi, M.K., Lucks, J.B. (2013) A modular strategy for engineering orthogonal chimeric RNA transcription regulators. *Nucleic Acids Res.*, **41**, 7577–7588.
28. Nielsen, A.A., Segall-Shapiro, T.H., Voigt, C.A. (2013) Advances in genetic circuit design: novel biochemistries, deep part mining, and precision gene expression. *Curr. Opin. Chem. Biol.*, **17**, 878–892.
29. Morbitzer, R., Romer, P., Boch, J., Lahaye, T. (2010) Regulation of selected genome loci using de novo-engineered transcription activator-like effector (TALE)-type transcription factors. *Proc. Natl Acad. Sci. USA*, **107**, 21617–21622.
30. Salis, H.M., Mirsky, E.A., Voigt, C.A. (2009) Automated design of synthetic ribosome binding sites to control protein expression. *Nat. Biotechnol.*, **27**, 946–950.
31. Blount, B.A., Weenink, T., Vasylechko, S., Ellis, T. (2012) Rational diversification of a promoter providing fine-tuned expression and orthogonal regulation for synthetic biology. *PLoS One*, **7**, e33279.

32. Sohka, T., Heins, R.A., Phelan, R.M., Greisler, J.M. *et al.* (2009) An externally tunable bacterial band-pass filter. *Proc. Natl Acad. Sci. USA*, **106**, 10135–10140.
33. Beerli, R.R., Barbas, C.F. III, (2002) Engineering polydactyl zinc-finger transcription factors. *Nat. Biotechnol.*, **20**, 135–141.
34. Garg, A., Lohmueller, J.J., Silver, P.A., Armel, T.Z. (2012) Engineering synthetic TAL effectors with orthogonal target sites. *Nucleic Acids Res.*, **40**, 7584–7595.
35. Purcell, O., Peccoud, J., Lu, T.K. (2013) Rule-based design of synthetic transcription factors in eukaryotes. *ACS Synth. Biol.*; doi: 10.1021/sb400134k.
36. Kaseniit, K.E., Perli, S.D., Lu, T.K. (2011) Designing extensible protein–DNA interactions for synthetic biology. Proceedings of the 2011 IEEE Biomedical Circuits and Systems Conference (BioCAS), pp. 349–352.
37. Nevins, J.R. (1991) Transcriptional activation by viral regulatory proteins. *Trends Biochem. Sci.*, **16**, 435–439.
38. Gossen, M., Bujard, H. (1992) Tight control of gene expression in mammalian cells by tetracycline-responsive promoters. *Proc. Natl Acad. Sci. USA*, **89**, 5547–5551.
39. Konermann, S., Brigham, M.D., Trevino, A.E., Hsu, P.D. *et al.* (2013) Optical control of mammalian endogenous transcription and epigenetic states. *Nature*, **500**, 472–476.
40. Topp, S., Gallivan, J.P. (2010) Emerging applications of riboswitches in chemical biology. *ACS Chem. Biol.*, **5**, 139–148.
41. Grilly, C., Stricker, J., Pang, W.L., Bennett, M.R. *et al.* (2007) A synthetic gene network for tuning protein degradation in *Saccharomyces cerevisiae*. *Mol. Syst. Biol.*, **3**, 127.
42. Bashor, C.J., Helman, N.C., Yan, S., Lim, W.A. (2008) Using engineered scaffold interactions to reshape MAP kinase pathway signaling dynamics. *Science*, **319**, 1539–1543.
43. Farzadfard, F., Perli, S.D., Lu, T.K. (2013) Tunable and multifunctional eukaryotic transcription factors based on CRISPR/Cas. *ACS Synth. Biol.*, **2**, 604–613.
44. van Hijum, S.A., Medema, M.H., Kuipers, O.P. (2009) Mechanisms and evolution of control logic in prokaryotic transcriptional regulation. *Microbiol. Mol. Biol. Rev.*, **73**, 481–509.
45. Courey, A.J., Jia, S. (2001) Transcriptional repression: the long and the short of it. *Genes Dev.*, **15**, 2786–2796.
46. Hahn, S., Young, E.T. (2011) Transcriptional regulation in *Saccharomyces cerevisiae*: transcription factor regulation and function, mechanisms of initiation, and roles of activators and coactivators. *Genetics*, **189**, 705–736.
47. Harmston, N., Lenhard, B. (2013) Chromatin and epigenetic features of long-range gene regulation. *Nucleic Acids Res.*, **41**, 7185–7199.
48. Gaszner, M., Felsenfeld, G. (2006) Insulators: exploiting transcriptional and epigenetic mechanisms. *Nat. Rev. Genet.*, **7**, 703–713.
49. Rao, S., Procko, E., Shannon, M.F. (2001) Chromatin remodeling, measured by a novel real-time polymerase chain reaction assay, across the proximal promoter region of the IL-2 gene. *J. Immunol.*, **167**, 4494–4503.
50. Lander, E.S., Linton, L.M., Birren, B., Nusbaum, C. *et al.* (2001) Initial sequencing and analysis of the human genome. *Nature*, **409**, 860–921.
51. Blattner, F.R., Plunkett, G. III, Bloch, C.A., Perna, N.T. *et al.* (1997) The complete genome sequence of *Escherichia coli* K-12. *Science*, **277**, 1453–1462.
52. Mewes, H.W., Albermann, K., Bahr, M., Frishman, D. *et al.* (1997) Overview of the yeast genome. *Nature*, **387**, 7–65.
53. Sadowski, I., Ma, J., Triezenberg, S., Ptashne, M. (1988) GAL4-VP16 is an unusually potent transcriptional activator. *Nature*, **335**, 563–564.
54. Urlinger, S., Baron, U., Thellmann, M., Hasan, M.T. *et al.* (2000) Exploring the sequence space for tetracycline-dependent transcriptional activators: novel mutations yield expanded range and sensitivity. *Proc. Natl Acad. Sci. USA*, **97**, 7963–7968.
55. Carson-Jurica, M.A., Schrader, W.T., O'Malley, B.W. (1990) Steroid receptor family: structure and functions. *Endocr. Rev.*, **11**, 201–220.
56. Christopherson, K.S., Mark, M.R., Bajaj, V., Godowski, P.J. (1992) Ecdysteroid-dependent regulation of genes in mammalian cells by a *Drosophila* ecdysone

receptor and chimeric transactivators. *Proc. Natl Acad. Sci. USA*, **89**, 6314–6318.
57. Wang, Y., O'Malley, B.W. Jr. Tsai, S.Y., O'Malley, B.W. (1994) A regulatory system for use in gene transfer. *Proc. Natl Acad. Sci. USA*, **91**, 8180–8184.
58. Braselmann, S., Graninger, P., Busslinger, M. (1993) A selective transcriptional induction system for mammalian cells based on Gal4-estrogen receptor fusion proteins. *Proc. Natl Acad. Sci. USA*, **90**, 1657–1661.
59. Louvion, J.F., Havaux-Copf, B., Picard, D. (1993) Fusion of GAL4-VP16 to a steroid-binding domain provides a tool for gratuitous induction of galactose-responsive genes in yeast. *Gene*, **131**, 129–134.
60. Weber, W., Fux, C., Daoud-el Baba, M., Keller, B. *et al.* (2002) Macrolide-based transgene control in mammalian cells and mice. *Nat. Biotechnol.*, **20**, 901–907.
61. Fussenegger, M., Morris, R.P., Fux, C., Rimann, M. *et al.* (2000) Streptogramin-based gene regulation systems for mammalian cells. *Nat. Biotechnol.*, **18**, 1203–1208.
62. Hartenbach, S., Daoud-El Baba, M., Weber, W., Fussenegger, M. (2007) An engineered L-arginine sensor of *Chlamydia pneumoniae* enables arginine-adjustable transcription control in mammalian cells and mice. *Nucleic Acids Res.*, **35**, e136.
63. Weber, W., Bacchus, W., Gruber, F., Hamberger, M. *et al.* (2007) A novel vector platform for vitamin H-inducible transgene expression in mammalian cells. *J. Biotechnol.*, **131**, 150–158.
64. Wieland, M., Fussenegger, M. (2012) Engineering molecular circuits using synthetic biology in mammalian cells. *Annu. Rev. Chem. Biomol. Eng.*, **3**, 209–234.
65. Pabo, C.O., Peisach, E., Grant, R.A. (2001) Design and selection of novel Cys2His2 zinc finger proteins. *Annu. Rev. Biochem.*, **70**, 313–340.
66. Maeder, M.L., Thibodeau-Beganny, S., Sander, J.D., Voytas, D.F. *et al.* (2009) Oligomerized pool engineering (OPEN): an 'open-source' protocol for making customized zinc-finger arrays. *Nat. Protoc.*, **4**, 1471–1501.
67. Pavletich, N.P., Pabo, C.O. (1991) Zinc finger-DNA recognition: crystal structure of a Zif268–DNA complex at 2.1 Å. *Science*, **252**, 809–817.
68. Greisman, H.A., Pabo, C.O. (1997) A general strategy for selecting high-affinity zinc finger proteins for diverse DNA target sites. *Science*, **275**, 657–661.
69. Beerli, R.R., Dreier, B., Barbas, C.F. III, (2000) Positive and negative regulation of endogenous genes by designed transcription factors. *Proc. Natl Acad. Sci. USA*, **97**, 1495–1500.
70. Lohmueller, J.J., Armel, T.Z., Silver, P.A. (2012) A tunable zinc finger-based framework for Boolean logic computation in mammalian cells. *Nucleic Acids Res.*, **40**, 5180–5187.
71. Sander, J.D., Maeder, M.L., Reyon, D., Voytas, D.F. *et al.* (2010) ZiFiT (Zinc Finger Targeter): an updated zinc finger engineering tool. *Nucleic Acids Res.*, **38**, W462–W468.
72. Fu, F., Sander, J.D., Maeder, M., Thibodeau-Beganny, S. *et al.* (2009) Zinc Finger Database (ZiFDB): a repository for information on C2H2 zinc fingers and engineered zinc-finger arrays. *Nucleic Acids Res.*, **37**, D279–D283.
73. Beerli, R.R., Schopfer, U., Dreier, B., Barbas, C.F. III, (2000) Chemically regulated zinc finger transcription factors. *J. Biol. Chem.*, **275**, 32617–32627.
74. Magnenat, L., Schwimmer, L.J., Barbas, C.F. III, (2008) Drug-inducible and simultaneous regulation of endogenous genes by single-chain nuclear receptor-based zinc-finger transcription factor gene switches. *Gene Ther.*, **15**, 1223–1232.
75. Ramirez, C.L., Foley, J.E., Wright, D.A., Muller-Lerch, F. *et al.* (2008) Unexpected failure rates for modular assembly of engineered zinc fingers. *Nat. Methods*, **5**, 374–375.
76. Zhang, F., Cong, L., Lodato, S., Kosuri, S. *et al.* (2011) Efficient construction of sequence-specific TAL effectors for modulating mammalian transcription. *Nat. Biotechnol.*, **29**, 149–153.
77. Boch, J., Scholze, H., Schornack, S., Landgraf, A. *et al.* (2009) Breaking the code of DNA binding specificity of TAL-type III effectors. *Science*, **326**, 1509–1512.
78. Mercer, A.C., Gaj, T., Sirk, S.J., Lamb, B.M. *et al.* (2013) Regulation of endogenous human gene expression by ligand-inducible TALE transcription factors. *ACS Synth. Biol.*; doi: 10.1021/sb400114p.

79. Maeder, M.L., Angstman, J.F., Richardson, M.E., Linder, S.J. *et al.* (2013) Targeted DNA demethylation and activation of endogenous genes using programmable TALE-TET1 fusion proteins. *Nat. Biotechnol.*, **31**, 1137–1142.
80. Mendenhall, E.M., Williamson, K.E., Reyon, D., Zou, J.Y. *et al.* (2013) Locus-specific editing of histone modifications at endogenous enhancers. *Nat. Biotechnol.*, **31**, 1133–1136.
81. Qi, L.S., Larson, M.H., Gilbert, L.A., Doudna, J.A. *et al.* (2013) Repurposing CRISPR as an RNA-guided platform for sequence-specific control of gene expression. *Cell*, **152**, 1173–1183.
82. Jinek, M., Chylinski, K., Fonfara, I., Hauer, M. *et al.* (2012) A programmable dual-RNA-guided DNA endonuclease in adaptive bacterial immunity. *Science*, **337**, 816–821.
83. Jiang, W., Bikard, D., Cox, D., Zhang, F. *et al.* (2013) RNA-guided editing of bacterial genomes using CRISPR-Cas systems. *Nat. Biotechnol.*, **31**, 233–239.
84. Mali, P., Yang, L., Esvelt, K.M., Aach, J. *et al.* (2013) RNA-guided human genome engineering via Cas9. *Science*, **339**, 823–826.
85. Cong, L., Ran, F.A., Cox, D., Lin, S. *et al.* (2013) Multiplex genome engineering using CRISPR/Cas systems. *Science*, **339**, 819–823.
86. Gilbert, L.A., Larson, M.H., Morsut, L., Liu, Z. *et al.* (2013) CRISPR-mediated modular RNA-guided regulation of transcription in eukaryotes. *Cell*, **154**, 442–451.
87. Liang, J.C., Bloom, R.J., Smolke, C.D. (2011) Engineering biological systems with synthetic RNA molecules. *Mol. Cell*, **43**, 915–926.
88. Chappell, J., Takahashi, M.K., Meyer, S., Loughrey, D. *et al.* (2013) The centrality of RNA for engineering gene expression. *Biotechnol. J.*, **8**, 1379–1395.
89. Chen, X., Ellington, A.D. (2009) Design principles for ligand-sensing, conformation-switching ribozymes. *PLoS Comput. Biol.*, **5**, e1000620.
90. Walton, S.P., Wu, M., Gredell, J.A., Chan, C. (2010) Designing highly active siRNAs for therapeutic applications. *FEBS J.*, **277**, 4806–4813.
91. Rinaudo, K., Bleris, L., Maddamsetti, R., Subramanian, S. *et al.* (2007) A universal RNAi-based logic evaluator that operates in mammalian cells. *Nat. Biotechnol.*, **25**, 795–801.
92. Leisner, M., Bleris, L., Lohmueller, J., Xie, Z. *et al.* (2010) Rationally designed logic integration of regulatory signals in mammalian cells. *Nat. Nanotechnol.*, **5**, 666–670.
93. Xu, Z.L., Mizuguchi, H., Ishii-Watabe, A., Uchida, E. *et al.* (2001) Optimization of transcriptional regulatory elements for constructing plasmid vectors. *Gene*, **272**, 149–156.
94. Helfer, S., Schott, J., Stoecklin, G., Forstemann, K. (2012) AU-rich element-mediated mRNA decay can occur independently of the miRNA machinery in mouse embryonic fibroblasts and *Drosophila* S2-cells. *PLoS One*, **7**, e28907.
95. Isaacs, F.J., Dwyer, D.J., Ding, C., Pervouchine, D.D. *et al.* (2004) Engineered riboregulators enable post-transcriptional control of gene expression. *Nat. Biotechnol.*, **22**, 841–847.
96. Lucks, J.B., Qi, L., Mutalik, V.K., Wang, D. *et al.* (2011) Versatile RNA-sensing transcriptional regulators for engineering genetic networks. *Proc. Natl Acad. Sci. USA*, **108**, 8617–8622.
97. Liu, C.C., Qi, L., Lucks, J.B., Segall-Shapiro, T.H. *et al.* (2012) An adaptor from translational to transcriptional control enables predictable assembly of complex regulation. *Nat. Methods*, **9**, 1088–1094.
98. Barrick, J.E., Breaker, R.R. (2007) The distributions, mechanisms, and structures of metabolite-binding riboswitches. *Genome Biol.*, **8**, R239.
99. Werstuck, G., Green, M.R. (1998) Controlling gene expression in living cells through small molecule–RNA interactions. *Science*, **282**, 296–298.
100. Harvey, I., Garneau, P., Pelletier, J. (2002) Inhibition of translation by RNA–small molecule interactions. *RNA*, **8**, 452–463.
101. Ellington, A.D., Szostak, J.W. (1990) In vitro selection of RNA molecules that bind specific ligands. *Nature*, **346**, 818–822.
102. Tuerk, C., Gold, L. (1990) Systematic evolution of ligands by exponential enrichment: RNA ligands to bacteriophage T4 DNA polymerase. *Science*, **249**, 505–510.
103. Topp, S., Gallivan, J.P. (2008) Random walks to synthetic riboswitches – a high-throughput selection based on cell motility.

ChemBioChem: Eur. J. Chem. Biol., **9**, 210–213.

104. Lynch, S.A., Gallivan, J.P. (2009) A flow cytometry-based screen for synthetic riboswitches. *Nucleic Acids Res.*, **37**, 184–192.
105. Desai, S.K., Gallivan, J.P. (2004) Genetic screens and selections for small molecules based on a synthetic riboswitch that activates protein translation. *J. Am. Chem. Soc.*, **126**, 13247–13254.
106. Suess, B., Hanson, S., Berens, C., Fink, B. *et al.* (2003) Conditional gene expression by controlling translation with tetracycline-binding aptamers. *Nucleic Acids Res.*, **31**, 1853–1858.
107. Grate, D., Wilson, C. (2001) Inducible regulation of the *S. cerevisiae* cell cycle mediated by an RNA aptamer–ligand complex. *Bioorg. Med. Chem.*, **9**, 2565–2570.
108. Topp, S., Gallivan, J.P. (2007) Guiding bacteria with small molecules and RNA. *J. Am. Chem. Soc.*, **129**, 6807–6811.
109. Thompson, K.M., Syrett, H.A., Knudsen, S.M., Ellington, A.D. (2002) Group I aptazymes as genetic regulatory switches. *BMC Biotech.*, **2**, 21.
110. Win, M.N., Smolke, C.D. (2007) A modular and extensible RNA-based gene-regulatory platform for engineering cellular function. *Proc. Natl Acad. Sci. USA*, **104**, 14283–14288.
111. Auslander, S., Ketzer, P., Hartig, J.S. (2010) A ligand-dependent hammerhead ribozyme switch for controlling mammalian gene expression. *Mol. Biosyst.*, **6**, 807–814.
112. Wieland, M., Fussenegger, M. (2010) Ligand-dependent regulatory RNA parts for synthetic biology in eukaryotes. *Curr. Opin. Biotechnol.*, **21**, 760–765.
113. Burgess-Beusse, B., Farrell, C., Gaszner, M., Litt, M. *et al.* (2002) The insulation of genes from external enhancers and silencing chromatin. *Proc. Natl Acad. Sci. USA*, **99** (Suppl. 4), 16433–16437.
114. Qi, L., Haurwitz, R.E., Shao, W., Doudna, J.A. *et al.* (2012) RNA processing enables predictable programming of gene expression. *Nat. Biotechnol.*, **30**, 1002–1006.
115. Li, X., Zhao, X., Fang, Y., Jiang, X. *et al.* (1998) Generation of destabilized green fluorescent protein as a transcription reporter. *J. Biol. Chem.*, **273**, 34970–34975.
116. Gottesman, S., Roche, E., Zhou, Y., Sauer, R.T. (1998) The ClpXP and ClpAP proteases degrade proteins with carboxy-terminal peptide tails added by the SsrA-tagging system. *Genes Dev.*, **12**, 1338–1347.
117. Egeler, E.L., Urner, L.M., Rakhit, R., Liu, C.W. *et al.* (2011) Ligand-switchable substrates for a ubiquitin-proteasome system. *J. Biol. Chem.*, **286**, 31328–31336.
118. Lim, W.A. (2010) Designing customized cell signalling circuits. *Nat. Rev. Mol. Cell Biol.*, **11**, 393–403.
119. Wei, P., Wong, W.W., Park, J.S., Corcoran, E.E. *et al.* (2012) Bacterial virulence proteins as tools to rewire kinase pathways in yeast and immune cells. *Nature*, **488**, 384–388.
120. Garg, R.K., Dixit, A., Yadav, P. (2008) *Basic Electronics*, Firewall Media.
121. Purcell, O., Savery, N.J., Grierson, C.S., di Bernardo, M. (2010) A comparative analysis of synthetic genetic oscillators. *J. R. Soc. Interface/R. Soc.*, **7**, 1503–1524.
122. Stricker, J., Cookson, S., Bennett, M.R., Mather, W.H. *et al.* (2008) A fast, robust and tunable synthetic gene oscillator. *Nature*, **456**, 516–519.
123. Fung, E., Wong, W.W., Suen, J.K., Bulter, T. *et al.* (2005) A synthetic gene-metabolic oscillator. *Nature*, **435**, 118–122.
124. Atkinson, M.R., Savageau, M.A., Myers, J.T., Ninfa, A.J. (2003) Development of genetic circuitry exhibiting toggle switch or oscillatory behavior in *Escherichia coli*. *Cell*, **113**, 597–607.
125. Prindle, A., Selimkhanov, J., Danino, T., Samayoa, P. *et al.* (2012) Genetic circuits in *Salmonella typhimurium*. *ACS Synth. Biol.*, **1**, 458–464.
126. Swinburne, I.A., Miguez, D.G., Landgraf, D., Silver, P.A. (2008) Intron length increases oscillatory periods of gene expression in animal cells. *Genes Dev.*, **22**, 2342–2346.
127. Tigges, M., Marquez-Lago, T.T., Stelling, J., Fussenegger, M. (2009) A tunable synthetic mammalian oscillator. *Nature*, **457**, 309–312.
128. Tigges, M., Denervaud, N., Greber, D., Stelling, J. *et al.* (2010) A synthetic low-frequency mammalian oscillator. *Nucleic Acids Res.*, **38**, 2702–2711.
129. Toettcher, J.E., Mock, C., Batchelor, E., Loewer, A. *et al.* (2010) A synthetic-natural

hybrid oscillator in human cells. *Proc. Natl Acad. Sci. USA*, **107**, 17047–17052.
130. Purcell, O., di Bernardo, M., Grierson, C.S., Savery, N.J. (2011) A multi-functional synthetic gene network: a frequency multiplier, oscillator and switch. *PLoS One*, **6**, e16140.
131. Del Vecchio, D., Ninfa, A.J., Sontag, E.D. (2008) Modular cell biology: retroactivity and insulation. *Mol. Syst. Biol.*, **4**, 161.
132. Jayanthi, S., Nilgiriwala, K.S., Del Vecchio, D. (2013) Retroactivity controls the temporal dynamics of gene transcription. *ACS Synth. Biol.*, **2**, 431–441.
133. Franco, E., Friedrichs, E., Kim, J., Jungmann, R. *et al.* (2011) Timing molecular motion and production with a synthetic transcriptional clock. *Proc. Natl Acad. Sci. USA*, **108**, E784–E793.
134. Kramer, B.P., Viretta, A.U., Daoud-El-Baba, M., Aubel, D. *et al.* (2004) An engineered epigenetic transgene switch in mammalian cells. *Nat. Biotechnol.*, **22**, 867–870.
135. Deans, T.L., Cantor, C.R., Collins, J.J. (2007) A tunable genetic switch based on RNAi and repressor proteins for regulating gene expression in mammalian cells. *Cell*, **130**, 363–372.
136. Muller, K., Engesser, R., Metzger, S., Schulz, S. *et al.* (2013) A red/far-red light-responsive bi-stable toggle switch to control gene expression in mammalian cells. *Nucleic Acids Res.*, **41**, e77.
137. Ellis, T., Wang, X., Collins, J.J. (2009) Diversity-based, model-guided construction of synthetic gene networks with predicted functions. *Nat. Biotechnol.*, **27**, 465–471.
138. Lou, C., Liu, X., Ni, M., Huang, Y. *et al.* (2010) Synthesizing a novel genetic sequential logic circuit: a push-on push-off switch. *Mol. Syst. Biol.*, **6**, 350.
139. Kobayashi, H., Kaern, M., Araki, M., Chung, K. *et al.* (2004) Programmable cells: interfacing natural and engineered gene networks. *Proc. Natl Acad. Sci. USA*, **101**, 8414–8419.
140. Wu, M., Su, R.Q., Li, X., Ellis, T. *et al.* (2013) Engineering of regulated stochastic cell fate determination. *Proc. Natl Acad. Sci. USA*, **110**, 10610–10615.
141. Hooshangi, S., Thiberge, S., Weiss, R. (2005) Ultrasensitivity and noise propagation in a synthetic transcriptional cascade. *Proc. Natl Acad. Sci. USA*, **102**, 3581–3586.
142. Shenoi, B.A. (2006) *Introduction to Digital Signal Processing and Filter Design*, John Wiley & Sons, Inc.
143. Greber, D., Fussenegger, M. (2010) An engineered mammalian band-pass network. *Nucleic Acids Res.*, **38**, e174.
144. Ham, T.S., Lee, S.K., Keasling, J.D., Arkin, A.P. (2008) Design and construction of a double inversion recombination switch for heritable sequential genetic memory. *PLoS One*, **3**, e2815.
145. Inniss, M.C., Silver, P.A. (2013) Building synthetic memory. *Curr. Biol.*, **23**, R812–R816.
146. Ferrell, J.E. Jr. (2002) Self-perpetuating states in signal transduction: positive feedback, double-negative feedback and bistability. *Curr. Opin. Cell Biol.*, **14**, 140–148.
147. Chickarmane, V., Troein, C., Nuber, U.A., Sauro, H.M. *et al.* (2006) Transcriptional dynamics of the embryonic stem cell switch. *PLoS Comput. Biol.*, **2**, e123.
148. Ng, R.K., Gurdon, J.B. (2008) Epigenetic inheritance of cell differentiation status. *Cell Cycle*, **7**, 1173–1177.
149. Haynes, K.A., Silver, P.A. (2011) Synthetic reversal of epigenetic silencing. *J. Biol. Chem.*, **286**, 27176–27182.
150. Ajo-Franklin, C.M., Drubin, D.A., Eskin, J.A., Gee, E.P. *et al.* (2007) Rational design of memory in eukaryotic cells. *Genes Dev.*, **21**, 2271–2276.
151. Friedland, A.E., Lu, T.K., Wang, X., Shi, D. *et al.* (2009) Synthetic gene networks that count. *Science*, **324**, 1199–1202.
152. Ham, T.S., Lee, S.K., Keasling, J.D., Arkin, A.P. (2006) A tightly regulated inducible expression system utilizing the fim inversion recombination switch. *Biotechnol. Bioeng.*, **94**, 1–4.
153. Siuti, P., Yazbek, J., Lu, T.K. (2013) Synthetic circuits integrating logic and memory in living cells. *Nat. Biotechnol.*, **31**, 448–452.
154. Ghosh, P., Pannunzio, N.R., Hatfull, G.F. (2005) Synapsis in phage Bxb1 integration: selection mechanism for the correct pair of recombination sites. *J. Mol. Biol.*, **349**, 331–348.
155. Bonnet, J., Subsoontorn, P., Endy, D. (2012) Rewritable digital data storage in live cells via engineered control of recombination directionality. *Proc. Natl Acad. Sci. USA*, **109**, 8884–8889.

156. Dorer, D.E., Nettelbeck, D.M. (2009) Targeting cancer by transcriptional control in cancer gene therapy and viral oncolysis. *Adv. Drug Delivery Rev.*, **61**, 554–571.
157. Weber, W., Fussenegger, M. (2012) Emerging biomedical applications of synthetic biology. *Nat. Rev. Genet.*, **13**, 21–35.
158. Folcher, M., Fussenegger, M. (2012) Synthetic biology advancing clinical applications. *Curr. Opin. Chem. Biol.*, **16**, 345–354.
159. Weber, W., Luzi, S., Karlsson, M., Sanchez-Bustamante, C.D. *et al.* (2009) A synthetic mammalian electro-genetic transcription circuit. *Nucleic Acids Res.*, **37**, e33.
160. Ye, H., Daoud-El Baba, M., Peng, R.W., Fussenegger, M. (2011) A synthetic optogenetic transcription device enhances blood-glucose homeostasis in mice. *Science*, **332**, 1565–1568.
161. Bacchus, W., Fussenegger, M. (2012) The use of light for engineered control and reprogramming of cellular functions. *Curr. Opin. Biotechnol.*, **23**, 695–702.
162. Purnick, P.E., Weiss, R. (2009) The second wave of synthetic biology: from modules to systems. *Nat. Rev. Mol. Cell Biol.*, **10**, 410–422.
163. Kramer, B.P., Fischer, C., Fussenegger, M. (2004) BioLogic gates enable logical transcription control in mammalian cells. *Biotechnol. Bioeng.*, **87**, 478–484.
164. Lienert, F., Torella, J.P., Chen, J.H., Norsworthy, M. *et al.* (2013) Two- and three-input TALE-based AND logic computation in embryonic stem cells. *Nucleic Acids Res.*, **41**, 9967–9975.
165. Anderson, J.C., Voigt, C.A., Arkin, A.P. (2007) Environmental signal integration by a modular AND gate. *Mol. Syst. Biol.*, **3**, 133.
166. Anderson, J.C., Clarke, E.J., Arkin, A.P., Voigt, C.A. (2006) Environmentally controlled invasion of cancer cells by engineered bacteria. *J. Mol. Biol.*, **355**, 619–627.
167. Auslander, S., Auslander, D., Muller, M., Wieland, M. *et al.* (2012) Programmable single-cell mammalian biocomputers. *Nature*, **487**, 123–127.
168. Kahan-Hanum, M., Douek, Y., Adar, R., Shapiro, E. (2013) A library of programmable DNAzymes that operate in a cellular environment. *Sci. Rep.*, **3**, 1535.
169. Moon, T.S., Lou, C., Tamsir, A., Stanton, B.C. *et al.* (2012) Genetic programs constructed from layered logic gates in single cells. *Nature*, **491**, 249–253.
170. Bonnet, J., Yin, P., Ortiz, M.E., Subsoontorn, P. *et al.* (2013) Amplifying genetic logic gates. *Science*, **340**, 599–603.
171. Daniel, R., Rubens, J.R., Sarpeshkar, R., Lu, T.K. (2013) Synthetic analog computation in living cells. *Nature*, **497**, 619–623.
172. Regot, S., Macia, J., Conde, N., Furukawa, K. *et al.* (2011) Distributed biological computation with multicellular engineered networks. *Nature*, **469**, 207–211.
173. Ferrell, J.E. Jr. (2009) Signaling motifs and Weber's law. *Mol. Cell*, **36**, 724–727.
174. Kaplan, S., Bren, A., Zaslaver, A., Dekel, E. *et al.* (2008) Diverse two-dimensional input functions control bacterial sugar genes. *Mol. Cell*, **29**, 786–792.
175. Tu, Y., Shimizu, T.S., Berg, H.C. (2008) Modeling the chemotactic response of *Escherichia coli* to time-varying stimuli. *Proc. Natl Acad. Sci. USA*, **105**, 14855–14860.
176. Nevozhay, D., Adams, R.M., Murphy, K.F., Josic, K. *et al.* (2009) Negative autoregulation linearizes the dose-response and suppresses the heterogeneity of gene expression. *Proc. Natl Acad. Sci. USA*, **106**, 5123–5128.
177. Koshland, D.E. Jr. (1996) The structural basis of negative cooperativity: receptors and enzymes. *Curr. Opin. Struct. Biol.*, **6**, 757–761.
178. Madar, D., Dekel, E., Bren, A., Alon, U. (2011) Negative auto-regulation increases the input dynamic-range of the arabinose system of *Escherichia coli*. *BMC Syst. Biol.*, **5**, 111.
179. Novotny, C.P., Englesberg, E. (1966) The L-arabinose permease system in *Escherichia coli* B/r. *Biochim. Biophys. Acta*, **117**, 217–230.
180. Schleif, R. (1969) An L-arabinose binding protein and arabinose permeation in *Escherichia coli*. *J. Mol. Biol.*, **46**, 185–196.
181. Doyle, M.E., Brown, C., Hogg, R.W., Helling, R.B. (1972) Induction of the ara operon of *Escherichia coli* B-r. *J. Bacteriol.*, **110**, 56–65.
182. Lobell, R.B., Schleif, R.F. (1990) DNA looping and unlooping by AraC protein. *Science*, **250**, 528–532.

183. Seabold, R.R., Schleif, R.F. (1998) Apo-AraC actively seeks to loop. *J. Mol. Biol.*, **278**, 529–538.
184. Bacchus, W., Fussenegger, M. (2013) Engineering of synthetic intercellular communication systems. *Metab. Eng.*, **16**, 33–41.
185. Mee, M.T., Wang, H.H. (2012) Engineering ecosystems and synthetic ecologies. *Mol. Biosyst.*, **8**, 2470–2483.
186. Kurjan, J. (1993) The pheromone response pathway in *Saccharomyces cerevisiae*. *Annu. Rev. Genet.*, **27**, 147–179.
187. Sprinzak, D., Lakhanpal, A., Lebon, L., Santat, L.A. *et al.* (2010) Cis-interactions between Notch and Delta generate mutually exclusive signalling states. *Nature*, **465**, 86–90.
188. Lum, L., Beachy, P.A. (2004) The Hedgehog response network: sensors, switches, and routers. *Science*, **304**, 1755–1759.
189. Attisano, L., Wrana, J.L. (2002) Signal transduction by the TGF-beta superfamily. *Science*, **296**, 1646–1647.
190. Keshet, Y., Seger, R. (2010) The MAP kinase signaling cascades: a system of hundreds of components regulates a diverse array of physiological functions. *Methods Mol. Biol.*, **661**, 3–38.
191. Logan, C.Y., Nusse, R. (2004) The Wnt signaling pathway in development and disease. *Annu. Rev. Cell Dev. Biol.*, **20**, 781–810.
192. Fiuza, U.M., Arias, A.M. (2007) Cell and molecular biology of Notch. *J. Endocrinol.*, **194**, 459–474.
193. Ortiz, M.E., Endy, D. (2012) Engineered cell–cell communication via DNA messaging. *J. Biol. Eng.*, **6**, 16.
194. Goni-Moreno, A., Amos, M., de la Cruz, F. (2013) Multicellular computing using conjugation for wiring. *PLoS One*, **8**, e65986.
195. Danino, T., Mondragon-Palomino, O., Tsimring, L., Hasty, J. (2010) A synchronized quorum of genetic clocks. *Nature*, **463**, 326–330.
196. Prindle, A., Samayoa, P., Razinkov, I., Danino, T. *et al.* (2012) A sensing array of radically coupled genetic 'biopixels'. *Nature*, **481**, 39–44.
197. Tabor, J.J., Salis, H.M., Simpson, Z.B., Chevalier, A.A. *et al.* (2009) A synthetic genetic edge detection program. *Cell*, **137**, 1272–1281.
198. Liu, C., Fu, X., Liu, L., Ren, X. *et al.* (2011) Sequential establishment of stripe patterns in an expanding cell population. *Science*, **334**, 238–241.
199. Tamsir, A., Tabor, J.J., Voigt, C.A. (2011) Robust multicellular computing using genetically encoded NOR gates and chemical 'wires'. *Nature*, **469**, 212–215.
200. Weber, W., Daoud-El Baba, M., Fussenegger, M. (2007) Synthetic ecosystems based on airborne inter- and intrakingdom communication. *Proc. Natl Acad. Sci. USA*, **104**, 10435–10440.
201. Bacchus, W., Lang, M., El-Baba, M.D., Weber, W. *et al.* (2012) Synthetic two-way communication between mammalian cells. *Nat. Biotechnol.*, **30**, 991–996.
202. Cardinale, S., Arkin, A.P. (2012) Contextualizing context for synthetic biology – identifying causes of failure of synthetic biological systems. *Biotechnol. J.*, **7**, 856–866.
203. Arkin, A.P. (2013) A wise consistency: engineering biology for conformity, reliability, predictability. *Curr. Opin. Chem. Biol.*, **17**, 893–901.
204. Ma, W., Trusina, A., El-Samad, H., Lim, W.A., Tang, C. (2009) Defining network topologies that can achieve biochemical adaptation. *Cell*, **138**, 760–773.
205. Chau, A.H., Walter, J.M., Gerardin, J., Tang, C., Lim, W.A. (2012) Designing synthetic regulatory networks capable of self-organizing cell polarization. *Cell*, **151**, 320–332.
206. Clancy, K., Voigt, C.A. (2010) Programming cells: towards an automated 'Genetic Compiler'. *Curr. Opin. Biotechnol.*, **21**, 572–581.
207. Rosenfeld, N., Elowitz, M.B., Alon, U. (2002) Negative autoregulation speeds the response times of transcription networks. *J. Mol. Biol.*, **323**, 785–793.
208. Zheng, Y., Sriram, G. (2010) Mathematical modeling: bridging the gap between concept and realization in synthetic biology. *J. Biomed. Biotechnol.*, **2010**, 541609.
209. Purcell, O., Jain, B., Karr, J.R., Covert, M.W. *et al.* (2013) Towards a whole-cell modeling approach for synthetic biology. *Chaos*, **23**, 025112.
210. Kimelman, A., Levy, A., Sberro, H., Kidron, S. *et al.* (2012) A vast collection

of microbial genes that are toxic to bacteria. *Genome Res.*, **22**, 802–809.

211. Cardinale, S., Joachimiak, M.P., Arkin, A.P. (2013) Effects of genetic variation on the *E. coli* host–circuit interface. *Cell Rep.*, **4**, 231–237.
212. Scott, M., Gunderson, C.W., Mateescu, E.M., Zhang, Z. *et al.* (2010) Interdependence of cell growth and gene expression: origins and consequences. *Science*, **330**, 1099–1102.
213. Pilpel, Y. (2011) Noise in biological systems: pros, cons, and mechanisms of control. *Methods Mol. Biol.*, **759**, 407–425.
214. Randall, A., Guye, P., Gupta, S., Duportet, X. *et al.* (2011) Design and connection of robust genetic circuits. *Methods Enzymol.*, **497**, 159–186.
215. Pedraza, J.M., van Oudenaarden, A. (2005) Noise propagation in gene networks. *Science*, **307**, 1965–1969.
216. Wang, B., Kitney, R.I., Joly, N., Buck, M. (2011) Engineering modular and orthogonal genetic logic gates for robust digital-like synthetic biology. *Nat. Commun.*, **2**, 508.
217. Becskei, A., Serrano, L. (2000) Engineering stability in gene networks by autoregulation. *Nature*, **405**, 590–593.
218. Alon, U. (2007) Network motifs: theory and experimental approaches. *Nat. Rev. Genet.*, **8**, 450–461.
219. Ghosh, B., Karmakar, R., Bose, I. (2005) Noise characteristics of feed forward loops. *Phys. Biol.*, **2**, 36–45.
220. Cobb, R.E., Si, T., Zhao, H. (2012) Directed evolution: an evolving and enabling synthetic biology tool. *Curr. Opin. Chem. Biol.*, **16**, 285–291.
221. Ozbudak, E.M., Thattai, M., Kurtser, I., Grossman, A.D. *et al.* (2002) Regulation of noise in the expression of a single gene. *Nat. Genet.*, **31**, 69–73.
222. van der Meer, J.R., Belkin, S. (2010) Where microbiology meets microengineering: design and applications of reporter bacteria. *Nat. Rev. Microbiol.*, **8**, 511–522.
223. Stocker, J., Balluch, D., Gsell, M., Harms, H. *et al.* (2003) Development of a set of simple bacterial biosensors for quantitative and rapid measurements of arsenite and arsenate in potable water. *Environ. Sci. Technol.*, **37**, 4743–4750.
224. Wackwitz, A., Harms, H., Chatzinotas, A., Breuer, U. *et al.* (2008) Internal arsenite bioassay calibration using multiple bioreporter cell lines. *Microb. Biotechnol.*, **1**, 149–157.
225. Lu, T.K., Bowers, J., Koeris, M.S. (2013) Advancing bacteriophage-based microbial diagnostics with synthetic biology. *Trends Biotechnol.*, **31**, 325–327.
226. Campbell, A. (2003) The future of bacteriophage biology. *Nat. Rev. Genet.*, **4**, 471–477.
227. Schofield, D.A., Sharp, N.J., Westwater, C. (2012) Phage-based platforms for the clinical detection of human bacterial pathogens. *Bacteriophage*, **2**, 105–283.
228. Saeidi, N., Wong, C.K., Lo, T.M., Nguyen, H.X. *et al.* (2011) Engineering microbes to sense and eradicate *Pseudomonas aeruginosa*, a human pathogen. *Mol. Syst. Biol.*, **7**, 521.
229. Duan, F., March, J.C. (2010) Engineered bacterial communication prevents *Vibrio cholerae* virulence in an infant mouse model. *Proc. Natl Acad. Sci. USA*, **107**, 11260–11264.
230. Chen, Y.Y., Jensen, M.C., Smolke, C.D. (2010) Genetic control of mammalian T-cell proliferation with synthetic RNA regulatory systems. *Proc. Natl Acad. Sci. USA*, **107**, 8531–8536.
231. Nettelbeck, D.M. (2008) Cellular genetic tools to control oncolytic adenoviruses for virotherapy of cancer. *J. Mol. Med.*, **86**, 363–377.
232. Hanna, N., Ohana, P., Konikoff, F.M., Leichtmann, G. *et al.* (2012) Phase 1/2a, dose-escalation, safety, pharmacokinetic and preliminary efficacy study of intratumoral administration of BC-819 in patients with unresectable pancreatic cancer. *Cancer Gene Ther.*, **19**, 374–381.
233. Zakeri, B., Lu, T.K. (2013) Synthetic biology of antimicrobial discovery. *ACS Synth. Biol.*, **2**, 358–372.
234. Walsh, C. (2003) Where will new antibiotics come from? *Nat. Rev. Microbiol.*, **1**, 65–70.
235. Lu, T.K., Koeris, M.S. (2011) The next generation of bacteriophage therapy. *Curr. Opin. Microbiol.*, **14**, 524–531.
236. Parsek, M.R., Singh, P.K. (2003) Bacterial biofilms: an emerging link to disease pathogenesis. *Annu. Rev. Microbiol.*, **57**, 677–701.
237. Lu, T.K., Collins, J.J. (2007) Dispersing biofilms with engineered enzymatic

bacteriophage. *Proc. Natl Acad. Sci. USA*, **104**, 11197–11202.

238. Lu, T.K., Collins, J.J. (2009) Engineered bacteriophage targeting gene networks as adjuvants for antibiotic therapy. *Proc. Natl Acad. Sci. USA*, **106**, 4629–4634.
239. Weber, W., Schoenmakers, R., Keller, B., Gitzinger, M. *et al.* (2008) A synthetic mammalian gene circuit reveals antituberculosis compounds. *Proc. Natl Acad. Sci. USA*, **105**, 9994–9998.
240. Boyle, P.M., Silver, P.A. (2012) Parts plus pipes: synthetic biology approaches to metabolic engineering. *Metab. Eng.*, **14**, 223–232.
241. Holtz, W.J., Keasling, J.D. (2010) Engineering static and dynamic control of synthetic pathways. *Cell*, **140**, 19–23.
242. Farmer, W.R., Liao, J.C. (2000) Improving lycopene production in *Escherichia coli* by engineering metabolic control. *Nat. Biotechnol.*, **18**, 533–537.
243. Zhang, F., Carothers, J.M., Keasling, J.D. (2012) Design of a dynamic sensor-regulator system for production of chemicals and fuels derived from fatty acids. *Nat. Biotechnol.*, **30**, 354–359.
244. Anesiadis, N., Cluett, W.R., Mahadevan, R. (2008) Dynamic metabolic engineering for increasing bioprocess productivity. *Metab. Eng.*, **10**, 255–266.
245. Anesiadis, N., Kobayashi, H., Cluett, W.R., Mahadevan, R. (2013) Analysis and design of a genetic circuit for dynamic metabolic engineering. *ACS Synth. Biol.*, **2**, 442–452.
246. Moser, F., Broers, N.J., Hartmans, S., Tamsir, A. *et al.* (2012) Genetic circuit performance under conditions relevant for industrial bioreactors. *ACS Synth. Biol.*, **1**, 555–564.
247. Chen, A.Y., Deng, Z., Billings, A.N., Seker, U.O. *et al.* (2014). Synthesis and patterning of tunable multiscale materials with engineered cells. *Nat. Materials* **13**, 515-523.
248. Nissim, L., Perli, S.D., Fridkin, A., Perez-Pinera, P. *et al.* (2014). Multiplexed and Programmable Regulation of Gene Networks with an Integrated RNA and CRISPR/Cas Toolkit in Human Cells. *Mol. Cell* **54**, 698-710.
249. Siuti, P., Yazbek, J., Lu, T.K. (2014). Engineering genetic circuits that compute and remember. *Nat. Protoc.* **9**, 1292-1300.
250. Purcell, O., Lu, T.K. (2014). Synthetic analog and digital circuits for cellular computation and memory. *Curr. Opin. Biotechnol.* **29C**, 146-155.
251. Roquet, N., Lu, T.K. (2014). Digital and analog gene circuits for biotechnology. *Biotechnol. J.* **9**, 597-608.

6
Synthetic Hybrid Biosensors

Apoorv Shanker[1], *Kangwon Lee*[2], *and Jinsang Kim*[3]
[1] *University of Michigan, Macromolecular Science and Engineering Center, 2300 Hayward Street, Ann Arbor, MI 48109, USA*
[2] *Korea Institute of Science and Technology (KIST), Center for Biomaterials, Biomedical Research Institute, Hwarangno 14-gil 5, Seongbuk-gu Seoul, 136-791 Republic of Korea*
[3] *University of Michigan, Departments of Materials Science and Engineering, Chemical Engineering, Biomedical Engineering, Chemistry and Macromolecular Science and Engineering Center, 2300 Hayward Street, Ann Arbor, MI 48109, USA*

Translational Nanomedicine. First Edition. Edited by Robert A. Meyers.

Keywords

Receptor
The biological recognition element used in the sensor.

Target
The analyte of interest; the one which is to be quantified.

Biorecognition
The process of identifying the target molecules through biocatalysis or bioaffinity reactions.

Transducer
A device which converts the chemical signal from the biorecognition event into a quantifiable physical signal.

Immobilization
The process of attaching receptor molecules to the transducer surface, without compromising its activity and selectivity.

Biosensors are analytical tools in which biological or biologically derived receptor molecules are used as recognition elements in conjunction with physicochemical transduction mechanisms. Biosensors can be classified according to the bio-recognition process or the transduction mechanism employed. Bioaffinity sensors involve affinity reactions between the receptor and the target, while biocatalytic sensors employ the specific catalysis of the target analyte by the biological molecule. Depending on the transduction mechanism, biosensors can be divided in four broad types: electrochemical; optical; piezoelectric; and thermal. Biosensors find application in fields ranging from clinical and point-of-care diagnosis, medicine and drugs, process industries, environmental monitoring to defense and biowarfare. Present biosensor research is focused on developing more compact and easy-to-use devices while retaining the efficiency and sensitivity.

1
Introduction

Biosensors are fascinating analytical tools that combine the specificity and sensitivity of biological processes with the physico-chemical transduction mechanism to provide bioanalytical measurements. The International Union of Pure and Applied Chemistry (IUPAC) defines a biosensor as a device that uses specific biochemical reactions mediated by isolated enzymes, immunosystems, tissues, organelles or whole cells to detect chemical compounds, usually by electrical, thermal or optical signals [1, 2]. In 1962, Leyland C. Clark was the first to elucidate the basic concept of the biosensor in his seminal report on "enzyme electrodes" [3], which he had built on his earlier invention of the oxygen electrode. Clark reasoned that the electrochemical detection of oxygen or hydrogen peroxide could be used for the analysis of a wide range of analytes that produce either oxygen or hydrogen peroxide on being acted upon by a specific enzyme(s) [4].

The field of biosensors can be divided into two broad categories of instrumentation: (i) sophisticated high-throughput laboratory instruments capable of delivering rapid and accurate measurements; and (ii) easy-to-use, portable devices for use by the non-specialists for decentralized, *in situ*, or home analysis [4]. Medical diagnostics – and in particular blood glucose sensors for diabetic patients – present the largest field of applications for biosensors, though they also find applications in food and process control, medicine, environmental monitoring, and defense and security. A biosensor may serve different analytical functions; in a clinical diagnosis it may just be required to determine whether a targeted analyte is above or below a certain threshold value, whereas in process control the sensor may be required to provide a continuous and precise feedback about the analyte. Hence, the sensor needs to be designed to meet the requirements of each and any application.

Based on the type of biological recognition process involved, biosensors can be allocated to two categories: (i) biocatalytic, which are typically based on the selective catalysis of biochemical reactions by enzymes; and (ii) bioaffinity, in which affinity interactions resulting in the formation of biocomplexes – such as antigen–antibody, the hybridization of complementary single-strand DNAs, protein–nucleic acids, and chemoreceptor–ligand – provide a very selective and sensitive mechanism for biosensing. Based on the transduction mechanism, biosensors can be broadly divided into four categories: electrochemical; optical; piezoelectric; and thermal. A combination of two transduction mechanisms can also be used to yield better sensitivity.

The aim of this chapter is to provide a basic knowledge of the various types of biosensor, and to outline the underlying principles and general design criteria, by providing specific examples.

2
Sensor Design

Figure 1 shows a general schematic of a biosensor, which will usually consist of three main components: the bioreceptor or recognition unit, which is used in conjunction with a transducer that converts the chemical information from the analyte–receptor interaction into an easily measurable and quantifiable signal which is then shown on a display unit. The

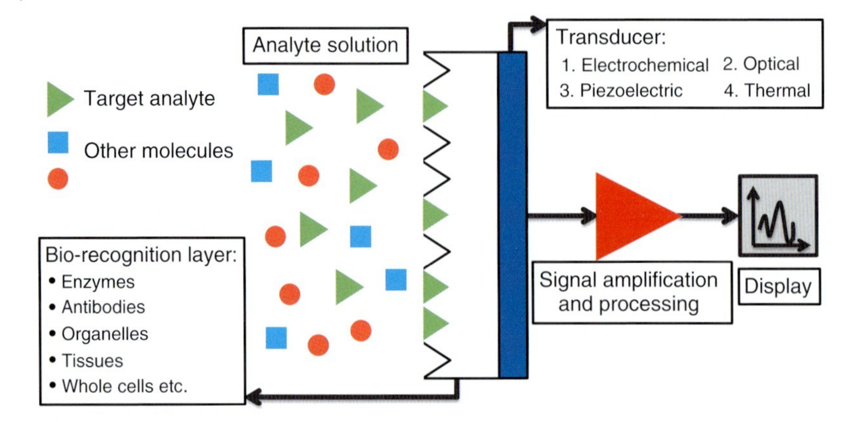

Fig. 1 Schematic representation of sensor design.

sensor design also incorporates an associated electronic circuit or signal processor. In a biosensor, enzymes, cell organelles, tissues, microorganisms, antibodies and nucleic acids are the commonly used bioreceptors. Biologically derived materials such as aptamers and apoenzymes, or biomimetic materials such as molecularly imprinted polymers (MIPs) can also be used as bioreceptors. The biological reaction must usually take place in close vicinity of the transducer, so that the transducer can pick up most of the chemical information from the receptor–analyte interaction.

In order to create a viable biosensor, the biorecognition unit must be properly attached to the transducer surface, without affecting the former's activity. This process, which is known as *immobilization*, is the most critical step in the fabrication of any biosensing device. The choice of immobilization method depends on several factors, including the nature of the biological component, the type of transducer, the physico-chemical properties, and the environment in which the sensor is intended to be used. The most commonly used immobilization methods used are adsorption, covalent binding, intermolecular crosslinking, matrix entrapment, and membrane entrapment.

- **Physical adsorption** utilizes a combination of van der Waals forces, hydrophobic interactions, H bonding and columbic interactions to immobilize biological elements on the surface of the transducer. Many substrates, such as cellulose, collodion, collagen, silica gel, glass, alumina and hydroxyapatite are known to adsorb biomolecules. However, the interaction forces between the substrate and the immobilized elements are weak, and the latter may tend to be released over a period of time, leading to sensor dysfunction [5, 6].
- **Covalent binding** involves the formation of covalent bonds between the certain reactive groups of the biological element which do not play a role in the biorecognition process and the substrate surface, which is modified to have functional groups. Generally, the nucleophilic functional groups present in the amino acid side chain, such as amine, carboxylic acid, imidazole, thiol, and hydroxyl are used for the coupling reaction. Coupling requires mild conditions such as low temperature, low ionic strength, and pH

in the physiological range. Covalent binding leads to a uniform surface coverage and helps to eliminate certain problems such as instability, aggregation, diffusion, and deactivation of the immobilized biocomponent [1].

- Bifunctional or multifunctional reagents such as glutaraldehyde, hexamethylene di-isocyanate, 1,5-difluoro 2,4-dinitrobenzene and bisdiazobenzidine-2,2'-disulfonic acid are used for immobilizing biocomponents through **intermolecular crosslinking**. The nonrigidity of the enzyme layer formed, the higher demands for amounts of biological material and the formation of multiple layers of enzyme, which negatively affects the activity, represent some of the disadvantages of this method. Moreover, larger diffusional barriers may delay interactions and increase the response time of the device [7].
- In **matrix entrapment**, the polymeric gel matrix precursors are polymerized in the presence of the biological elements to be entrapped. The most commonly used gels are polyacrylamide, polyvinyl alcohol, polycarbonate, cellulose acetate, starch, alginate, and silica gel. Matrix entrapment is usually not the preferred method of immobilization as it may lead to possible delays in response time due to a diffusional barrier to the analyte and the leakage of biological species during sensor operation, resulting in a loss of bioactivity.
- In **membrane entrapment**, enzyme solutions, cell suspensions or tissue slices can simply be encapsulated in analyte-permeable preformed membranes on the electrochemical transducer. Self-assembled monolayers (SAMs) and bilayer lipid membranes (BLMs) can also be used to encapsulate biological molecules and bind them to the transducer surface. A sol–gel method is used to immobilize biological molecules in ceramics, glasses, and other inorganic materials. Bulk modification of the entire electrode, for example, enzyme-modified carbon paste or graphite epoxy resin [8], magnetic interactions [9], and biotin–avidin binding [10, 11] are also effective methods of immobilization.

3
Electrochemical Biosensors

In the simplest of terms, an *electrochemical biosensor* can be defined as one which transduces or converts a biological event into a measurable, reproducible, and discrete electronic signal. Electrochemical biosensors combine the electrochemical transducers' sensitivity with the specificity of biological recognition processes involving biological elements such as enzymes, proteins, nucleic acids, antibodies, cells, or tissues [12]. Electrochemical biosensors provide easy fabrication, ease of operation, portability, and entail low costs in manufacturing. An electrochemical study of a reaction will yield a measurable current (amperometry), a measurable charge accumulation or potential (potentiometry), a change in the conductivity of the medium (conductometry), or changes in capacitance and/or resistance of the medium (impedance spectroscopy). As electrochemistry is a surface phenomenon, electrochemical transduction does not require large sample volumes, and this provides for the effective miniaturization of biosensor devices. Electrochemical transduction helps in the speedy, continuous, real-time and inexpensive monitoring of many components in clinical laboratories and industries [13].

3.1 Amperometric Biosensors

The amperometric technique involves applying a fixed potential to the working electrode versus a reference electrode, and measuring the current produced as a result of the electrochemical reduction or oxidation occurring at the working electrode. This current is proportional to the concentration of the electroactive product, which is in turn proportional to the nonelectroactive substrate in the sample. However, the electrolysis current is limited by mass transfer rates. Amperometric biosensors provide an additional selectivity because the oxidation or reduction potential used is normally characteristic of a particular analyte [12]. Additionally, the fixed potential used in amperometry results in a negligible charging current, which minimizes the background signal. Simplicity and a low limit of detection make amperometric transduction a good choice for biocatalytic and bioaffinity sensors [14].

The electrochemical cell usually consists of a three electrodes system: (i) a working electrode made from conductive inert metals such as Pt, Au, or graphite, and at which the biochemical reaction involving the target analyte occurs catalyzed, in most cases by enzymes which are immobilized on the electrode surface; (ii) a counterelectrode, which is usually a Pt wire; and (iii) a reference electrode against which the potential measurement is made. However, if the current density is low ($<\mu A\,cm^{-2}$), a two-electrode system without the reference electrode can also be used; in fact, such a system is generally preferred in disposable sensors as long-term stability of the reference electrode is not required and the costs are lower [12].

The electrodes can easily be miniaturized to micron size, or even to nanometer size [15, 16], which results in low sample volume requirements for detection of the analyte. Recently, screen-printed electrodes (SPEs) with patterned microelectrodes have gained in popularity because of their low cost, ease, and speed of mass production [17]. Disposable SPEs have been used in immunochemical sensors and glucose sensors [18]. Interdigitated array electrodes consisting of two pairs of working electrodes made from parallel metal finger strips interdigitated and separated by insulating materials may serve as another good amperometric transducer [12, 19].

Amperometric biosensors are typically based on enzyme electrodes. The simplest design of an amperometric biosensor is the direct detection of either the increase of an enzymatically produced electroactive species or the decrease of a substrate of a redox enzyme. A typical example of this design is a glucose sensor based on using glucose oxidase (GOx) as the biorecognition element. The increase in concentration of the product, H_2O_2, or the decrease in concentration of the cosubstrate, O_2, is electrochemically monitored in order to quantify the glucose concentration [3, 20–23]. Such sensors are termed *"first-generation" biosensors* (Fig. 2). Unfortunately, the reproducibility of these biosensors is dependent on the concentration of oxygen, while the electrode potential is prone to interference [24].

The use of artificial redox mediators was introduced to overcome the operational problems associated with their first-generation counterparts. Cass *et al.* developed the first amperometric glucose biosensor based on ferrocene, a redox mediator [25]; such biosensors are termed *"second-generation" biosensors* (Fig. 3). Artificial redox mediators are small, soluble

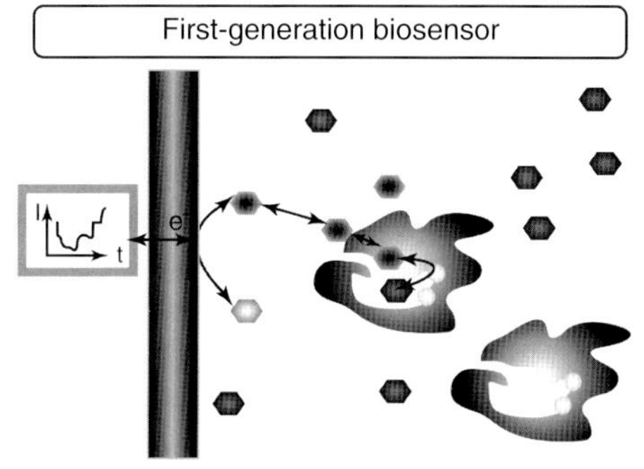

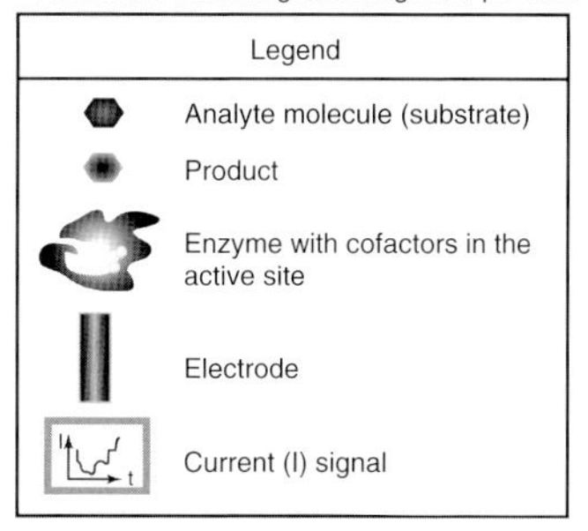

Fig. 2 "First-generation" amperometric biosensors. Reprinted with permission from Ref. [24]; © 2011, John Wiley & Sons.

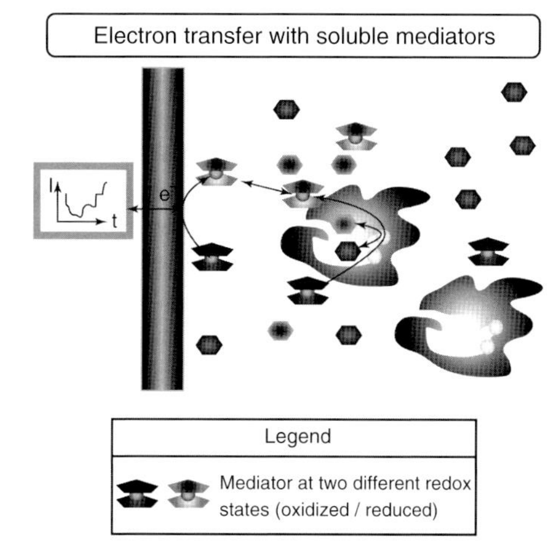

Fig. 3 "Second-generation" amperometric biosensors. Reprinted with permission from Ref. [24]; © 2011, John Wiley & Sons.

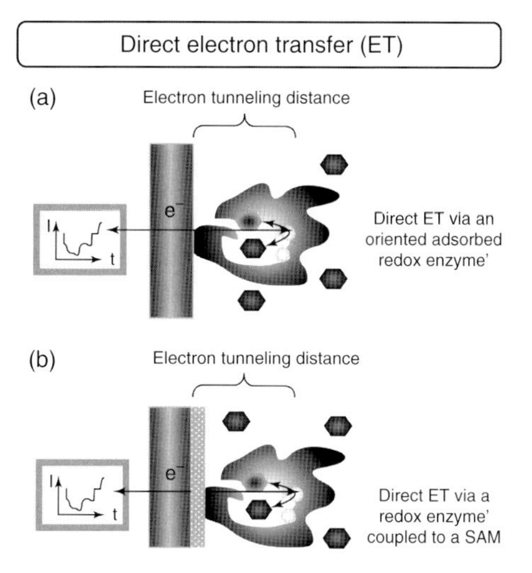

Fig. 4 "Third-generation" amperometric biosensors. Reprinted with permission from Ref. [24]; © 2011, John Wiley & Sons.

molecules capable of undergoing rapid and reversible redox reactions that shuttle electrons between the active site of the enzyme and the electrode surface. These sensors are prone to leakage of free diffusing redox mediators from the electrode surface, which adversely affects their long-term operational stability. However, this does not affect their successful application in one-shot devices, such as those for the self-monitoring of glucose [13].

A better biosensor architecture can be realized through the immobilization of a redox enzyme on the electrode surface, in such a way that a direct electron transfer is made possible between the active site of the enzyme (where the catalytic reactions occur) and the transducer. Such a design obviates the use of freely diffusing redox mediators, and biosensors based on this design principle are termed *"third-generation" biosensors* (Fig. 4). A reagentless biosensor architecture can be realized by co-immobilizing the enzyme and the mediator at the electrode surface [24]. Third-generation biosensors have a greater stability and can be used for repeated measurements or continuous monitoring, as neither the enzyme nor the mediator needs to be added. Consequently, such sensors are self-contained and the cost of each measurement is reduced [12].

3.2 Potentiometric Biosensors

Potentiometric biosensors are based on the principle of measurement of charge accumulation at the working electrode compared to a reference electrode, under the conditions of negligible or zero current, and are governed by the Nernst equation:

$$E = \frac{E^\circ \pm RT}{n\mathrm{F}\ \log(a_\mathrm{i})} \tag{1}$$

where E is the measured potential of the cell, E° is the standard cell potential at temperature T, R is the universal gas constant, n is the number of moles of electrons transferred in the cell reaction, F is Faraday's constant (~96 500 C mol^{-1}), and a_i is the chemical activity of the species i.

A typical potentiometric device set-up consists of a reference and one working electrode in contact with the sample solution. Common electrodes used for potentiometric quantification are the glass pH electrodes and ion-selective electrodes (ISEs) for ions such as K^+, Na^+, and Ca^{+2} [26]. These sensors can be converted into biosensors by using biological elements such as enzymes capable of catalyzing reactions that involve analyte molecules to produce ions for which the sensor is designed. An immobilized enzyme layer adjacent to the working electrode catalyzes a biological reaction involving the analyte in which ionic species are either consumed or produced. A local equilibrium is established at the sensor interface and the membrane potential, developed due to the difference in concentration of the ions across the membrane, is measured. The ISE generates an electrical signal in response to the change in concentration of ionic species. Currently, three types of ISE are used in potentiometric biosensors:

- Glass electrodes for cations: these are made from a very thin hydrated glass membrane as the sensing element. A transverse electrical potential is developed due to the concentration-dependent competition between cations for specific binding sites. The selectivity of a glass electrode is determined by the composition of the glass. A common glass electrode is shown in Fig. 5.
- Gas electrodes, which are the usual glass pH electrodes coated with hydrophobic gas-permeable polypropylene or Teflon membranes selective for gases such as CO_2, NH_3, and H_2S. The diffusion of gases through the membrane causes a change in the pH of the sensing solution between the membrane and the electrode, which is then determined.
- Solid-state electrodes which consist of a thin membrane of a specific ion conductor made from a mixture of Ag_2S and AgX, where X is a halide anion.

Although, amperometric transduction – given its good sensitivity and low limit of detection – is favored in the case of glucose biosensors, the details of a number of potentiometric biosensors have been reported [27, 28]. Potentiometric biosensors are known for their simplicity of operation, and their continuous measurement capability makes them interesting for environmental applications, especially for monitoring heavy metals and pesticides [13]. With limits of detection as low as 10^{-8} to 10^{-11} M, potentiometric biosensors are suitable for measuring low concentrations in small sample volumes as they do not chemically influence the sample [29].

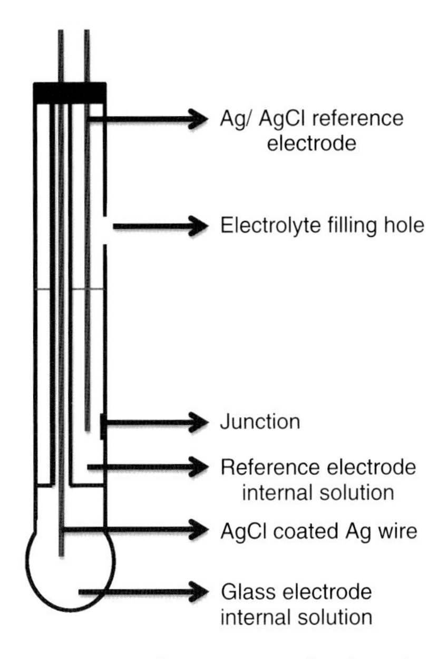

Fig. 5 Typical arrangement of a glass electrode.

The ion-selective field effect transistor (ISFET), a type of potentiometric device, was developed by Bergeveld and successfully combines solid-state integrated circuit (IC) technology with ISEs. The chemical-sensitive property of the glass membrane electrode is used in conjunction with the impedance-converting characteristics of the metal oxide semiconductor field effect transistor (MOSFET), in which the metal electrode (gate) is removed and its function is taken over by the sample solution under study (Fig. 6). This modified FET is capable of detecting changes in ion concentration when the gate is exposed to a solution containing ions. Many biosensors based on ISFET have been described since the first report of the enzymatic-modified ISFET (enzyme field-effect transistor; EnFET) for the determination of penicillin [30]. ISFET-based enzyme biosensors can also be used to detect and quantify heavy metal ions and organic pollutants, through the inhibitory action of such species on enzyme activity [31].

These biosensors have many advantages over other types of biosensor, notably miniaturization, high sensitivity, low cost and multianalyte detection potential [32]. Unfortunately, however, they still suffer from a variety of fundamental and technological problems, such as the impurity of the

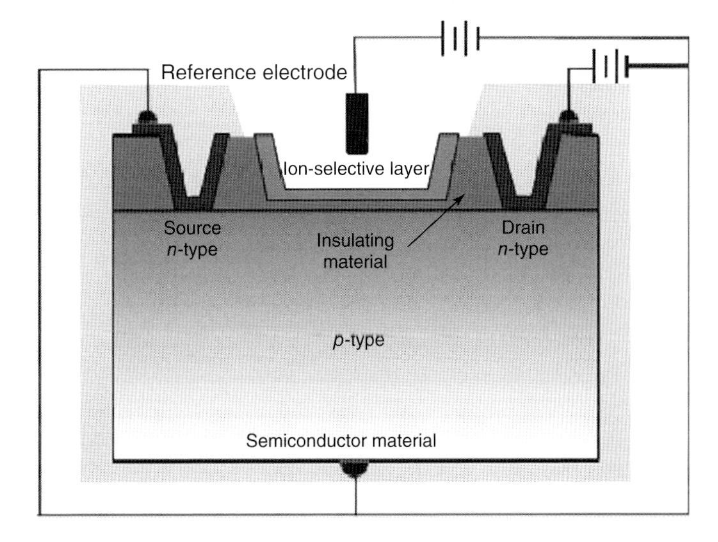

Fig. 6 Typical arrangement of an ISFET device.

semiconductor layer and instability of the functional groups in the sensing layer [33]. On the other hand, ISFETs can be directly incorporated into the electronic signal processing circuitry [13], and this can lead to their integration in microsystems such as micro-total-analysis-systems (μ-TAS) and lab-on-a-chip (LoC) [33].

3.3 Conductometric Biosensors

Conductometric techniques rely on measurements of the change in electrical conductivity of the sample solution due to the production of charged species, such as ions and electrons, during the course of a biochemical reaction catalyzed by an enzyme. For example, urease – which catalyzes the production of ionic species – can be used in combination with conductometric transduction. Conductance measurements have a lower sensitivity compared to other techniques [34], as conductance is sensitive to temperature, faradaic processes, double-layer charging, and concentration polarizations [13]. These effects can be minimized, however, by using an alternating current voltage for measurements [34]. This will result in a higher limit of detection and reduces potential interferences from variations in the ionic strength of the samples [13]. Conductometric techniques can be used to create inexpensive and disposable sensors; however, in order to obtain reliable measurements the ionic strength of the sample solution should undergo a significant change.

Conductometric biosensors have been used for environmental monitoring, as they provide an easy-to-use, accurate, selective, fast and cheap alternative to conventional methods of heavy metal determination, such as gas and liquid chromatography, spectrophotometry, and chemical and physical techniques which are time-consuming and require expensive instruments and skilled personnel. The presence of heavy metals can be determined using thin-film interdigitated planar conductometric electrodes, with enzymes such as GOx, butyric oxidase and urease having been used to detect Ag^+, Hg^{2+} and Pb^{2+} [13]. Conductometric biosensors have also been used to monitor the presence of organic pollutants and pesticides in the environment [35, 36].

4 Optical Biosensors

4.1 Conjugated Polymer-Based Biosensors

Conjugated polymers (CPs) are π-conjugated polymeric compounds in which the backbone is composed of alternating saturated and unsaturated bonds, while the backbone atoms are sp^2-hybridized [37–41]. These sp^2 hybrid orbitals, which are bonded through σ bonds with the remaining out-of-plane P_z orbitals overlapping with the neighboring P_z orbitals, provide the movement of free electrons. Therefore, the p-orbital overlap is the origin of the emissive and conductive properties of CPs and provides unique optoelectronic properties under certain conditions. For example, CPs are highly conductive under chemically doped conditions and are good candidates for flexible electronic materials. The unique optoelectronic property of CPs has attracted much attention for use as effective optical transducers, with CPs emerging as the active materials for various applications including light-emitting diodes (LEDs), field effect transistors (FETs), light-emitting electrochemical cells (LECs), polymer actuators, plastic lasers, batteries,

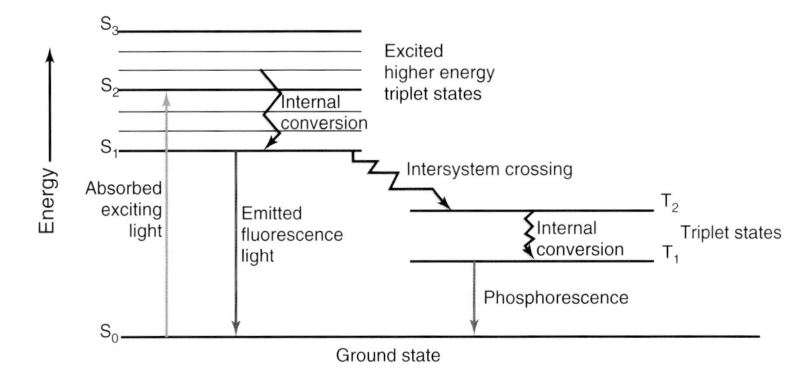

Fig. 7 Jablonski diagram.

photovoltaic cells, and biomaterials for sensory applications.

Conjugated polyelectrolytes (CPEs) are π-CPs that have charged (anionic or cationic) side chains [42–46]. In this case, ionic groups such as sulfonate, carboxylate, phosphate and quaternary ammonium ion are introduced into the chemical structures of the CPs to change their polarity. These ionic functional groups usually prevent the CPs from aggregating in water, and also control their solubility. In particular, CPEs may be good candidates for biological applications because the excellent water solubility of the CPs is essential for their homogeneous use in aqueous media. The ability to control the water-solubility of CPs might be considered as a sensing mechanisms for CPEs to be exploited as biosensory materials [47]. A change in water solubility by adding the target analytes, and the subsequent conformational change of the CPs, lead to alterations in the fluorescence wavelength and intensity of CPs. Recently, hydrophobic CPs have also been prepared as nanoparticles in an aqueous environment and used as sensory materials [48, 49].

CPs are largely classified based on how they release the energy absorbed from excitation. A CP molecule in an excited state can lose either an emission of radiation (as fluorescence and phosphorescence) or radiationless transition, such as the intersystem crossing shown in the Jablonski diagram (Fig. 7). The emission of radiation from the lowest vibrational level of the excited state S_1 to any of the vibrational levels of the ground state S_0 is termed *fluorescence* (fluorescence lifetime: 10^{-9} to 10^{-7} s). Although the population of triplet states by direct absorption from the ground state is insignificant, a more efficient process exists for the population of triplet states from the lowest excited state in CPs (intersystem crossing). If intersystem crossing has occurred, and the initial spin state is different from the final energy levels, then the emission energy may change; this is termed *phosphorescence*. Once intersystem crossing has occurred, the molecule undergoes a singlet–triplet process within the lifetime of an excited singlet state (10^{-8} s), and the life time of a triplet state is much longer than that of an excited state (ca. 10^{-4} to 10 s). This could be a good indication to judge the type of emission in CPs. However, fluorescence in CPs is usually statistically much more likely than phosphorescence, unless vibrational coupling between the excited singlet state and a triplet state causes intersystem crossing (this usually could occur at a very low temperature, <80 K) [39]. A similar type of emission termed *delayed fluorescence* has been identified which normally follows the fluorescence characteristic emission spectrum; however, the lifetime of delayed fluorescence in CPs is slightly shorter than that of phosphorescence as it is caused by a recombination of geminate electron hole pairs rather than triplet–triplet annihilation.

Fluorescence from CPs is very sensitive to any environmental changes around CPs. The optical properties of CPs undergo dramatic changes such as fluorescence amplification, quenching, or nonradiative energy transfer when the light is absorbed [50], and therefore the provision of mechanisms for optical changes in CPs allows their implementation in sensing applications. This appealing property of CPs provides a highly sensitive transduction mechanism by signal amplification of CP fluorescence, and also explains various detection modes. The signal-amplifying model of CPs was proposed by Swager and colleagues in 1995 [51, 52]. When a target analyte binds locally to a receptor on a CP repeat unit, the entire conjugated backbone is affected due to its one-dimensional wire-like optoelectronic property, such that the fluorescence of the entire polymer chain is altered. The wiring of chemosensory molecules in series provides a universal method by which to obtain signal amplification relative to single-molecule systems. CPs are "molecular wires," as the key feature of a CP is that it can harness extended electronic communication and transport. However, the terms amplification and sensitivity enhancement only indicate when a single event – the binding of an analyte – in a supramolecular polyreceptor system produces a response larger than that afforded by a similar interaction in an analogous small monoreceptor system.

The photophysical properties of CPs are strongly related to their polymer structure, whether in solution and/or in solid state [50]. Changes in the chemical nature, effective conjugation length, intramolecular conformation and intermolecular packing will each have an influence on the color and intensity of fluorescence. The emission wavelength can be modulated through the design of backbone structure of CPs and by changing the charge density around the CP backbone. Side-chain modification along the backbone, using either electron-rich or electron-deficient functional groups, provides the emission wavelength. The effective conjugation length is also a critical factor in determining wavelength, with long chains generally showing a longer wavelength emission. However, the fluorescence wavelength of CPs is not influenced further when the conjugation length of CP exceeds the "effective" conjugation length. Rather, the shorter lifetime of CPs and the exciton mobility – which may be limited by conformational disorder in solution – will prevent diffusion throughout the entire length of high-molecular-weight CPs.

Several detection modes in CPs have been actively developed for the sensing of chemical or biomolecules, including fluorescence turn-on (amplification) and turn-off (quenching), fluorescence color change, and visible color change (Fig. 8). In the turn-on mechanism, the fluorescence signal of CPs is inherently excellent but is completely and partially quenched due to the change in electron density along the CP backbone as a result of conformation changes or intramolecular packing. Target binding to CPs, and the associated conformational rearrangement of CPs or unpacking among the backbones, perturb the electronic state along the CP

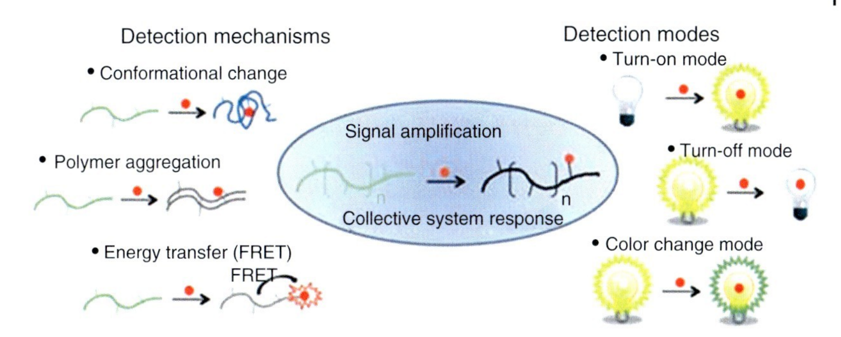

Fig. 8 Conjugated polymer based biosensors: detection mechanisms and modes. (*http://pubs.rsc.org/en/Content/ArticleLanding/2010/AN/c0an00239a*) Reproduced with permission from Ref. [50]; © 2010, Royal Society of Chemistry.

backbone and induce an enhancement in fluorescence. Although the polymer may be soluble in water, the fluorescence quantum efficiency of CPs in aqueous solution may be low due to their limited water-solubility and the resultant polymer aggregation. The use of a surfactant to improve CO solubility in water can also provide an improvement in signal turn-on, without affecting target binding and signal transduction by surfactants.

Another interesting turn-on sensor has been developed as a colorimetric CP sensor by using polythiophene derivatives. These CPs provide a color change and signal enhancement when a target is bound to the receptor such that their conformation is altered. As an example, cationic polythiophene derivatives will form a duplex with negatively charged single-strand DNA molecules, which results in polymer aggregation and, hence, fluorescence quenching. When a target single-strand DNA molecule is hybridized to the complementary receptor DNA, the DNA/DNA/polymer triplex will be less planar than when in the duplex conformation, and so will have a shorter conjugation length and different absorption characteristics. Target molecules detected using this system vary from small DNA molecules to large protein molecules.

In a turn-off system, a variety of mechanisms can result in quenching, such as Forster or Dexter energy transfer, static quenching, complex-formation between polymers and a target, and collisional quenching [53, 54]. For these reasons, the fluorescence quenching of CPs is often dependent on environmental factors such as temperature and pressure. In most CP-based quenching systems, target binding induces the electronic state of CPs by an intermolecular aggregation of the polymer chains. Such aggregation is often due to hydrophobic effects induced by target analytes, and the addition of surfactants or a change in temperature can prevent CP aggregation. Quenching that occurs upon interaction with a specific molecular biological target forms the basis of active optical contrast agents for molecular imaging.

In a similar way, the fluorescence resonance energy transfer (FRET)-induced detection mode also begins with the fluorescence quenching of CPs that normally are used as energy donors [55–57]. FRET (also known as Forster energy transfer) is a dynamic quenching mechanism because energy transfer occurs while the CP donor is in an excited state. When a target analyte labeled with fluorescence acceptor molecules is bound to a target receptor and is located in the proximity of CPs (usually within 10 nm), the CP as the donor chromophore will transfer energy to an acceptor chromophore through dipole–dipole coupling. The efficiency of FRET depends on many physical parameters, including the distance between the donor and the acceptor, the spectral overlap of the donor emission spectrum and the acceptor absorption mechanism, and the relative orientation of the donor emission dipole moment and acceptor absorption dipole moment. The dominant factor among these CP-based sensors is the distance between the donor and the acceptor, because the efficiency of this energy transfer is inversely proportional to the sixth power of the distance between the CP and the acceptor dye. Consequently, FRET-based CPs have been used as potent tools to measure distances and to detect molecular interactions in a number of systems, and are widely applied in biology and chemistry.

4.2 Surface Plasmon Resonance-Based Biosensors

Surface plasmon resonance (SPR) was first observed by Wood in 1902, but a complete explanation of the phenomenon was not provided until 1968, by Otto [58–63]. Since the first application of SPR-based sensors to biomolecular interaction monitoring by Liedberg *et al.* in 1983, the phenomenon of SPR has served as a fascinating detection tool for biosensor applications [64]. When polarized light is shone through a prism on a sensor chip, on top of which is a thin film of metal (usually gold or silver); the film will act as a mirror and reflect the light. On changing the angle of incidence, the intensity of the reflected light will pass through a minimum. At the angle of incidence when this occurs, the light will excite the surface plasmon so as to induce SPR and cause a dip in the intensity of the reflected light. The angle at which the maximum loss of reflected light intensity occurs is termed the *resonance angle* or *SPR angle*. The SPR angle depends on the optical refractive indices of the media at both sides of the metal. The SPR conditions can be changed, and the shift of the SPR angle is suited to provide information on the kinetics of target adsorption on the metal surface. A schematic of a biosensor device based on SPR is shown in Fig. 9.

4.2.1 SPR Principles

SPR can be used to monitor changes in the refractive index in the near vicinity of the metal surface. When the refractive index changes, the angle at which the intensity minimum is observed will also change [66]. Hence, SPR not only provides an excellent means of measuring the difference between two states, but can also be used to monitor the intensity change in time-lapse. SPR sensors function in only a very limited vicinity or fixed volume at the metal surface, as the exponential decay of the evanescent field intensity in a typical SPR-based sensor presents practical blindness at distances beyond 600 nm from its surface. A process occurring within the first few tens of nanometers from the metal surface will result in a few-fold higher response than the same process performed at a distance of a few hundred nanometers. A signal observed at the penetration depth of the electromagnetic field is termed the *evanescent field*, and does not exceed a few hundred nanometers. The penetration depth of the evanescent field is a function of the wavelength of the incident light. In order to provide selectivity for the SPR sensor, its surface must be modified with ligands that are suitable for capturing the target compounds and which are

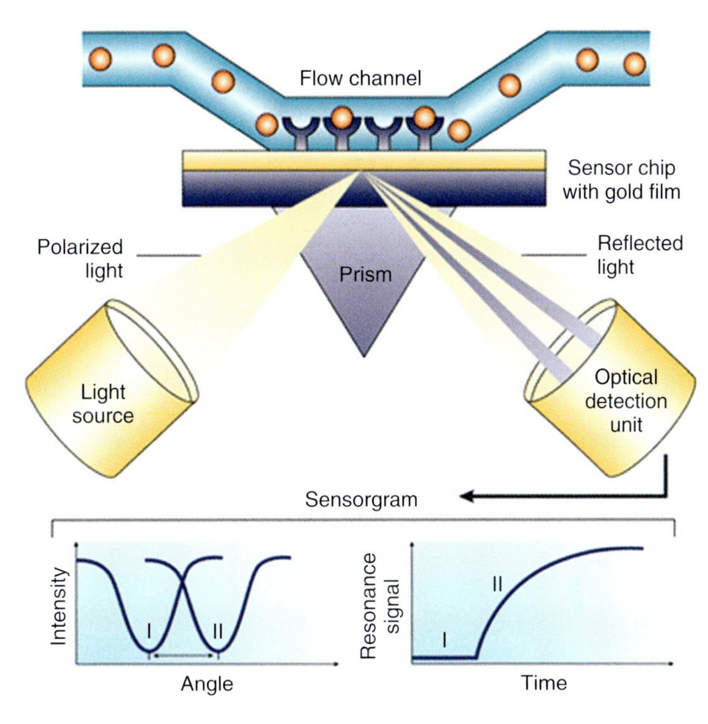

Fig. 9 Schematic of surface plasmon resonance (SPR)-based biosensor. Reproduced with permission from Ref. [65]; © 2002, Macmillan Publishers Ltd.

permanently immobilized on the sensor surface.

SPR-based sensor applications are associated with specific properties: (i) field enhancement; (ii) surface plasmon (SP) coherence length; and (iii) the phase jump of the reflected light upon SP excitation. In field enhancement, calculation of the electric field transmission coefficient based on Fresnel's equation for the interface shows that the electric field at the high refractive index side of the metal can be much smaller than that at the low index side of the metal layer. At very close to the SPR angle, the intensity can be enhanced by a factor of more than 30, a circumstance which explains much of the remarkable sensitivity that the SPR condition has for a changing dielectric environment. The metal thickness is also critical for SPR phenomena, and affects the biosensor's efficiency; for example, with an excitation wavelength of 700 nm the field enhancement is normally maximized when the gold layer is about 50 nm thick, but is decreased as the gold layer become either thicker or thinner.

SP coherence length implies that the field intensity of SPR decays with a characteristic distance $1/2k_x''$, where the metal's dielectric constant is complex and a complex propagation constant is $k_x = k_x''$ (real parts) $+ jk_x''$ (imaginary parts). For gold or silver (the most frequently used metals in sensory applications), the imaginary part of the dielectric constant increases with decreasing wavelength, and the SP propagation length decreases accordingly. Hence, the SP propagation length will become longer with increasing wavelengths used for the sensor studies. Finally, phase jump refers to the reflection event at an interface, that is generally accompanied by a phase jump of the reflected field. The phase of the reflected electric field undergoes a relatively large change around the SPR dip, and this is critical for sensing purposes. However, the absolute values found are of limited validity due to the complicated experimental set-up, though a phase measurement will provide an order-of-magnitude better sensitivity.

4.2.2 Surface Chemistry in SPR Technique

Detection processes in SPR between a target and a receptor are critical to provide a thorough understanding of all processes in living organisms [60]. Fast, selective and quantitative analyses, without a need for labels for optical biosensors, is the key to this situation. However, an elegant direct detection technique for label-free targets in bare SPR substrates provides a nonspecific binding of other components, as well as the desired specific targets of a biomolecular interaction. Another issue when using SPR detection without a relevant surface modification is the irreversible binding of numerous proteins and other biomolecules, as this can result in a failure to completely regenerate chip surfaces for reuse. In fact, after being used only a couple of times, the SPR substrates will be only partially regenerated, and 90% or more of their activity will be lost. Thus, it is clear that the surface energy and surface charge of the SPR chips must both be carefully modulated to obtain not only a minimum false signal but also maximum detection yields. In addition to a high signal-to-noise ratio, the ability to retrieve the biological activity of an immobilized ligand is essential when creating successful SPR-based sensors.

The transport of targets to the SPR chip surface by convection and diffusion has a profound effect on the signal. A correctly selected surface nanoarchitecture is extremely important to control the amount of immobilized ligand. The steric hindrance of target binding sites via a chemical immobilization process has a strong effect on the affinity of the ligand towards the target molecules. A sufficient spacing between the ligands can help to minimize any steric problems by controlling the ligand densities. The binding rate and equilibrium constants can normally be determined based on a novel distribution analysis. A detailed characterization of the distribution of binding properties provides a useful tool for optimizing surface modifications to achieve an effective functionalization of biosensor surfaces with uniform high-affinity binding sites, and also for studying immobilization processes and surface properties. The surface charge may also influence the interaction kinetics and the extent of nonspecific interactions.

Controlling the hydrophilicity can achieve protection of the sensitive biomolecular ligands by the appropriate selection of a functional group. Nonspecific binding can be prevented by introducing a bioinert layer, while the most popular functional group is the carboxylate ion. The optimum thickness of the bioinert layer is based on the exponentially decaying strength of the evanescent field. A thickness >10 nm would reduce the sensitivity of the binding signal, whereas a thickness <1 nm would usually cause an inhomogeneous coating of the metal surface. Hence, the preferred thickness of the bioinert layer, as an adhesion-linking layer, should be 2–5 nm.

4.2.3 Surface Plasmon Fluorescence Technique

Among the various sensing principles proposed for biosensor studies, surface plasmon fluorescence spectroscopy has, in particular, found widespread application and has demonstrated its potential for the sensitive detection of targets in several examples. SPR provides a label-free detection principle, as only the presence of bound analytes will slightly alter the optical architecture at the sensor surface probed by the surface plasmon mode propagating along this metal/dielectric interface. Another reason for such rapid growth is that surface plasmon fluorescence-based detection principles present attractive sensitivities for the *in situ* and real-time monitoring of biological targets. In addition, several facile surface-modification protocols are available to obtain the required functionalization of the sensor surface for label-free detections.

As the intensity profile normal to the metal/dielectric interface decreases exponentially in the direction into the metal, this suggests that the analyte molecules must be brought as close to the metal surface as possible in order to place their chromophores into the highest possible optical field. Metal-enhanced fluorescence is wavelength- as well as environment-dependent; notably, the environment (e.g., the solvent) can affect the enhancement factors [67]. Other effects on chromophores in the excited state close to a metal surface such as gold, and which can lead to a quenching of fluorescence, should also be considered. The fluorescence subsequently decreases with close contact, most likely due to FRET at very close contact. At intermediate distances, however, an efficient back-coupling of the excitation energy from the vibrational relaxed–excited state of the chromophore to the metal substrate becomes the driving force for the excitation of a red-shifted plasmon mode that can re-radiate via the prism at its respective resonance angle; this effect can be used to enhance the fluorescence emission. Fluorescence emitted directly from chromophores is sufficiently separated from the substrate surface, but can still be enhanced within the enhanced optical field of surface plasmon mode. This combination of field enhancement and fluorescence detection has been applied to a range of chemical and biosensing studies [68–80].

4.3 Surface-Enhanced Raman Spectroscopy-Based Biosensors

Surface-enhanced Raman spectroscopy (SERS) or surface-enhanced Raman scattering is a surface-sensitive technique that enhances Raman scattering by molecules absorbed on rough metal surfaces [81]. Since the discovery in 1974 by Fleischmann *et al.*, that a high-intensity Raman scattering of small molecules could be achieved on an electrochemically roughened silver surface, the field of SERS has expanded dramatically due to improvements in technique that have resulted from advances in nanotechnology and improved instrumental capabilities [82]. Today, the SERS technique is becoming widespread and is encountering new and exciting horizons in analytical chemistry, biology and biotechnology, forensic science, and also in the study of artistic objects. Although the exact mechanism of SERS remains a matter of debate, and the mechanisms proposed experimentally have not been straightforward, two primary theories have persisted: (i) an electromagnetic theory based on the excitation of localized surface plasmons; and (ii) a chemical theory based on the formation of charge–transfer complexes [83–88]. As the chemical theory applies only to species which have formed a chemical bond with the surface, it cannot explain the observed signal enhancement in all cases. In contrast, the electromagnetic theory can be applied even to those cases where the specimen is absorbed only physically to the surface. Although the electromagnetic theory of enhancement can be applied regardless of the molecule being studied, it does not fully illustrate the magnitude of the enhancement observed in many molecules which have lone pair electrons and are bound to the surface. In this situation, the enhancement mechanism cannot be solely explained by involving surface plasmons. The chemical mechanism involves charge transfer between the chemically adsorbed species and the metal surface; in this case a spectroscopic transition – which takes place in the ultraviolet range and where the metal acts as a charge-transfer intermediate – can be excited by visible light.

An increase in the Raman signal on metal surfaces occurs due to enhancements in the electric field provided by the surface (Fig. 10). The light incident on the surface can excite a variety of phenomena in the surface, but the complexity of this condition can be simplified by surfaces with features much smaller than the wavelength of the light. For this explanation, one useful approximation to solve enhancement numerically – and which has been widely used in the literature – is the *electrostatic approximation*. In this case, the problem can be solved as in electrostatics, and the approximation corresponds then to ignoring the presence of the wave vector k. Therefore, the applied electric field does not have a wavelength; rather, it is a uniform field oscillating up and down with frequency. Although this approximation fails in many cases, it is not too difficult to imagine that the electrostatic approximation functions well when the size of the object is much smaller than the wavelength. This means that the electrostatic approximation will be valid mostly for objects of typical sizes in the range of about 10 nm or smaller. Another factor that affects the intensity in enhancement is the shape of the features.

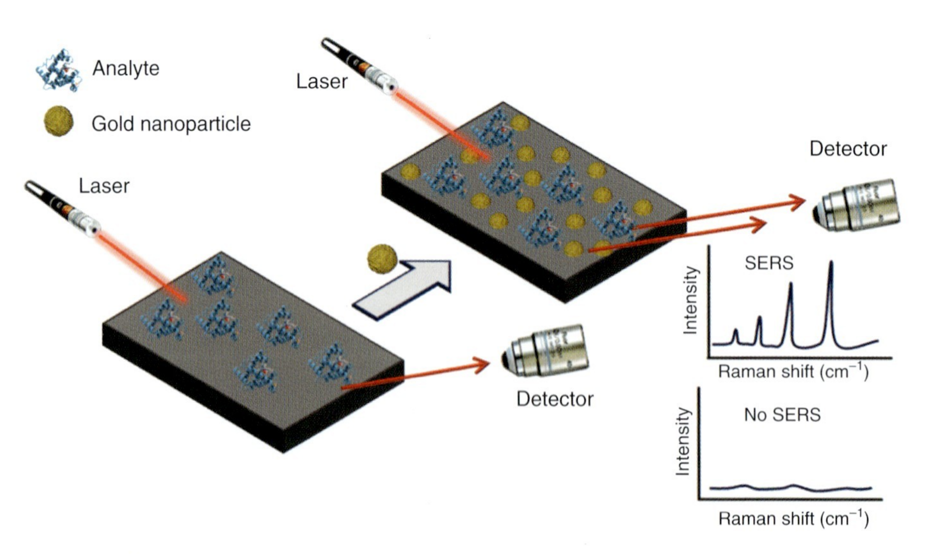

Fig. 10 Surface-enhanced Raman spectroscopy (SERS)-based biosensor.

Objects with different shapes will have different resonances, and more than one resonance condition associated with a given shape. The local field intensity enhancement factor at two different wavelengths is strongly position-dependent in most cases, and the direction of the polarization is vertical. Intensity enhancement in more complicated shapes than in the simplest cases of a cylinder or sphere can also be very high in some circumstances.

In electrostatic approximations for the calculation of field enhancement, size is not really important because the local field intensity enhancement factor will be the same as in the approximation. However, size is important if the objects are in the range of typical dimensions of ~30 to 100 nm. Generally, localized SPR red shifts as the size increases. SPR are also strongly damped as the size increases, mostly as a result of increased radiation losses. This results in a broadening of the resonance, such that the latter will eventually disappear for large sizes (typically 100 nm for dipolar localized surface plasmon in sphere, but possibly larger sizes for other geometries).

The choice of metal in SERS experiments is also critical for improving the results obtained. Generally, it is clear that silver outperforms gold, an advantage that can be tracked down to the higher absorption of gold at the frequencies where resonance occurs. However, the red shift induced by object interaction and shape and size effects can push the resonance in gold to the wavelength region (>600 nm). In this case, gold may be as good as silver, especially for bioapplications. Many biological applications of the techniques are based on near-infrared lasers (typical examples being diode lasers at ~750 or ~830 nm), and gold will be probably the most preferred plasmonic substrate.

5
Piezoelectric Biosensors

Piezoelectric biosensors are sensing devices which couple the bioaffinity recognition processes between the biological probe molecule and the target analyte molecule with the acoustic wave-based transduction mechanism, better known as the *piezoelectric effect*. The demonstration of a linear relationship between the mass adsorbed onto the surface of the piezoelectric crystal and its resonant frequency [89] and the development of suitable oscillator circuits for their operation in liquid [90], led to their application in the field of biological sensing. Piezoelectric crystals can be combined with interfacial chemistry for the immobilization of biorecognition elements and macro- and micro-fluidic systems which enable a controlled contact of the analyte solution with the receptors [91], so as to yield efficient biosensing devices.

Certain solid materials (especially crystals lacking a center of symmetry) that demonstrate charge accumulation upon the application of mechanical stress or internal mechanical strain when subjected to an external electric field are said to exhibit the "piezoelectric effect." In general, a mechanical stress that originates from a change in the mass of the adsorbed film on the piezoelectric crystal changes the resonance frequency of the crystal; this relationship is given by:

$$\Delta f = \frac{-2f_0 \Delta m}{A\sqrt{\rho\mu}} \qquad (2)$$

where Δf is the change in resonant frequency (in Hz); f_0 is the resonant frequency of the crystal (in MHz); Δm is the change in mass (in g); A is the piezoelectrically active area of the crystal, between the electrodes (in cm^2); ρ is the density of the crystal (in

$g\,cm^{-3}$); and μ is the shear modulus of the crystal (in $g\,cm^{-2}.s$).

The change Δf of the resonant frequency f_0 of the piezoelectric crystal is directly proportional to the change in mass, Δm. More specifically, the density, viscosity, elasticity, electric conductivity and dielectric constant of the sensing element can also undergo changes and, in turn, affect the piezoelectric transducer [92]. For an acoustically thin film, the mass change affects the transducer response [93, 94], whereas for an acoustically thick film, the film's viscous and elastic properties and geometric features also make significant contributions. Generally, the change in mass is central to the application of piezoelectric transducers in biosensors. However, there are instances when the ability of such transducers to quantify changes in shear modulus and viscosity has been exploited to fabricate efficient biosensors to study lipids and membranes. Figure 11 shows a typical piezoelectric transducer; here, the biosensing layer with the immobilized bioreceptors can be fabricated over the transducer surface. Piezoelectric biosensors have been employed in the label-free detection of a wide array of analytes ranging from proteins, oligonucleotides and DNAs, antigens, small molecules to viruses and bacteria. They have also been used widely to study protein–protein, protein–DNA, protein–peptide, peptide–peptide interactions, as well as interactions of carbohydrates with proteins, lipids, and other carbohydrates.

In general, for biosensor applications the piezoelectric material should be capable of operation in the liquid media. For efficient operation in the liquid medium, the acoustic waves must be either shear horizontally polarized, or their phase speed should be lower than the speed of sound propagation in the liquid [95]. The dielectric constant of the material should match that of the medium in which the device is to be used, in order to prevent a capacitive short-circuit of the electric field at the interdigital transducers (IDTs) [96]. Quartz, lithium niobate ($LiNbO_3$), potassium niobate ($KNbO_3$), lithium tantalate ($LiTaO_3$), and langasite (lanthanum gallium silicate) are some of the most commonly used piezoelectric crystals for the fabrication of acoustic devices. Knowledge of the specific properties of these different types of acoustic wave devices, such as the mechanical displacement of acoustic waves, the spatial distribution of mechanical and electrical fields, susceptibility to spurious coupling modes, and the sensitivity to

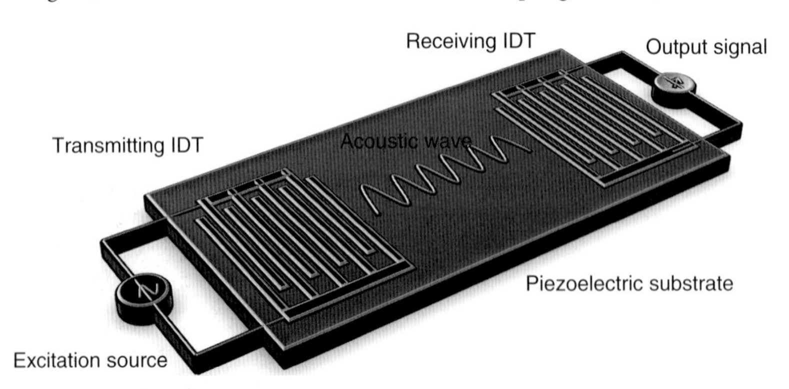

Fig. 11 Schematic of a piezoelectric transducer.

temperature and pressure, is also important for the design of an efficient biosensor [92].

Depending on the acoustic wave-guiding mechanism, acoustic wave devices can be divided into three categories: (1) bulk acoustic wave (BAW) devices, in which the wave propagates unguided through the volume of the material; (2) surface acoustic wave (SAW) devices, in which the wave propagates, guided or unguided, along a single surface of the material; and (3) acoustic plate mode (APM) devices, in which the acoustic waves are guided by reflection from multiple surfaces [93]. The SAW and APM devices can be grouped together as surface-generated acoustic wave (SGAW) devices.

5.1 Bulk Acoustic Wave (BAW) Sensors

The BAW sensors, better known as thickness shear mode (TSM) devices or quartz crystal microbalance (QCM) devices, have traditionally been the choice of transducers for biosensors [97]. In the QCM bulk wave devices, the acoustic wave travels unguided through the entire volume of the piezoelectric substrate, resulting in vibration of the complete substrate. The displacement is maximized at the surface of the crystal, which makes the devices sensitive to surface interactions [98]. A typical BAW device consists of a piezoelectric crystal sandwiched between two electrodes that are generally produced by vapor-depositing Au or Pt onto the electrode surface. An electric field applied between the electrodes results in the mechanical oscillation of a standing shear wave across the bulk of the crystal at its natural resonant frequency. The frequency of the vibration depends on the properties of the crystal (size, density, cut, and shear modulus), and also on the properties of the phases adjacent to it [99]. This frequency changes when the target analyte molecules become attached to the bioactive layer that has been immobilized on the piezoelectric substrate, and this constitutes the output signal from the sensing device pertaining to the analyte. The sensitivity of these devices is limited by the thickness of the piezoelectric crystal. For higher sensitivity, a higher resonant frequency is required; this can be achieved by reducing the thickness of the crystal, but thinner crystals are more fragile and difficult to handle. The QCM devices have been well investigated for the past 50 years, and have subsequently become a mature, commercially available, robust, and affordable technology [100, 101]. Typically, frequencies between 5 and 30 MHz are used.

The BAW sensors can generally be operated in two ways. In the first (*"dip-and-dry"*) method, the reaction between the analyte and the immobilized biorecognition element takes place in the solution phase, while the analysis and quantification occur under the gas phase [99]. The method involves measuring the vibrational frequency of the piezoelectric quartz crystal (PQC) before dipping the device in the analyte solution for a stipulated time. The device is then rinsed to remove any nonspecifically bound molecules, dried, and the vibrational frequency is measured again. Shons *et al.* [102] described the first PQC biosensor in 1972, while Grande *et al.* [103] discussed some of the complications of dip-and-dry methods, such as solvent retention. Unfortunately, the dip-and-dry method does not provide any real-time analysis. The second method involves solution-phase sensing for which the contact cell is configured in a flow or batch mode and a peristaltic or syringe pump is used to introduce the test solution into the cell [99]. Solution-phase sensing allows for real-time analysis.

Previously, BAW sensors have been used for the detection of viruses, bacteria and other cells. Lee *et al.* [104] demonstrated sensitivity comparable to an enzyme-linked immunosorbent assay (ELISA) for the detection of cattle bovine ephemeral fever virus. The application of QCM biosensors in microbiology can be categorized into three areas: the direct detection of a microbe or spore; the detection of an associated antigen or toxin; and the study and characterization of biofilm formed by a microbe [91]. Biosensors for a large number of bacteria, and for the toxins produced by them, have been reported. QCM sensors have been used successfully to monitor and quantify some key processes in biofilm formation and colonization in real time. QCM sensors have also been used to determine proteins, small molecules such as drugs, hormones and pesticides, and nucleic acids.

5.2 Surface-Generated Acoustic Wave (SGAW) Sensors

SAW and APM devices can be grouped together as SGAW devices as both involve the generation and detection of acoustic waves at the surface of the piezoelectric crystal by means of IDT [94]. Since, the acoustic wave is confined to the surface of the crystals, these devices are not affected by the crystal thickness [98]. SAW devices operate at higher frequencies than BAW devices which, in principle, may lead to higher sensitivities because the acoustic wave penetration depth in the adjacent media is reduced [105]. In a typical configuration, an electrical signal is converted at the input IDT into a polarized transverse acoustic wave traveling parallel to the substrate surface. The amplitude and/or velocity of the wave are affected by any coupling reaction at the surface. The output IDT at the opposite end picks up the acoustic wave and converts it back to an electrical signal; any attenuation of the wave is then reflected in the output signal (Fig. 12). Depending on the piezoelectric substrate material, the crystal cut, the positioning of IDTs on the substrate, plate thickness and wave guide mechanism, different operational modes of SGAW such as shear horizontal surface acoustic wave (SH-SAW), surface transverse wave (STW), Love wave, shear horizontal acoustic plate mode (SH-APM), and layer-guided acoustic plate mode (LG-APM) can be achieved [95]. For a better understanding of these SGAW modes, an excellent review is provided in Ref. [95]. SGAW devices can be manufactured using IC microfabrication or central metallica-oxen semiconductor (CMOS) techniques, which allows for the integration of a signal processing unit in the sensor architecture itself [95, 106].

The basic set-up of a SGAW-based biosensor consists of a piezoelectric transducer with an immobilized biospecific layer coupled to a driving electronic circuit and integrated with a sample flow mechanism driven by a peristaltic or syringe pump. Different SGAW techniques have been used according to the sensitivity and operational requirements. Love wave sensors are the most sensitive, with an operating range of 80–300 MHz and mass sensitivity of 150–500 $cm^2\,g^{-1}$ [108, 109]. When Gizeli *et al.* [110] reported the first biosensor based on Love waves in 1992, the device consisted of a quartz crystal with poly(methylmethacrylate) (PMMA) wave guide layer and immunoglobulin G (IgG) immobilized on the surface as the probe. Among others, a STW device is reported to have a sensitivity of 100–200 $cm^2\,g^{-1}$ and an operating frequency of 30–300 MHz [108, 109]. Respective values for SH-APM are 20–50 $cm^2\,g^{-1}$[93, 108, 109] and

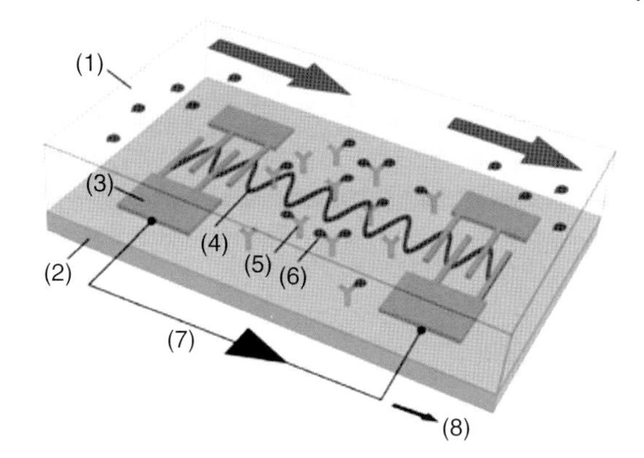

Fig. 12 Surface-generated acoustic wave (SGAW) sensor set-up. The arrows at the top indicate the flow of the liquid sample (1) in which the sensor is immersed. The elements of the SGAW biosensor are a piezoelectric crystal (2), IDTs (3), the surface acoustic wave (4), and immobilized antibodies (5) corresponding to the analyte molecules (6) in the sample. The driving electronics (7) operate the SAW biosensor and generate changes in the output signal (8) as the analyte binds to the sensor surface. Reproduced with permission from Ref. [107]; © 2008, Springer Science and Business Media.

25–200 MHz, and for LG-APM are 20–40 $cm^2\ g^{-1}$ [108, 109] and 25–200 MHz.

6 Thermal Biosensors

Thermal biosensors function by monitoring the change in temperature due to the enthalpy changes associated with any biochemical reaction, and as such are independent of the optical or electrochemical properties of the biocatalyst (usually enzyme), substrate or product. The invention of the enzyme thermistor (ET), which couples flow injection analysis (FIA) with an immobilized biocatalyst and a heat-sensing element [111], led to a surge in investigations into the design and applications of such biosensors. Their versatility and superior operational stability make thermal biosensors useful for such diverse applications as clinical analysis, food analysis, industrial process monitoring and environmental monitoring.

Any biochemical reaction is accompanied by the evolution or absorption of heat. The change in temperature (ΔT) of the system can be defined in terms of the enthalpy change associated with the reaction and the heat capacity of the system as: total heat evolved or absorbed during the reaction is given by:

$$\Delta T = \frac{-n_p \cdot \Delta H}{C_S} \tag{3}$$

where n_p is the total number of moles of the product, ΔH is the molar enthalpy change associated with the reaction and C_S is the total heat capacity of the system.

The enthalpy changes associated with biochemical enzymatic reactions are usually in the range of 10 to 200 $kJ\ mol^{-1}$, and are adequate to determine substrate concentrations at clinically interesting levels [112]. The molar enthalpy changes for some common enzyme-catalyzed reactions are listed in Table 1. The total enthalpy change is the sum of enthalpy changes associated with individual reactions. Thus, the measurement can be improved

Tab. 1 Molar enthalpy changes for some common enzyme reactions [112]

Enzyme	*Substrate*	*Enthalpy change (ΔH; $kJ\ mol^{-1}$)*
Catalase	Hydrogen peroxide	100
Cholesterol oxidase	Cholesterol	53
Glucose oxidase	Glucose	80–100
Hexokinase	Glucose	75[a]
Lactate dehydrogenase	Sodium pyruvate	62
NADH dehydrogenase	NADH	225
β-Lactamase	Penicillin G	115[a]
Trypsin	Benzoyl-L-arginine amide	29
Urease	Urea	61
Uricase	Urate	49

a) In Tris buffer (protonation enthalpy: -47.5 $kJ\ mol^{-1}$).

by coimmobilizing two enzymes – for example, oxidases with catalases – using a high-protonation enthalpy buffer such as tris(hydroxymethyl) aminomethane (TRIS) in the case of proton-producing biochemical reactions [112], using organic solvents which have lower heat capacities than aqueous solvents [113], or by the enzymatic recycling of the substrate where the net enthalpy change in each cycle adds to the overall enthalpy change [114, 115]. An enthalpy change of 100 $kJ\ mol^{-1}$ is sufficient for the detection of analyte concentrations down to 5 $\mu mol\ l^{-1}$.

The conventional thermometric device consists of a working column with the enzyme immobilized on a supporting matrix, and a thermal transducer (usually a thermistor) placed in the vicinity of the column. A schematic of a flow injection analysis enzyme thermistor (FIA-ET), a commonly used thermal biosensor, is shown in Fig. 13. The broad design includes an external jacket for insulation, a working column with immobilized enzymes, an indirect placement of the thermistor to prevent fouling, a heat exchanger prior to the working column to avoid temperature fluctuation, and a peristaltic pump to drive the buffer and analyte solution through the system. Nonspecificity is inherent in calorimetry, since all enthalpy changes in the reaction contribute to the measurement. However, this problem can be overcome by having a split-flow arrangement in which the test solution passes through two different columns. Typically, an active column and a reference column containing only the support matrix or, in some cases inactivated enzyme, is often used to minimize the effects of nonspecific enthalpies arising from nonenzymatic reactions [112]. On the other hand, the fabrication of miniaturized thermometric biosensors has become possible due to advances in the field of IC technology and the micromachining of liquid filters, microvalves and micropumps [113]. Miniaturized devices are suitable for portable use because of their high sensitivity, small size, modest buffer consumption, and good operational stability [112, 116–119].

Enzymes are the most commonly used biorecognition element in a thermal biosensor; however, in cases where the isolation of a pure enzyme is not possible, whole

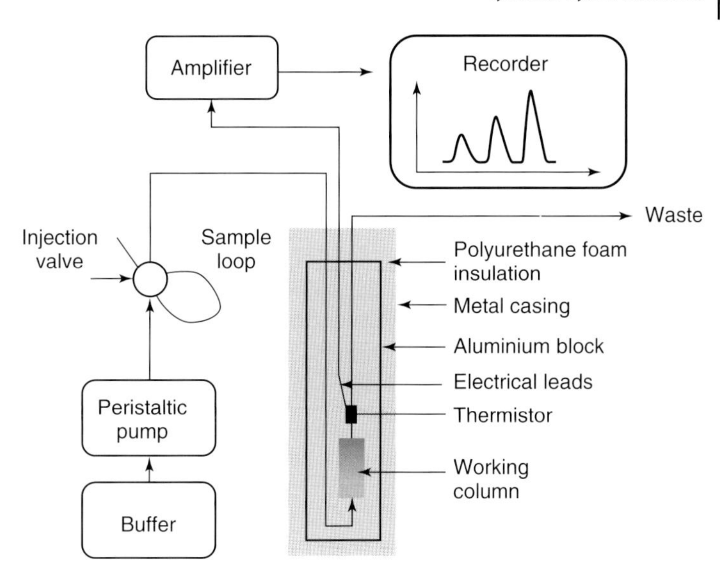

Fig. 13 Schematic of an enzyme thermistor. Reproduced with permission from Ref. [112]; © 2012, Elsevier.

cells, organelles or tissue slices present a good alternative, although they may lack specificity and may also respond to some interfering compounds; MIPs may represent an alternative choice of receptor. Correct immobilization of the biorecognition element on the support matrix is important to maintain a good catalytic activity of the former. The supporting matrix chosen should also be mechanically stable to withstand physical stress, to allow good flow properties [112], and should not interfere with the enzymatic reactions. Controlled pore glass (CPG), Sepharose CL-6B or CL-4B, Eupergit [112], reticulated vitreous carbon (for hybrid thermal-electrochemical sensors) [115, 120–122] and ceramic hydroxyapatite [123] are the most commonly used matrix materials. In general, a large excess of enzyme is immobilized on the support matrix to ensure correct operational stability.

Enzyme thermistors (ETs) have been used to determine a wide range of analytes, such as ethanol, glucose, oxalate, ascorbate, cellobiose and sucrose, and penicillin. Thermal biosensors have been used for the selective measurement of fructose in the presence of glucose [124] and to determine levels of urea in adulterated milk [125]. ETs have also been used for clinical monitoring; for example, a semi-continuous blood glucose monitoring ET device [126] and a cholesterol ET sensor [127] have been described. ETs have also been used for off-line as well as on-line monitoring of bioprocesses such as fermentation [128–130]. Heavy-metal ions such as Hg^{2+}, Cu^{2+} and Ag^{+} can be monitored using ETs via their inhibitory actions on urease, while pesticides can be determined by their inhibitory actions on the enzymes acetylcholinesterase and butylcholinesterase. Apoenzyme-based ETs can be used to monitor heavy-metal ions up to submillimolar levels [112]; an example is that of Cu^{2+} concentrations in human blood sera, which were measured using immobilized ascorbate oxidase [131] or galactose oxidase [132].

MIP-based thermistors have been used for the label-free characterization of MIP binding and catalysis [133, 134]. Thermometric transduction has also been coupled to an ELISA to yield a thermometric enzyme-linked immunosorbent assay (TELISA) [135] that can be used to determine the presence of hormones, antibodies and other biomolecules in complex matrices such as fermentation broth, blood samples and hybridoma cell media [112]. Although the sample capacity and sensitivity of TELISA is lower than for other established techniques (such as radioimmunoassay), it offers a faster monitoring and can be employed where rapid results are desired.

7 Microarrays

A microarray is a high-throughput, two-dimensional screening array that is located on a glass slide or a silicon thin-film cell and can be used to assay large quantities of biological materials. The concept and methodology of microarrays were introduced in 1983; since then the technologies of DNA, protein, peptide, tissue, cellular, carbohydrate and even phenotype microarrays have become highly sophisticated and the most used worldwide [136].

7.1 DNA Microarray

A microarray usually contains picomoles of a specific sequence as probes, and this enables many genetic tests to be executed in parallel, simultaneously. As such, DNA microarrays have dramatically accelerated many types of investigation [137–143]. The core principle of DNA (or RNA) microarrays is based on a hybridization between complementary base pairs by strong hydrogen bonds; the total number of fluorescently labeled target sequences that will bind to a probe sequences will depend on the amount of target sample and thus provide quantitative information relating to the target (Fig. 14). The amount of target samples to be detected is generally limited, however, and additional steps such as the polymerase chain reaction (PCR) and target labeling with fluorescent dyes can be burdensome steps. Consequently, much effort has been made to increase the detection signal by using highly fluorescent probing materials such as CPs or inorganic quantum dots, and to develop label-free detection methods in microarrays in order to avoid a cumbersome labeling step [55, 56, 144–150].

Whilst a "traditional" solid-state array involves a collection of orderly microscopic spots called *features*, each with thousands of probes attached onto a surface, an "alternative" bead array is a collection of microscopic polystyrene beads, each with a specific probe and a mixture of two or more dyes which do not interfere with the fluorescence of the dyes used on the target sequence. Depending on the number of probes, the types of scientific questions being asked and the cost, DNA microarrays can be manufactured in different ways. Some of the most-often used technique for DNA microarray manufacture include printing with fine-pointed pins onto glass slides, photolithography using pre-prepared masks or dynamic micro-mirror devices, ink-jet printing, and electrochemistry on microelectrode arrays. DNA probes can also be synthesized directly onto a microarray substrate (*in situ*) or attached (spotted) via surface engineering by a covalent bonding such as silane, lysine, or amide chemistry.

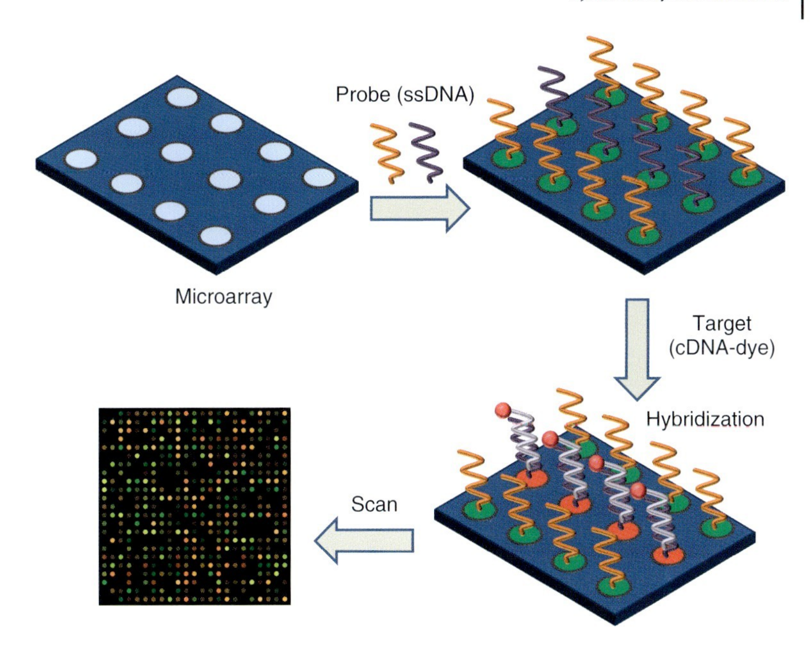

Fig. 14 DNA microarray.

Applications of DNA microarrays include gene expression profiling, comparative genomic hybridization, GeneID, chromatin immuneprecipitation on chip, DamID, single nucleotide polymorphism (SNP) detection, exon (junction) arrays, fusion gene microarray, and tiling arrays.

7.2
Protein Microarray

A protein microarray (protein chip) is a high-throughput method used to track the interaction of large numbers of proteins in parallel, and to determine their function [151–157]. One critical disadvantage of DNA microarrays lies in the fact that the quantity of mRNA in the cell often does not reflect the expression level of the corresponding proteins because proteins – unlike DNA or mRNA – are functional in cell response. Protein microarrays have enabled research groups to study the biological interactions at the cell level. The protein technology was relatively easy to develop as it is based on previously developed DNA microarray technology. Similar to the DNA microarray, the chip consists of a support surface such as a glass slide, nitrocellulose membrane or bead, while the probe molecules are typically labeled with fluorescent dyes. Nowadays, protein microarrays have replaced cumbersome techniques such as two-dimensional gel electrophoresis or chromatography, which are not suited to the analysis of low-abundance proteins and are time-consuming and costly. Protein microarrays can be applied to proteomics, protein functional analysis, antibody characterization, disease treatment development such as antigen-specific therapies for autoimmunity, cancer, allergies, for diagnostics such as tests for antigen–antibody interaction, the discovery of new biomarkers, and the monitoring of disease states.

The surfaces of protein microarrays must meet the sophisticated requirements of immobilizing protein probes, notably to prevent protein denaturation and to provide a relevant surface polarity at which the binding reaction can occur. There is also a need to prevent the nonspecific binding of other proteins, and to minimize the creation of false signals from the background noise. Immobilizing agents vary from layers of inorganic aluminum or gold to organic polymers, polyacrylamide gels, or small functional moieties such as amines, aldehyde and epoxy. Occasionally, thin-film technologies such as physical vapor deposition (PVD) and chemical vapor deposition (CVD) are also used to apply the coating to the support surface. Protein array methods include ink-jetting, robotic spotting, piezoelectric spotting, a drop-on-demand, and photolithography [158–162]. The probe molecules may be antigens, antibodies, aptamers, protein-mimicking peptides, or full-length proteins. Recently, an *in-situ*, on-chip synthesis of proteins directly from DNA, using cell-free expression systems called DAPA (DNA array to protein array), PISA (protein in situ array) or NAPPA (nucleic acid programmable protein array), was introduced as the proteins in an arrayal surface are highly sensitive and easily deteriorate, whereas DNA molecules are more stable over time and better suited to long-term storage.

The detection methods employed included fluorescence labeling, as well as affinity, photochemical or radioisotope tagging. For label-free detection, SPR, carbon nanowire sensors (where detection occurs via changes in electronic conductance) and microelectromechanical system (MEMS) cantilevers can be used. However, these systems are ill-suited for high-throughput screening and need to undergo further development before their future use.

8 Conclusions

During recent years the field of biosensors has undergone an evolutionary phase with ever-increasing demands for efficient, sensitive and robust sensors in the fields of clinical diagnostics, medicine and drugs, process control, and environmental monitoring. In this chapter, attention has been focused on the most basic principles of biosensory sciences, and has hopefully proved valuable to the reader. Electrochemical biosensors are the most widely used sensing devices, given their high efficiency and ease of operation. Optical techniques such as SPR and SER have also found widespread use in research and development, while biosensors based on conducting polymers are beginning to open up a new field of hand-held biosensors that involve internal transduction mechanisms and can respond to analyte recognition through color changes. While acoustic resonance devices seem to be a good choice for bioaffinity sensors, with some noteworthy advances having been made in this field, thermal biosensors do not yet appear to have made any serious practical impact.

During the past decade, the development of biosensors has been greatly spurred by advancements made in the materials sciences. For example, nanostructured metal oxides have been shown to provide an effective immobilization of biomolecules with desired orientation and conformation, resulting in better sensing characteristics [163]. While silver and gold nanoparticles are widely used as electrochemical labels in amperometric immunoassays, metal quantum dots have been used as multilabels for affinity reactions [4]. Carbon nanotubes (CNTs) have also proved to be a material of choice for electrode fabrication, due to their semiconductive behavior and high porosity

[164]. Indeed, amperometric biosensors comprising a CNT-modified electrode have shown an enhanced reactivity of NADH and hydrogen peroxide at the electrode [4]. Graphene, with its one atom-thick single graphitic layer, has attracted much scientific interest due to its unique physicochemical properties such as high surface area and excellent thermal and electric conductivities. Given its excellent electron transport properties and high surface area, functionalized graphene is expected to help in the direct electron transfer between the electrode substrate and enzymes, and thus aid in the design of mediator-free biosensors with potentially better sensing parameters. Whilst conducting polymers have been used successfully as a material for the immobilization of biomolecules, as well as providing enhanced electron transfer properties, they have also been shown to function as stand-alone sensors as their emission properties are influenced by their molecular environment.

With the principles of biosensing and transduction mechanism having been well established, attention in the field of biosensors has now been focused on miniaturization, and this has resulted in smaller, more sensitive, and more easily affordable devices. Microchip technology has helped to concentrate electronic circuits onto a single chip through embedded ICs. However, in spite of the technological innovations and improvements, the miniaturization of these devices poses technical challenges as they lack sensitivity, long-term stability and robustness for their intended applications. Nonetheless, newer technologies such as silicon microsensors, fiber-optic biosensors and cell-on-chip sensors are increasingly being investigated [165].

Further developments in the field of biosensors will most likely be effected by the emergence of personalized medicine, as escalating healthcare costs continue to force the development of a new generation of wearable, integrated and less-invasive sensors [4]. In addition to healthcare and clinical diagnostics, industrial processes and environmental monitoring will also continue to press for more efficient, sensitive and robust biosensors. Moreover, with the increasing risks of biological and chemical warfare, security and biodefense will also require new, innovative and efficient biosensors to meet these challenges.

Acknowledgments

The authors acknowledge the financial support from the Converging Research Center Program funded by the Ministry of Science, ICT and Future Planning (Project No. 2013K000314).

References

1. Monosik, R., Stred'ansky, M., Sturdik, E. (2012) Biosensors-classification, characterization and new trends. *Acta Chim. Slovacaa*, **5**, 109–120.
2. McNaught, A.D., Wilkinson, A. (1997) *IUPAC. Compendium of Chemical Technology*, 2nd edn (the "Gold Book"). Blackwell Scientific Publications, Oxford.
3. Clark, L.C., Lyons, C. (1962) Electrode systems for continuous monitoring in cardiovascular surgery. *Ann. N. Y. Acad. Sci.*, **102**, 29–45.
4. Turner, A.P.F. (2013) Biosensors: sense and sensibility. *Chem. Soc. Rev.*, **42** 3184–3196.
5. Bartlett, P.N. (2008) *Bioelectrochemistry Fundamentals, Experimental Techniques and Applications*, John Wiley & Sons, Ltd, West Sussex.
6. Mikkelsen, S.R., Corton, E. (2004) *Bioanalytical Chemistry*. John Wiley & Sons, Inc., Hoboken, NJ.
7. Collings, A.F., Caruso, F. (1997) Biosensors: recent advances. *Rep. Prog. Phys.*, **60**, 1397–1445.

8. Gorton, L. (1995) Carbon paste electrodes modified with enzymes, tissues, and cells. *Electroanalysis*, **7**, 23–45.
9. Gao, X., Ge, S., Cai, Q., Zeng, K., Grimes, C.A. (2008) Kinetic study on the interaction between tannin and bovine serum albumin with a wireless magnetoelastic biosensor. *Sens. Actuators, B: Chem.*, **129**, 929–933.
10. De Palma, R., Liu, C., Barbagini, F., Reekmans, G. *et al.* (2007) Magnetic particles as labels in bioassays: interactions between a biotinylated gold substrate and streptavidin magnetic particles. *J. Phys. Chem. C*, **111**, 12227–12235.
11. Barhoumi, H., Maaref, A., Cosnier, S., Martelet, C., Jaffrezic-Renault, N. (2008) Urease immobilization on biotinylated polypyrrole-coated ChemFEC devices for urea biosensor development. *IRBM*, **29**, 192–201.
12. Ronkainen, N.J., Halsall, H.B., Heineman, W.R. (2008) Electrochemical biosensors. *Chem. Soc. Rev.*, **39**, 1747–1763.
13. Arya, S.K., Singh, S.P., Malhotra, B.D. (2008) Electrochemical Techniques in Biosensors, in: *Handbook of Biosensors and Biochips*, John Wiley & Sons, Ltd, Chichester.
14. Halsall, H.B., Heineman, W.R. (1990) Electrochemical immunoassay: an ultrasensitive method. *J. Int. Fed. Clin. Chem.*, **2**, 179–187.
15. Jiang, T., Halsall, H.B., Heineman, W.R., Giersch, T. *et al.* (1995) Capillary enzyme immunoassay with electrochemical detection for the determination of Atrazine in water. *J. Agric. Food Chem.*, **43**, 1098–1104.
16. Wightman, R.M., Wipf, D.O. (1989) Voltammetry at Ultramicroelectrodes, in: Bard, A.J. (Ed.), *Electroanalytical Chemistry*, vol. 15, Marcel Dekker, New York, pp. 267–353.
17. Ronkainen-Matsuno, N.J., Thomas, J.H., Halsall, H.B., Heineman, W.R. (2002) Electrochemical immunoassay moving into the fast lane. *Trends Anal. Chem.*, **21**, 213–225.
18. van Emon, J.M. (Ed.) (2007) *Immunoassay and Other Bioanalytical Methods*, CRC Press, Boca Raton, FL.
19. Niwa, O., Morita, M., Tabei, H. (1990) Electrochemical behavior of reversible redox species at interdigitated array electrodes with different geometries: consideration of redox cycling and collection efficiency. *Anal. Chem.*, **62**, 447–452.
20. Updike, S.J., Hicks, G.P. (1967) The enzyme electrode. *Nature*, **214**, 986–988.
21. Guilbault, G.G., Lubrano, G.J. (1973) Enzyme electrode for amperometric determination of glucose. *Anal. Chim. Acta*, **64**, 439–455.
22. Harrison, D.J., Turner, R.B.F., Baltes, H.P. (1988) Characterization of perfluorosulfonic acid polymer coated enzyme electrodes and a miniaturized integrated potentiostat for glucose analysis in whole blood. *Anal. Chem.*, **60**, 2002–2007.
23. Shimizu, Y., Morita, K. (1990) Microhole array electrode as a glucose sensor. *Anal. Chem.*, **62**, 1498.
24. Borgmann, S., Schulte, A., Neugebauer, S., Schuhmann, W. (2011) Amperometric Biosensors, in: Alkire, R.C., Kolb, D.M., Lipkowski, J. (Eds), *Advances in Electrochemical Science and Engineering: Bioelectrochemistry*, Vol. 13 Wiley-VCH Verlag GmbH and Co. KGaA, Weinheim, pp. 1–83.
25. Cass, A.G., Davis, G., Francis, G.D., Hill, H.A.O. *et al.* (1984) Ferrocene-mediated enzyme electrode for amperometric determination of glucose. *Anal. Chem.*, **56**, 667–671.
26. Schuhmann, W., Bonsen, E.M. (2007) Biosensors, in: *Encyclopedia of Electrochemistry*, Wiley-VCH Verlag GmbH & Co. KGaA, Weinheim.
27. Kormos, F., Sziraki, L., Tarische, I. (2000) Potentiometric biosensor for urinary glucose level monitoring. *Lab. Rob. Autom.*, **12**, 291–295.
28. Psychoyios, V.N., Nikoleli, G.-P., Tzamtzis, N., Nikolelis, D.P. *et al.* (2013) Potentiometric cholesterol biosensor based on ZnO nanowalls and stabilized polymerized lipid film. *Electroanalysis*, **25**, 367–372.
29. Grieshaber, D., MacKenzie, R., Voros, J., Reimhult, E. (2008) Electrochemical biosensors – sensor principles and architectures. *Sensors*, **8**, 1400–1458.
30. Caras, S., Janata, J. (1980) Field effect transistor sensitive to penicillin. *Anal. Chem.*, **52**, 1935–1937.
31. Dzyadevych, S., Soldatkin, A., Korpan, Y., Arkhypova, V. *et al.* (2003) Biosensors based on enzyme field-effect transistors for determination of some substrates and inhibitors. *Anal. Bioanal. Chem.*, **377**, 496–506.

32. Ulber, R., Scheper, T. (1998). Enzyme Biosensors Based on ISFETs, in: Mulchandani, A., Rogers, K.R. (Eds), *Enzyme and Microbial Biosensors Techniques and Protocols* Vol. **6**, Methods in Biotechnology, Humana Press, Totowa, NJ, pp. 35–50.
33. Lee, C.-S., Kim, S., Kim, M. (2009) Ion-sensitive field-effect transistor for biological sensing. *Sensors*, **9**, 7111–7131.
34. Mohanty, S.P., Kougianos, E. (2006) Biosensors a tutorial review. *IEEE Potentials*, **25**, 35–40.
35. Dzydevich, S.V., Shuga, A.A., Soldatkin, A.P., Hendji, A.M.N. *et al.* (1994) Conductometric biosensors based on cholinesterases for sensitive detection of pesticides. *Electroanalysis*, **6**, 752–758.
36. Anh, T.M., Dzyadevych, S.V., Van, M.C., Renault, N.J. *et al.* (2004) Conductometric tyrosinase biosensor for the detection of diuron, atrazine and its main metabolites. *Talanta*, **63**, 365–370.
37. McQuade, D.T., Pullen, A.E., Swager, T.M. (2000) Conjugated polymer-based chemical sensors. *Chem. Rev.*, **100** (7), 2537–2574.
38. Bunz, U.H.F. (2000) Poly(arylene ethynylene)s: syntheses, properties, structures, and applications. *Chem. Rev.*, **100** (4), 1605–1644.
39. Hertel, D., Romanovskii, Y.V., Schweitzer, B., Scherf, U. *et al.* (2001) Spectroscopy of conjugated polymers: phosphorescence and delayed fluorescence. *Macromol. Symp.*, **175**, 141–150.
40. Thomas, S.W. III, Joly, G.D., Swager, T.M. (2007) Chemical sensors based on amplifying fluorescent conjugated polymers. *Chem. Rev.*, **107** (4), 1339–1386.
41. Herland, A., Inganas, O. (2007) Conjugated polymers as optical probes for protein interactions and protein conformations. *Macromol. Rapid Commun.*, **28** (17), 1703–1713.
42. Zhu, C., Liu, L., Yang, Q., Lv, F. *et al.* (2012) Water-soluble conjugated polymers for imaging, diagnosis, and therapy. *Chem. Rev.*, **112** (8), 4687–4735.
43. Ho, H.A., Najari, A., Leclerc, M. (2008) Optical detection of DNA and proteins moth cationic polythiophenes. *Acc. Chem. Res.*, **41** (2), 168–178.
44. Liu, Y., Ogawa, K., Schanze, K.S. (2009) Conjugated polyelectrolytes as fluorescent sensors. *J. Photochem. Photobiol., C: Photochem. Rev.*, **10** (4), 173–190.
45. Jiang, H., Taranekar, P., Reynolds, J.R., Schanze, K.S. (2009) Conjugated polyelectrolytes: synthesis, photophysics, and applications. *Angew. Chem. Int. Ed.*, **48** (24), 4300–4316.
46. Wang, D., Gong, X., Heeger, P.S., Rininsland, F. *et al.* (2002) Biosensors from conjugated polyelectrolyte complexes. *Proc. Natl Acad. Sci. USA*, **99** (1), 49–53.
47. Lee, K., Kim, H.J., Kim, J. (2012) Design principle of conjugated polyelectrolytes to make them water-soluble and highly emissive. *Adv. Funct. Mater.*, **22** (5), 1076–1086.
48. Tuncel, D., Demir, H.V. (2010) Conjugated polymer nanoparticles. *Nanoscale*, **2** (4), 484–494.
49. Li, K., Liu, B. (2012) Polymer encapsulated conjugated polymer nanoparticles for fluorescence bioimaging. *J. Mater. Chem.*, **22** (4), 1257–1264.
50. Lee, K., Povlich, L.K., Kim, J. (2010) Recent advances in fluorescent and colorimetric conjugated polymer-based biosensors. *Analyst*, **135** (9), 2179–2189.
51. Zhou, Q., Swager, T.M. (1995) Fluorescent chemosensors based on energy migration in conjugated polymers: the molecular wire approach to increased sensitivity. *J. Am. Chem. Soc.*, **117** (50), 12593–12602.
52. Swager, T.M. (1998) The molecular wire approach to sensory signal amplification. *Acc. Chem. Res.*, **31** (5), 201–207.
53. Xie, D.P., Parthasarathy, A., Schanze, K.S. (2011) Aggregation-induced amplified quenching in conjugated polyelectrolytes with interrupted conjugation. *Langmuir*, **27** (19), 11732–11736.
54. Harrison, B.S., Ramey, M.B., Reynolds, J.R., Schanze, K.S. (2000) Amplified fluorescence quenching in a poly(*p*-phenylene)-based cationic polyelectrolyte. *J. Am. Chem. Soc.*, **122** (35), 8561–8562.
55. Lee, K., Rouillard, J.M., Kim, B.G., Gulari, E. *et al.* (2009) Conjugated polymers combined with a molecular beacon for label-free and self-signal-amplifying DNA microarrays. *Adv. Funct. Mater.*, **19** (20), 3317–3325.
56. Lee, K., Rouillard, J.M., Pham, T., Gulari, E. *et al.* (2007) Signal-amplifying conjugated polymer-DNA hybrid chips. *Angew. Chem. Int. Ed.*, **46** (25), 4667–4670.

57. Bolinger, J.C., Traub, M.C., Brazard, J., Adachi, T. *et al.* (2012) Conformation and energy transfer in single conjugated polymers. *Acc. Chem. Res.*, **45** (11), 1992–2001.
58. Wood, R.W. (1902) On a remarkable case of uneven distribution of light in a diffraction grating spectrum. *Philos. Mag.*, **4** (19-24), 396–402.
59. Otto, A. (1968) Excitation of nonradiative surface plasma waves in silver by method of frustrated total reflection. *Z. Phys.*, **216** (4), 398–410.
60. Schasfoort, R.B.M., Tudos, A.J. (2008) *Handbook of Surface Plasmon Resonance*. Royal Society of Chemistry, Cambridge, vol. xxi.
61. Zeng, S.W., Yong, K.T., Roy, I., Dinh, X.Q. *et al.* (2011) A review on functionalized gold nanoparticles for biosensing applications. *Plasmonics*, **6** (3), 491–506.
62. Rich, R.L., Myszka, D.G. (2007) Higher-throughput, label-free, real-time molecular interaction analysis. *Anal. Biochem.*, **361** (1), 1–6.
63. Hiep, H.M., Endo, T., Kerman, K., Chikae, M. *et al.* (2007) A localized surface plasmon resonance based immunosensor for the detection of casein in milk. *Sci. Technol. Adv. Mater.*, **8** (4), 331–338.
64. Liedberg, B., Nylander, C., Lundstrom, I. (1983) Surface-plasmon resonance for gas-detection and biosensing. *Sens. Actuators*, **4** (2), 299–304.
65. Cooper, M.A. (2002) Optical biosensors in drug discovery. *Nat. Rev. Drug Discov.*, **1**, 515–528.
66. Raether, H. (1987) *Surface Plasmons on Smooth and Rough Surfaces and on Gratings*. Springer-Verlag, Berlin, New York, vol. x.
67. Zhang, Y., Dragan, A., Geddes, C.D. (2009) Wavelength dependence of metal-enhanced fluorescence. *J. Phys. Chem. C*, **113** (28), 12095–12100.
68. Lee, K., Hahn, L.D., Yuen, W.W., Vlamakis, H. *et al.* (2011) Metal-enhanced fluorescence to quantify bacterial adhesion. *Adv. Mater.*, **23** (12), H101–H104.
69. Pompa, P.P., Martiradonna, L., Torre, A.D., Sala, F.D. *et al.* (2006) Metal-enhanced fluorescence of colloidal nanocrystals with nanoscale control. *Nat. Nanotechnol.*, **1** (2), 126–130.
70. Geddes, C.D., Gryczynski, I., Malicka, J., Gryczynski, Z. *et al.* (2003) Metal-enhanced fluorescence: potential applications in HTS. *Comb. Chem. High Throughput Screen.*, **6** (2), 109–117.
71. Malicka, J., Gryczynski, I., Lakowicz, J.R. (2003) DNA hybridization assays using metal-enhanced fluorescence. *Biochem. Biophys. Res. Commun.*, **306** (1), 213–218.
72. Pugh, V.J., Szmacinski, H., Moore, W.E., Geddes, C.D. *et al.* (2003) Submicrometer spatial resolution of metal-enhanced fluorescence. *Appl. Spectrosc.*, **57** (12), 1592–1598.
73. Geddes, C.D., Parfenov, A., Lakowicz, J.R. (2003) Photodeposition of silver can result in metal-enhanced fluorescence. *Appl. Spectrosc.*, **57** (5), 526–531.
74. Geddes, C.D., Parfenov, A., Roll, D., Uddin, M.J. *et al.* (2003) Fluorescence spectral properties of indocyanine green on a roughened platinum electrode: metal-enhanced fluorescence. *J. Fluoresc.*, **13** (6), 453–457.
75. Aslan, K., Geddes, C.D. (2006) Microwave-accelerated metal-enhanced fluorescence (MAMEF): application to ultra-fast and sensitive clinical assays. *J. Fluoresc.*, **16** (1), 3–8.
76. Aslan, K., Lakowicz, J.R., Geddes, C.D. (2005) Metal-enhanced fluorescence using anisotropic silver nanostructures: critical progress to date. *Anal. Bioanal. Chem.*, **382** (4), 926–933.
77. Aslan, K., Gryczynski, I., Malicka, J., Matveeva, E. *et al.* (2005) Metal-enhanced fluorescence: an emerging tool in biotechnology. *Curr. Opin. Biotechnol.*, **16** (1), 55–62.
78. Aslan, K., Lakowicz, J.R., Szmacinski, H., Geddes, C.D. (2005) Enhanced ratiometric pH sensing using SNAFL-2 on silver island films: metal-enhanced fluorescence sensing. *J. Fluoresc.*, **15** (1), 37–40.
79. Li, H., Chen, C.Y., Wei, X., Qiang, W. *et al.* (2012) Highly sensitive detection of proteins based on metal-enhanced fluorescence with novel silver nanostructures. *Anal. Chem.*, **84** (20), 8656–8662.
80. Geddes, C.D., Lakowicz, J.R. (2002) Metal-enhanced fluorescence. *J. Fluoresc.*, **12** (2), 121–129.
81. Schlücker, S. (Ed.) (2011) *Surface-Enhanced Raman Spectroscopy: Analytical, Biophysical*

and Life Science Applications, Wiley-VCH Verlag GmbH, Weinheim.

82. Fleischm, M., Hendra, P.J., Mcquilla, A.J. (1974) Raman-spectra of pyridine adsorbed at a silver electrode. *Chem. Phys. Lett.*, **26** (2), 163–166.
83. Kahl, M., Voges, E. (2000) Analysis of plasmon resonance and surface-enhanced Raman scattering on periodic silver structures. *Phys. Rev. B*, **61** (20), 14078–14088.
84. Shalaev, V.M., Sarychev, A.K. (1998) Nonlinear optics of random metal-dielectric films. *Phys. Rev. B*, **57** (20), 13265–13288.
85. Smith, E., Dent, G. (2005) *Modern Raman Spectroscopy: A Practical Approach*. John Wiley & Sons, Inc., Hoboken, NJ, vol. x.
86. Kneipp, K., Moskovits, M., Kneipp, H. (2006) *Surface - enhanced Raman Scattering: Physics and Applications*. Springer, Berlin, New York, vol. xvii.
87. Lombardi, J.R., Birke, R.L., Lu, T.H., Xu, J. (1986) Charge-transfer theory of surface enhanced Raman-spectroscopy – Herzberg–Teller contributions. *J. Chem. Phys.*, **84** (8), 4174–4180.
88. Lombardi, J.R., Birke, R.L. (2008) A unified approach to surface-enhanced Raman spectroscopy. *J. Phys. Chem. C*, **112** (14), 5605–5617.
89. Sauerbrey, G. (1959) Verwendung von Schwingquarzen zur Wagung dunner Schichten und zur Mikrowagung. *Z. Phys.*, **155**, 206–222.
90. Nomura, T., Okuhara, M. (1982) Frequency shifts of piezoelectric quartz crystals immersed in organic liquids. *Anal. Chim. Acta*, **142**, 281–284.
91. Cooper, M.A., Singleton, V.T. (2007) A survey of the 2001 to 2005 quartz crystal microbalance biosensor literature: applications of acoustic physics to the analysis of biomolecular interactions. *J. Mol. Recognit.*, **20**, 154–184.
92. Lec, R.M. (2001) Piezoelectric biosensors: recent advances and applications. Presented at Proceedings of the 2001 IEEE International Frequency Control Symposium and PDA Exhibition, 2001.
93. Ballantaine, D.S., White, R.M., Martin, S.J., Ricco, A.J. *et al.* (1997) *Acoustic Wave Sensors: Theory, Design and Physico-Chemical Applications*. Academic Press, San Diego, CA.
94. Lucklum, R., Behling, C., Hauptmann, P. (1999) Role of mass accumulation and viscoelastic film properties for the response of acoustic-wave-based chemical sensors. *Anal. Chem.*, **71**, 2488–2496.
95. Rocha-Gaso, M.A.-I., March-Iborra, C., Montoya-Baides, A., Arnau-Vives, A. (2009) Surface generated acoustic wave biosensors for the detection of pathogens: a review. *Sensors*, **9**, 5740–5769.
96. Rupp, S., von Schickfus, M., Hunklinger, S., Eipel, H. *et al.* (2008). A shear horizontal surface acoustic wave sensor for the detection of antigen-antibody reactions for medical diagnosis. *Sens. Actuators B*, **134**, 225–229.
97. Kogai, T., Yotsuda, H. (2006) 3F-3 Liquid Sensor using SAW and SH-SAW on Quartz. IEEE Ultrason Symposium, 2–6 October, Vancouver, BC, Canada, pp. 552–555.
98. Gronewold, T.M.A. (2007) Surface acoustic wave sensors in the bioanalytical field: Recent trends and challenges. *Anal. Chim. Acta*, **603**, 119–128.
99. Bunde, R.L., Jarvi, E.J., Rosentreter, J.J. (1998) Piezoelectric quartz crystal biosensors. *Talanta*, **46**, 1223–1236.
100. Janshoff, A., Galla, H.J., Steinem, C. (2000) Piezoelectric mass sensing devices as biosensors – an alternative to optical biosensors? *Angew. Chem. Int. Ed.*, **39**, 4005–4032.
101. Andle, J.C., Vetelino, J.F. (1994) Acoustic wave biosensors. *Sens. Actuators A*, **44**, 167–176.
102. Shons, A., Dorman, F., Najarian, J. (1972) An immunospecific microbalance. *J. Biomed. Mater. Res.*, **6**, 565–570.
103. Grande, L.H., Geren, C.R., Paul, D.W. (1988) Detection of galactosyltransferase using chemically modified piezoelectric quartz. *Sens. Actuators*, **14**, 387–403.
104. Lee, Y.-G., Chang, K.-S. (2005) Application of a flow type quartz crystal microbalance immunosensor for real time determination of cattle bovine ephemeral fever virus in liquid. *Talanta*, **65**, 1335–1342.
105. Smith, J.P., Hinson-Smith, V. (2006) Commercial SAW sensors move beyond military and security applications. *Anal. Chem.*, **78**, 3505–3507.
106. Tigli, O., Zagnhloul, M.E. (2007) A novel SAW device in CMOS: design, modeling and fabrication. *IEEE Sens. J.*, **7**, 219–227.

107. Länge, K., Rapp, B.E., Rapp, M. (2008) Surface acoustic wave biosensors a review. *Anal. Bioanal. Chem.*, **391**, 1509–1519.
108. Grate, J.W., Stephen, J.M., Richard, M.W. (1993). Acoustic wave microsensors. *Anal. Chem.*, **65**, 940A.
109. Ferrari, V., Lucklum, R. (2008) Overview of Acoustic Wave Microsensors, in: Arnau, A. (Ed.), *Piezoelectric Transducers and Applications* (2nd edn), Springer, Berlin, Heidelberg, pp. 39–59.
110. Gizeli, E., Goddard, N.J., Lowe, C.R., Stevenson, A.C. (1992) A Love plate biosensor utilising a polymer layer. *Sens. Actuators B: Chem.*, **6**, 131–137.
111. Danielsson, B., Mosbach, K. (1986) Theory and Applications of Calorimetric Sensors, in: Turner, A.P.F., Karube, I., Wilson, G.S. (Eds), *Biosensors: Fundamentals and Applications*, Oxford University Press, Oxford, pp. 575–595.
112. Yakovleva, M., Bhand, S., Danielsson, B. (2013) The enzyme thermistor – A realistic biosensor concept. A critical review. *Anal. Chim. Acta*, **766**, 1–12.
113. Danielsson, B., Flygare, L., Velev, T. (1989) Biothermal analysis performed in organic solvents. *Anal. Lett.*, **22**, 1417–1428.
114. Scheller, F., Siegbahn, N., Danielsson, B., Mosbach, K. (1985) High-sensitivity enzyme thermistor determination of L-lactate by substrate recycling. *Anal. Chem.*, **57**, 1740–1743.
115. Xie, B., Tang, X., Wollenberger, U., Johansson, G. *et al.* (1997) Hybrid biosensor for simultaneous electrochemical and thermometric detection. *Anal. Lett.*, **30**, 2141–2158.
116. Xie, B., Ramanathan, K., Danielsson, B. (2000) Mini/micro thermal biosensors and other related devices for biochemical/clinical analysis and monitoring. *Trends Anal. Chem.*, **19**, 340–349.
117. Danielsson, B., Hedberg, U., Rank, M., Xie, B. (1992) Recent investigations on calorimetric biosensors. *Sens. Actuators B: Chem.*, **6**, 138–142.
118. Shimohigoshi, M., Yokoyama, K., Karube, I. (1995) Development of a bio-thermochip and its application for the detection of glucose in urine. *Anal. Chim. Acta*, **303**, 295–299.
119. Xie, B., Danielsson, B., Norberg, P., Winquist, F. *et al.* (1992) Development of a thermal micro-biosensor fabricated on a silicon chip. *Sens. Actuators B: Chem.*, **6**, 127–130.
120. Xie, B., Khayyami, M., Nwosu, T., Larsson, P.-O. *et al.* (1993) Ferrocene-mediated thermal biosensor. *Analyst*, **118**, 845–848.
121. Ramanathan, K., Jonsson, B.R., Danielsson, B. (2001) Sol-gel based thermal biosensor for glucose. *Anal. Chim. Acta*, **427**, 1–10.
122. Khayyami, M., Pérez Pita, M.T., Peña Garcia, N., Johansson, G. *et al.* (1998) Development of an amperometric biosensor based on acetylcholine esterase covalently bound to a new support material. *Talanta*, **45**, 557–563.
123. Salman, S., Soundararajan, S., Safina, G., Satoh, I. *et al.* (2008) Hydroxyapatite as a novel reversible in situ adsorption matrix for enzyme thermistor-based FIA. *Talanta*, **77**, 490–493.
124. Bhand, S.G., Soundararajan, S., Surugiu-Wärnmark, I., Milea, J.S. *et al.* (2010) Fructose-selective calorimetric biosensor in flow injection analysis. *Anal. Chim. Acta*, **668**, 13–18.
125. Mishra, G.K., Mishra, R.K., Bhand, S. (2010) Flow injection analysis biosensor for urea analysis in adulterated milk using enzyme thermistor. *Biosens. Bioelectron.*, **26**, 1560–1564.
126. Carlsson, T., Adamson, U., Lins, P.-E., Danielsson, B. (1996) Use of an enzyme thermistor for semi-continuous blood glucose measurements. *Clin. Chim. Acta*, **251**, 187–200.
127. Raghavan, V., Ramanathan, K., Sundaram, P.V., Danielsson, B. (1999) An enzyme thermistor-based assay for total and free cholesterol. *Clin. Chim. Acta*, **289**, 145–158.
128. Danielsson, B., Mosbach, K. (1979) Determination of enzyme activities with the enzyme thermistor unit. *FEBS Lett.*, **101**, 47–50.
129. Mattiasson, B., Danielsson, B. (1982) Calorimetric analysis of sugars and sugar derivatives with aid of an enzyme thermistor. *Carbohydr. Res.*, **102**, 273–282.
130. Mandenius, C., Danielsson, B., Mattiasson, B. (1981) Process control of an ethanol fermentation with an enzyme thermistor as a sucrose sensor. *Biotechnol. Lett.*, **3**, 629–634.

131. Satoh, I. (1991) An apoenzyme thermistor microanalysis for zinc(II) ions with use of an immobilized alkaline phosphatase reactor in a flow system. *Biosens. Bioelectron.*, **6**, 375–379.
132. Satoh, I. (1990) Calorimetric biosensing of heavy metal ions with the reactors containing the immobilized apoenzymes. *Ann. N. Y. Acad. Sci.*, **613**, 401–404.
133. Rajkumar, R., Katterle, M., Warsinke, A., Moehwald, H. *et al.* (2008) Thermometric MIP sensor for fructosyl valine. *Biosens. Bioelectron.*, **23**, 1195–1199.
134. Lettau, K., Katterle, M., Warsinke, A., Scheller, F.W. (2008) Sequential conversion by catalytically active MIP and immobilized tyrosinase in a thermistor. *Biosens. Bioelectron.*, **23**, 1216–1219.
135. Birnbaum, S., Bulow, L., Hardy, K., Danielsson, B. *et al.* (1986) Automated thermometric enzyme immunoassay of human proinsulin produced by *Escherichia coli*. *Anal. Biochem.*, **158**, 12–19.
136. Chang, T.W. (1983) Binding of cells to matrices of distinct antibodies coated on solid-surface. *J. Immunol. Methods*, **65** (1-2), 217–223.
137. Maskos, U., Southern, E.M. (1992) Oligonucleotide hybridizations on glass supports a novel linker for oligonucleotide synthesis and hybridization properties of oligonucleotides synthesised in situ. *Nucleic Acids Res.*, **20** (7), 1679–1684.
138. Augenlicht, L.H., Kobrin, D. (1982) Cloning and screening of sequences expressed in a mouse colon tumor. *Cancer Res.*, **42** (3), 1088–1093.
139. Augenlicht, L.H., Wahrman, M.Z., Halsey, H., Anderson, L. *et al.* (1987) Expression of cloned sequences in biopsies of human colonic tissue and in colonic carcinoma cells induced to differentiate in vitro. *Cancer Res.*, **47** (22), 6017–6021.
140. Kulesh, D.A., Clive, D.R., Zarlenga, D.S., Greene, J.J. (1987) Identification of interferon-modulated proliferation-related cDNA sequences. *Proc. Natl Acad. Sci. USA*, **84** (23), 8453–8457.
141. Lashkari, D.A., DeRisi, J.L., McCusker, J.H., Namath, A.F. *et al.* (1997) Yeast microarrays for genome wide parallel genetic and gene expression analysis. *Proc. Natl Acad. Sci. USA*, **94** (24), 13057–13062.
142. Schena, M., Shalon, D., Davis, R.W., Brown, P.O. (1995) Quantitative monitoring of gene expression patterns with a complementary DNA microarray. *Science*, **270** (5235), 467–470.
143. Sassolas, A., Leca-Bouvier, B.D., Blum, L.J. (2008) DNA biosensors and microarrays. *Chem. Rev.*, **108** (1), 109–139.
144. Liu, B., Bazan, G.C. (2006) Synthesis of cationic conjugated polymers for use in label-free DNA microarrays. *Nat. Protoc.*, **1** (4), 1698–1702.
145. Lee, K., Maisel, K., Rouillard, J.M., Gulari, E. *et al.* (2008) Sensitive and selective label-free DNA detection by conjugated polymer-based microarrays and intercalating dye. *Chem. Mater.*, **20** (9), 2848–2850.
146. Sun, C., Gaylord, B.S., Hong, J.W., Liu, B. *et al.* (2007) Application of cationic conjugated polymers in microarrays using label-free DNA targets. *Nat. Protoc.*, **2** (9), 2148–2151.
147. Liu, B., Bazan, G.C. (2005) Methods for strand-specific DNA detection with cationic conjugated polymers suitable for incorporation into DNA chips and microarrays. *Proc. Natl Acad. Sci. USA*, **102** (3), 589–593.
148. Geho, D., Lahar, N., Gurnani, P., Huebschman, M. *et al.* (2005) Pegylated, streptavidin-conjugated quantum dots are effective detection elements for reverse-phase protein microarrays. *Bioconj. Chem.*, **16** (3), 559–566.
149. Liang, R.Q., Li, W., Li, Y., Tan, C.Y. *et al.* (2005) An oligonucleotide microarray for microRNA expression analysis based on labeling RNA with quantum dot and nanogold probe. *Nucleic Acids Res.*, **33** (2), e17.
150. Han, M., Gao, X., Su, J.Z., Nie, S. (2001) Quantum-dot-tagged microbeads for multiplexed optical coding of biomolecules. *Nat. Biotechnol.*, **19** (7), 631–635.
151. Schena, M. (2005) *Protein Microarrays*, vol. **xv**, Jones and Bartlett, Sudbury, MA.
152. Kambhampati, D. (2004) *Protein Microarray Technology*, vol. **xxxii**, Wiley-VCH Verlag GmbH, Weinheim.
153. Templin, M.F., Stoll, D., Schrenk, M., Traub, P.C. (2002) Protein microarray technology. *Trends Biotechnol.*, **20** (4), 160–166.
154. Mitchell, P. (2002) A perspective on protein microarrays. *Nat. Biotechnol.*, **20** (3), 225–229.

155. Hall, D.A., Ptacek, J., Snyder, M. (2007) Protein microarray technology. *Mech. Ageing Dev.*, **128** (1), 161–167.
156. Talapatra, A., Rouse, R., Hardiman, G. (2002) Protein microarrays: challenges and promises. *Pharmacogenomics*, **3** (4), 527–536.
157. Fung, E.T., Thulasiraman, V., Weinberger, S.R., Dalmasso, E.A. (2001) Protein biochips for differential profiling. *Curr. Opin. Biotechnol.*, **12** (1), 65–69.
158. Calvert, P. (2001) Inkjet printing for materials and devices. *Chem. Mater.*, **13** (10), 3299–3305.
159. Zhu, H., Snyder, M. (2003) Protein chip technology. *Curr. Opin. Chem. Biol.*, **7** (1), 55–63.
160. Delehanty, J.B. (2004) Printing functional protein microarrays using piezoelectric capillaries. *Methods Mol. Biol.*, **264**, 135–143.
161. Shigeta, K., He, Y., Sutanto, E., Kang, S. *et al.* (2012) Functional protein microarrays by electrohydrodynamic jet printing. *Anal. Chem.*, **84** (22), 10012–10018.
162. Zaugg, F.G., Wagner, P. (2003) Drop-on-demand printing of protein biochip arrays. *MRS Bull.*, **28** (11), 837–842.
163. Solanki, P.R., Kaushik, A., Agrawal, V.V., Malhotra, B.D. (2011) Nanostructured metal oxide-based biosensors. *NPG Asia Mater.*, **3**, 17–24.
164. Sarma, A.K., Vatsyayan, P., Goswami, P., Minteer, S.D. (2009) Recent advances in material science for developing enzyme electrodes. *Biosens. Bioelectron.*, **24**, 2313–2322.
165. Thusu, R. (2010) Strong Growth Predicted for Biosensors Market. Available at: *http://www.sensorsmag.com/specialty-markets/medical/strong-growth-predicted-biosensors-market-7640* (accessed 17 April 2013).

Part III
Pharmaceutical Delivery

Translational Nanomedicine. First Edition. Edited by Robert A. Meyers.

7
Carbon Nanotubes for Enhanced Biopharmaceutical Delivery

Harikrishna Rallapalli[1,2] *and Bryan Ronain Smith*[2]
[1]*New York University Langone Medical Center, Center for Biomedical Imaging, 660 1st Avenue, Room 420, New York, NY 10016, USA*
[2]*Stanford University, Clark Center, 318 Campus Drive, E-150, Stanford, CA 94305, USA*

Keywords

Carbon nanotubes
Cylindrically arranged carbon atoms formed by rolled sheets of graphene with all sp^2 bonds. A member of the fullerene family.

Translational Nanomedicine. First Edition. Edited by Robert A. Meyers.

Biopharmaceuticals
Therapeutic compounds extracted from biological sources, or synthesized using biological methods.

Functionalization
The process of modifying nanoparticles by means of chemical reaction or surface adhesion, including adding functional groups, to produce a desired effect.

SWNTs
Single-walled carbon nanotubes.

Delivery
The transport of a compound to its target site within the body.

Carbon nanotubes (CNTs) can be applied as versatile biopharmaceutical delivery systems due to their high drug-loading capacity, excellent cell-penetrating ability, and customizable surface chemistry. Biopharmaceuticals are potent therapeutics when applied correctly, but are challenged by significant drawbacks. The coupling of biopharmaceuticals to CNTs can mitigate these issues, thus additively or synergistically improving their performance. Improved therapeutic effects, enhanced (targeted) delivery to specific tissue types, highly controlled release profiles, and even the visualization of *in-vivo* biodistribution comprise just a few examples of the benefits of such complexes. Furthermore, novel complexes have been tailored to combat cancer and other difficult-to-treat diseases, with promising success. In this chapter, the application of CNTs for the delivery of biopharmaceuticals to disease sites *in vitro* and *in vivo* will be discussed.

1
Introduction

Since their discovery in 1991 by Sumio Iijima, carbon nanotubes (CNTs) have been applied as versatile tools for diverse applications ranging from efficient electronics and high-strength materials for construction to many biological and medical uses [1]. In this chapter, the application of CNTs for the delivery of biopharmaceuticals to disease sites *in vitro* and *in vivo* will be discussed. Although, the early stages of CNT-mediated biopharmaceutical delivery are still being witnessed, ample investigations have suggested that this approach has tremendous potential due to the unique biological and modular properties displayed by CNTs. In many ways, CNTs are analogous to a "tactical rifle" – modifications can be made to both tools according to the needs of the user and the task – making them more lethal to their targets.

Biopharmaceuticals are therapeutic compounds extracted from biological sources or synthesized using biological methods. They are in many ways completely distinct from artificially synthesized pharmaceuticals, and often come with many added benefits. While traditional pharmaceuticals are screened from a list of potential compounds with desired effects, biopharmaceuticals are generally tailored to treat specific biological defects. Quite often, biopharmaceutical mechanisms interact with the same pathways as "typical" disease function, making them highly specific treatments. Their high specificity reduces the risk of side effects, making biopharmaceuticals typically safer than traditional pharmaceuticals [2, 3].

The combination of CNTs and biopharmaceuticals consistently outperforms separate biopharmaceutical treatment or nanoparticle treatment methods, for several reasons. CNT functionalization makes the nanoparticles more biocompatible, allows targeted delivery [4, 5], and eases delivery into the cells. Intrinsically, CNTs are able to enter cells [6, 7] through multiple pathways, and are able to carry a high density of drug to the target [8, 9]. This is accomplished by attachment to the surface, attachment to an encapsulating polymer, or interior loading of the CNT. In contrast to the delivery of biopharmaceuticals alone, conjugation to CNTs reduces some of the inherent drawbacks of certain biopharmaceuticals. For example, whilst small interfering RNA (siRNA) can serve as an excellent anticancer therapeutic, pure siRNA alone easily cannot easily cross the cell membranes due to electrostatic interactions, as both siRNA and the lipid bilayer are negatively charged. The conjugation of siRNA to functionalized carbon nanotubes (f-CNTs) significantly increases their delivery into cells due to CNT uptake through mechanisms other than endocytosis [10]. On the organism scale, CNTs can also act as a vehicle to help overcome various biological barriers in the body to reach target sites better than therapeutic molecules and biopharmaceuticals alone.

2 Nanotube Background

Two distinct varieties of CNT exist today: single-walled nanotubes (SWNTs) and multi-walled nanotubes (MWNTs). SWNTs are composed of single sheets of graphene rolled into a cylinder, while MWNTs are formed from multiple concentric SWNTs. The SWNTs and MWNTs may vary widely in terms of their lengths and diameter, as well as their functionalization (discussed in detail in Sect. 2.3). These physical differences have led to very different biological responses being observed in living subjects [11–14]. For any therapeutic application, it is, of course, necessary that the base nanoparticle is biocompatible. While this point is beyond the scope of this chapter, many recent reviews have explored links between the chemistry, length and other physico-chemical parameters and CNT toxicity [13–15]. Moreover, the distinctive properties of both SWNTs and MWNTs are important not only for biocompatibility, but they have also been exploited for therapeutic applications and biopharmaceutical delivery.

In the past, CNTs were primarily produced in small batches but with high defect rates. The CNT size was difficult to control, and early experiments were difficult to reproduce due to batch-to-batch inconsistencies in physical properties. Today, the diameter, length and mechanical properties can each be adjusted precisely during synthesis, leading to an ability to create biocompatible CNT batches for medical applications. In addition, batches are no longer volume-limited, allowing single, large batches to be used across experiments. The ability to control CNT formation has increased in the past 20 years, producing safer, easier-to-handle CNTs for both *in-vitro* and *in vivo* experimentation [16].

Many synthesis methods have been developed for SWNTs and MWNTs, each generating a unique variant of CNT with several tunable properties that can be adjusted based on the intended application. After synthesis, the CNT can be functionalized for a variety of purposes.

2.1 Carbon Nanotube Synthesis

CNTs were initially produced by arc-discharge evaporation [1]. This method requires a high voltage to be applied across two graphite electrodes in an inert gas atmosphere, and SWNTs and MWNTs can be selectively synthesized based on the electrode properties. If the electrodes are made from pure graphite rods, then MWNTs will accumulate on the cathode. However, if a metal catalyst is applied in addition to the pure graphite, then SWNTs can be collected in the residue from the chamber. Arc-discharge synthesis methods are limited in the quantity of product they can produce with respect to the initial amount of graphite [17]. Early, large scale arc-discharge methods could produce CNTs ranging from 2 to 20 nm in diameter, with a nanotube yield of 25% [18]. In addition, multiple purification steps are necessary in order to remove any unwanted synthesis byproducts and excess catalyst. Although more high-throughput synthesis methods now exist, many groups have made recent advancements in the performance and safety of arc-discharge synthesis [19]. Today, fully automated arc-discharge methods have been developed to manufacture MWNTs without the need for metal catalysts [20].

Laser ablation synthesis methods have several similar limitations to the arc-discharge synthesis, but can produce more pure products. Laser ablation involves a high-intensity laser impinging upon a carbon target that is held in a fixed 1200 °C oven while released carbon atoms are accelerated towards a copper collection plate. This method can be used to produce both types of CNT: MWNTs are produced simply by the impinging radiation, while SWNTs are produced in the presence of a metal catalyst [17]. Several factors affect the characteristics and quantity of CNTs produced by laser ablation, including the amount/type of metal catalyst, the type of inert gas, and the power/wavelength of the laser. CNTs produced by this method are 90% more pure than those produced by arc-discharge, and have diameter variances of only 1.0–1.6 nm [21, 22]. However, like arc-discharge synthesis, this method is limited in terms of production volume; indeed, the need to scale-up production of CNTs has motivated the development of synthesis methods with minimal physical synthesis limitations.

Modern CNT synthesis methods typically employ chemical vapor deposition (CVD). Essentially, the controlled decomposition of a gas-phase carbon source has the potential to produce CNTs with little impurity and easy-to-control growth [23]. In this process, a metal catalyst "stage" is placed into a controlled-atmosphere chamber and heated to approximately 700 °C. A gas-phase carbon source and a mixed gas are then introduced into the chamber, and carbon is deposited on the catalyst as the gas comes into contact with the plate [24]. The CNT diameters are correlated with the metal catalyst particle size, and the yield is dependent on the metal support compounds [23]. Although this method does not share the same limitations in production volume, all three methods discussed here require the use of metal catalysts – some of which are cytotoxic – to produce SWNTs. The implications of this feature for medical applications will be discussed in Sect. 2.2.

2.2 Cytotoxicity of Carbon Nanotubes

In order for CNTs to be useful for biopharmaceutical delivery, it is critical that their cytotoxicity and long-term toxicity be understood. For that reason – and the fact that pristine CNTs are significantly more cytotoxic than many other nanoparticles – some recent insights are provided into the debate on CNT toxicity. Pristine CNTs are cytotoxic primarily because of their exquisite ability to enter cells [25] and their subsequent potential interactions with DNA and the cytoskeleton. Currently, three mechanisms have been proposed for cell entry, namely nanopenetration, endocytosis, and phagocytosis:

- CNT nanopenetration is a passive process by which CNTs diffuse across cell membranes, most likely by "spearing" them with their very small cross-sectional diameter (1–30 nm).
- Endocytosis is an active process by which CNTs are incorporated into a vesicle for internalization after interaction with the cell surface [25, 26].
- Specialized immune cells known as phagocytes may take up CNTs through a process termed phagocytosis.

All three pathways are represented schematically in Fig. 1 [27]. Recent discoveries regarding the energy dependence of CNT uptake have confirmed that endocytosis is the primary mechanism for transport across cell membranes [26].

After internalization, CNTs may cause toxicity in cells by activating several pathways simultaneously, with the more significant effects involving DNA damage [25]. Changes in cell cycle stage, the generation of apoptotic signals and DNA strand breaks have all been reported in mesothelial cells exposed to pristine SWNTs at approximately 25 $\mu g\,cm^{-2}$ [28]. In addition, certain cell types are more susceptible to damage as the dose increases. A correlation between increased dose and increased oxidative stress has been reported *in vivo* in the liver and lung, but interestingly no significant correlation was found in the spleen [29]. Intriguingly, mesothelial cancer cells generally experience an increased phosphorylation of extracellular signal-regulated kinases (ERK) and p38-kinase with respect to their normal counterpart, but are less susceptible to DNA damage than the latter [30].

The physical dimensions of CNTs affect their cell internalization dynamics. A negative correlation between CNT length and macrophage uptake has been reported both *in vivo* and *in vitro* [25]. Short CNTs (<0.22 μm) were found in macrophages four weeks after their subcutaneous injection into rats, whereas longer CNTs (>0.80 μm) were still free-floating in the extracellular milieu. As free CNTs can cause inflammation and local cell death, the toxicity effects of longer CNTs must be considered before their use in medical applications [25].

CNT diameter also affects cellular uptake, and is a significant factor when considering drug loading and surface modification. Jin *et al.* have reported that the endocytosis rate of SWNTs is governed by thermodynamic entropic contributions, and that the rate reaches a maximum when SWNTs display a radius of 25–30 nm (i.e., when clusters at the surface of the cell reach such radii) based on *in-vitro* experiments and computational models; larger or smaller diameters result in significant decreases in simulated uptake [11]. While computational models are not now widely used in the field of medical nanotechnology, such models will become increasingly important as the numbers of nanoparticles rapidly proliferate. The choice and optimization of nanoparticles will then likely eventually be subject to *in-silico* models for major transitions, for example, *in vitro* to preclinical animal models, and preclinical models to patient use in the clinic.

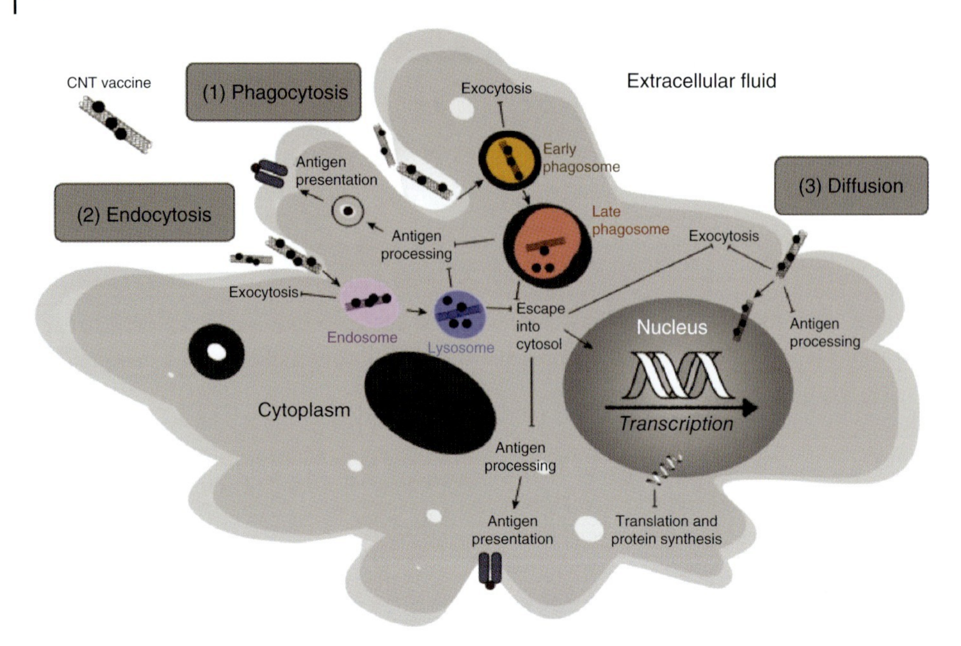

Fig. 1 Mechanisms of carbon nanotube (CNT) uptake by cells. (1) Phagocytes may recognize foreign bodies conjugated to the CNT surface and take up the complex. (2) Functional groups on the CNT surface may interact with cell-surface proteins, resulting in endocytosis. (3) The nanopenetrating ability of CNTs allows them to enter the cell through passive diffusion. (Reproduced with permission from Ref. [27]; © 2014, Elsevier.)

Metal catalysts are commonly used in SWNT synthesis, and are required for the CVD synthesis of any CNT type. They are also incorporated into CNT products at concentrations as high as 40% of the total CNT mass. Catalysts often contain metals such as iron, cobalt, nickel, yttrium [31], and arsenic, some of which can produce free radicals *in vivo.* Small iron oxide nanoparticles, which are commonly used as a catalyst, are known to be biocompatible based on a long history of iron oxide use in humans [25, 32]. However, with other catalysts, free radical production may be a concern, especially during CNT uptake by macrophages [25]. Transition metals can come into contact with superoxide species produced when the enzyme NADPH-oxidase is activated in macrophages, and produce highly reactive hydroxyl groups that have the potential to bind to proteins, lipids and DNA, causing significant damage to the cell [25]. For medical applications of CNTs, purification protocols have been developed and refined to remove any catalyst material from the CNT product by several orders of magnitude, though not without losses in CNT yield [33, 34]. Although these purification procedures leave only trace amounts of metal catalysts, this most likely leads (at least in part) to the minimal toxicity observed in various animal models for optimized CNT formulations [35–37].

Whilst the toxicity of these trace compounds has not been proven in humans, animal studies have provided critical insights. Indeed, many small animal studies have demonstrated a lack of toxicity of correctly functionalized f-CNTs of appropriate length and diameter [35–37]. However, if not correctly modified, the CNTs may have unavoidable structural defects that can have negative effects on cell health [38]. For example, charge density abnormalities caused by five- or seven-membered ring defects in the CNT surface chemistry can cause undesirable nucleic acid binding, or may weaken the bonds between nucleic acid functional groups and the CNT surface. When the CNTs are in cells, such active chemistries can lead to deleterious interactions and interference with DNA [25]. Both, the arc-discharge and laser ablation methods leave the CNT ends "closed," whereas CVD synthesis produces "open" ends. Consequently, CNTs produced using CVD may have "dangling" highly reactive species at each end, which can lead to oxidative stress without further modification [25].

The most common routes for introducing CNTs into the body are intravenous injection, inhalation, and transdermal injection. In general, inhalation and transdermal injections of CNTs result in a distinct biodistribution and pathological effects [39, 40]. Many descriptions have been made of the effects of inhalation of both SWNTs and MWNTs. Based on rodent studies, the consensus is that the inhalation of these structures generally causes adverse effects, but that the results are heavily dependent on CNT functionalization and length [41]. In one study, Ma-Hock *et al.* described the long-term effects of inhaled nanotube exposure in rats [42] when, after 90 days, the inhaled MWNTs had not produced any adverse systemic effects but had caused an increased lung weight, local inflammation, and blood neutrophilia in the case of high inhaled NT concentrations ($2.5\,mg\,m^{-3}$). Previous conflicting reports of lung toxicity in early studies of long-term CNT inhalation were also summarized [42]. For example, in 2001 Huczko *et al.* reported no significant lung toxicity after four weeks of exposure in guinea pigs [43], but Lam *et al.* subsequently (in 2004) disputed this claim, reporting inflammation and granuloma formation after single-dose exposure [44]. In 2008, Shvedova *et al.* showed that inhaled SWNTs would cause oxidative stress, collagen deposition and fibrosis in mice [45]. These apparently conflicting results were most likely due to differences in the manufacture, functionalization and/or dose levels across studies, as well as differences in the species tested. Recent advances in CNT aerosolization testing have led to many studies being conducted in quantitative toxicity measurements [46].

Similar aerosolization advancements have begun to resolve issues with experimental consistency across studies [47]. In a provocative study, Poland *et al.* reported that the pathological effects of CNTs introduced into the abdominal cavity may be similar to those of asbestos, although others have reported that asbestos-like effects are due almost solely to the length/high aspect ratio, chemical functionalization, and bundling activity of CNTs [48–50]. When toxic, MWNTs are reported to produce local inflammation and granuloma in mice, and may have the capacity to cause mesothelioma after intraperitoneal injection. Yet, when correctly functionalized and at the appropriate lengths, minimal toxicity is observed [48–50]. Due to an ability to control the dose precisely, the intravenous injection of CNTs is preferred over inhalation for whole-body, drug-delivery applications; nevertheless, and particularly for studies of toxicity in small animals, inhalation has historically been the more common approach for introducing CNTs into the body [15, 51]. The safe clinical use of CNTs through either inhalation or intravenous injection has been discussed in detail by Saito *et al.* [15], with intravenous injection, using CNTs of small to medium length and appropriate functionalization (e.g., with amphiphilic polymer coating [35]) having been shown to be acutely and chronically safe (for at least three to four months) in rodent models [35, 52]. For these reasons, the intravenous route is the preferred method for nanomedicine applications, including drug delivery.

2.3 Functionalization of Carbon Nanotubes

By design, the surface functionalization of CNTs typically results in an increased biocompatibility, increased circulation times, increased cell- or receptor-specific targeting, and an increased drug-carrying capacity depending on the nature of the attached functional groups. This functionalization is key for medical applications such as biopharmaceutical delivery.

Functionalization can be subdivided into either noncovalent (nondestructive) or covalent (destructive) methods, and further classified based on application. While Mittal has provided an extensive treatment of CNT surface modification methods, the present chapter will detail functionalization methods that are particularly applicable to biopharmaceutical delivery [53]. Currently, oxidation, cycloaddition, and the noncovalent attachment of polymers are three of the most commonly used approaches to produce CNTs for medical applications in living subjects:

- *Oxidation of CNTs:* Oxidation is a common method of functionalization as it is used to help covalently link molecules of interest to CNT sidewalls and end caps. Oxidizing agents, such as nitric acid, are used to produce carboxylic acid groups on defect sites on the CNT sidewall and opened caps. Although oxidized CNTs are generally water-soluble, their aggregation characteristics prevent them from being used *in vivo* directly. To improve the biocompatibility of oxidized CNTs, linker molecules such as polyethylene glycol (PEG) can be covalently attached to oxidized sites [54].
- *Cycloaddition of CNTs:* Cycloaddition is the process of creating a chemical "ring" at the CNT surface, without producing defects on the surface or carboxyl groups at the nanotube ends. The ring provides residual groups, which can be replaced with PEG, proteins, drugs, or certain positively charged moieties. Both, oxidation and cycloaddition are covalent methods of functionalization which can disrupt optical properties of CNTs, thereby limiting certain therapeutic and diagnostic mechanisms of pristine SWNTs [54].
- *Polymer attachment:* Pristine CNTs are not water-soluble and therefore cannot be injected. Moreover, they are cytotoxic unless modified with amphiphilic moieties on the pure graphitic surface. Noncovalent methods to attach water-soluble compounds to CNT surfaces have been developed to easily suspend CNTs in aqueous solutions. Analogous to many phospholipid bilayers, the hydrophobic CNT surface interacts with a hydrophobic segment of a polymer, which is connected to a hydrophilic segment of the polymer that is thermodynamically most stable when presented on the outermost layer of the system in aqueous solvents. Weak van der Waals forces and hydrophobic effects hold the polymer to the CNT surface, preventing nonspecific binding of proteins and other biomolecules to the CNT. Such polymers are commercially available and have been demonstrated to perform well

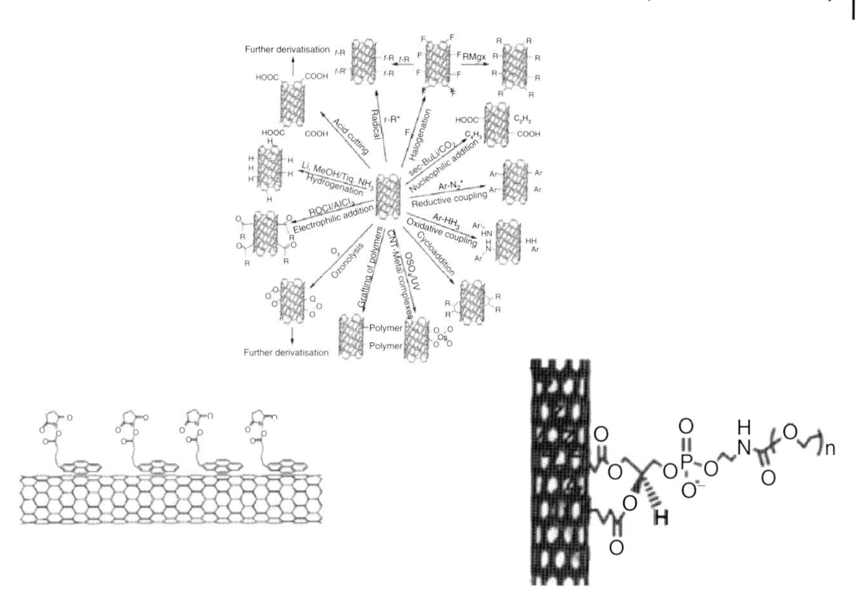

Fig. 2 CNT surface functionalization methods. (a) Examples of direct covalent addition of functional groups and general mechanism for addition of chemical groups. (Reproduced with permission from Ref. [55]; © 2010, Royal Society of Chemistry); (b) Noncovalent functionalization method primarily applied for peptide immobilization. 1-Pyrenebutanoic acid, succinimidyl ester is irreversibly adsorbed onto the hydrophobic surface of CNTs in organic solvents. Proteins then form amide bonds with the exposed succinimidyl ester groups. (Reproduced with permission from Ref. [56]; © 2001, American Chemical Society); (c) Distearoylphosphatidylethanolamine (DSPE) copolymer-based addition of polyethylene glycol (PEG), a molecule commonly used to enhance CNT solubility in water and increase blood circulation time. This approach applies noncovalent hydrophobic interactions of the lipid group to link to the CNT and the PEG group for aqueous solubility. (Reproduced with permission from Ref. [57]; © 2008, Nature Biotechnology.)

in biosensing, suspension stability, and *in-vivo* applications [54].

Liu *et al.* have listed the critical characteristics of noncovalent functionalization coatings for CNTs. The coating should be biocompatible and nontoxic, resist detachment from the CNT surface, have a low critical micelle concentration (CMC) in order to create a stable coating on the CNT surface after application, and have functional groups capable of interacting with biologically active molecules or application-specific compounds [54]. These points are all critical for biopharmaceutical applications of CNTs, and the biopharmaceutical-specific functionalization chemistries will be discussed as each particular chemistry is presented. A general overview of CNT surface functionalization is provided in Fig. 2.

2.4 Pharmacokinetics of Carbon Nanotubes

Although the *in-vivo* kinetics of f-CNTs are still not completely understood, significant efforts have recently been made to understand their short- and long-term biodistributions. This is due in part to the fact that so many different lengths, diameters and functionalization distributions have been tested by different groups [58]. Nevertheless, several characteristics are common across most studies: (i) the CNT's physical properties, particularly functionalization, play a critical role in biodistribution across all time points; (ii) clearance via the urine is comparable to the clearance of larger-diameter nanoparticles (most likely due in part to the large diameter of CNT aggregates); (iii) accumulation is apparent in the liver, spleen, and often the bladder; and (iv) covalent functionalization often promotes a higher stability of CNTs *in vivo* with respect to noncovalent functionalization. Yang *et al.* have provided a detailed overview of the *in-vivo* kinetics of f-CNTs [58].

3 Biopharmaceutical Delivery Using Carbon Nanotubes

Biopharmaceuticals can be broadly categorized into three groups, namely antibodies, nucleic acids, and peptides/proteins (Table 1). Antibodies and peptides/proteins are separated because antibodies are specific to particular antigen epitopes and tend to be larger, whereas peptides/proteins are a general class of macromolecule. The common functionalization methods, recent applications in living subjects, and advantages for each group are discussed in the following sections.

3.1 Peptides/Proteins

Gene expression and cell signaling are commonly accomplished through proteins. The cell membrane can house a variety of proteins serving several purposes, including chemical channels, signaling systems, or pumping systems to build an electrochemical gradient. For therapeutics, membrane-bound proteins that specifically bind to small amino acid sequences (that are related to a particular disease) are of interest. These proteins can be exploited as targets for drug-delivery systems, or they can be blocked by means of a blocker/inhibitor. Conjugating proteins to CNTs can serve both purposes: (i) a drug can be carried by the CNT with a protein-targeting agent; or (ii) high densities of inhibitor (which in many cases is also a protein) can be loaded onto the CNT for delivery [59]. In many instances, the protein conjugated to CNTs can serve as both a targeting agent and as an inhibitor for the membrane-bound protein.

Tab. 1 Biopharmaceutical delivery using carbon nanotubes: an overview.

Biopharmaceutical	*Function*	*Section*
Peptide/protein		3.1
	Targeted drug delivery	3.1.1
	Adjuvant	3.1.2
Antibodies		3.2
	Targeted drug delivery	3.2.1
	Targeting imaging	3.2.2
Nucleic acids		3.3
	Gene silencing	3.3.1

Many functionalization methods for peptide–CNT systems exist, including nonspecific adsorption [60], noncovalent functionalization [61], and covalent functionalization (see Sect. 2.3). Nonspecific adsorption is possible due to the interaction of hydrophobic groups on the protein or polymer with the hydrophobic CNT surface or other hydrophobic scaffolds [62]. Noncovalent functionalization involves the addition of linker molecules with hydrophobic and hydrophilic ends to the CNT surface; the hydrophobic end interacts with the CNT surface, while the hydrophilic end is free to bind protein and provide solubility in an aqueous environment [63].

Noncovalent functionalization is similar to nonspecific adsorption, with the caveat that noncovalent functionalization contains a linker molecule between the nanotube and the biomolecule.

Recent applications of peptides in biopharmaceutical delivery systems can be broadly classified into either targeting or adjuvant approaches. In practice, peptides are more commonly used as targeting agents for CNTs, most likely due to their considerably smaller size. This is logical considering that proteins are often 3–5 nm or more in hydrodynamic diameter; as SWNTs are on the order of 2 nm in diameter, molecules the size of proteins could in many cases adversely modulate some of the properties that make SWNTs and other CNTs so useful in biopharmaceutical delivery.

3.1.1 Peptides and Proteins as Targeting Agents

Peptide- and protein-f-CNTs are efficient drug-delivery systems and imaging agents, mainly because of their targeting ability and high loading capacity due to their extremely high surface area-to-volume ratio.

The use of Arginine-Glycine-Aspartic Acid (RGD) functionalized SWNTs, or other short peptide sequences for receptor-specific targeting, has been explored *in vitro* [64] and *in vivo* [65]. RGD selectively binds the $\alpha_v\beta_3$ integrin, a receptor protein that is upregulated on many tumor cell types (e.g., U87MG cells), downregulated on other cells (e.g., MCF-7 cells), and overexpressed on tumor blood vessels [5, 65]. In practice, an RGD ring – cyclic RGD – is typically used for its increased $\alpha_v\beta_3$ integrin affinity. Overexpression can be exploited by conjugating RGD to SWNTs to increase homing and efficacy at the target site. Once targeted, therapy can be achieved by incorporating doxorubicin (DOX), a well-characterized chemotherapeutic agent, into RGD–SWNTs to significantly increase binding to U87MG cells with respect to $\alpha_v\beta_3$ integrin-deficient MCF-7 cells [64].

A way to selectively target particular subsets of immune cells has long been sought [66]. In a novel application of SWNTs in immune nanomedicine, SWNTs were shown to be taken up with very high selectivity into a monocyte subset [67]; the SWNTs could then be used to deliver molecules to modulate the function of these monocytes. A novel tumor "homing" mechanism was also characterized, showing how RGD-conjugation of SWNTs increased monocyte trafficking into tumor regions after the monocytes had engulfed the SWNTs. In particular, RGD–SWNT-laden circulating monocytes extravasated into tumor regions more frequently than plain or nonspecific control groups [67]. These results suggest that certain peptides can be conjugated to drug carriers for the specific delivery of not only nanoparticles, but also cells carrying such nanoparticles into tumors.

Peptide-conjugated SWNTs can be used as tools for targeted photothermal therapy. SWNTs are inherently excellent absorbers of near-infrared (NIR) excitation, but they must be targeted in order to produce heating at the appropriate site. In order to prevent the irradiation of a nonspecifically targeted region, SWNTs are coupled with a repeating amino acid Lys-Phe-Lys-Ala (KFKA) structure via β-sheet wrapping. This polymer is engineered for water-solubility, binding to tumor tissue by electrostatic interaction, and the ability to be further modified as a carrier moiety for drug delivery. Functionalization was confirmed using atomic force microscopy, transmission electron microscopy, and molecular modeling [68]. The system was tested both *in vitro* and *in vivo* in mice. A high-efficiency, concentration-correlated heat production resulted from the NIR irradiation of peptide-carrying SWNTs in water. Pure peptide was found to cause no cell death below 10 μg ml^{-1} concentrations. *In-vitro* testing confirmed that NIR excitation caused localized cell death. SWNT-$(KFKA)_7$ (a system wherein SWNTs are conjugated with a KFKA sequence which repeats seven times) were injected directly into tumor tissue and subjected to NIR irradiation for 30 s. Localized heating was confirmed using infrared thermometry, with very small temperature increases outside of SWNT localization. A statistically significant suppression of tumor growth was induced by SWNT introduction and ablation [68]. In the next steps, this system should be tested using systemic administration and in a larger animal model more representative of human disease, as this result is promising if it can be successfully administered via a more clinically applicable route.

In addition to radiative therapy methods, peptide f-CNTs can be used to specifically deliver therapeutics to a target site. For example, a multifunctional CNT system which employs asparagine-glycine-arginine (NGR) peptide as a targeting agent has been validated *in vivo*. This system is then loaded with tamoxifen (TMX), an anticancer drug, exploiting the unparalleled loading capacity of CNTs [69]. In addition to chemical therapeutics, the SWNT–peptide retains the useful optical properties of pristine SWNTs. This allows photothermal ablation therapy, which can be used in combination with TMX. In a recent study, peptide-laden SWNTs were taken up into tumors more readily than their pristine counterparts. Although the full NGR/TMX–SWNT system showed cytotoxic effects alone, the combined effects of TMX and NIR irradiation increased the *in vitro* cytotoxicity 1.5-fold. *In-vivo* biodistribution assays confirmed a specific uptake of complete NGR/TMX-SWNT system by tumors, whereas other variants accumulated in the clearance organs. In addition, significant reductions in tumor volume were apparent at 14 days post-injection of NGR/TMX–SWNTs due to the excellent delivery and uptake characteristics of the system [69].

Extracellular receptors may be relatively easy targets for biopharmaceutical delivery systems, but intracellular targets are much more difficult to reach. A method to repair damage to mitochondrial DNA by means of peptide-conjugated CNTs has recently been developed. Mitochondrial targeting can be accomplished via known endogenous peptide markers conjugated to nanoparticles; this is known as the mitochondrial targeting system (MTS). *In vitro*, peptide-conjugated CNTs were significantly more effective in penetrating cell membranes than the peptide alone. Also *in vitro*, MTS–MWNTs aggregate around mitochondrial membranes, which is due to an interaction with the translocator inner membrane/translocator outer membrane protein complex. Importantly, the MTS–MWNT conjugates were also found not to be cytotoxic. The findings of these studies suggest that a need exists to use targeting peptides when delivering therapeutic agents to mitochondria, rather than nonspecific delivery into the cytosol [70].

3.1.2 Peptides and Proteins as Adjuvants

CNT-bound peptides can also serve as adjuvants, which are molecules that produce a localized immune response in the host. A resumé of recent progress in this area of technology is now presented.

SWNTs have been employed as peptide carriers to boost weak immune

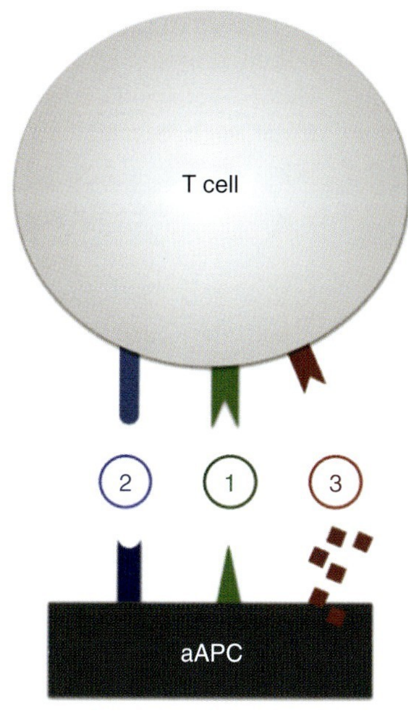

(1) Antigen recognition: interaction of antigen stimuli with receptors on T cells. Examples of T cell stimuli include physiological molecules (i.e., MHCs) and antibodies specific to the CD3 complex (i.e., anti-CD3)

(2) Co-stimulation: interaction of co-stimulatory molecules with ligands on T cells

(3) Cytokine signaling: interaction of cell signaling molecules with receptors expressed by T cells

Fig. 3 CNTs can be applied as artificial antigen-presenting cells (aAPCs) in order to produce or boost an immune response to an antigen. (Reproduced with permission from Ref. [71]; © 2011, Elsevier.)

responses. In a recent study, it was found that peptide-laden SWNTs were actively engulfed by immune cells *in vitro*, producing a marked increase in immune response over the pure peptide antigen. In this paradigm, tumor-specific peptide antigens could theoretically "prime" the immune system against tumor subtypes, potentially serving as quasi-vaccines, as shown in Fig. 3 [71].

It is also now possible to induce an immune response to haptens, which are low-molecular-weight compounds that are too small to be presented on major histocompatibility complex (MHC) II molecules. Other methods of immunization against haptens involve conjugation to nanoparticles, and the introduction of these systems into the body. Both, SWNTs and MWNTs were conjugated with azoxystrobin, an antifungal with the broadest known spectrum of activity. When these systems were injected into mice and rabbits for *in-vivo* immunization, the data acquired showed the CNTs to be extremely efficient antigen carriers and to produce an immune response to the hapten complex at a 10-fold lower concentration than did the pure complex alone. Moreover, when CNTs of varying length were tested, the results showed that shorter CNTs would produce a significantly increased immune response in comparison to longer CNTs of similar diameter. To further illustrate the amplification of immune response to CNT antigen carriers, the total CNT–protein complex dose was diluted to one-tenth of the pure protein complex dose. Subsequently, whilst the pure complex produced no significant immune response, the CNT–protein complex, when inoculated, produced significant increases in blood immunoglobulin (Ig) G concentrations [72].

The delivery in this application was to cells of the immune system (e.g., T cells) present in the circulation rather than to a particular region of the body. Consequently, a major feature of this delivery system would be for the presented molecules to circulate for as long as possible in order to enable their maximal exposure to the immune cells.

3.2 Antibodies

Antibodies are a subset of proteins that can serve as versatile biomolecules generated by B cells and are capable of recognizing a wide variety of unique chemical signals or epitopes present on the surface of foreign bodies. By definition, antibodies (or Igs) are large, Y-shaped proteins deployed by the immune system to detect and neutralize antigens. Variable portions at the tips of the structure serve as chemical complements to certain antigens, allowing them to tag the foreign body for phagocytosis, degranulation, or digestion by cytotoxic natural killer (NK) cells. Because of their exquisite sensitivity and specificity for unique antigens, as well as their high binding affinity, antibodies have applications as immunoassay tools, immunization promoters, and targeting agents. Furthermore, it emerges that many antibodies can in fact have therapeutic effects, as is the case with the monoclonal antibody Herceptin® [73]. This is in addition to any drug or other molecules that the CNT can carry into the cell once it is targeted (as will be discussed later). Therefore, the conjugation of antibodies to CNTs as a targeting agent (as well as a potential drug) has great potential to amplify their therapeutic effects.

Recent advances in CNT chemistry have led to the synthesis of antibody f-CNTs primarily for use as targeted drug-delivery vehicles. The covalent functionalization of antibodies to CNTs may be optimal, as the potential loss of the targeting agent would lead to nonspecific drug delivery. To date, three such covalent strategies have been reported:

- A simple oxidation reaction where carboxylic acid groups are appended to defect sites and "open" ends. The antibodies are coupled to the carboxyl groups by amino groups on an amino acid side chain, such as lysine, available on the antibody [74].
- A strategy which also follows standard oxidation techniques – the oxidized CNT is attached to pyrrolidine or groups at the defect sites, and the antibody binds the ring through an amidation reaction.
- A strategy which requires thiolated antibodies and maleimide f-CNT. The complete system consists of a pyrrolidine ring coupled to a maleimide group, which binds the thiolated antibody.

All three methods have been well-characterized and have been validated both *in vitro* and *in vivo* [74].

Currently, two major categories of CNT immunotherapy exist, each with its own benefits. Passive immunotherapy involves the introduction of antibodies or complements as targeting agents for drug delivery vehicles that can then treat the disease site. Active systems boost the host immune response to a specific antigen by introducing pathogen fragments or other adjuvants. Though active systems do not house antibodies themselves, they are designed to drive antibodies and primary immune cells to antigen sites, which has been previously discussed (see Sect. 3.1.2). The targeting and inhibitory action of antibody f-CNTs will be discussed in the following section, beginning with recent *in-vitro* applications and leading to the delivery of these systems in living subjects.

3.2.1 Antibodies as Targeting Agents for Drug Delivery

Multidrug resistance (MDR) is a major problem in clinical oncology today. In time, MDR essentially nullifies many treatment approaches because cancer cells adapt to resist the effects of previously administered drugs. Fortunately, targeted drug delivery can counteract MDR effects by enhancing the local drug concentration at the target, and particularly by enabling the delivery of multiple drugs simultaneously (this is a key advantage of nanoparticle delivery systems, such as CNTs). As an example, P-glycoprotein (P-gp) antibody-functionalized, water-soluble SWNTs were used to target the membrane-bound P-gp on MDR cells. Anti-P-gp was covalently linked to oxidized SWNTs, and the SWNTs were loaded with DOX. In addition to passive diffusion of the drug, it is possible to irradiate the *in-vitro* culture of MDR cells and SWNTs with NIR light in order to boost drug release. The proposed mechanism for attachment (corroborated by fluorescence imaging) is an initial interaction between anti-P-gp functionalized on the SWNT and membrane-bound P-gp, followed by endocytosis by the MDR cells where the drug is released. This enhances the time of contact between the drug and the target cells. This method has been confirmed to produce a 2.4-fold increase in targeted cytotoxicity with respect to free DOX *in vitro* [75], and indicates that the antibodies in this case were effective targeting moieties. It also enabled the delivery of both the antibody and the loaded drug to overcome the drug resistance of the cells, allowing them to be destroyed.

3.2.2 Antibodies as Targeting Agents for Imaging

f-CNTs are useful not only as imaging agents but also as drug carriers. Targeted imaging agents have the potential to serve also as drug carriers in the future, provided that the drug and imaging agent share similar steric and chemical properties; these types of agent are known as "theranostics." It is also important to recognize the prevalence and distribution of antibody targets on the cell of interest, in order to understand the binding and internalization mechanisms and thereby provide a better design of the CNT vehicle for optimal uptake [76]. In the context of imaging agents, antibodies bound to f-CNTs can increase accumulation at a target site, or increase cellular uptake.

Imaging agents and drug-delivery systems are intrinsically connected when working with f-CNTs. When specific surface proteins are upregulated – as often occurs in cancer cells compared to healthy cells – these proteins can be ideal targets for specific drug-delivery vehicles. Her2+ breast cancers are particularly aggressive and are generally less responsive to gene therapy, but display increased Her2 receptor expression on the cell surface. A SWNT–antibody complex which was produced for targeting Her2-positive breast cancer cells *in vitro* [76] consisted of a biotinylated SWNT functionalized to bind a chromophore, a secondary antibody, and primary antibody to the Her2 receptor (Fig. 4). The full complex was seen to bind to the Her2 receptor on BT-474 cells with remarkable specificity, as quantified by Raman imaging and confirmed in control immunofluorescence studies [76].

It is also possible to slow the rate of endocytosis by cooling the cells to 15 °C, and then to obtain confocal Raman spectroscopy images in three dimensions to image the delivery process. The "Raman effect" is an apparent shift in the frequency of monochromatic incident light that inelastically scatters after contact with a

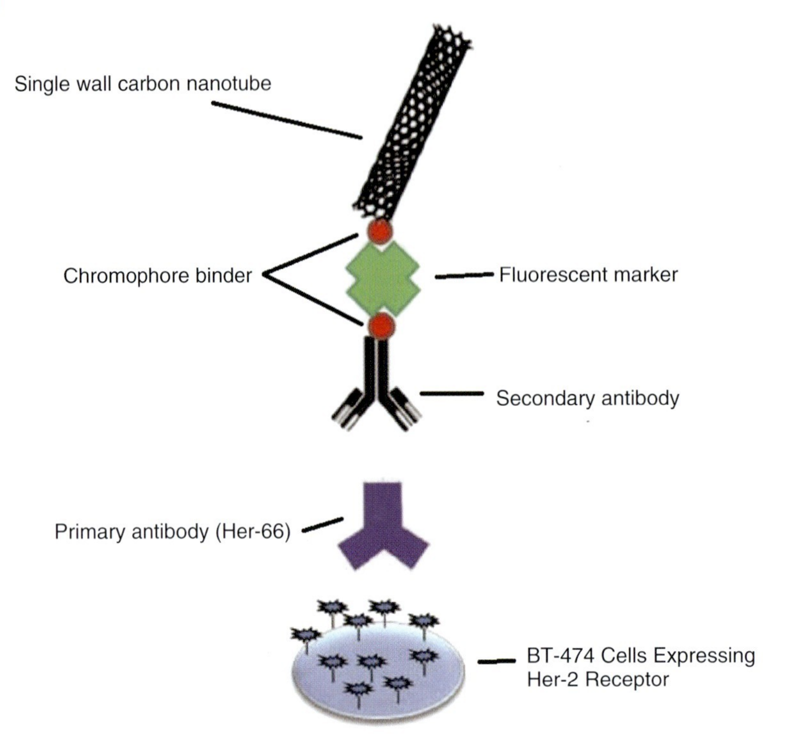

Fig. 4 Schematic diagram of SWNT–antibody complex for Her-2 receptor targeting to the cell surface *in vitro*. The Her-2 distribution on BT-474 cells was evaluated using Her-66 anti-Her-2 monoclonal antibodies. The biotinylated secondary antibody binds the primary antibody and is complexed with a fluorescent marker (FITC) and SWNTs via a chromophore binder (NeutrAvidin-FITC bridge). (Reproduced with permission from Ref. [76]; © 2014, Royal Society of Chemistry.)

target. The frequency shift is associated with the vibrational and rotational energy states of the target molecules and their bonds, which in turn makes it possible to determine the contents of the target by analyzing the frequency shifts of the scattered light [76]. Precision Raman imaging is possible with CNTs due to the unique chemical structure of the C–C bonds within the SWNTs, which makes them easily and very specifically detectable by Raman imaging in a background of carbon (i.e., the milieu of a living subject such as a mouse) [77]. The imaging results showed that SWNTs are localized at the cell membrane, suggesting the start of endocytosis. These results, in combination with data from controls, suggested that antibody-mediated cell binding caused a significant increase in SWNT binding to the cell membrane, which in turn potentiated SWNT internalization by the cell [76].

Antibodies do not need to be directly attached to CNTs in order to serve as targeting agents. A novel two-step system was shown to guide SWNTs to tumor regions by "priming" sites of interest with antibodies, as shown in Fig. 5. The SWNTs were conjugated to oligonucleotide sequences and fluorophores [78], antibodies were employed as targeting agents for specific tumor cell receptors, and complementary

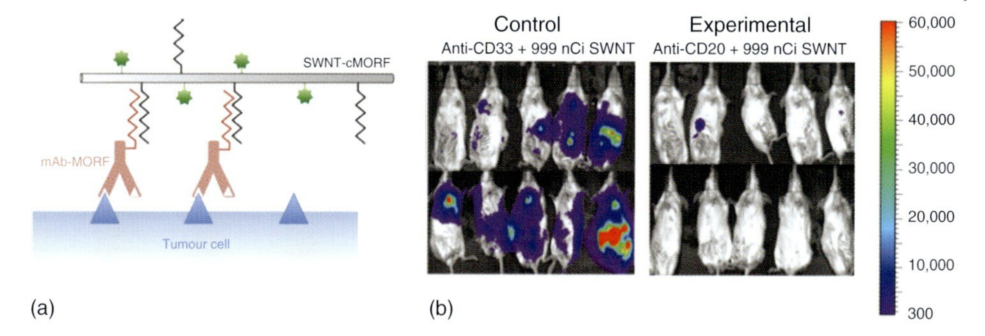

Fig. 5 Imaging agents can be coupled to biopharmaceutical-conjugated CNTs to monitor their *in-vivo* kinetics and therapeutic effect. (a) Schematic representation of two-step SWNT delivery system. Tumor-specific antigens are represented as blue triangles. Antibodies (mAb) with morpholino (MORF) oligonucleotide sequence are shown in red. SWNTs with bound complementary morpholino (cMORF) oligonucleotide sequences and radiotherapeutic agents are shown in gray; (b) Bioluminescence images demonstrating specific $CD20^+$ tumor volume reduction using antibody-conjugated CNTs. Control and experimental conditions are anti-CD33 (Ab for cell-surface markers are not present on this tumor cell type) and anti-CD20 (Ab for cell-surface markers are overexpressed on this tumor cell type) respectively. Both conditions have the same radiotherapeutic dose of 999 nCi. (Reproduced with permission from Ref. [78]; © 2013, Nature Nanotechnology.)

oligonucleotides were covalently bound to the constant regions of the antibody. The SWNT system was loaded with alpha- and gamma-radiation-emitting moieties to serve as a therapeutic agent, and with fluorophores for *in-vivo* tracking. The antibody system was injected intravenously 24 h before the SWNT system was introduced, in order to allow binding to the tumor sites. The combined Ab–SWNT system can be considered as self-assembling because the binding occurs on the target, not during synthesis. This method showed a five- to 10-fold reduction in toxicity over preconjugated systems, which was an amazing achievement. The imaging results indicated dose-dependent therapeutic effects, motivating further exploration to use this method for drug delivery that applies SWNT–Ab systems in a "sandwich assay" technique that is more commonly associated with immunoassays [78].

A MWNT–antibody system to measure levels of prostate stem cell antigen (PSCA) has recently been developed. The purpose of this system is to target cancers that upregulate PSCA, and also to be used as a targeted ultrasound contrast agent *in vitro* and *in vivo* [79]. The hyperechogenicity of MWNTs which was only recently described is somewhat surprising, given the typical size range of many echogenic materials for ultrasonic applications (often $\geq 1\,\mu m$ in diameter, such as microbubbles). Ultrasonic imaging using smaller materials such as MWNTs has great potential, as these agents can reach sites that are inaccessible to larger materials such as microbubbles, including extravascular spaces and cells. Moreover, the high drug-loading capacity of MWNTs will allow for future anticancer drug delivery and real-time tracking. This DOX-loaded system did not cause any adverse hemolytic effects over a large range of drug concentrations ($0–400\,\mu g\,ml^{-1}$) either *in vitro* or *in vivo*, due to complexation with the CNTs [79]. Neither were any organ systems significantly affected after long-term exposure to the MWNT system. However, after an 18-day exposure period the rate of tumor growth was significantly decreased when DOX and PSCA monoclonal antibodies were conjugated to MWNTs with respect to other system variants (i.e., controls) [79].

3.3 Nucleic Acids

Nucleic acids are large macromolecules which store genetic information and are composed of nucleotide bases and a ribose sugar backbone. Ribosomes translate messenger RNA (mRNA) into proteins, which are then packed and processed. This mechanism can be exploited to express exogenous genes or to halt unwanted gene expression by means of engineered nucleic acids delivered through the cell membrane using f-CNTs. The CNTs are able to traverse cell membranes more readily than pure sequences of nucleic acids, and can provide nucleic acid stability against endogenous enzymes, making them a key part of an ideal transport mechanism.

Three overarching methods for nucleotide functionalization of CNTs have been employed in recent studies;

1) Polyethyleneimine (PEI) functionalization was first demonstrated by Liu *et al.* as the first noncovalent method to immobilize DNA on CNTs. PEI is composed of a high density of amines, which readily bind to nucleic acids. The polymer is grafted onto CNTs by a polymerization reaction in the presence of amine-f-CNTs [80].
2) Amino-functionalization of CNTs by the cycloaddition of a pyrrolidine ring bearing a terminal amine group is commonly used to bind plasmid DNA. Pantarotto *et al.* have reported that the addition of this group markedly increases CNT solubility in aqueous solution, as well as enabling covalent immobilization of the plasmid DNA [81].
3) Carboxyl-functionalization is another form of covalent attachment of nucleotides to CNTs; however, this method requires the amino-modification of nucleic acids to overcome electrostatic repulsion forces between the negatively charged nucleic acids and cell membranes [82].

3.3.1 Nucleic Acids as Gene-Silencing Agents

A major focus in the field of nucleic acid delivery has involved the enhancement of siRNA delivery into the cytoplasm [83]. The conjugation of siRNA to CNTs results in four critical benefits when used for treatments in living subjects: (i) ease of entering the cytoplasm; (ii) ability to add targeting ligands for intracellular organelles; (iii) potential therapeutic loading; and (iv) exploitable optical properties for simultaneous cytotoxic and silencing effects. While siRNA has been featured, it is likely that clustered regularly interspersed short palindromic repeats (CRISPR) systems, which only recently have displayed an incredible ability to perform precision gene editing in mammalian cells, will require delivery systems such as can be provided by nanoparticles (including nanotube-based approaches) in the near future.

In the field of cancer therapy, siRNA has been applied to silence the gastrin-releasing peptide receptor (*GRP-R*) gene, which produces a cell-surface receptor (GRP-R) that is commonly upregulated on neuroblastomas. GRP-R is known to promote the growth of epithelial cells and the formation of abnormal tissue growths [83]. The assay was tested *in vitro* and the results confirmed by using NIR fluorescence imaging of siRNA-treated cells and Western blotting for GRP-R. *In-vivo* testing entailed the subcutaneous injection of 1.7 μM SWNT–siRNA into a murine xenograft model, whereby such treatment was found to be effective in limiting the number of mitotic cells, leading to a decreased tumor growth [84]. Conjugation to CNTs thus improves the delivery of siRNA to intracellular targets, facilitating therapy in living subjects.

Although siRNA alone is a useful anticancer therapeutic, a combination of gene silencing and photothermal irradiation can lead to a significant further reduction in tumor volume [85]. In this case, PEI–SWNTs were conjugated with siRNA and NGR, a peptide sequence that can selectively recognize the tumor neovasculature [85]. The system was tested both *in vitro* and *in vivo*. An analysis of cell growth inhibition and transfection efficiency after CNT system incubation with cells and exposure to NIR irradiation occurred *in vitro*. The antitumor activities of fluorescein isothiocyanate (FITC)-labeled SWNT–PEI, SWNT–PEI/NGR, SWNT–PEI/siRNA, and SWNT–PEI/siRNA/NGR systems were tested in human prostate cancer cell line (PC-3) tumor-laden mice. Cellular uptake into the cytoplasm was confirmed using fluorescence microscopy, imaging the FITC label [85]. Nanotube–siRNA complexes were red fluorescence-labeled and cultured with FITC-labeled cells, and imaged at 1, 2, and 4 h time points. Whilst the siRNA-conjugated SWNT–PEI did not completely enter the cell after 4 h, the NGR-conjugated SWNT–PEI showed complete cell entry within 4 h, suggesting that the use of targeting peptides may be useful in nucleic acid delivery. Cell viability was significantly decreased following the NIR irradiation of PC-3 cells exposed to SWNT– PEI/siRNA/NGR.

CdSe quantum dots were used to fluorescently label SWNTs for an *in-vivo* biodistribution analysis, though it should be noted that the size of these labels has a greater likelihood of affecting the unlabeled SWNT biodistribution compared to smaller molecular fluorophores. Both, SWNT–PEI/NGR/quantum dot (QD) and SWNT–PEI/siRNA/QD showed fluorescence activity in tumor regions, with SWNT–PEI/NGR/QD producing a significantly higher signal than SWNT–PEI/QD, suggesting that NGR is important for an enhanced cellular uptake of SWNTs. The targeting peptide may increase the interaction time between CNTs and the cell membrane, thus promoting endocytosis of the complex. Significant reductions in tumor size were noted in SWNT–PEI/siRNA and SWNT–PEI/siRNA/NGR conditions, but not in SWNT–PEI, SWNT–PEI/NGR, or saline conditions. This suggests that siRNA played a critical role in tumor size reduction. Subsequent laser irradiation of all SWNT–PEI/X conditions showed significant reductions in tumor size; the most significant difference being noted in the SWNT–PEI/NGR condition, and again suggesting the power of peptide conjugation as a critical piece in an integrated targeted drug-delivery system [85].

In addition to the innate high-loading capacity of CNTs, it is possible to add scaffolds to CNT surfaces to increase the delivery of siRNA. The effects of branched siRNA scaffold moieties conjugated to MWNTs have been characterized. Branched polymers, or dendrons, can multiply drug-loading capacities but may also lead to steric hindrance and possibly impact SWNT biodistribution. Hence, trade-offs must be considered and typically tested empirically, depending on the particular application. Multiple conjugation

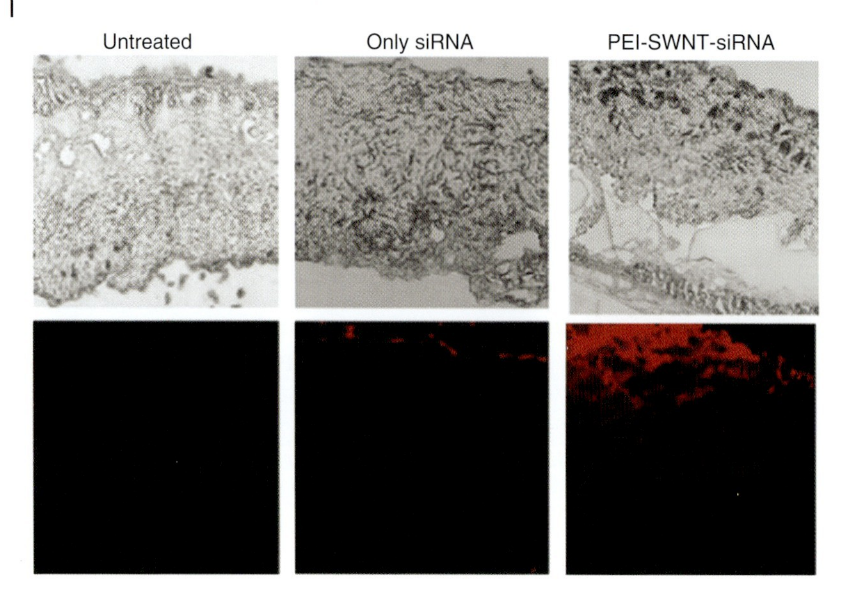

Fig. 6 Topical delivery of siRNA into skin in a murine model. Top row: Bright-field microscopy images of cryosectioned skin samples. Bottom row: Fluorescence images of cryosectioned skin samples. siRNA is Cy-3 fluorescence-labeled, and each condition is tested 4 h after siRNA solution application. The untreated condition shows no siRNA signal, as expected. The siRNA-only condition shows minimal siRNA signal at the surface of the skin. The siRNA–SWNT complex shows both a deeper skin penetration and an increased overall fluorescence signal. (Reproduced with permission from Ref. [87]; © 2014, Elsevier.)

methods, including novel "click chemistry" protocols have been used for appending dendrons and siRNA. Several branching chemistry methods were also employed, and the results characterized *in vitro* [86]. The ability to complex siRNA was inversely proportional to the degree of dendron branching in click chemistry-produced branching [86]. This effect may be due to a nonspecific binding of the long alkyl chains with the CNT surface, which hinders siRNA binding to the chain. For example, long chains may wrap around the CNT, producing unwanted conformations and reducing the exposed siRNA binding sites on the branch. However, the loading capacity can be increased by increasing the charge ratio (nitrogen/phosphate), partially mitigating this effect. Two variations of this complex are significantly more effective at delivering siRNA into cells than siRNA alone. The dendron-conjugated MWNTs produce longer-lasting silencing effects than cationic liposomes, though the liposomes have better gene-silencing abilities at early time points [86]. This may be due to faster interaction mechanisms between liposomes and cell membranes than peptide-conjugated CNTs, though MWNT–siRNA complexes display a high stability and an improved siRNA delivery at much longer time points than liposomes. Therefore, MWNT–siRNA exhibit ideal characteristics for controlled, steady-state delivery of the siRNA therapeutic into cells.

The topical delivery of biopharmaceuticals can serve as a highly efficient method for introducing therapeutics into superficial tumor sites such as skin (melanoma is the most common and invasive). Yet, these therapeutics typically require a carrier for adequate delivery. A topical application method for siRNA delivery into melanoma using SWNTs was recently developed. Although modified SWNTs have been successfully applied as transdermal drug delivery vehicles in the past, siRNA delivery was previously hindered by an insufficient drug loading to achieve significant effects. In order to mitigate this issue, siRNA was bound noncovalently to CNTs via specialized scaffolding proteins. Modified PEI-functionalized SWNTs for noncovalent binding to siRNA and an enhanced loading capacity were employed [87]. Binding of siRNA to the PEI–SWNT complex was confirmed using nuclear magnetic resonance (NMR) spectroscopy. Results from murine model experiments showed significant decreases in tumor size as early as 15 days post-injection, an increased siRNA tissue penetration (see Fig. 6), and a consistently small tumor size and weight up to 25 days post application [87]. The study results indicated that a topical application of siRNA-conjugated SWNTs may be a viable treatment mechanism for superficial tumors, due to the unique ability of SWNTs to interact with cell membranes and deliver their cargoes.

4 Conclusions

CNTs are versatile platforms for drug delivery, especially when conjugated to biopharmaceuticals. These versatile nanocarriers support efficient biopharmaceutical delivery through mechanisms distinct from other nanoparticles. Such mechanisms are made possible by the unique, quasi-one-dimensional structure, ultra-high surface area-to-volume ratio and facile chemistries of the CNTs, which allows them to be more efficient and specific to their target sites. The innate ability of CNTs to penetrate cells and to bind a wide range of compounds, in addition to their optical properties, makes them ideal tools for applications as diverse as immune system boosting, disease detection, and drug treatment for cancer and other diseases.

References

1. Iijima, S. (1991) Helical microtubules of graphitic carbon. *Nature*, **354** (6348), 56–58.
2. Zhu, J. (2012) Mammalian cell protein expression for biopharmaceutical production. *Biotechnol. Adv*, **30** (5), 1158–1170.
3. Steinberg, F.M., Raso, J. (1998) Biotech pharmaceuticals and biotherapy: an overview. *J. Pharm. Pharm. Sci.*, **1** (2), 48–59.
4. Smith, B.R., Kempen, P., Bouley, D., Xu, A. *et al.* (2012) Shape matters: intravital microscopy reveals surprising geometrical dependence for nanoparticles in tumor models of extravasation. *Nano Lett*, **12** (7), 3369–3377.
5. Smith, B.R., Zavaleta, C., Rosenberg, J., Tong, R. *et al.* (2013) High-resolution, serial intravital microscopic imaging of nanoparticle delivery and targeting in a small animal tumor model. *Nano Today*, **8** (2), 126–137.
6. Cai, D., Mataraza, J.M., Qin, Z.H., Huang, Z. *et al.* (2005) Highly efficient molecular delivery into mammalian cells using carbon nanotube spearing. *Nat. Methods*, **2** (6), 449–454.
7. Kam, N.W., Dai, H. (2005) Carbon nanotubes as intracellular protein transporters: generality and biological functionality. *J. Am. Chem. Soc*, **127** (16), 6021–6026.
8. Liu, Z., Chen, K., Davis, C., Sherlock, S. *et al.* (2008) Drug delivery with carbon nanotubes for in vivo cancer treatment. *Cancer Res*, **68** (16), 6652–6660.
9. Madani, S.Y., Naderi, N., Dissanayake, O., Tan, A. *et al.* (2011) A new era of cancer treatment: carbon nanotubes as drug delivery tools. *Int. J. Nanomedicine*, **6**, 2963–2979.
10. Varkouhi, A.K., Foillard, S., Lammers, T., Schiffelers, R.M. *et al.* (2011) SiRNA delivery

with functionalized carbon nanotubes. *Int. J. Pharm*, **416** (2), 419–425.
11. Jin, H., Heller, D.A., Sharma, R., Strano, M.S. (2009) Size-dependent cellular uptake and expulsion of single-walled carbon nanotubes: single particle tracking and a generic uptake model for nanoparticles. *ACS Nano*, **3** (1), 149–158.
12. Sato, Y., Yokoyama, A., Shibata, K., Akimoto, Y. *et al.* (2005) Influence of length on cytotoxicity of multi-walled carbon nanotubes against human acute monocytic leukemia cell line THP-1 in vitro and subcutaneous tissue of rats in vivo. *Mol. Biosyst*, **1** (2), 176–182.
13. Kostarelos, K., Bianco, A., Prato, M. (2009) Promises, facts and challenges for carbon nanotubes in imaging and therapeutics. *Nat. Nanotechnol*, **4** (10), 627–633.
14. Kostarelos, K. (2008) The long and short of carbon nanotube toxicity. *Nat. Biotechnol*, **26** (7), 774–776.
15. Saito, N., Haniu, H., Usui, Y., Aoki, K. *et al.* (2014) Safe clinical use of carbon nanotubes as innovative biomaterials. *Chem. Rev*, **114** (11), 6040–6079.
16. Collins, P.G. (2009) Defects and disorder in carbon nanotubes, in *Oxford Handbook of Nanoscience and Technology: Frontiers and Advances*, Oxford University Press, Oxford, pp. 31–81.
17. Thostenson, E.T., Ren, Z., Chou, T.-W. (2001) Advances in the science and technology of carbon nanotubes and their composites: a review. *Compos. Sci. Technol*, **61** (13), 1899–1912.
18. Ebbesen, T.W., Ajayan, P.M. (1992) Large-scale synthesis of carbon nanotubes. *Nature*, **358** (6383), 220–222.
19. Hedmer, M., Isaxon, C., Nilsson, P.T., Ludvigsson, L. *et al.* (2014) Exposure and emission measurements during production, purification, and functionalization of arc-discharge-produced multi-walled carbon nanotubes. *Ann. Occup. Hyg*, **58** (3), 355–379.
20. Yousef, S., Khattab, A., Osman, T.A., Zaki, M. (2013) Effects of increasing electrodes on CNTs yield synthesized by using arc-discharge technique. *J. Nanomater*, **2013**, 9.
21. Guo, T., Nikolaev, P., Thess, A., Colbert, D.T. *et al.* (1995) Catalytic growth of single-walled nanotubes by laser vaporization. *Chem. Phys. Lett*, **243** (1-2), 49–54.
22. Thess, A., Lee, R., Nikolaev, P., Dai, H. *et al.* (1996) Crystalline ropes of metallic carbon nanotubes. *Science*, **273** (5274), 483–487.
23. Eftekhari, A., Jafarkhani, P., Moztarzadeh, F. (2006) High-yield synthesis of carbon nanotubes using a water-soluble catalyst support in catalytic chemical vapor deposition. *Carbon*, **44** (7), 1343–1345.
24. Sayangdev, N., Ishwar, K.P. (2008) A model for catalytic growth of carbon nanotubes. *J. Phys. D Appl. Phys*, **41** (6), 065304.
25. Firme, C.P., Bandaru, P.R. III (2010) Toxicity issues in the application of carbon nanotubes to biological systems. *Nanomedicine*, **6** (2), 245–256.
26. Yaron, P.N., Holt, B.D., Short, P.A., Lösche, M. *et al.* (2011) Single wall carbon nanotubes enter cells by endocytosis and not membrane penetration. *J. Nanobiotechnol*, **9**, 45.
27. Fadel, T.R., Fahmy, T.M. (2014) Immunotherapy applications of carbon nanotubes: from design to safe applications. *Trends Biotechnol*, **32** (4), 198–209.
28. Jacobsen, N.R., Pojana, G., White, P., Møller, P. *et al.* (2008) Genotoxicity, cytotoxicity, and reactive oxygen species induced by single-walled carbon nanotubes and C(60) fullerenes in the FE1-Muta™ mouse lung epithelial cells. *Environ. Mol. Mutagen*, **49** (6), 476–487.
29. Lacerda, L., Singh, R., Al-Jamal, K.T., Turton, J., *et al.* (2008) Dynamic imaging of functionalized multi-walled carbon nanotube systemic circulation and urinary excretion. *Adv. Mater*, **20** (2), 225–230.
30. Pacurari, M., Yin, X.J., Zhao, J., Ding, M. *et al.* (2008) Raw single-wall carbon nanotubes induce oxidative stress and activate MAPKs, AP-1, NF-kappaB, and Akt in normal and malignant human mesothelial cells. *Environ. Health Perspect*, **116** (9), 1211–1217.
31. Jakubek, L.M., Marangoudakis, S., Raingo, J., Liu, X. *et al.* (2009) The inhibition of neuronal calcium ion channels by trace levels of yttrium released from carbon nanotubes. *Biomaterials*, **30** (31), 6351–6357.
32. Gupta, A.K., Gupta, M. (2005) Synthesis and surface engineering of iron oxide nanoparticles for biomedical applications. *Biomaterials*, **26** (18), 3995–4021.
33. Xu, Y.-Q., Peng, H., Hauge, R.H., Smalley, R.E. (2004) Controlled multistep purification

of single-walled carbon nanotubes. *Nano Lett*, **5** (1), 163–168.

34. Ebbesen, T.W., Ajayan, P.M., Hiura, H., Tanigaki, K. (1994) Purification of nanotubes. *Nature*, **367** (6463), 519.
35. Schipper, M.L., Nakayama-Ratchford, N., Davis, C.R., Kam, N.W. *et al.* (2008) A pilot toxicology study of single-walled carbon nanotubes in a small sample of mice. *Nat. Nanotechnol*, **3** (4), 216–221.
36. Kayat, J., Gajbhiye, V., Tekade, R.K., Jain, N.K. (2011) Pulmonary toxicity of carbon nanotubes: a systematic report. *Nanomed. Nanotechnol. Biol. Med*, **7** (1), 40–49.
37. Kolosnjaj-Tabi, J., Hartman, K.B., Boudjemaa, S., Ananta, J.S. *et al.* (2010) In vivo behavior of large doses of ultrashort and full-length single-walled carbon nanotubes after oral and intraperitoneal administration to Swiss mice. *ACS Nano*, **4** (3), 1481–1492.
38. Bussy, C., Pinault, M., Cambedouzou, J., Landry, M.J. *et al.* (2012) Critical role of surface chemical modifications induced by length shortening on multi-walled carbon nanotubes-induced toxicity. *Part. Fibre Toxicol*, **9** (1), 46–60.
39. Liu, Y., Zhao, Y., Sun, B., Chen, C. (2012) Understanding the toxicity of carbon nanotubes. *Acc. Chem. Res*, **46** (3), 702–713.
40. Mostafalou, S., Mohammadi, H., Ramazani, A., Abdollahi, M. (2013) Different biokinetics of nanomedicines linking to their toxicity; an overview. *Daru*, **21** (1), 14.
41. Zhang, Y., Deng, J., Zhang, Y., Guo, F. *et al.* (2013) Functionalized single-walled carbon nanotubes cause reversible acute lung injury and induce fibrosis in mice. *J. Mol. Med. (Berl.)*, **91** (1), 117–128.
42. Ma-Hock, L., Treumann, S., Strauss, V., Brill, S. *et al.* (2009) Inhalation toxicity of multiwall carbon nanotubes in rats exposed for 3 months. *Toxicol. Sci*, **112** (2), 468–481.
43. Huczko, A., Lange, H., Całko, E., Grubek-Jaworska, H. *et al.* (2001) Physiological testing of carbon nanotubes: are they asbestos-like? *Fullerene Sci. Technol*, **9** (2), 251–254.
44. Lam, C.W., James, J.T., McCluskey, R., Hunter, R.L. (2004) Pulmonary toxicity of single-wall carbon nanotubes in mice 7 and 90 days after intratracheal instillation. *Toxicol. Sci*, **77** (1), 126–134.
45. Shvedova, A.A., Kisin, E., Murray, A.R., Johnson, V.J. *et al.* (2008) Inhalation vs. aspiration of single-walled carbon nanotubes in C57BL/6 mice: inflammation, fibrosis, oxidative stress, and mutagenesis. *Am. J. Physiol. Lung Cell. Mol. Physiol*, **295** (4), L552–L565.
46. Shaughnessy, P., Adamcakova-Dodd, A., Altmaier, R., Thorne, P.S. (2014) Assessment of the aerosol generation and toxicity of carbon nanotubes. *Nanomaterials*, **4** (2), 439–453.
47. Kasai, T., Umeda, Y., Ohnishi, M., Kondo, H. *et al.* (2015) Thirteen-week study of toxicity of fiber-like multi-walled carbon nanotubes with whole-body inhalation exposure in rats. *Nanotoxicology*, **9** (4), 413–422.
48. Ali-Boucetta, H., Nunes, A., Sainz, R., Herrero, M.A. *et al.* (2013) Asbestos-like pathogenicity of long carbon nanotubes alleviated by chemical functionalization. *Angew. Chem. Int. Ed*, **52** (8), 2274–2278.
49. Donaldson, K., Murphy, F.A., Duffin, R., Poland, C.A. (2010) Asbestos, carbon nanotubes and the pleural mesothelium: a review of the hypothesis regarding the role of long fibre retention in the parietal pleura, inflammation and mesothelioma. *Part. Fibre Toxicol*, **7**, 5.
50. Poland, C.A., Duffin, R., Kinloch, I., Maynard, A. *et al.* (2008) Carbon nanotubes introduced into the abdominal cavity of mice show asbestos-like pathogenicity in a pilot study. *Nat. Nanotechnol*, **3** (7), 423–428.
51. Madani, S.Y., Mandel, A., Seifalian, A.M. (2013) A concise review of carbon nanotube's toxicology. *Nano Rev.*, **4**, 4. doi: 10.3402/nano.v4i0.21521
52. Liu, Z., Davis, C., Cai, W., He, L. *et al.* (2008) Circulation and long-term fate of functionalized, biocompatible single-walled carbon nanotubes in mice probed by Raman spectroscopy. *Proc. Natl Acad. Sci. USA*, **105** (5), 1410–1415.
53. Mittal, V. (2011) Carbon nanotubes surface modifications: An overview, in *Surface Modification of Nanotube Fillers*, Wiley-VCH Verlag GmbH & Co. KGaA, Weinheim, pp. 1–23.
54. Liu, Z., Tabakman, S., Welsher, K., Dai, H. (2009) Carbon nanotubes in biology and medicine: in vitro and in vivo detection, imaging and drug delivery. *Nano Res*, **2** (2), 85–120.

55. Wu, H.-C., Chang, X., Liu, L., Zhao, F. *et al.* (2010) Chemistry of carbon nanotubes in biomedical applications. *J. Mater. Chem*, **20** (6), 1036–1052.
56. Chen, R.J., Zhang, Y.G., Wang, D.W., Dai, H.J. (2001) Noncovalent sidewall functionalization of single-walled carbon nanotubes for protein immobilization. *J. Am. Chem. Soc*, **123** (16), 3838–3839.
57. Chen, Z., Tabakman, S.M., Goodwin, A.P., Kattah, M.G. *et al.* (2008) Protein microarrays with carbon nanotubes as multicolor Raman labels. *Nat. Biotechnol*, **26** (11), 1285–1292.
58. Yang, S.T., Luo, J., Zhou, Q., Wang, H. (2012) Pharmacokinetics, metabolism and toxicity of carbon nanotubes for biomedical purposes. *Theranostics*, **2** (3), 271–282.
59. Moore, T.L., Grimes, S.W., Lewis, R.L., Alexis, F. (2014) Multilayered polymer-coated carbon nanotubes to deliver dasatinib. *Mol. Pharm*, **11** (1), 276–282.
60. Malmsten, M. (2013) Inorganic nanomaterials as delivery systems for proteins, peptides, DNA, and siRNA. *Curr. Opin. Colloid Interface Sci*, **18** (5), 468–480.
61. Witus, L.S., Rocha, J.-D.R., Yuwono, V.M., Paramonov, S.E. *et al.* (2007) Peptides that non-covalently functionalize single-walled carbon nanotubes to give controlled solubility characteristics. *J. Mater. Chem*, **17** (19), 1909–1915.
62. Balavoine, F., Schultz, P., Richard, C., Mallouh, V. *et al.* (1999) Helical crystallization of proteins on carbon nanotubes: a first step towards the development of new biosensors. *Angew. Chem. Int. Ed*, **38** (13-14), 1912–1915.
63. Chen, R.J., Bangsaruntip, S., Drouvalakis, K.A., Kam, N.W. *et al.* (2003) Noncovalent functionalization of carbon nanotubes for highly specific electronic biosensors. *Proc. Natl Acad. Sci. USA*, **100** (9), 4984–4989.
64. Sun, H., She, P., Lu, G., Xu, K. *et al.* (2014) Recent advances in the development of functionalized carbon nanotubes: a versatile vector for drug delivery. *J. Mater. Sci*, **49** (20), 6845–6854.
65. Smith, B.R., Cheng, Z., De, A., Koh, A.L. *et al.* (2008) Real-time intravital imaging of RGD-quantum dot binding to luminal endothelium in mouse tumor neovasculature. *Nano Lett*, **8** (9), 2599–2606.
66. Weissleder, R., Pittet, M.J. (2008) Imaging in the era of molecular oncology. *Nature*, **452** (7187), 580–589.
67. Smith, B.R., Ghosn, E.E., Rallapalli, H., Prescher, J.A. *et al.* (2014) Selective uptake of single-walled carbon nanotubes by circulating monocytes for enhanced tumour delivery. *Nat. Nanotechnol*, **9** (6), 481–487.
68. Hashida, Y., Tanaka, H., Zhou, S., Kawakami, S. *et al.* (2014) Photothermal ablation of tumor cells using a single-walled carbon nanotube-peptide composite. *J. Control. Release*, **173**, 59–66.
69. Chen, C., Hou, L., Zhang, H., Zhu, L. *et al.* (2013) Single-walled carbon nanotubes mediated targeted tamoxifen delivery system using asparagine-glycine-arginine peptide. *J. Drug Target*, **21** (9), 809–821.
70. Battigelli, A., Russier, J., Venturelli, E., Fabbro, C. *et al.* (2013) Peptide-based carbon nanotubes for mitochondrial targeting. *Nanoscale*, **5** (19), 9110–9117.
71. Villa, C.H., Dao, T., Ahearn, I., Fehrenbacher, N. *et al.* (2011) Single-walled carbon nanotubes deliver peptide antigen into dendritic cells and enhance IgG responses to tumor-associated antigens. *ACS Nano*, **5** (7), 5300–5311.
72. Parra, J., Abad-Somovilla, A., Mercader, J.V., Taton, T.A. *et al.* (2013) Carbon nanotube-protein carriers enhance size-dependent self-adjuvant antibody response to haptens. *J. Control. Release*, **170** (2), 242–251.
73. Izumi, Y., Xu, L., di Tomaso, E., Fukumura, D. *et al.* (2002) Tumour biology: herceptin acts as an anti-angiogenic cocktail. *Nature*, **416** (6878), 279–280.
74. Venturelli, E., Fabbro, C., Chaloin, O., Ménard-Moyon, C. *et al.* (2011) Antibody covalent immobilization on carbon nanotubes and assessment of antigen binding. *Small*, **7** (15), 2179–2187.
75. Li, R., Wu, R., Zhao, L., Wu, M. *et al.* (2010) P-glycoprotein antibody functionalized carbon nanotube overcomes the multidrug resistance of human leukemia cells. *ACS Nano*, **4** (3), 1399–1408.
76. Bajaj, P., Mikoryak, C., Wang, R., Bushdiecker, D.K. II, *et al.* (2014) A carbon nanotube-based Raman-imaging immunoassay for evaluating tumor targeting ligands. *Analyst*, **139** (12), 3069–3076.

77. Zavaleta, C., de la Zerda, A., Liu, Z., Keren, S. *et al.* (2008) Noninvasive Raman spectroscopy in living mice for evaluation of tumor targeting with carbon nanotubes. *Nano Lett*, **8** (9), 2800–2805.
78. Mulvey, J.J., Villa, C.H., McDevitt, M.R., Escorcia, F.E. *et al.* (2013) Self-assembly of carbon nanotubes and antibodies on tumours for targeted amplified delivery. *Nat. Nanotechnol*, **8** (10), 763–771.
79. Wu, H., Shi, H., Zhang, H., Wang, X. *et al.* (2014) Prostate stem cell antigen antibody-conjugated multiwalled carbon nanotubes for targeted ultrasound imaging and drug delivery. *Biomaterials*, **35** (20), 5369–5380.
80. Liu, Y., Wu, D.C., Zhang, W.D., Jiang, X. *et al.* (2005) Polyethylenimine-grafted multiwalled carbon nanotubes for secure noncovalent immobilization and efficient delivery of DNA. *Angew. Chem. Int. Ed*, **44** (30), 4782–4785.
81. Pantarotto, D., Singh, R., McCarthy, D., Erhardt, M. *et al.* (2004) Functionalized carbon nanotubes for plasmid DNA gene delivery. *Angew. Chem. Int. Ed*, **43** (39), 5242–5246.
82. Bates, K., Kostarelos, K. (2013) Carbon nanotubes as vectors for gene therapy: past achievements, present challenges and future goals. *Adv. Drug Deliv. Rev*, **65** (15), 2023–2033.
83. Benya, R.V., Kusui, T., Pradhan, T.K., Battey, J.F. *et al.* (1995) Expression and characterization of cloned human bombesin receptors. *Mol. Pharmacol*, **47** (1), 10–20.
84. Qiao, J., Hong, T., Guo, H., Xu, Y.Q. *et al.* (2013) Single-walled carbon nanotube-mediated small interfering RNA delivery for gastrin-releasing peptide receptor silencing in human neuroblastoma, in *NanoBiotechnology Protocols* (eds S.J. Rosenthal and D.W. Wright), Humana Press, pp. 137–147.
85. Wang, L., Shi, J., Zhang, H., Li, H. *et al.* (2013) Synergistic anticancer effect of RNAi and photothermal therapy mediated by functionalized single-walled carbon nanotubes. *Biomaterials*, **34** (1), 262–274.
86. Battigelli, A., Wang, J.T., Russier, J., Da Ros, T. *et al.* (2013) Ammonium and guanidinium dendron-carbon nanotubes by amidation and click chemistry and their use for siRNA delivery. *Small*, **9** (21), 3610–3619.
87. Siu, K.S., Chen, D., Zheng, X., Zhang, X. *et al.* (2014) Non-covalently functionalized single-walled carbon nanotube for topical siRNA delivery into melanoma. *Biomaterials*, **35** (10), 3435–3442.

8
Cholesterol in Nanobiotechnology

Philipp Schattling, Yan Zhang, Boon M. Teo, and Brigitte Städler
Aarhus University, Interdisciplinary Nanoscience Center (iNano), Gustav Wieds Vej 14, 8000 Aarhus, Denmark

Translational Nanomedicine. First Edition. Edited by Robert A. Meyers.

Keywords

Cholesterol
An essential component of mammalian cells that is largely found in cell membranes. The biosynthesis occurs mainly in the liver. Cholesterol refers to the class of steroids.

Liposomes
Spherical vesicles which consist of two or more lipid bilayer membranes. Liposomes feature a similar structure to natural cell membranes, and thus are suitable to mimic biological cell membranes. The architecture allows both hydrophilic and lipophilic compounds to be entrapped.

Nanobiotechnology
An interdisciplinary field which combines nanotechnology and biotechnology, with the aim of mimicking processes in Nature and to make them feasible on an industrial scale.

Polymer
A macromolecule which is formed by the linkage of a large number of smaller molecules (monomers) in a polymerization reaction.

Self-assembly
Describes the process, in which an initially disordered system composed of pre-existing particles or molecules forms an ordered system, without external direction but as a consequence of the intrinsic properties of the particles or molecules.

Cell mimicry
Engineered assemblies that can mimic the (specific) functions of a biological cell.

Biosensor
A device in which a biologically derived recognition is coupled to a transducer for the purpose of quantitative development of a complex biochemical parameter.

1

Cholesterol is an essential and natural component of animal cells. The human body contains approximately 140 g of cholesterol, which is found mainly in cellular membranes. Since its first recognition during the late eighteenth century, cholesterol has evolved from an alcohol-soluble side fraction of human gallstones into a compound with an extraordinary impact on nanobiotechnology. Such remarkable development is based on two key facts: (i) the intrinsic chemical and physical properties of the steroid skeleton; and (ii) the advent of modern polymerization techniques which enable the design of sophisticated polymeric architectures with significant impact on nanomedicine. The contributions made by cholesterol in the fields of liposomal drug delivery, biosensing and cell mimicry will be discussed in this chapter, with emphasis placed on the design and strategic access to cholesterol-containing polymer structures.

1
Introduction

Successful concepts in nanomedicine are often inspired by Nature, with advances in nanobiotechnology having benefited from the entire spectra, starting with the creation of individual natural building blocks (e.g., peptides, DNA, lipids) and extending to their (self-)assembly into nanodevices capable of mimicking entire functional systems. The combination of natural building blocks with synthetic materials has proven to further enlarge the available toolbox to assemble (nano)materials. Cholesterol represents just one such example where natural building blocks can serve as indispensable components for diverse applications in nanomedicine.

Although, historically, cholesterol was first recognized as a discrete substance in the late eighteenth century, the name cholesterol was introduced only at the start of the twentieth century. The structure of cholesterol was elucidated in multiple steps; first, it was shown to be an alcohol [1] and later, more specifically a secondary alcohol [2], while the presence of the double-bond was confirmed in 1868. Further details on these aspects are available in the reviews by Windaus [3, 4].

In Nature, cholesterol is an essential component of mammalian cells, occurring largely in the cell membranes but with a great abundance of cellular cholesterol located in the plasma membrane [5, 6]. Two ways have been identified by which cells can ensure a sufficient cholesterol availability: (i) most nucleated cells can synthesize cholesterol via the mevalonic acid pathway in the endoplasmic reticulum; or (ii) cholesterol can be internalized by cells from lipoproteins. The ratio of synthesized cholesterol to dietary acquired cholesterol has been estimated as 7:3 [7]. Cells can transport cholesterol to the required locations in the membranes in a variety of ways [8–10]. The major importance of cholesterol is its ability to modulate the physico-chemical properties of the cell membrane, that is, to affect the membrane's permeability and fluidity. Cholesterol also modulates the activity of membrane proteins (e.g., of G-protein-coupled receptors [11]), the distribution of membrane proteins, or the lipid organization within the membrane, and is also involved in signal transduction and endocytosis due to its presence in lipid rafts [6] and caveolae [12]. The connection between cholesterol and human health was identified at an early stage, and serum total cholesterol levels are considered an important parameter in this context [13], including the assessment of risk of developing cardiovascular problems [14] or dementia/Alzheimer disease [15]. According to the World Health Organization, one-third of all ischemic heart disease is attributable to high cholesterol levels, causing three million deaths annually and 30 million disability-adjusted life-years. In Europe and the USA in particular, more than 50% of men and women suffer from raised cholesterol levels, and from this perspective it is hardly surprising that in modern society cholesterol is recognized in a negative context. Yet, despite this drawback – which is rather a society issue than of the sterol itself – cholesterol is well recognized as an excellent compound in the biomaterials sciences, with applications in drug delivery, nanomedicine, and biosensing.

The aim of this chapter is to provide an overview of the importance of cholesterol in nanobiotechnology, initially by outlining the chemistry of the molecule, and the special features of which contribute greatly to the applications described below. Details of the fundamental aspects are followed by a summary of relevant cholesterol-modifying approaches and the synthesis of cholesterol-containing polymers. Cholesterol is an essential factor whereby liposomes can be used as carriers in drug-delivery systems or as components in biosensors, and in this context it can be regarded as either a stabilizer or as an anchoring unit, and the importance of cholesterol in these contexts will be illustrated by recent examples. The ultimate aim of the chapter is to show that cholesterol is not only a biologically important molecule, but it is also an essential player in the field of nanobiotechnology, contributing to diverse applications aimed at improving human health.

2
The Chemistry of Cholesterol

Cholest-5-en-3β-ol – better known under its trivial name cholesterol – features a hydroxyl group at the C-3 position, and thus refers to the class of sterols as a subclass of steroids (Fig. 1a) [16]. The structure of cholesterol mainly comprises of the three structural domains, explaining the importance of cholesterol: (i) Polar hydroxyl group, allowing hydrogen bonding and chemical modification of cholesterol; (ii) the steroid motif, composed of four jointed rings, provides a rigid planar skeleton; and (iii) flexible aliphatic side chain.

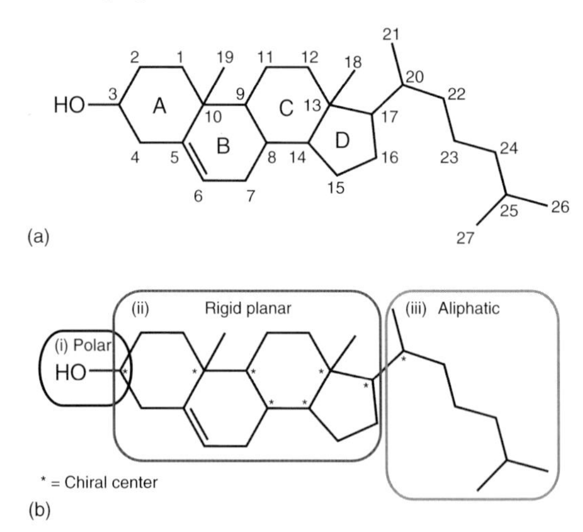

Fig. 1 (a) Chemical structure of cholesterol. The rings of the cyclopentan[α]phenanthrene skeleton are assigned using the letters A to D. The carbon atoms are numbered according to (a), starting with the A ring. At position C-10 and C-13 methyl groups are present, whereas the *iso*-octyl side chain is linked at position C-17; (b) Illustration

three regions (Fig. 1b): (i) the hydroxyl group connected to the (ii) rigid, nonpolar tetracyclic cyclopenta[α]phenanthrene skeleton [17], which exhibits methyl groups at the C-10 and C-13 positions; and finally (iii) a flexible, nonpolar *iso*-octyl side chain linked at the C-17 position [16, 18]. These structural features make cholesterol an amphipathic compound, exhibiting both hydrophilic and lipophilic properties, while the rigid skeleton in combination with the flexible tail induces the intrinsic ability for orientational ordering.

2.1
Properties of Cholesterol

Since the nineteenth century, cholesterol has been used frequently as a steroid model system for testing novel chemical reactions. As the most abundant member of the family of sterols, and due to its importance in the biosynthesis of steroid hormones, bile acids and vitamin D, cholesterol is readily available and relatively inexpensive [19]. Hence, a huge library of pathways towards sophisticated and efficient chemical modifications of cholesterol has been compiled and is available in excellent reviews on this topic [19–21]. Within the scope of the present chapter, simple derivatization strategies of cholesterol are introduced, enabling its conjugation to (bio)molecules and polymers.

In this context, the hydroxyl group in the 3β position of the cholesterol A ring (Fig. 1b) is of primary interest, since its nucleophilic character can easily be utilized in nucleophilic substitution reactions to conjugate polymerizable groups onto cholesterol, or cholesterol to preformed polymers, respectively. The most common strategy is to use the commercially available cholesteryl chloroformate. Notwithstanding, a potential synthetic access to the chloroformate derivative could proceed via the reaction of cholesterol with phosgene. Conversion towards the chloroformate formally renders the reactivity of the former nucleophilic hydroxyl group electrophilic, and thus enables conjugation to any nucleophile (e.g., amino- or hydroxyl groups) to yield carbamates or carbonates, respectively. In general, high yields were achieved when auxiliary bases such as triethylamine (TEA) or 4-dimethylaminopyridine (DMAP) were added to the reaction under ice-cooled anhydrous conditions. Following this approach, the primary amino functionalities of 3′-*O*-(6-aminohexyl)-uridine [22] or poly(ethylenimine) (PEI) [23] could be coupled with cholesterol. In another study (and also an example of the carbonate-conjugation of cholesterol), the alcohol function of hydroxylethyl (meth)acrylate (HEMA) was employed [24]. It was further shown that the reactivity of the chloroformate was sufficiently high, and that even aromatic hydroxyl groups could be converted [25].

The cholesterol hydroxyl group can also participate in esterification reactions. In general, functional acid chlorides were normally used, as the reaction proceeds rapidly and with high yields. In this manner, 5-norbornene-2-carboxylic acid chloride was converted with cholesterol [26]. Similarly, 4,4′-azobis(4-cyanovaleric acid) was transferred with thionyl chloride into its acid chloride derivative and subsequently conjugated to cholesterol [27]. Especially in the context of conducting a subsequent polymerization, the acylation of cholesterol with acryloyl [28, 29] or methacryloyl chloride [30–32] was widely used. Steglich esterification, a mild and efficient strategy to form ester bonds between alcohols and carboxylic acids, is particularly suitable for the formation of steric-demanding esters, such as ester-conjugation between

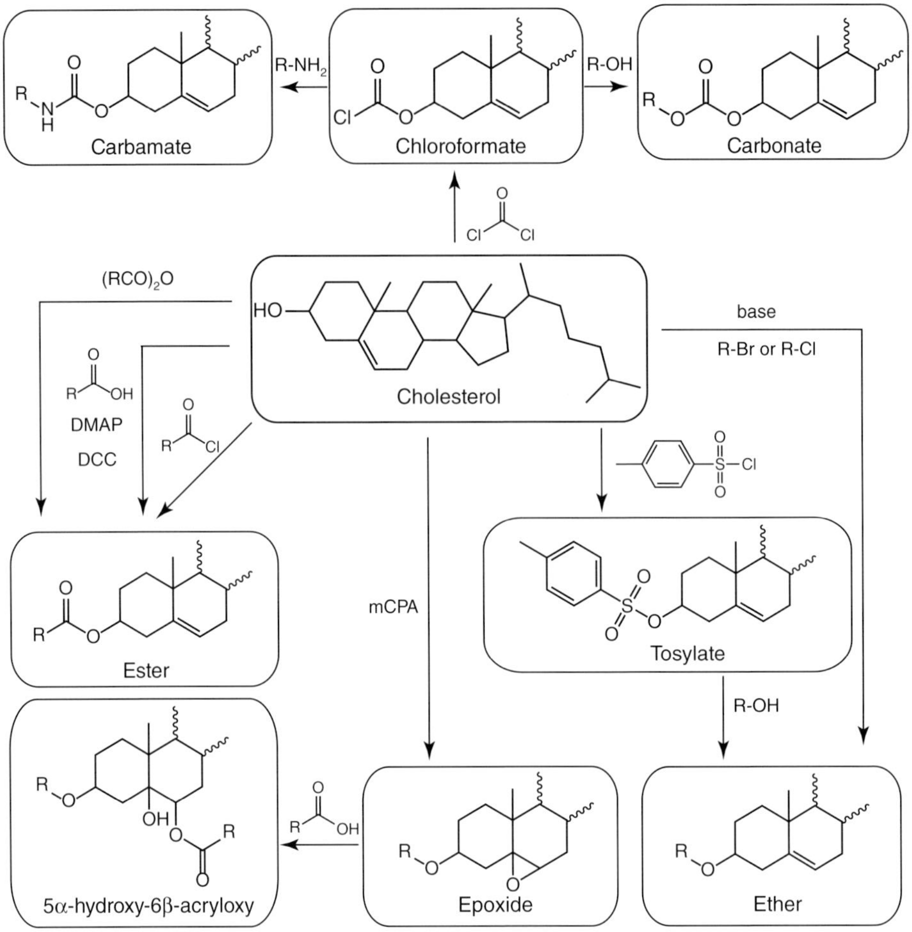

Fig. 2 Schematic illustration of strategies to chemically modify cholesterol. Amines and carbonates are accessible utilizing the chloroformate derivative of cholesterol (top), etherification via tosylation of cholesterol or base-supported conversion with alkyl halides (right), esterification via acid chlorides, Steglich approach or anhydrides (left), modification of the C=C double bond present in ring B by epoxidation and subsequent conversion with carboxylic acids (below).

cholesterol and carboxylic acids [33, 34]. This mechanism proceeds via an *O*-acylisourea intermediate with dicyclohexylcarbodiimide (DCC) in the presence of DMAP (Fig. 2).

It was shown, that the nucleophilic character of the cholesterol hydroxyl group provided the possibility of a nucleophilic attack of succinic anhydride and, as a result, the cyclic structure of the anhydride was opened and gave a cholesterol ester, featuring a terminal carboxylic acid [35].

Strategies whereby cholesterol is conjugated via carbonyl functions are advantageous in terms of biodegradability, as esters and carbonates are both generally susceptible to hydrolysis under physiological conditions [36]. Nevertheless, ether formation plays a major role in the conjugation of sugar or oligo ethylene glycol to cholesterol. The Williamson ether synthesis is considered as the traditional approach for synthesizing ether, where an alkoxide ion ("deprotonated hydroxyl group") reacts with a primary alkyl halide. In this fashion, cholesterol was deprotonated by sodium hydride and added to bromoxylene norbornene, yielding the corresponding ether [37]. Unfortunately, this method often competes with the base-catalyzed elimination reaction, especially when steric-demanding or tertiary halides were used. In an alternative approach, two alcohols can be combined via an ether bond by employing *p*-toluene sulfonyl chloride, whereby the sulfonyl chloride compound tosylates one alcohol which, in turn, reacts with the second alcohol. By applying this tool, short and long oligo ethylene glycol spacers could be conjugated to cholesterol [38, 39]. Starting with tosylated cholesterol, several 3-cholesteryl 6-(glycosylthiohexyl) ethers were obtained via thiol alkylation with the corresponding 1-thioaldoses in three steps [40]. In another elegant approach, trichloroacetimidate derivatives of acylated galactose were successfully conjugated to cholesterol via glycosylation [41, 42].

When considering the structure of cholesterol, the double bond in the B ring represents another feature which can be utilized for cholesterol conjugation. This approach provides the possibility of adding extra functionalities by further modification of the hydroxyl group. When employing this strategy, the etherification of cholesterol with alkyl halides was performed as a first step, followed by epoxidation of the double bond, using *meta*-chloroperoxybenzoic acid (mCPA). The epoxy function formed was opened by an excess amount of acrylic acid, leading to 3β-alkoxy-5α-hydroxy-6β-acrylatecholestanes [43]. Similarly, the carbon double bond of cholesteryl acetate was epoxidized and converted with methacrylic acid [44, 45].

2.2 Polymerization Method

Currently, ten million tons of polymers are produced each year via conventional radical polymerization, and consequently free-radical polymerization has asserted itself as the most important (radical) polymerization technique in both academia and industry [46]. The reasons for this are threefold: (i) it allows the polymerization of a wide variety of monomers (e.g., (meth)acrylates, (meth)acrylamide, styrenes, vinylic compounds); (ii) it demonstrates a high tolerance against many unprotected functional groups (hydroxyl-, carboxy-, amino-, and sulfonate); and (iii) it further features an excellent compatibility with reaction conditions (bulk, solution, or emulsion). Yet, despite these advantages, the free-radical approach is not suitable for designing complex polymeric architectures or well-defined polymers – at least not on a level similar to ionic polymerization techniques [47].

The advent of controlled polymerization techniques opened new avenues to novel polymeric materials, with narrow molecular weight distributions and an ability to control both macromolecular architecture and copolymer composition. From a mechanistic point of view, controlled radical polymerization strategies proceed in similar fashion to conventional free-radical polymerization, but employ a reversible activation process [48]. In this context, for controlled radical chain polymerization the term reversible-deactivation radical polymerization (RDRP) is also used [49].

The basics of controlled radical polymerization techniques, which are mainly used for the synthesis of cholesteryl-containing polymers, are briefly summarized in the following sections, though further comprehensive reviews on the corresponding methods are cited in the individual subsections.

2.2.1 Reversible Addition-Fragmentation Chain Transfer Polymerization (RAFT)

RAFT (reversible addition fragmentation chain transfer) polymerization proceeds in the presence of a so-called RAFT agent, the chain-transfer agent (CTA). The CTAs are mainly comprised of thiocarbonylthio compounds (dithioester and trithiocarbonate) or xanthates [50, 51], and behave like ideal CTAs. The reactivity of the C=S double bond towards the radical addition can be tailored by the substituent Z (Fig. 3). The propagating radical chain is added to the CTA at an early stage of the polymerization, and the formed intermediate radical (**1**) fragments into a polymeric CTA and a new propagating radical. A rapid establishment of the addition–fragmentation equilibrium between the active propagating radical and the dormant species ensures an equal probability for all chains to grow, yielding low-disperse polymers [46, 52–55].

2.2.2 Atom Transfer Radical Polymerization (ATRP)

Similar to the RAFT mechanism, the control in atom transfer radical polymerization (ATRP) is the subject of an equilibrium between a propagating radical and a dormant species [47]. This activation–deactivation process occurs between alkyl halides (dormant species) and transition metal complexes; the exchange mechanism is explained formally in Fig. 4. The transition from dormant to active species involves a reversible redox process with a concomitant abstraction of the halogen, during which the transition metal complex undergoes a one-electron oxidation in its higher oxidation state. As a consequence, growing radical species are

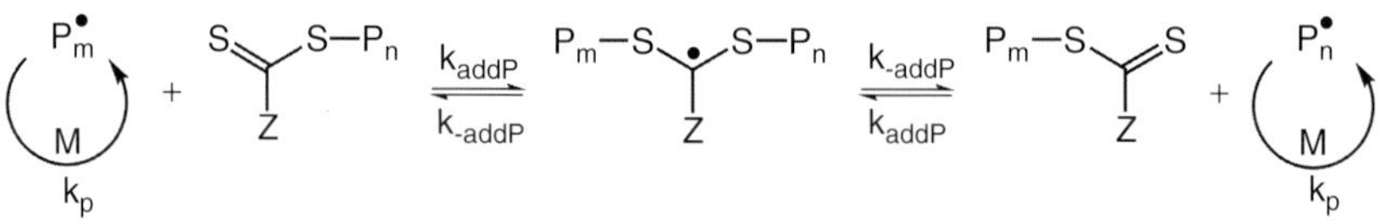

$$P_m^\bullet + S{=}C(Z){-}S{-}P_n \underset{k_{-addP}}{\overset{k_{addP}}{\rightleftharpoons}} P_m{-}S{-}\dot{C}(Z){-}S{-}P_n \underset{k_{addP}}{\overset{k_{-addP}}{\rightleftharpoons}} P_m{-}S{-}C(Z){=}S + P_n^\bullet$$

2   **1**   **3**

Fig. 3 Mechanism of the reversible addition fragmentation chain transfer (RAFT) polymerization. Note that only the addition–fragmentation equilibrium is shown, whereas the initiation and termination are omitted.

$$P_n{-}X + Mt^m/L \underset{k_{deact}}{\overset{k_{act}}{\rightleftharpoons}} P_n^\bullet + X{-}Mt^{m+1}/L$$

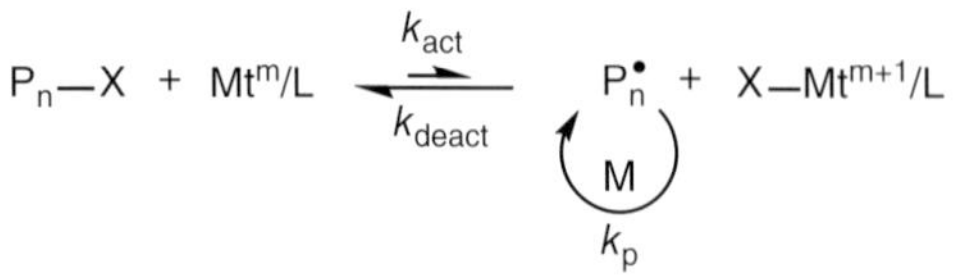

Fig. 4 Compact description of the mechanism of atom transfer radical polymerization (ATRP), highlighting the equilibrium between propagating radical and dormant species (initiation and termination are omitted).

formed intermittently, and these are rapidly deactivated again by the halide-coordinated transition metal complex [56, 57].

2.3 Polymer End Group Functionalization

In the case of a polymer chain being sufficiently short, the nature of the polymer end group has a remarkable impact on the properties of the entire polymer [58–60]. Consequently, controlling the end group of a polymer is highly desirable, and therefore a variety of strategies has been developed over the past decades that allows for the transformation, modification and utilization of both ends of a polymer chain [58–61].

The architecture of a polymer chain can be subdivided into three compartments (Fig. 5):

- The α-position, which is defined by the nature of the radical initiator and forms during the initiation step.
- The polymer backbone, which is formed during the chain growth mechanism (in radical and ionic polymerization) and basically contains the polymerized monomer sequence.
- The ω-position, which is implemented into the polymer during the termination step [62]. (It should be noted that in the literature the terms ω and α are often used in an inconsistent fashion.)

The introduction of cholesterol moieties as end groups in the polymer backbone can be realized by two prevalent strategies: (i) a direct introduction by utilizing cholesterol-prefunctionalized initiators, respectively CTAs; or (ii) indirectly by implementing a post-polymerization modification of pre-existing end groups. In the case of post-modification, highly efficient and specific reactions are needed in order to ensure a high degree of conversion on the one hand, and to prevent a potential modification of the reactive sites that are

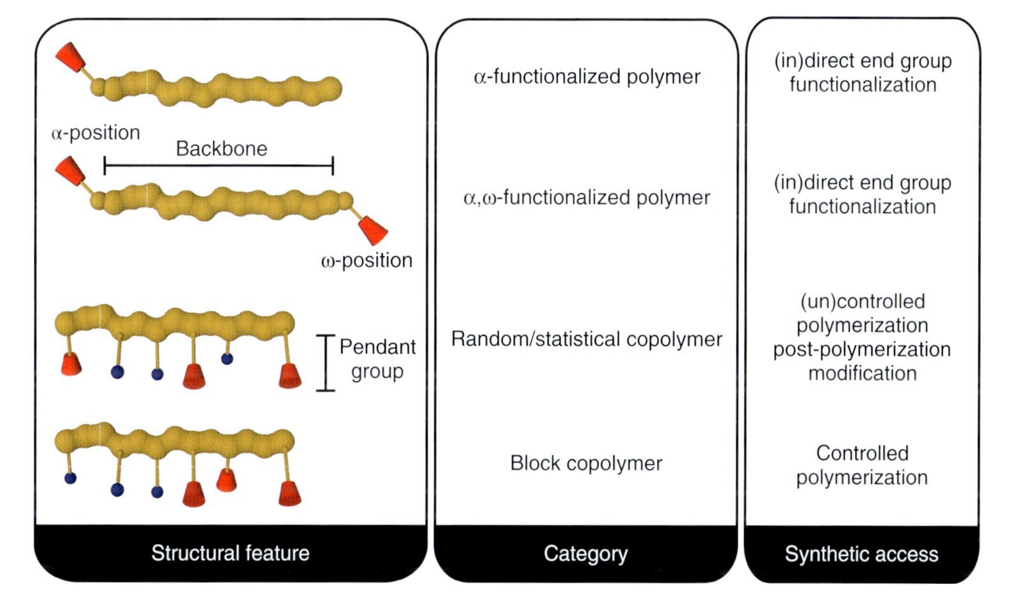

Fig. 5 The different sites in the polymer architecture that can be modified with cholesterol moieties and the different synthetic strategies capable of yielding the corresponding architecture are shown.

present in the polymer backbone on the other hand.

2.3.1 Direct Introduction

Although the mechanisms of the above-mentioned radical polymerization techniques differ considerably, they each have a preceding initiation step in common. As a consequence, the initiator (in free-radical polymerization and ATRP) or a fragment of the CTA (in RAFT polymerization) remains at the α-position, and for the controlled radical polymerization methods at the ω-position of each polymer chain. In this context, the local resolution implemented into the initiation step can be exploited so as to functionalize exclusively one or both ends of the polymer chains; this serves as an ideal approach for end-capping polymers with cholesterol units.

The use of cholesterol-modified radical initiators in a free-radical polymerization emerged as a simple way to design cholesterol end-functionalized polymers, and some selected examples of cholesterol-modified initiators are shown in Fig. 6. For example, 4,4′-azobis(4-cyanovaleric acid) was successfully labeled with cholesterol [27]. The resulting 4,4′-azobis(4-cyano-1-cholesteryl pentanoate) was further used to initiate the polymerization of sodium 2-(acrylamide)-2-methylpropane-sulfonate (AMPS), yielding a linear α-cholesteryl functional poly(AMPS) polymer [27]. In contrast to the relatively simple introduction of a single cholesterol functionality on one end, the functionalization on both ends is more challenging as the termination step must be employed. In free-radical polymerization the termination occurs rapidly but less specifically, and proceeds mainly via recombination or disproportionation reactions. In order to gain control over the nature of the ω end group, efficient prefunctionalized CTAs need to be utilized; subsequently, thiol groups proved to be near-ideal CTAs, as a consequence of both the weakness of the S-H bond and the high reactivity of the thiyl radical. Both of these characteristics feature large chain-transfer constants, which in turn leads to a low molecular weight of the polymer but without any significant change in the rate of polymerization [63]. Taking these considerations into account, it is possible to modify one or both ends of a polymer chain, depending if only one or both mechanistic steps are instrumentalized. As an example, *N*-(2-hydroxypropyl)methacrylamide (HPMA) was polymerized in a free-radical fashion, utilizing 4,4′-azobis-(4-cyano-1-cholesteryl pentanoate) as the initiator [64]. When the polymerization was carried out in the presence of 2-mercaptoethanol, α-cholesterol-functionalized poly(HPMA) was obtained, whereas both ends of the polymer were functionalized when thiocholesterol was utilized as the CTA instead.

Besides free-radical polymerization, controlled radical polymerizations methods have proven to be suitable strategies to synthesize cholesterol end-capped polymers. In RAFT polymerization, the R fragment of the CTA is regioselectively placed at the α-end of each polymer chain. As an example, 4-cyano-4-methyl-trithiopentanoic acid was converted with cholesterol (**6**), and the CTA obtained was further used to place the cholesteryl residual at the α-position of a poly(ethylene glycol) acrylate polymer [33]. The introduction of cholesterol groups at the end of polymeric chains does not necessarily lead to an α-mono-functionalized polymer; for example, a bis-cholesteryl functional CTA was designed (**7**) and used to polymerize α,α-bis-functional poly(*N*,*N*-dimethylmethacrylamide) and poly(HPMA) [34]. In another study, bis-cholesteryl functional reactive ester

Free radical polymerization

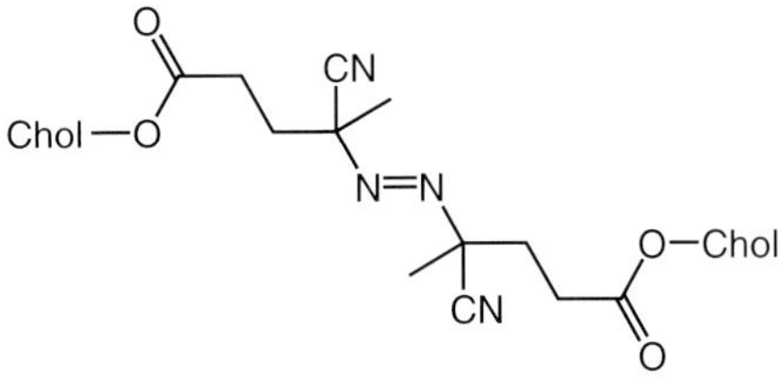

4 [27, 63]

Ring-opening polymerization

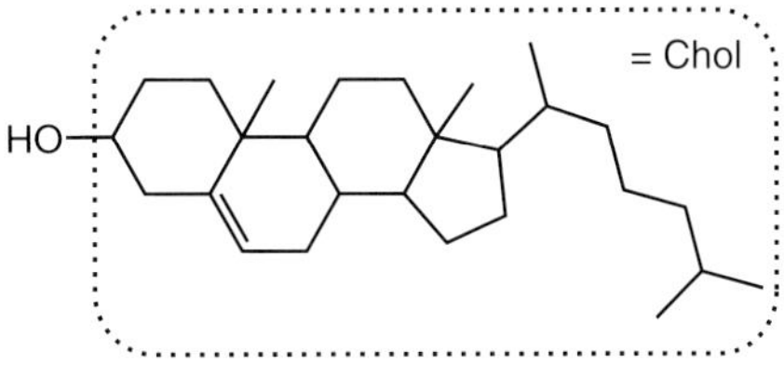

= Chol

5 [70-74]

Reversible addition-fragmentation chain transfer polymerization

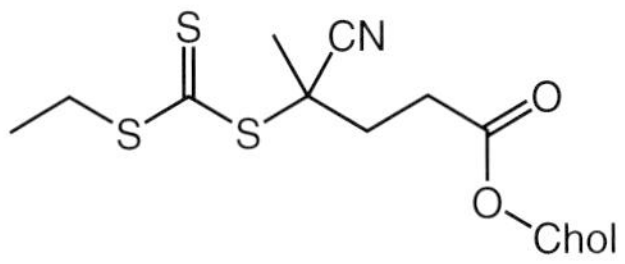

6 [33]

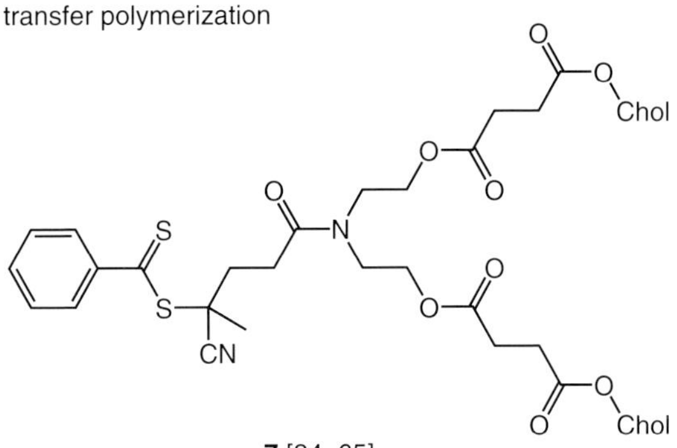

7 [34, 65]

Atomic transfer radical polymerization

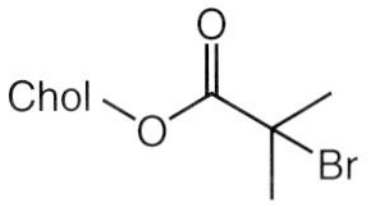

8 [66, 67]

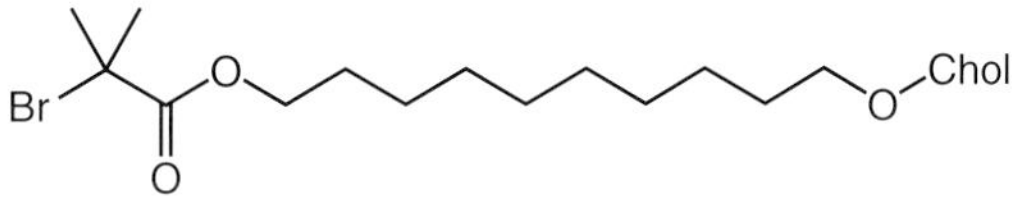

9 [68, 69]

Fig. 6 Cholesterol-modified initiators, which enable end group functionalization of polymers with cholesterol moieties by direct attachment of the steroid during the initiation.

polymers, based on pentafluorophenyl acrylate and pentafluorophenyl methacrylate, were synthesized using the same CTA. These polymers could be used in a post-polymerization modification such that polymers were obtained which were not accessible via conventional polymerization techniques [65].

ATRP is equally applicable to the synthesis of cholesteryl end-capped polymers, the main factor being a matter of the initiator design. Cholesteryl functionalization was achieved via a nucleophilic substitution of 2-bromoisobutyryl bromide with cholesterol (**8**), after which the modified ATRP initiator was used to polymerize methyl methacrylate (MMA) [66] and various oligo(ethylene glycol) (meth)acrylates [67]. In another report, a 10-cholesteryloxydecanol-based ATRP macroinitiator (**9**) was used to polymerize 2-(methacryloyloxy)-ethyl phosphorylcholine (MPC) [68, 69].

The cholesteryl end group functionalization of degradable polymers such as poly(lactide) (PLA) can be realized via ring-opening polymerization (ROP). The hydroxyl group of the cholesterol (**5**) possessed sufficient nucleophilicity to ring open the corresponding monomers, without the need of additional catalyst, and this approach proved suitable for the polymerization of PLA [70], poly(glycolide), and their copolymerization, yielding poly(lactide-*co*-glycolide) [71]. Use of the hydroxyl group of cholesterol to polymerize polycarbonates from trimethylene carbonate [72] or 2,2-dimethyltrimethylene carbonate [73] was also reported. When using cholesterol, the ROP approach could be extended to ε-caprolactone, when the catalyst tin(II) octoate ($Sn(Oct)_2$) was added [73, 74].

2.3.2 Indirect Introduction

The synthesis of telechelic polymers precedes the indirect approach. The term telechelic defines polymers with reactive end groups, which are able to participate in further reactions such as block copolymerization or chemical modifications. It should be noted that the end group reactivity originates from the initiator, termination or from the CTAs, but not from the monomers, as in polycondensations or polyadditions [62]. The indirect introduction is advantageous when the desired functionality may interfere with the growing species during the polymerization step, or when the functionality reduces the reactivity of the initiator, in such a way that polymerization does not occur. Following this strategy, the ω-hydroxyl end group of ROP-polymerized PLA was exploited for conversion with cholesteryl chloroformate [75].

The indirect approach becomes increasingly more interesting when RAFT is considered. As noted above, during RAFT polymerization the monomer is inserted between the R and Z groups of the corresponding CTA, yielding heterotelechelic polymers. The nature of the R group can easily be controlled by the design of the CTA, whereas the retaining Z group remains reactive towards radicals or nucleophiles [76]. As a consequence, the end group reactivity can easily be controlled by the choice of CTA used in the RAFT process, and this enables the synthesis of both symmetrically and unsymmetrically end-capped polymers. For instance, poly(*N*-isopropylacrylamide) (PNIPAM) was polymerized via RAFT polymerization, utilizing a symmetrical bis(dithioester) CTA. (A bis(dithioester) is a compound that bears two functional groups each capable of reversible addition fragmentation chain transfer.) In a subsequent aminolysis, the end groups of the α,ω-isobutyldithiocarbo-PNIPAM thus obtained were converted into mercapto groups, and this in turn reacted with cholest-5-en-3β-yl 6-iodohexyl ether via a nucleophilic reaction to yield α,ω-di(cholest-5-en-3β-yl 6-oxyhexylthio) PNIPAM [77]. In another study, the same approach was applied to design end-capped cholesterol polymers of poly(vinylpyrrolidone) (PVP). In one example, α-amino-PVP was converted with cholesteryl chloroformate [78], while in the same study, a disulfide end-capped PVP polymer was synthesized and conjugated to thiocholesterol via a thiol–disulfide exchange reaction. In a similar manner, thiocholesterol was conjugated to a α-pyridyldisulfide-ω-bovine serum albumin-poly(oligo(ethylene glycol) acrylate) [79].

2.4 Homopolymers

Clearly, the simplest method of obtaining cholesterol-containing polymers involves the homopolymerization of a monomer bearing a cholesterol as pendant group. Free-radical polymerization of the monomer cholesteryl acrylate (CHA) (**10**) in the bulk represents only one possibility [28, 80]. While the free-radical polymerization of CHA led to quite large polydispersities, modern controlled radical polymerization techniques such as RAFT yielded a narrower molecular weight distribution and allowed additional chain extension (see Sect. 2.6) [29]. A ring-opening metathesis polymerization of norbornene monomers, modified with cholesterol and a variety of different spacers in between the norbornene motif and cholesterol (**11**), was also reported [26]. A rhodium-catalyzed polymerization of polyacetylenes, bearing cholesterol pendant groups (**12**), was conducted in order to characterize their liquid crystalline properties [81]. Unfortunately, homopolymers which feature cholesterol pendant groups suffer from serious solubility issues in water, and hence their applications in a biomedical context are limited. The chemical structures of the homopolymers discussed above are illustrated in Fig. 7.

2.5 Random and Statistical Copolymers

During radical copolymerization, two (or more) monomers simultaneously participate in the chain growth of the polymer. The reactivity, which is expressed by the so-called copolymerization parameter [82, 83] of each monomer and of its corresponding radical species, determines the way in which the monomers are incorporated into the polymer. The choice of monomer pair is crucial, as this affects the architecture of the resulting polymer, and in this regard radical copolymerization can lead to the formation of random, alternating, or block copolymers. It should be noted that, in an extreme example, an unfortunate choice may even inhibit the entire polymerization. Nonetheless, these above-mentioned features lead to the copolymerization approach being a convenient and simple tool for incorporating cholesteryl pendant groups along a polymer chain at various molar percentages. In this context, Fig. 8 provides the chemical structures of the acrylic-based copolymers, while Fig. 9 summarizes the discussed nonacrylic copolymers. Copolymers of *N*-isopropylacrylamide (NIPAM) and CHA (poly(CHA-*co*-NIPAM)) (**13**) were synthesized in different molar ratios [84].

CHA was successfully copolymerized with a long (oligoethylene glycol) methacrylate (**14**) [85], while the free-radical

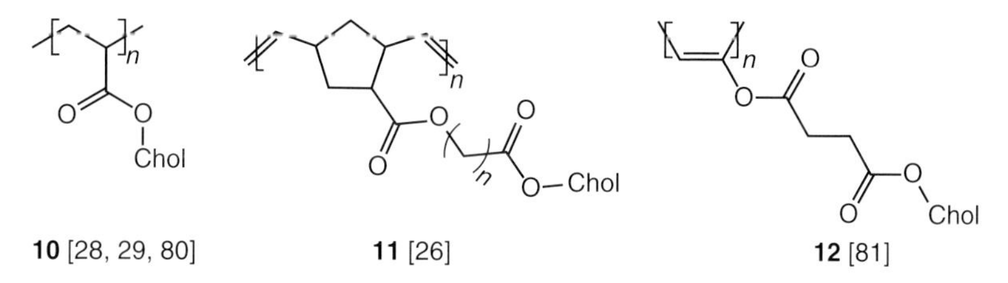

Fig. 7 Homopolymers bearing cholesterol as a pendant group.

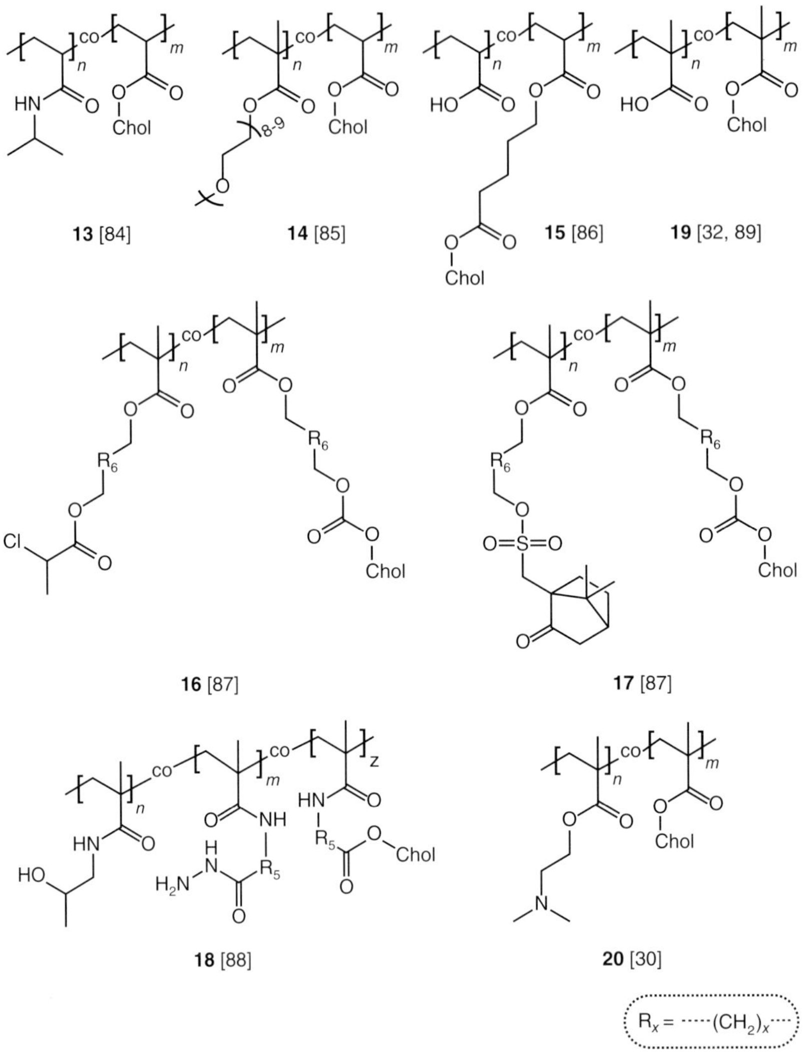

Fig. 8 Acrylic-based random copolymers bearing cholesterol moieties as pendant groups.

copolymerization of acrylic acid (AA) with 5-acrylolyloxypentyl cholesterate (Ch5A) yielded random poly(Ch5A-*co*-AA) copolymers (**15**) with cholesteryl molar fractions ranging from 1% to 100% [86]. In another example, cholesteryl 8-(methacryloyloxy)octyl carbonate was copolymerized with both diastereomers of 8-(methacry1oyloxy)octyl-2-chloropropionate (**16**) or with both diastereomers of 8-(methacryloyloxy)octyl-camphorsulfonate (**17**) [87]. HPMA simultaneously copolymerized with 6-methacrylamido hexanoyl hydrazine (MAHH) and cholest-5en-3β-yl 6-methacrylamido hexanoate (CMAH) at different feed ratios to yield the corresponding poly(CMAH-*co*-MAHH-*co*-HPMA) terpolymers (**18**) [88]. The copolymerization of methacrylic acid (MA) with cholesteryl methacrylate (CMA)

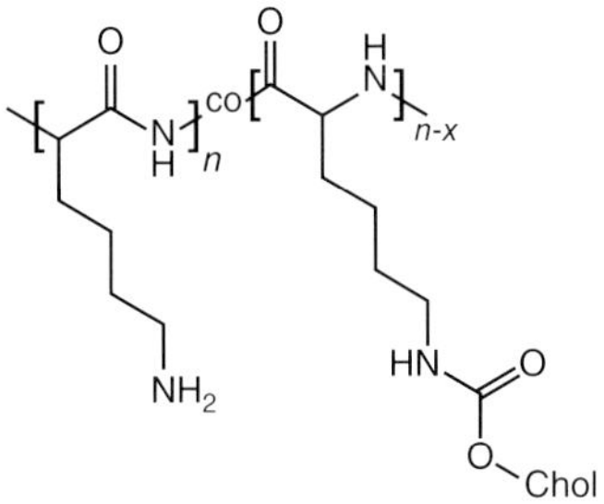

23 [93]

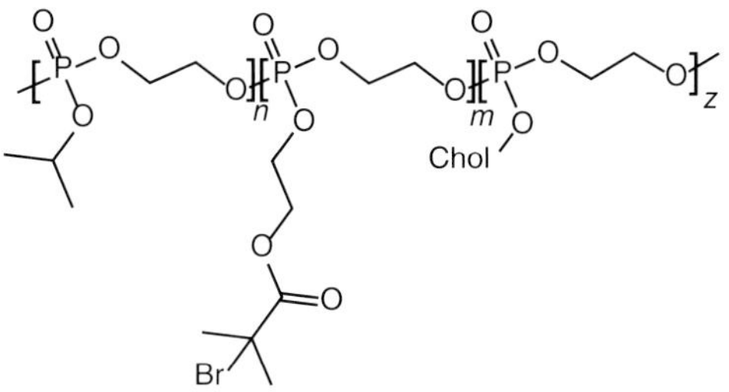

21 [90]

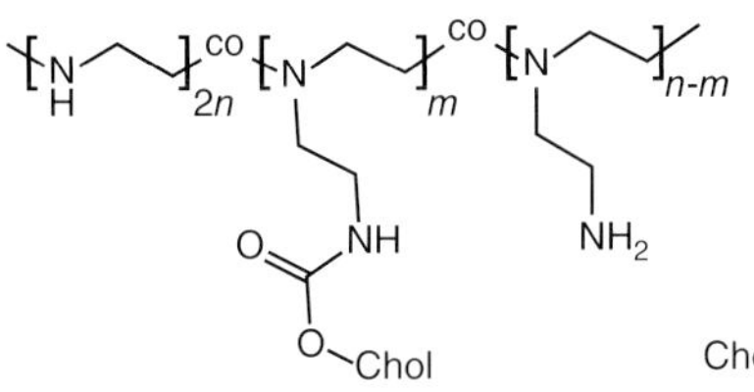

24 [23]

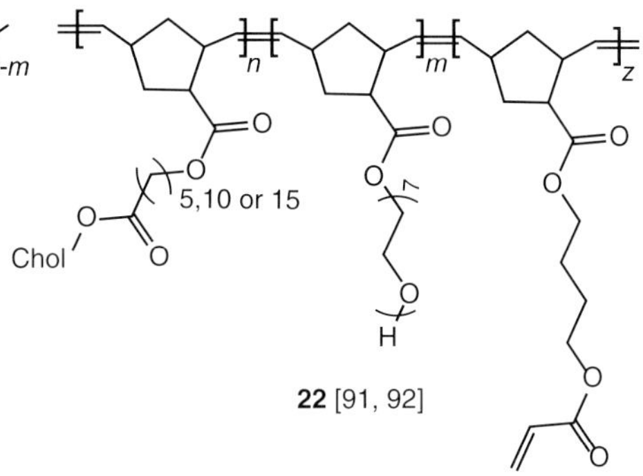

22 [91, 92]

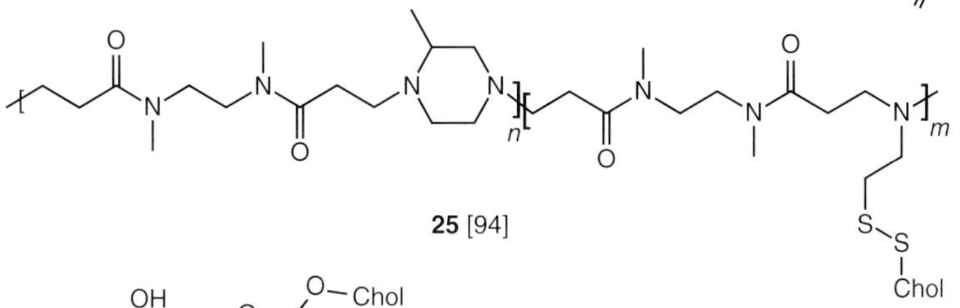

25 [94]

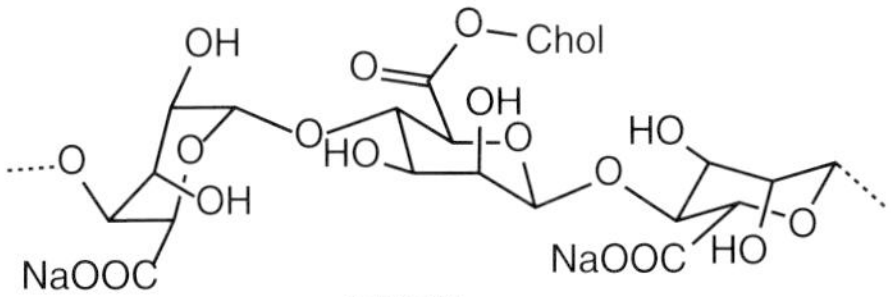

26 [95]

Fig. 9 Chemical structures of cholesterol-containing random copolymers, which were not synthesized via radical polymerization techniques.

in the presence of the CTA 4-cyano-4-(dodecylsulfanylthiocarbonyl)sulfanyl pentanoic acid (**19**) represents a very good example that random copolymers can be obtained via RAFT-mediated copolymerization [32]. A similar strategy with the same CTA was followed with the copolymerization of *N,N*-dimethylamino ethyl methacrylate (DMAEMA) with CMA, yielding poly(DMAEMA-*co*-CMA) (**20**) [30] or poly(MA-*co*-CMA) (poly(methacrylic acid-*co*-cholesteryl methacrylate)) (**19**), when CMA was copolymerized with *tert*-butyl methacrylate [89].

Copolymers containing cholesteryl pendant groups could also be obtained using ROP methods. In one example, cholesteryl derivatives of an cyclic phosphate were terpolymerized with 2-isopropyl-2-oxo-1,3,2-dioxaphospholane and 2-(2-oxo-1,3,2-dioxaphosphoroyloxyethyl-2-bromoisobutyrate) (OPBB) (**21**), utilizing ROP. Interestingly, the OPBB repeating unit was considered as the ATRP initiating group, such that MPC was subsequently grafted from the OPBB sites [90]. In another study, cholesteryl-containing side-chain liquid crystalline networks were obtained via a ring-opening metathesis polymerization of 5-[*n*-(cholesteryloxycarbonyl)-alkyloxycarbonyl]bicyclo[2.2.1]hept-2-ene, poly(ethylene glycol)-functionalized norbornene, and 5-(acryloyl butoxycarbonyl)bicyclo[2.2.1] hept-2-ene (**22**) [91, 92].

Alternatively, post-polymerization modification techniques can be applied to homopolymers or copolymers in order to introduce cholesterol units. One prime example depicts the post-functionalization of poly(L-lysine) (PLL) with 4-nitrophenol cholesteryl formate (**23**) [93]. Further, cholesteryl chloroformate could be converted with free amino groups present in poly(ethylenimine-*co*-*N*-2-(aminoethyl)ethylenimine) copolymers (**24**) [23]. In another example, poly(amido amine) (PAAm) was copolymerized via the Michael addition of 2-methyl-piperazine and cystamine with 2,2-bis(acrylamido)acetic acid; the polymeric network obtained contained disulfide crosslinks which were further modified with 2,2′-dithiopyridine. A subsequent thiol–disulfide exchange reaction with thiocholesterol yielded at random cholesterol-containing PAAm copolymers (**25**) [94]. In fact, natural polymers are also receptive for post-functionalization with cholesterol; for example, cholesterol was linked covalently via a catalyzed DCC coupling to alginate (**26**) [95].

2.6 Block Copolymers

In contrast to the above-discussed strategies for introducing cholesteryl functionalities into polymers, free-radical polymerization is not suitable for designing block copolymers. In general, the synthesis of block architectures in polymers includes two separate polymerization steps. Polymerization of the first block requires preservation of the functionality capable of being utilized for polymerization. The halogen in ATRP, the CTA in RAFT, or an ion-pair in ionic polymerization number among these groups, whereas the termination step in free-radical polymerization leads to unreactive end groups. Block copolymers are particularly fascinating as they are able, potentially, to undergo phase separation and self-assembly into nanostructured architectures.

The block copolymers described here are summarized in Fig. 10 (diblock copolymers) and Fig. 11 (special examples of block copolymers). Ring-opening metathesis polymerization represents a technique that enables block copolymerization; for example, 5-(4-(2-exo-[2-(2-(-2 [2-(2-methoxy-ethoxy)-ethoxy]-ethoxy)-ethoxy)-ethoxy]-ethoxymethyl)-benzyloxy)-bicyclo[2.2.1]hept-2-ene [poly(ethylene oxide) (PEO) monomer] and the corresponding cholesterol derivative could be successfully block-copolymerized (**27**) [37]. Initially, the PEO monomer was polymerized using a first-generation Grubbs catalyst, and this was followed by polymerization of the cholesterol block. The three-membered rings of cyclopropane derivatives are well known for their ability to be opened by anions. In this regard, a thiol-functionalized PEG polymer was utilized as an anionic macroinitiator to initiate the anionic ROP of 4-(cholesteryl)butyl ethyl cyclopropane-1,1-dicarboxylate [96].

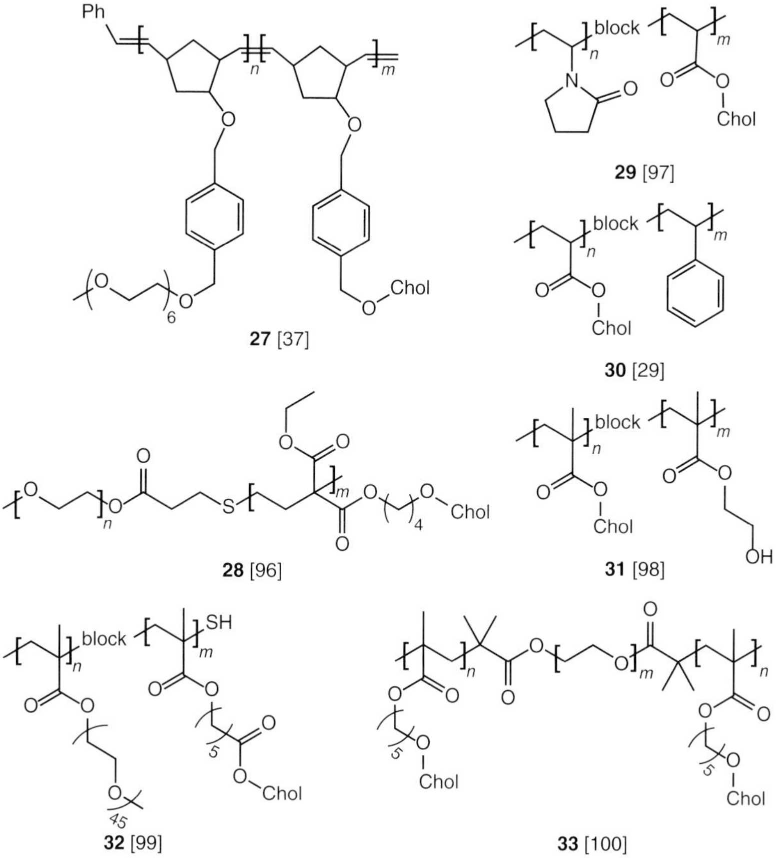

Fig. 10 Collection of block copolymers, featuring cholesterol pendant groups.

The copolymers obtained (**28**) were able to self-organize into vesicles with wavy membranes or as nanoribbons with twisted and folded structures, depending on the size of the cholesterol block.

In comparison to anionic polymerization, controlled radical polymerization techniques feature a much simpler processability. As a consequence, block copolymer structures having one block made from cholesterol derivatives synthesized by RAFT are much more common. For instance, when a short block of CHA was copolymerized with a macro-CTA made from PVP the cholesterol-containing block of the poly(VP-*b*-CHA) obtained (**29**) could be used to anchor liposomes, while the PVP block was employed to interact with other polymers [97]. A RAFT-mediated chain extension could also be performed with CHA as the primary block; for example, in the presence of the monomer styrene (St), use of the CHA macro-CTA led to poly(CHA-*b*-St) (**30**) [29]. Similarly, CMA was RAFT-polymerized with *S*-1-dodecyl-*S*′-(α,α′-dimethyl-α″-acetic acid) tricarbonate as the CTA, and thereafter the CMA block was extended with trimethylsilyl (TMS)-protected hydroxyethyl methacrylate (HEMP). Subsequent acid cleavage of the TMS group yielded the amphiphilic block copolymer

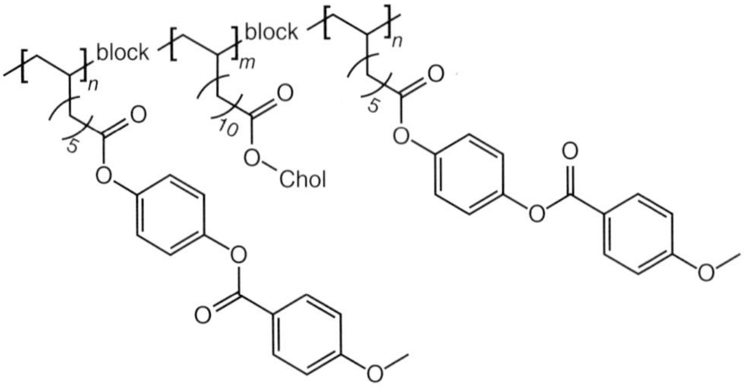

34 [101]

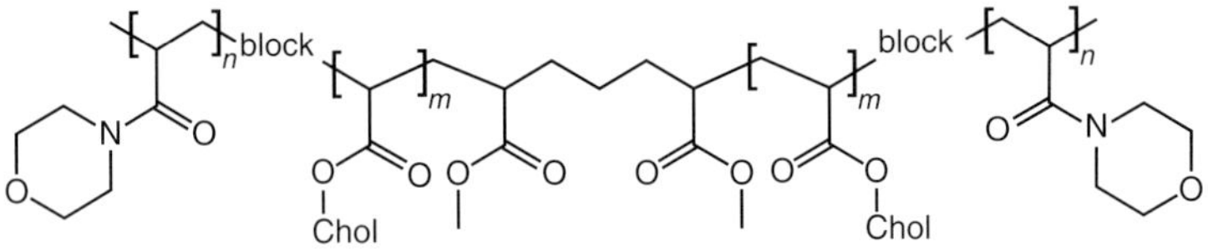

35 [102]

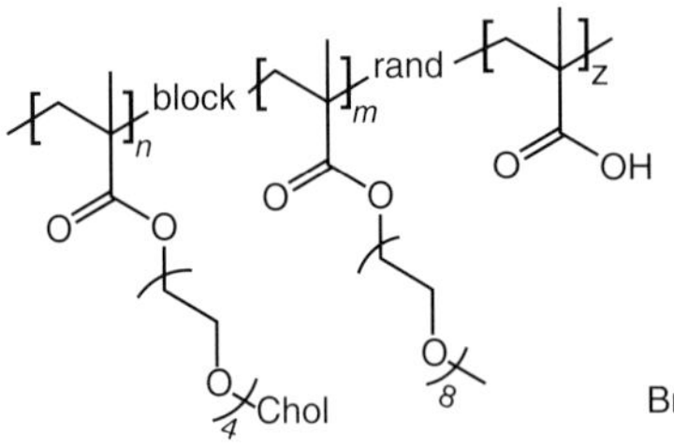

36 [103]

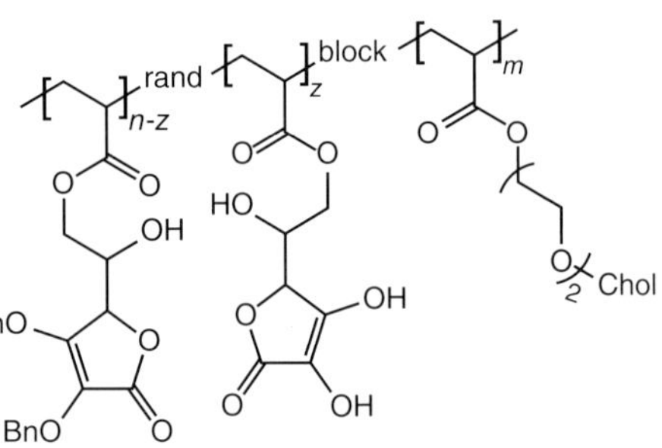

37 [38]

Fig. 11 Chemical structure of triblock copolymers bearing cholesterol moieties.

poly(CMA-*b*-HEMA) (**31**) [98]. Brush-like block copolymers (**32**) which comprised PEG-functionalized methacrylate in one block and CMA as the second block were realized using RAFT polymerization [99]. Interestingly, the dithioester end group of the copolymer was subsequently converted into a thiol group by aminolysis and exploited to graft the polymer onto gold nanoparticles. When an α,ω-dithioester end-functionalized PEG macro-CTA was used for the polymerization of 5-cholesteryloxypentyl methacrylate, a ABA triblockcopolymer was obtained (**33**) [100]. In similar fashion, triblock copolymers were accessible in which cholesteryl-11-acryloyloxyundecanoate (CAU) was located between two blocks of 4-(6-acryloyloxycaproyloxy)phenyl 4-methoxybenzoate (APMB) (**34**) [101]. The first step of the RAFT polymerization was carried out with CAU in the presence of the di-*tert*-butyl trithiocarbonate as CTA, followed by a symmetrical chain extension with APMB. In another example, a bis-functional RAFT agent was used for the copolymerization of CHA and *N*-acryloyl morpholine (NAM). The obtained ABA-triblockcopolymer (**35**), with CHA as the middle block, exhibited fascinating self-assembling properties that resulted in polymersomes [102]. Notably, mixed random/block architectures can also be obtained by RAFT polymerization; these types of structure can be realized when a macro-CTA comprising a copolymer is obtained from a simple copolymerization of two monomers, followed by chain extension of a third monomer. In this manner, ethylene glycol methacrylate (EGMA) was randomly copolymerized with methyl acrylate (MA) via RAFT polymerization. From the formed poly(EGMA-*co*-MA)-macro-CTA, an arm of cholesteryl methacryloyl tetraethylene glycol carbonate (Chol-TEGMA) was grown, yielding poly(EGMA-*co*-MA)-*b*-poly(Chol-TEGMA) (**36**) [103]. Alternatively, the above-mentioned RAFT-mediated block copolymerization can be performed, followed by the chemical modification of one block such that a random/block mixed polymer structure was obtained. An example of this was the copolymerization of benzyl-protected ascorbyl acrylate (BnAbA) with cholesteryl diethylene glycol acrylate in the presence of 4-cyanopentanoic acid as CTA. After copolymerization, the BnAbA block of the block copolymer was partially hydrogenolized, leading to a random distribution of the BnAbA and AbA in one block connected to the cholesteryl-containing block (**37**) [38].

3 Self-Assembly of Cholesterol

The three structural domains, as previously outlined in Fig. 1, reveal that cholesterol is equipped with the intrinsic property to self-assemble, a prime example of which is the formation of gallstones in the human gallbladder.

Cholesterol gallstones consist of plate-like cholesterol monohydrate crystals which form in a complex crystallization process that involves a sequence of metastable intermediate structures, starting from a cholesterol-supersaturated bile solution and filaments. The filaments develop in the first stage and are replaced by helical ribbons which, in turn, pass over into the tubules. In the final stage, the tubules fracture to produce plate-like crystals (Fig. 12) [104]. Notably, the helical ribbons produced featured only one of two distinct pitch angles, regardless of their diameter. High-pitch helices ($53.7 \pm 0.8°$) form initially but disappear as crystallization proceeds, whereas low-pitch helices ($11.1 \pm 0.5°$) evolve independently, driving the crystallization process forward [105, 106]. The low-pitch angle helices are of particular interest, as they have spring constants ranging from 0.5 to 500 pN μm^{-1}, and the springs are able to elongate from 1 to 100 μm [106]. These properties can be exploited for measuring forces between nanodimensional biological objects, and this led to their suggested application as mesoscopic springs [107]. In order to be able to choose the cholesterol ribbon with the desired spring constant from a polydisperse solution, Khaykovich *et al.* evaluated an explicit formula for the spring constant of any selected helical ribbon as a function of its length, width and radius, and found a quadratic relationship between the radius and thickness [107].

4 Self-Assembly of Cholesterol-Modified Polymers

As a consequence of the structural features associated with the steroid skeleton, the attachment of cholesterol moieties – either

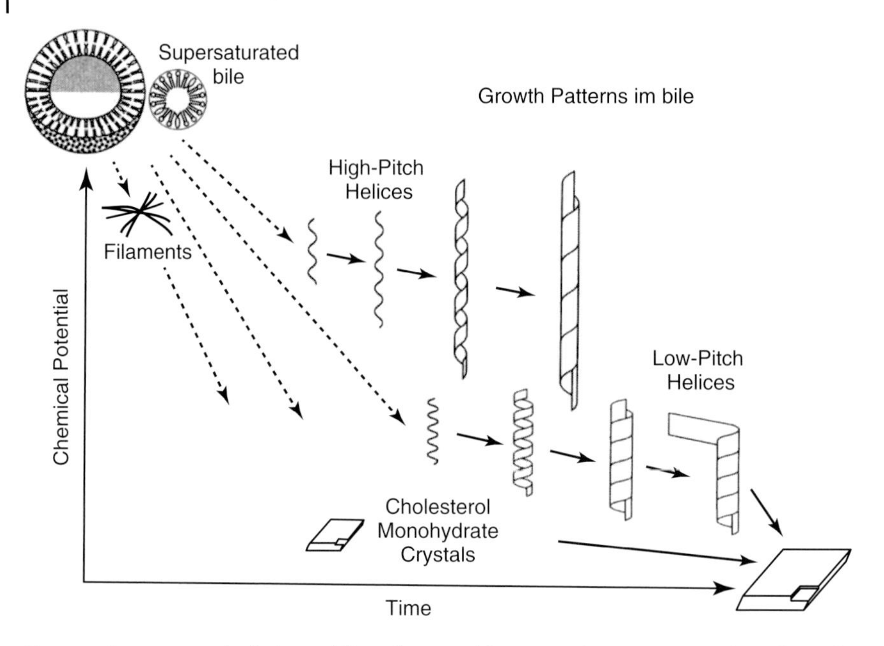

Fig. 12 Sequence and relative stability of metastable intermediate structures plotted as a function of time after supersaturation of bile. Reprinted with permission from Ref. [105]; © 1993, National Academy of Sciences, USA.

as pendant group or as end group – can induce self-assembly properties to a water-soluble polymeric compound. The driving force for the formation of well-defined nanostructured morphologies in water is explained on the one hand by the additional hydrophobic gain of the cholesteryl unit, and on the other hand by the mesogenic character and ability of cholesterol to form liquid crystalline phases [108–111].

In biomedical applications, amphiphilic-rendered polysaccharides such as cholesterol-modified mannan [112–114], pullan [115, 116], or chitosan, have attracted significant attention due to their excellent biodegradability and biocompatibility, and their ability to entrap hydrophobic drugs within self-assembled nanoparticles. Various intermolecular and intramolecular hydrogen bonds are considered to be responsible for the high rigidity of chitosan, resulting in chitosan's poor water solubility under physiological conditions (pH 7.4) [117]. In an effort to circumvent this limitation, Yu *et al.* reported on the synthesis of cholesteryl-modified glycol chitosan (CHGC) [118]. The glycol modification of the primary hydroxyl function of chitosan significantly enhanced the water solubility at all pH values, without losing its biodegradable or biocompatible properties, while the cholesterol moieties enabled the self-aggregation of CHGC into nanoparticles. When these nanoparticles were investigated in respect of their drug load and delivery, using the anti-cancer drug doxorubicin, it was found that compared to the non-encapsulated doxorubicin the nanoparticles exhibited not only a delayed blood clearance after intravenous administration but also a remarkable pH-dependent release of the drug, with drug release at pH 5.5 being much faster than at pH 6.5 or 7.4. This observed behavior was especially important as the biological environment of tumor tissue often shows weak acidic properties. Further, the enhanced pharmacokinetic parameters, by means of a prolonged circulation time, resulted in a greater inhibition of tumor growth, such that the CHGC nanoparticles represented a promising carrier system for cancer therapy.

An alternative approach for tackling the water solubility issue of chitosan, as reported by Zhang and coworkers, involved increasing the hydrophilicity of chitosan by modifying it with carboxymethyl groups [119–123]. Such carboxymethylation inhibited the intermolecular hydrogen bonding and thus increased the flexibility of the polysaccharide chain, which supported the compound's solubility in water. After increasing the overall hydrophilicity of chitosan, cholesterol was conjugated with different degrees of substitution (DS) via the chitosan amino group onto the polysaccharide, yielding in cholesterol-modified *O*-carboxymethyl chitosan (CCMC) derivatives. This partial gain in hydrophobicity enabled the amphiphilic chitosan to self-aggregate, and the observed self-association was attributed to strong interactions between the cholesterol moieties, forming hydrophobic microdomains [121]. In this context, the DS was found to crucially influence both the nanoparticle size and surface properties. It was also shown that the carboxymethyl modification in the CCMC significantly supported the self-aggregation, leading to narrow disperse nanoparticles, whereas chitosan modified only with cholesterol (CHCS) yielded irregular spherical structures [122]. The CCMCs and CHCS were investigated as potential drug carriers for paclitaxel (a hydrophobic anticancer drug) [123] and epirubicin (an anthracycline anticancer drug) [121], respectively. The drug release rates exhibited pH-dependent behaviors as a consequence of the protonation of residual amino groups in the chitosan backbone at low pH values. The increasing repulsive forces altered the swelling behavior of the nanoparticles and, as a consequence, increased the release rate. The paclitaxel-loaded CCMC particles were compared in a biodistribution assay with paclitaxel dissolved in Cremophor EL in ethanol, and revealed an increased uptake of the former nanoparticles in plasma, spleen and liver, but a decrease in heart and kidney [123]. In a further study, when the interaction of bovine serum albumin (BSA) with CCMC was investigated [119, 120], interaction of the self-aggregated CCMC nanoparticles with BSA proteins was found to cause a slight change in the BSA secondary structure. This unfolding process was confirmed using circular dichroism, revealing a decrease in α-helix content accompanied by an increase in β-strand content. Moreover, interaction of the BSA protein with CCMC nanoparticles increased the protein's stability against denaturation, as can be induced by denaturants such as urea.

In another report, stable hollow polymer nanocapsules were obtained by dialyzing PLA and cholesterol-modified dextran (Chol-Dex) dissolved in dimethylsulfoxide (DMSO) against water [124]. The explanation given for the assembly mechanism was that the shell had formed as a consequence of the Chol-Dex polymer acting like a surfactant, with the appended cholesterol units anchoring onto the swollen PLA. Simultaneously, in the shell interior, the amphiphilic Chol-Dex polymer chains became entangled with hydrophobic PLA and were further phase-separated and assembled into the innermost surface, leading to the observed sandwich-like structure of Chol-Dex/PLA/Chol-Dex. The capsules obtained were proposed to have potential applications as a drug carrier or a biomembrane.

As a consequence of extensive cellular uptake by phagocytes, the use of (polymer-based) micelles or vesicles suffers from a short circulation life time, which is an important consideration for increasing the selective accumulation at sites of enhanced vascular permeability [125]. In order to overcome this major problem, PEG-based (co-)polymers or the incorporation of PEG onto micelle surfaces are often used, due to their "stealth" properties against protein recognition [126, 127]. In this context, a combination of the advantages of both PEG and cholesterol is often applied, leading to self-assembled nanoparticles with applications in drug delivery [124, 128–135]. Jia *et al.* synthesized liquid crystalline (LC) block copolymers which consisted of a PEG block and a cholesterol-based smectic LC block, which was composed of a poly(cholesteryl acryloyloxy ethyl carbonate) block [129]. Subsequently it was found that, depending on the hydrophilic/hydrophobic weight ratio and on the choice of the cosolvent mixture, polymer aggregates with different morphologies were formed. As a result, for all sets of block copolymers, only micrometer-sized complex aggregates (or just precipitation) were obtained in tetrahydrofuran (THF), whereas distinct and well-defined micellar aggregates were formed in a dioxane/water cosolvent mixture, ranging from rod-like vesicles to nanofibers. Based on these results, the shape of the micellar aggregates can be tailored by the correct adjustment of the block size, by adjusting the ratio of the hydrophilic/hydrophobic balance. Cholesterol-modified PEG can also be utilized to form hydrogels; for example, Manakker *et al.* prepared eight-arm, star-shaped PEG polymers where each arm was derivatized with either cholesterol or β-cyclodextrin (βCD) [133]. The cholesterol and βCD were able to form reversible host–guest inclusion complexes such that the combination of both star-shaped polymers resulted in a self-assembly of tight viscoelastic hydrogels. A rheological analysis showed that the storage and loss modulus of the network deviated significantly from the Maxwell model (which described the viscoelastic properties), and the gel also exhibited a thermosensitive behavior. As a consequence of the slow overall chain relaxation of the polymer, viscoelastic properties were observed at room temperature, whereas a liquid-like behavior was recognized at higher temperatures due to an increased dissociation of the βCD–cholesterol complexes and concomitant faster chain relaxation.

Based on the above-mentioned case studies, cholesterol moieties induce self-formation into nanoparticles or nanoaggregates, respectively. In this context Nagahama *et al.* observed a rather unique property evoked by cholesterol [130] when they synthesized a star-shaped, partially cholesterol-substituted eight-arm poly(ethyleneglycol-*b*-L-lactide) polymer. At 20 °C, this amphiphilic polymer self-assembled into well-defined spherical object of about 15 μm diameter, after which a constant increase in temperature resulted in a second self-organized transition to form a continuous, fiber-like structure. Moreover, the direct injection of a 10 wt% solution of the star-shaped polymer into a phosphate-buffered saline (PBS) solution that had been preheated to 37 °C led to an instantaneous gelation and a hydrogel with a high mechanical strength.

Hosta-Rigau *et al.* showed that an amphiphilic ABA triblock copolymer, with poly(CHA) assembled in between two poly(NAM) blocks, self-assembles into polymersomes (Fig. 13a, panel i) [102].

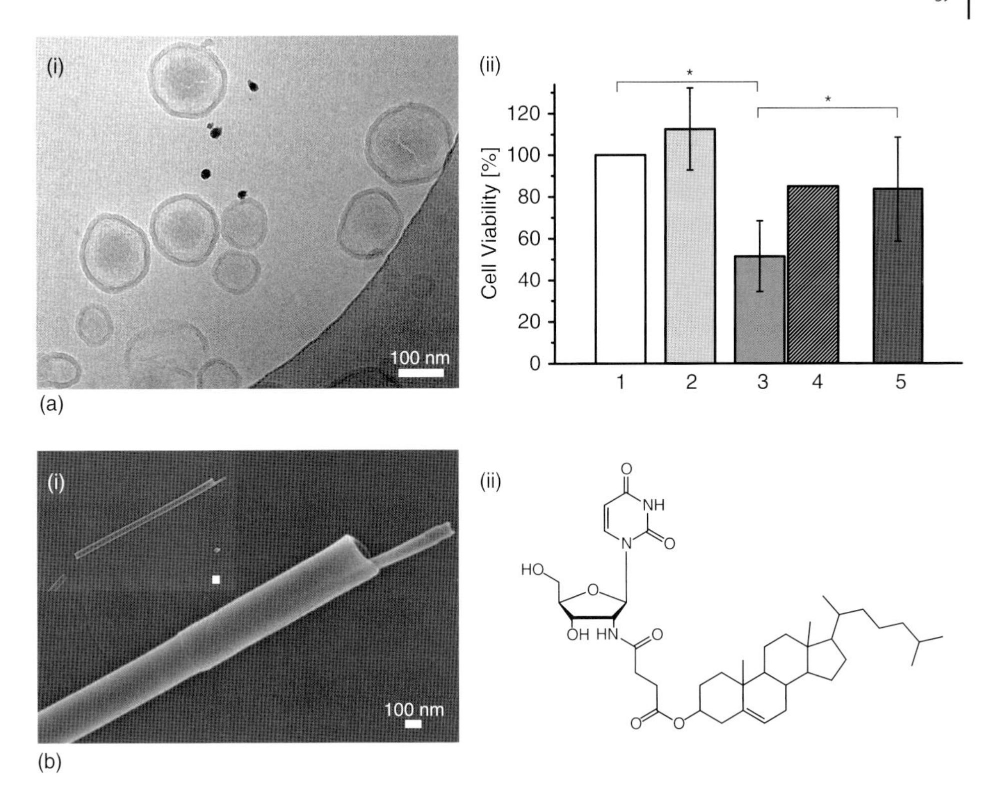

Fig. 13 (a) Utilizing the phase-inversion method, polymer 35 self-assembled into polymersomes. (i) Cryo-transmission electron microscopy image of the obtained polymersomes; (ii) Myoblast cell viability of cells grown on microstructured (μS) PVA (1); μS PVA with empty polymersomes (2) and with TC-loaded polymersomes (3); theoretical cell viability of myoblast cells, when exposed to TC in solution (4); μS PVA containing free TC (5). Reprinted with permission from Ref. [102]; (b) (i) Scanning electron microscopy image of self-organized microtubule composed of 2′-*N*-(2-(cholesteryl)-succinyl)-2′-deoxy-2′-aminouridine and DOPC. The inset shows the microtubule with a lower magnification (scale bar = 300 nm); (ii) Chemical structure of the cholesteryl-modified nucleoside 2′-deoxy-2′-aminouridine. Reprinted with permission from Ref. [136].

These artificial vesicles were investigated in the context of surface-mediated drug delivery (SMDD) by loading the polymersomes with the cytotoxic drug thiocoraline (TC) and mixing them into a poly(vinyl alcohol) (PVA) hydrogel matrix. Compared to the TC-free hydrogels, the TC-loaded polymersome/PVA functional coatings caused a significant decrease in the viability of adhering myoblast cells (Fig. 13a, panel ii). In addition, cell–drug interaction was shown to be more efficient when the drug was delivered from the surface than when administered from solution, thereby demonstrating its potential as a biomaterial in SMDD applications.

In another study, cholesterol was linked via a succinyl spacer to the 2′-position of the nucleoside 2′-deoxy-2′-aminouridine. When the modified nucleoside was mixed with the unsaturated phospholipid dioleoylphosphatidylcholine (DOPC), the formation of open-ended microtubes that were 20–40 μm long and exhibited an outer diameter of 2–3 μm was observed (Fig. 13b) [136].

5 Cholesterol and Liposomes

Liposomes are artificially prepared spherical vesicles composed of one (or more) lipid bilayer membrane(s) which has a similar structure to a natural cell membrane and can mimic the behavior of biological cell membranes [137, 138]. Glycerophospholipids, sphingolipids and cholesterol are the three predominant components involved in the preparation of liposomes [109]. In combination with other lipids, cholesterols are interspersed in the lipid bilayer membrane by orienting the steroid ring parallel to the hydrocarbon chains of the lipids and the hydroxyl group encountering the aqueous phase [139]. The subsequent incorporation and adjustment of the cholesterol content of the lipid bilayer can effectively increase the lipid membrane's mechanical strength [140], reduce the passive permeability of water, small molecules and gases [141, 142], and regulate membrane fluidity and the phase behavior of membranes [143, 144]. Further details on this topic are available in reviews produced by Rog *et al.* [145] or Ohvo-Rekila [139]. To date, cholesterol as a building block in liposomes has been employed in a wide range of biomedical applications.

5.1 Cholesterol as a Stabilizer

5.1.1 Drug Delivery

Liposomes are among the preferred carriers in the field of drug delivery for diverse therapeutic cargoes (e.g., antiviral [146], antibacterial [147], and anticancer drugs [148], or enzymes [149]) due to their excellent properties such as ease of preparation, colloidal size, controllable surface and membrane properties, large carrying capacity, and biocompatibility. The incorporation of cholesterol into liposomes is beneficial for their stability in the blood circulation upon systemic application [150], and also improves the liposomes' suboptimal control over cargo retention and release [151]. The first cholesterol-containing liposomal formulation of the antitumor drug doxorubicin (Doxil®) was approved by the Food and Drug Administration (FDA) in 1995, for medical use in oncology [152]. Following this, several cholesterol-containing liposomal formulations such as daunorubicin (DaunoXome®) for blood cancer [153], cytarabine (Depocyt®) for neoplastic meningitis and lymphomatous meningitis [154], and morphine sulfate (DepoDur®) for pain [155], were approved by the FDA [156–158], or are currently undergoing clinical trials [159, 160]. In addition to these prominent examples, all of which reached the pharmaceutical market, liposomal therapeutics containing cholesterol remain the subject of intensive research, and three recent examples have been selected at this point to illustrate the relevance of cholesterol in this respect.

Venegas *et al.* prepared liposomal combretastatin A-4 phosphate (CA4P) formulations by using liposomes with different cholesterol contents [161]. CA4P is a tubulin-binding drug that disrupts microtubule formation, and shows a preference for binding to the tubulin of endothelial cells of tumor blood vessels such that cell proliferation and viability is reduced. When Venegas and coworkers studied the dependence of the cholesterol content of liposomal CA4P formulations on MCF-7 breast cancer cells, the cytotoxic effect was found to vary significantly (with up to 50% changes in cytotoxicity) in line with the liposomal cholesterol content, but in a well-defined and alternating manner. Cytotoxicity was maximal when the liposomal cholesterol content was 20.0, 22.2, 25.0, 26.0, 33.3, 40.0, and 50 mol.%, and the cholesterol did not simply serve as a membrane-stabilizing agent but also modulated the release of liposomal CA4P. This was in accordance with the concept of a sterol superlattice, using small cholesterol molar fraction increments (0.5 mol.%) (Fig. 14a and b).

Yang *et al.* investigated the ability of cationic liposomes composed of 3β-[*N*-(*N*′,*N*′-dimethylaminoethane) carbamoyl] cholesterol and cholesterol to carry plasmid DNA into 293 T cells [162]. In this case, liposomes containing more than 66.7% cholesterol in the formulation exhibited a calorimetric transition caused by anhydrous cholesterol domain at about 41 °C. Moreover, the cholesterol-containing liposome carriers showed more stable particle sizes, a low turbidity, and a high activity for transfecting cells in the presence of high concentrations of fetal bovine serum (50%), due primarily to the neutral domain formation by increasing the molar ratios of cholesterol in the formulation, as well as the relatively low cytotoxicity.

Alhajlan *et al.* developed a new type of cholesterol-containing liposomal formulation of clarithromycin (an antibiotic) with different surface charges [163], and investigated the efficacy and safety of the formulation against clinical isolates of *Pseudomonas aeruginosa* from the lungs of cystic fibrosis (CF) patients. The encapsulation efficiencies of the drug

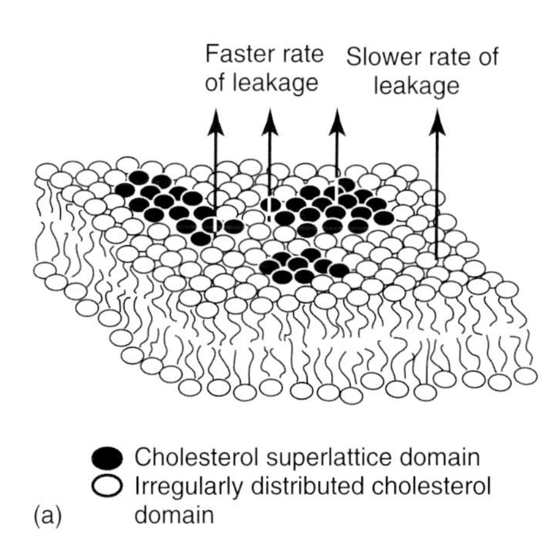

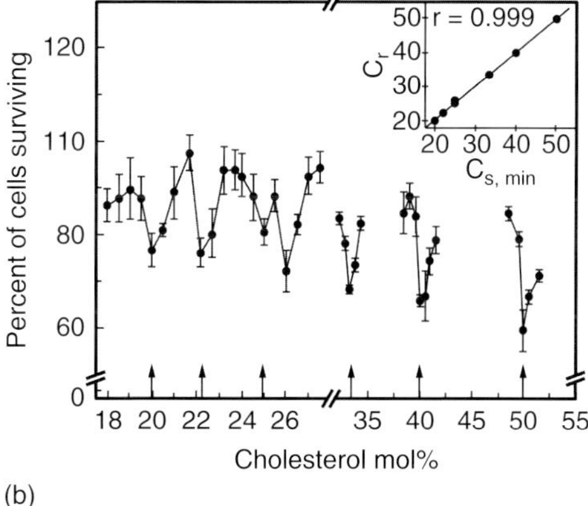

Fig. 14 (a) Schematic illustrating the concept that the rate of spontaneous release of entrapped CA4P varies with the lipid packing maintained in different membrane regions; (b) Dependence of the mole fraction of liposomal cholesterol and MCF-7 cells surviving in microplate wells after 90 min incubation with liposomal CA4P. Error bars represent the standard derivation from six independent samples. When treated with liposomal CA4P the number of surviving cells reached a local minimum at 20.0, 22.2, 25.0, 26.0, 33.3, 40.0, and 50.0 mol.% cholesterol. The theoretically predicted critical mole fractions for the maximal sterol superlattice formation (C_r) were indicated by arrows. Inset: correlation between theoretical C_r and observed critical cholesterol mole fraction for minimal cell survival (C_s, min), where r is the correlation coefficient. Reprinted with permission from Ref. [161].

ranged from 5.7% to 30.4%, while liposomal clarithromycin reduced bacterial growth within the biofilm by 3 to 4 log units, caused a significant attenuation of virulence factor production, and also reduced bacterial twitching, swarming and swimming motilities. Compared to the free drug, the clarithromycin-entrapped liposomes were significantly less cytotoxic.

5.1.2 **Biosensing**

The use of cholesterol-stabilized liposomes plays also an important role in analytical science [164]. Cholesterol-containing liposomes have been used in immobilized liposome chromatography (ILC), in liposome capillary electrophoresis (LCE), in biosensors, or in liposome immunosorbent assays (LISAs). These different techniques are briefly introduced in the following sections, together with some recent examples where cholesterol has proven to be an integral part for the success of the set-up.

ILC [165], which has been used successfully for studying solute–membrane interactions since the 1990s [166], benefits from the fact that liposomes are used in medical and pharmaceutical research as models for mimicking the structure and function of biological cell membranes. These models are often used with the aim of understanding the interaction between cell membranes and different molecules (e.g., drugs, proteins, and peptides) [167], specifically to predict drug adsorption *in vivo* by estimating the membrane partitioning coefficient of the tested drug. Cholesterol is particularly important for predicting a drug's intestinal adsorption, as the small-intestine brush border membrane is composed of phosphatidylcholine, phosphatidylserine, phosphatidylethanolamine, and cholesterol [168].

LCE [167] is a specific type of capillary electrophoresis [169] in which liposomes are used as either a coating material [170, 171] or a carrier [164, 167]. In order to develop a stable biological membrane coating for LCE for use in membrane interaction studies, Lindén *et al.* examined the effect of cholesterol on the stability of dipalmitoylphosphatidylcholine (DPPC) and sphingomyelin (SM) coatings [172]. For this, liposomes were first prepared from DPPC/SM with and without cholesterol or red blood cells ghost lipids, and then flushed through the capillary. The results of the coating stability confirmed that the cholesterol content in the liposomes affected the stability of the coatings. Solutions of DPPC with 0–30 mol.% cholesterol and SM in different ratios provided a stable coating, which more closely resembled natural membranes. Tiala *et al.* attached phospholipid liposomes covalently to iminoaldehyde-coated fused silica capillaries and applied them for the separation of model steroids by open-tubular capillary electrochromatography [173]. This allowed investigations to be made of reducing the Schiff's base formed with sodium borohydride, and also the effect of liposome composition on the stability of the coating. When cholesterol was present in the liposomes it enhanced the stability of the capillaries by making the coatings more rigid, which in turn resulted in lower retention factors for all of the model steroids studied. Li *et al.* developed a method for screening monoamine oxidase (MAO) inhibitors by using capillary electrophoresis based on the interaction of the MAO and its substrate kynuramine (Kyn) [174]. For this, bioactive proteoliposomes were first prepared by reconstituting phosphatidylcholine/cholesterol liposomes and MAO, and then used as the pseudostationary phase of a capillary electrophoresis system to mimic the interaction between the enzyme and its substrate. The relative migration time ratios of Kyn in the presence of rasagiline were found to be larger than in the presence of *N*-prolmrgyl-*R*-2-heptylamine, a known MAO inhibitor, and it was concluded that the interaction between Kyn and MAO was weakened with an increase of the inhibitors.

Biosensors are self-contained integrated devices capable of providing specific quantitative or semi-quantitative analytical information. They usually possess a biological recognition element (biochemical receptor) that is retained in direct spatial contact with a transduction element [175]. Liposomes can not only encapsulate various signal markers including enzymes [176], salts [177], DNA [178–180], electrochemical [181, 182], and chemiluminescent markers [183], but they can also serve as cell membrane models [184]. Therefore, liposomes are important candidates as building blocks for biosensors. As cholesterol helps to control the fluidity and permeability of the lipid membrane, cholesterol-containing liposomes are used to ensure a stable entrapment of the molecules required for detection, and four interesting examples of this that have been developed during the past three years are described.

Voccia *et al.* developed a biosensor for the detection of microRNAs (miRNAs) by using biotin-labeled cholesterol-containing liposomes as nanointerfaces to amplify the primary miRNA-sensing events [185]. MiRNAs are candidates for diagnostic and prognostic clinical biomarkers [186]. In this case, thiolated DNA capture probes were immobilized onto gold electrode surfaces (the biosensing platform) and a biotinylated RNA was prepared by hybridizing with specific capture probes. Biotin-labeled liposomes were used as a functional tether for the enzyme, as these are capable of carrying a large number of enzyme molecules. When the biosensing platform was incubated with the enzymatic label, the liposomes were found to alter the interfacial properties of the electrodes so as to increase the electron transfer resistance. As a consequence, a twofold increase in the target/blank signal ratio was observed by using electrochemical impedance spectroscopy (EIS). These preliminary results showed that liposomes coupled to EIS can be used for further investigations of the development of biosensors for miRNA detection.

Guan *et al.* developed a novel tyrosinase (Tyr) biosensor based on liposomal bioreactors and chitosan (CS) nanocomposites for the detection of phenolic compounds that occur abundantly as pollutants in food and the environment, and are inherently toxic (Fig. 15a, panel i) [187]. The liposomal microreactor was assembled by encapsulating the enzyme Tyr in L-α-phosphatidylcholine liposomes. Porins were then embedded into the lipid membrane, which forms a channel that allows the substrate to be freely transported but does not allow the enzyme to escape due to size limitations. When the glassy carbon electrode (GCE) was alternately immersed in CS and Tyr-loaded liposomes to assemble bilayer films, it was found that the biosensor could detect phenol over a broad linear range from 0.25 nM to 25 μM (Fig. 15a, panel ii). The detection limit was estimated as 0.091 nM, and the biosensor exhibited good repeatability and stability.

As the use of microarray-based immunoassays is often limited by their sensitivity, Ruktanonchai *et al.* investigated the use of liposome encapsulation to increase the signal in an antibody microarray platform [188]. This was

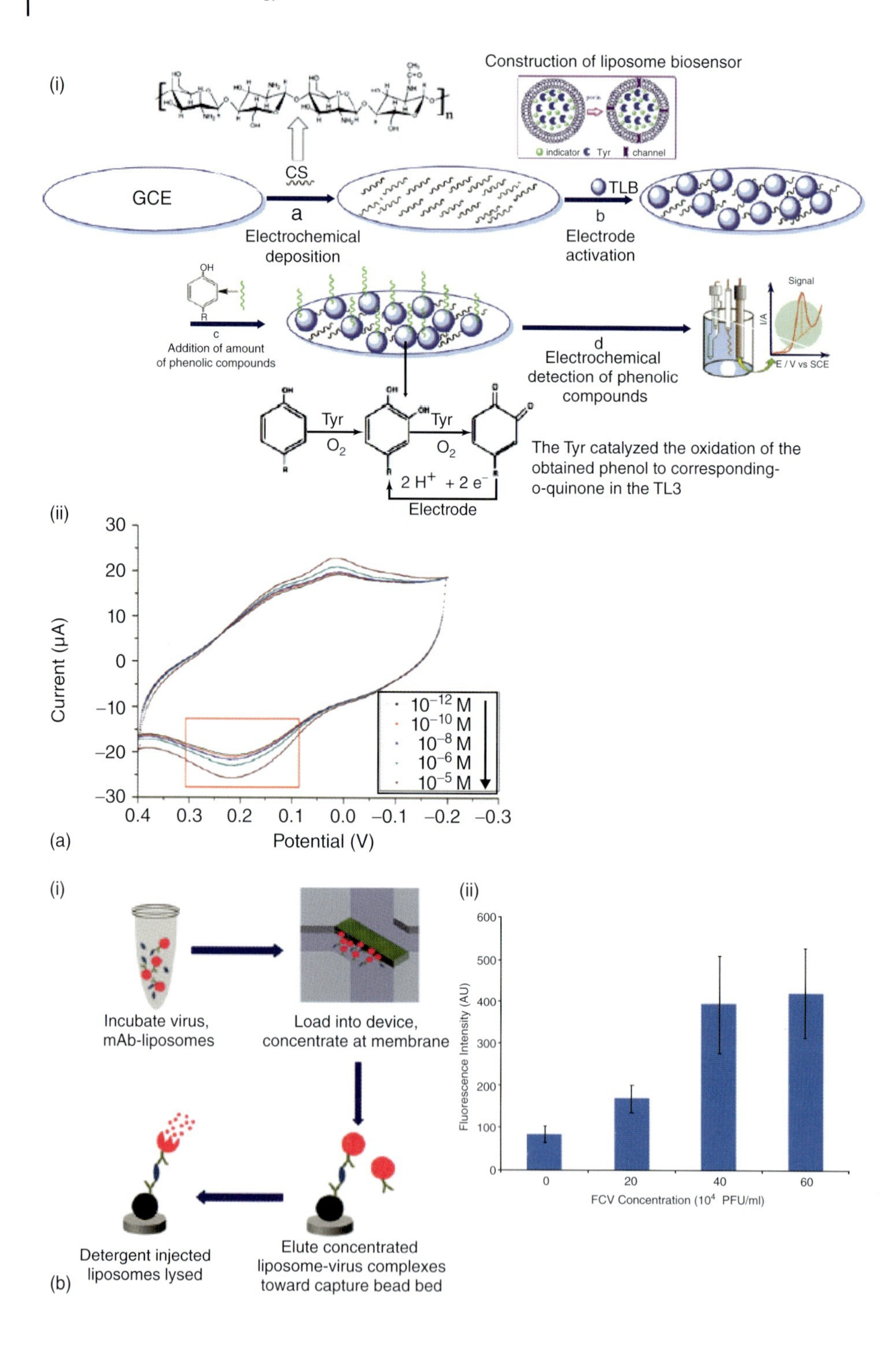
(i)
Construction of liposome biosensor
indicator Tyr channel
GCE
CS
a
Electrochemical deposition
TLB
b
Electrode activation
c
Addition of amount of phenolic compounds
d
Electrochemical detection of phenolic compounds
Signal
I/A
E / V vs SCE
Tyr
O2
2 H+ + 2 e−
Electrode
The Tyr catalyzed the oxidation of the obtained phenol to corresponding-o-quinone in the TL3
(ii)
Current (μA)
Potential (V)
10−12 M
10−10 M
10−8 M
10−6 M
10−5 M
(a)
(i)
Incubate virus, mAb-liposomes
Load into device, concentrate at membrane
Elute concentrated liposome-virus complexes toward capture bead bed
Detergent injected liposomes lysed
(b)
(ii)
Fluorescence Intensity (AU)
FCV Concentration (10⁴ PFU/ml)

Fig. 15 (a) (i) Schematic illustration of the liposome-based biosensor for electrochemical analysis of phenolic compounds; (ii) Response of the biosensor to different phenol concentrations. Reprinted with permission from Ref. [187]; (b) (i) Schematic of assays for virus (FCV) detection. The assay begins by loading the device with anti-FCV polyclonal antibody-labeled protein A superparamagnetic beads to create a capture bed and incubating the anti-FCV monoclonal antibody-labeled fluorescent liposomes with FCV. The sample is concentrated at the nanoporous membrane, and eluted toward the capture bead bed. Detergent is injected to lyse the liposomes, and the released fluorescent dye is used for quantification; (ii) FCV detection after electrokinetic concentration. FCV samples were incubated with anti-FCV-coupled liposomes for 2 h at the indicated concentrations and analyzed as described in panel (i). Reprinted with permission from Ref. [190].

demonstrated by detection of the food-borne pathogen *Listeria*. In a plate-trapped antigen enzyme-linked immunosorbent assay (ELISA), horseradish peroxidase (HRP)-loaded liposomes increased the signal ninefold compared to controls. The limits of detection of the HRP-encapsulated liposomes were 6.4×10^5 and 5.5×10^6 colony-forming units per milliliter in the sandwich ELISA and antibody array, respectively. Furthermore, when the chromogen 4-chloro-1-naphthol substrate was used for signal development in the antibody array, the results could be observed with the naked eye.

Subsequently, Román-Pizarro *et al.* developed on-flow microcontainers with liposomes containing magnetic gold nanoparticles and an enzymatic substrate (4-methylumbelliferyl-phosphate) for reagent preconcentration in a flow-injection method to determine alkaline phosphatase (ALP) activity [189]. These studies confirmed the value of using functionalized magnetic nanoparticles entrapped in liposomes for the automatic determination of enzymatic activity using a flow system, and ALP as a model analyte to check the applicability of this approach. The method has been applied to determine ALP activity in milk samples, as this is usually accepted as a rapid validation of the pasteurization of milk products. The method was found to improve the limit of detection (1.9×10^{-3} U l^{-1}) and precision in the range of 0.7–2.4%.

In LISA (the counterpart to ELISA) [191] the cholesterol-containing liposomes encapsulate large amounts of marker and receptor molecules and, when lysed, release detectable markers that can be monitored. Edwards and Baeumner developed a magnetically influenced cholesterol-containing liposome formulation encapsulating the sulforhodamine B dye [192]. In this case, the magnetic properties were added by incorporating a ferromagnetic metal oxide–oleic acid complex into the lipid bilayer of liposomes. In the magnetic field, those liposomes incorporating the iron oxide–oleic acid complex were drawn towards the binding surface of this heterogeneous sandwich hybridization assay format. The application of a magnetic field helps to overcome mass transfer limitations, yielding increased binding interactions of liposomes with surface-captured DNA or RNA target molecules, leading in turn to an increased assay sensitivity, reduced reagent concentrations, and reduced assay times.

Connelly *et al.* used cholesterol-containing liposomes for virus detection by applying an immunoassay sandwich approach where the reporting antibodies were tagged to liposomes [190]. For this, an integrated device was developed to detect environmentally relevant viruses, using feline calicivirus (FCV) as a model organism for human norovirus (Fig. 15b, panel i). The limit of detection for FCV of the integrated device was 1.6×10^5 plaque-forming units ml^{-1}, which was an order of magnitude lower than that achieved using the microfluidic biosensor without preconcentration (Fig. 15b, panel ii).

Shukla *et al.* prepared anti-*Salmonella* immunoglobulin G (IgG)-tagged liposomes (immunoliposomes) by conjugating purified anti-*Salmonella* IgG to cholesterol-containing liposomes [193]. These authors proposed an improved immunoassay method by using these immunoliposomes to detect *Salmonella typhimurium*; ultimately, this technique allowed a detection level 10^3 cells ml^{-1} *Salmonella*, compared to 10^4 cells ml^{-1} with ELISA [194].

5.2 Cholesterols as Anchors

5.2.1 Cholesterol as an Anchor for PEGylation

A key milestone in the design of liposomal formulations for drug-delivery systems is the creation of stable, intact liposomes for *in-vitro* and *in-vivo* applications. Liposomes are cleared from the systemic circulation via the reticuloendothelial system (RES), mainly in the liver and spleen [195]. The development of long-circulating liposomes capable of escaping RES uptake has been regarded as a major drive in the advancement of liposome technology. The stability of liposomes *in vivo* is determined mainly by their surface properties, and therefore the surface engineering of liposomes towards stealth characteristics is of paramount importance.

Among the many polymer coatings that endow liposomes with stealth properties, PEG is by far the best representative, due to its various attractive properties. Notably, PEG is nonionic, low-fouling, possesses high solubility in both aqueous and organic solvents [196], and is also characterized by its excellent biocompatibility, lack of toxicity and low immunogenicity [197, 198]. PEGylated liposomes have been shown to remain in the blood for periods up to 100-fold longer than conventional liposomes, and to show a greater accumulation in solid tumors based on their enhanced permeation and retention effects [199]; consequently, several successful PEGylated liposomes have been developed for clinical use.

PEG has also been used as a linker for the lipid conjugation of a variety of targeting molecules such as folate [200–202], transferrin [203], peptides [204, 205], and antibodies [206, 207]. In comparison to the direct conjugation of PEG to lipids in the bilayer, PEG linker-based conjugation allows for the steric stabilization of liposomes by preventing them from aggregating and fusing, and also facilitates cellular receptor recognition [202]. Cholesterol and phospholipids have been used successfully to anchor PEG to the liposome membrane; in particular, cholesterol-based PEG has been shown to be the more successful anchoring moiety [196]. Additionally, cholesterol improves hydration of the lipid head group, stabilizes the liposome membrane, and also aids in the retention of hydrophilic drugs. Due to the lipophilicity of cholesterol and their structural compatibility with phospholipids, cholesterol-based PEGs are easily incorporated into the membranes. Cholesterol-based PEGs can also maintain the fluidity of the membranes, increase the stability of the bilayer, and effectively control drug permeability of the liposomal bilayer [150, 208, 209]. For example, long-circulating cisplatin-loaded liposomes modified with cholesterol–PEG have been shown to exhibit prolonged circulation times. In fact, at 12 h after the administration of such modified cisplatin-loaded liposomes, blood levels of the drug were maintained more than fourfold higher than when unmodified liposomes were used [150].

In a recent study, a new type of multifunctional liposome (LP) consisting of a cell-penetrating peptide, octaarginines (R8), and a cholesterol-anchored reduction-sensitive PEG coating (CL), were designed to overcome kinetics barriers due to high PEGylation *in vivo*, so as to provide an improved tumor targeting (Fig. 16a) [210]. When these modified liposomes were tested both *in vitro* and *in vivo* in the presence of cysteine (Cys), the cellular uptake of CL-R8-LP was comparable to that of R8-LP, indicating that the outer reduction-sensitive PEG layer had been efficiently detached from the surface of the liposomes (Fig. 16b). It was also shown that, at 24 h after liposomal administration, the amount of modified liposomes which had accumulated in the tumors of treated mice was significantly higher than after the injection of unmodified liposomes. In this case, the R8-liposomes were mainly accumulated in the liver, spleen and lung, while the distribution of control formulations in these organs was decreased significantly (Fig. 16c).

Although the results of several studies have pointed favorably towards the use of cholesterol-based PEG as anchors for liposomes, it is worth noting that liposomes modified with cholesterol-PEG exhibit a shorter prolongation of circulation than do PEG-phospholipids. It has been

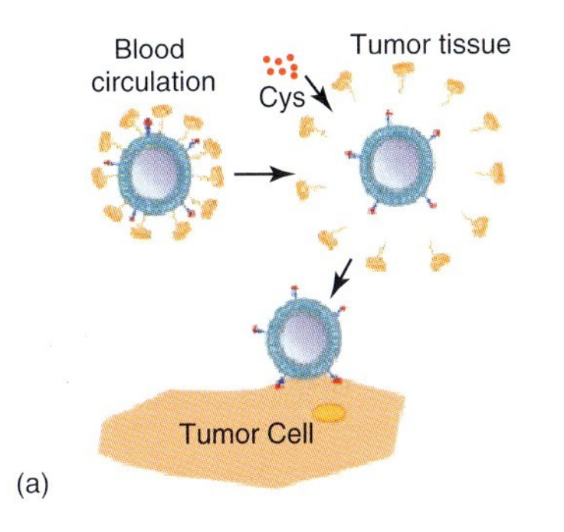

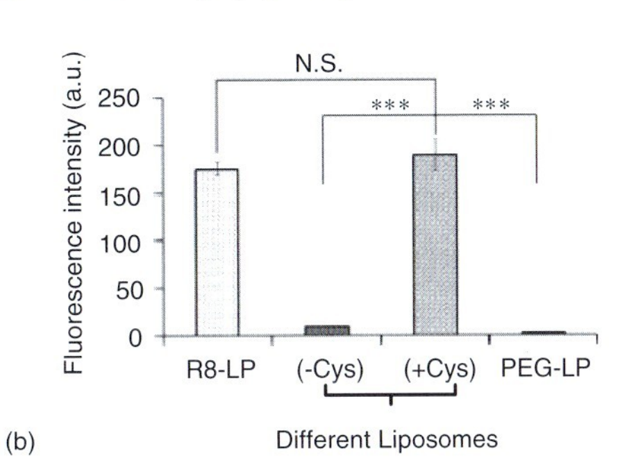

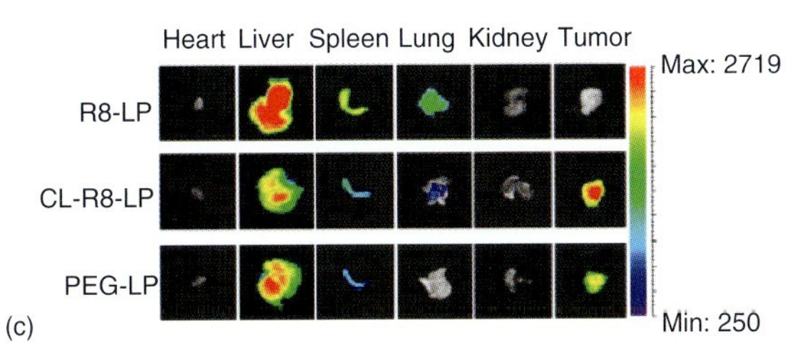

Fig. 16 (a) A cholesterol-anchored reduction-sensitive PEG coating for liposomes equipped with a cell-penetrating peptide R8. The PEG coating can be detached from the liposomes in the presence of cysteine to facilitate uptake by tumor cells; (b) Fluorescence intensity of C26 cells measured by flow cytometry after incubation with different types of fluorescently labeled liposomes for 4 h. These results show that the PEG shell is successfully detached in the presence of cysteine; (c) *Ex-vivo* images of C26 tumor-bearing mice at 24 h after injection of fluorescently labeled liposomes. Reprinted with permission from Ref. [210]; © 2014, Elsevier.

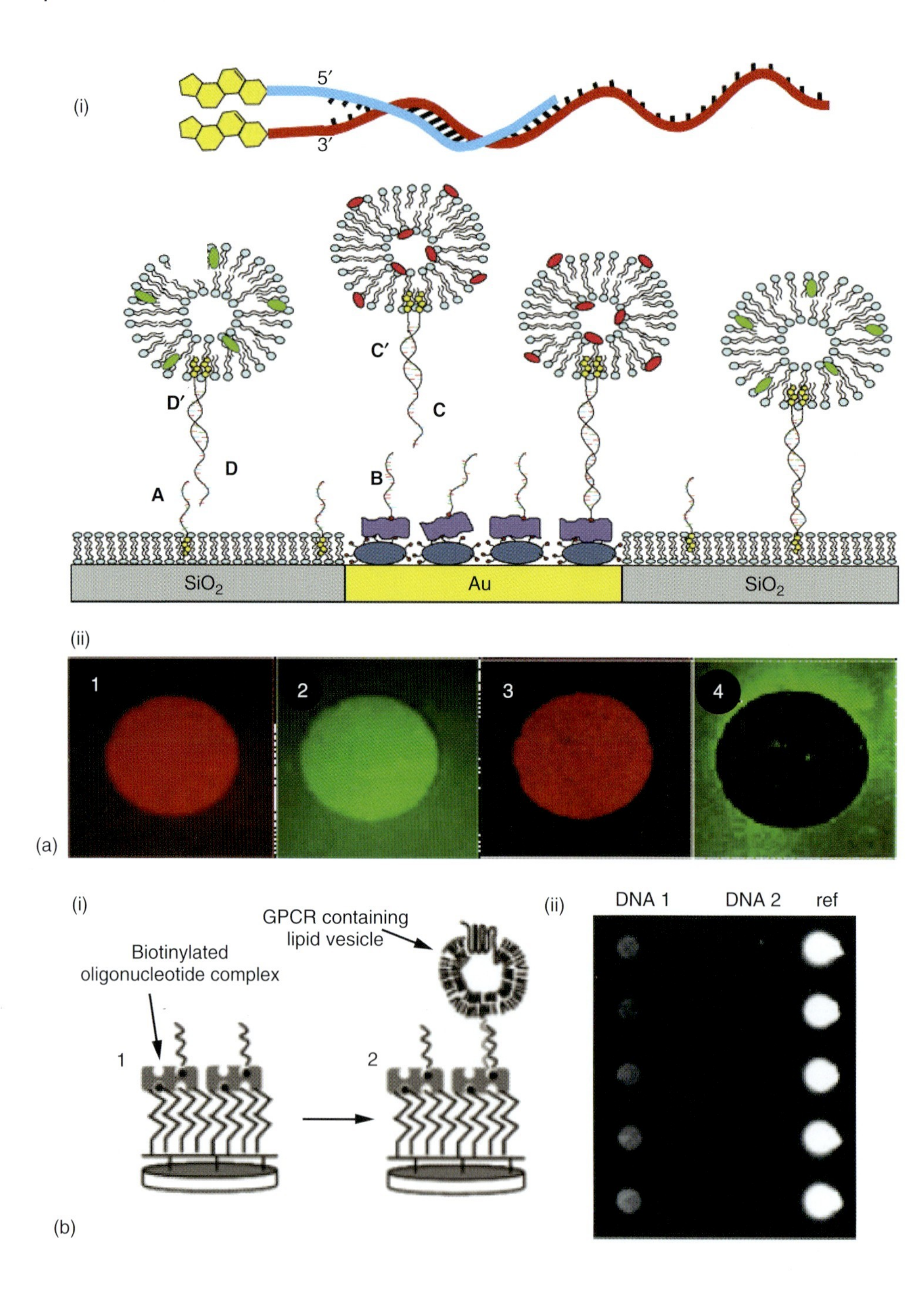
(i)
5′
3′
D′
D
A
C′
C
B
SiO2
Au
SiO2
(ii)
1
2
3
4
(a)
(i)
GPCR containing lipid vesicle
Biotinylated oligonucleotide complex
1
2
(ii)
DNA 1
DNA 2
ref
(b)

hypothesized that this difference in circulation duration between these types of liposome is most likely due to the reduced conformational flexibility of PEG, as the cholesterol anchor could be inserted more deeply into the liposome membrane [211]. A deeper presence of the PEG chain within the membrane has also been suggested as a reason why cholesterol-based PEG tends to disturb the lipid bilayer and cause a rapid release of the trapped therapeutic cargo, particularly at a high cholesterol-PEG density.

5.2.2 Cholesterol as an Anchor to Surface-Immobilized Liposomes: Biosensing

In the field of biosensing applications, liposomes have great potential as signal amplifiers as a result of their excellent properties for encapsulating signal markers such as dyes [212], enzymes [176], DNA [178, 179], and chemiluminescent markers [213]. Liposomes can also be used as cell membrane models [184], and have attracted enormous attention in the field of cytologic research. The concept of a liposome-based biosensor in this particular field lies with its similarity in structure and components with a cell membrane, that they can be used to monitor the simulated physiological processes of cells to detect interactions between a receptor and ligand. As noted above, the ability to surface-engineer liposomes to suit specific recognition functions with a range of different types of analytes makes liposomes ideal candidates in bioanalysis. More detailed discussions of liposome-based biosensors are available in recent reviews [175, 214, 215].

One approach for the development of membrane-based sensors using liposomes includes the use of intact liposomes arrayed on various substrates by using linkages such as avidin–biotin [216, 217], disulfide bonds [218], and chemical linkages between histidines [219]. The use of these linkages has limitations, however, in that different liposomes cannot be immobilized to predefined areas on the substrates [109]. To overcome this issue, Niemeyer and colleagues introduced the method of anchoring liposomes to substrates via DNA directed coupling [220], where the liposome array approach was based on DNA hybridization. In this case, capture DNA segments were immobilized on regular functional micro zones, after which the DNA-tagged liposomes were hybridized with the immobilized complementary DNA to form a regular array of liposomes. Similar attempts have been made to anchor liposomes to substrates using cholesterol-modified DNA that can insert spontaneously into the hydrophobic interior of lipid bilayers (Fig. 17a, panel i)

Fig. 17 (a) (i) Illustration of the bivalent cholesterol/DNA hybrid (above), obtained by modification with cholesterol in the 3′- and 5′-ends of a 15-mer DNA and a 30-mer DNA, respectively. Schematic illustration of the site-selective and sequence-specific sorting of differently DNA-modified lipid vesicles on a low-density cDNA array. (ii) Micrographs (1) and (2) were obtained by exposing the DNA-modified substrate to a mixture of red-labeled and green-labeled liposomes tagged using monovalent cholesterol–DNA. Micrographs (3) and (4) show an identical experiment, but for a liposome mixture using bivalent coupled cholesterol–DNA. Reprinted with permission from Ref. [223]; (b) (i) Schematic representation of the immobilization: (1) Immobilization of the preformed NA/bDNA complex on a PLL-g-PEG (12.5% or 50% of the PEG chains were terminated by biotin molecules) -modified surface; (2) liposomes containing G protein-coupled receptors (GPCRs), tagged with complementary oligonucleotides modified with a cholesterol moiety were captured onto the surface; (ii) Site-specific immobilization of liposomes containing GPCRs using cholesterol-DNA. Reprinted with permission from Ref. [225].

[218, 221, 222]. Although cholesterol-based anchoring of DNA is rather weak, it can be further strengthened by using bivalent cholesterol DNA coupling [221], and in this way red and green liposomes were sorted from a mixture of liposomes (Fig. 17a, panel ii) [223, 224]. This method also allowed for ligand binding to liposome arrays equipped with G protein-coupled receptors (Fig. 17b), paving the way for the development of this approach into a self-sorting biosensing platform [225].

5.2.3 Cholesterol as an Anchor to Surface-Immobilized Liposomes: Substrate-Mediated Drug Delivery (SMDD)

SMDD, with applications in drug-eluting implants, is a simple and versatile approach for delivering therapeutic compounds to their site of action in close vicinity. The method relies on the controlled trapping, retention and release of therapeutic molecules from thin polymer films. While many reports have been made on the successful incorporation of therapeutic compounds within films assembled via the layer-by-layer technique, and their subsequent delivery to adhering cells [226, 227], there remain several limitations to this approach. In particular, the trapping of small hydrophobic molecules or fragile proteins within the films still poses a challenge as a result of uncontrolled leakage and loss of functionality. In attempts to circumvent these problems, cyclodextrins, micelles, polymersomes and/or liposomes have been used as drug carriers embedded within the polymer films [228]. Different methods have also been proposed to stably embed liposomes within polymer films, without their displacement or rupture. Stabilization employing electrostatic interactions [229, 230] and noncovalent anchoring using cholesterol- [32] or oleic acid- [231] modified polymers have all proven to be promising strategies of immobilizing liposomes within polymer films. Lynge *et al.* were the among first to embed liposomes within multilayered polymer films, and demonstrated an interaction of these composite films with adhering mammalian cells (Fig. 18a, panel i) [232]. The successful assembly of these composite coatings relied on the previously demonstrated ability of cholesterol-modified poly(methacrylic acid) (PMA_c) to anchor the liposomes firmly to polymer layers due to the strong hydrophobic interaction between cholesterol and the lipid bilayer [32]. These composite coatings exhibited a sustained delivery of fluorescent lipids to adhering cells, with a much higher association of the fluorescent lipids to hepatocytes than to myoblasts (Fig. 18a, panel ii). Further, as a proof of concept, when the liposomes were loaded with a hydrophobic cytotoxic compound a significant decrease in viability of the adhering cells was observed. In a follow-up study, the same group combined the sequential deposition of polyelectrolytes with poly(dopamine) (PDA), and assembled films in which the liposomes were either trapped within a PDA matrix or embedded within polymer layers [233]. In this way, it was possible to demonstrate control over the amount of liposomes immobilized and the dependence of the cell mean fluorescence (CMF) of myoblasts adhering to films depending on the separation layers between the two liposome-deposition steps. Separation layers containing nondegradable building blocks (e.g., poly(styrene sulfonate)) led to a lower CMF compared to the separation layers where both components were (bio)degradable, and showed that the composition of the film could be used to "steer" the interaction with the cells.

Recently, lipogels – liposomes containing surface-adhering crosslinked PVA

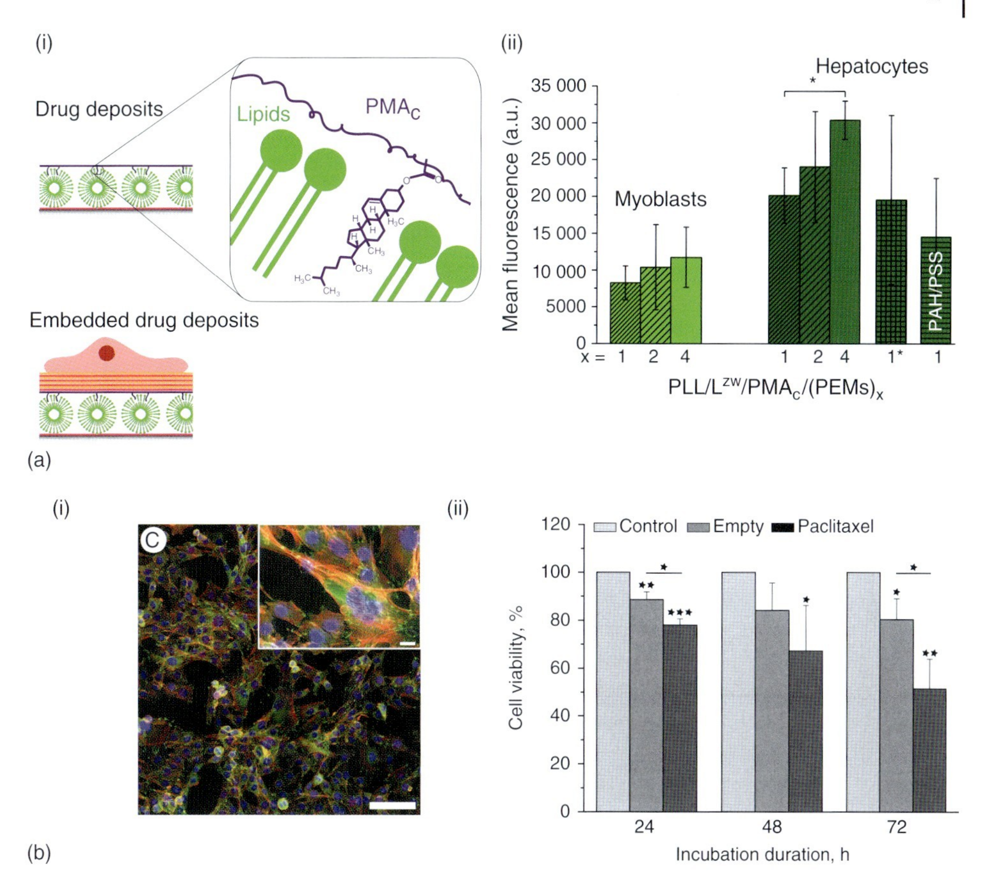

Fig. 18 (a) Liposome-loaded polymer films. (i) Schematic illustration showing the assembly of liposome-loaded polymer films using cholesterol-modified poly(methacrylic acid) (PMA_c) to anchor the liposomes to the polymers; (ii) Mean fluorescence of myoblasts and hepatocytes after attachment for 24 h to PLL/fluorescent liposomes (L^{ZW})/PMA_c/polyelectrolyte multilayers $(PEM)_x$ films (x = 1, 2 or 4, while 1* represents hepatocytes adhering to these coatings after the sample was incubated in cell media for 24 h) containing fluorescently labeled liposomes. Degradable capping layers ($(PEM)_x = ((PLL/PMA_c)_3)_x$-PLL) were compared to nondegradable capping layers ($(PEM)_x$ = ((poly(allylamine hydrochloride) $(PAH)/PSS)_3)_1$) for hepatocytes. Reprinted with permission from Ref. [232]; (b) Liposome-loaded hydrogel. (i) Confocal laser scanning microscopy image of myoblasts adhering to lipogels for 24 h Scale bars = 100 and 20 μm (inset) (red = actin filament; blue = DAPI; green = focal adhesion points); (ii) Cell viability of myoblast cells adhering to lipogels containing paclitaxel-loaded liposomes compared to pristine lipogels after 24, 48, and 72 h. Reprinted with permission from Ref. [234].

hydrogels – were also assembled [234]. It emerged that these liposomes had to be modified with PVA in order to remain nonaggregated when mixed with the PVA solution. To this end, thiocholesterol was embedded into the liposomes, which were then mixed with a solution of PVA equipped with a terminal thiol in a form which is activated towards thiol – disulfide exchange, in order to successfully add PVA chains to the liposomes. In this way, it was shown that cells could adhere to the lipogels when

PLL or PDA was present in the hydrogels (Fig. 18b, panel i). The cells were also able to uptake/associate with fluorescent lipids trapped in the liposomes. Encapsulating the cytotoxic compound paclitaxel within the liposomes led to a successful reduction in cell viability of the adhering myoblasts compared to controls (Fig. 18b, panel ii).

5.2.4 **Cholesterol as an Anchor to Surface-Immobilized Liposomes: Colloidal System towards Cell Mimicry**

The living cell is one of the most complex systems in Nature, and has continuously inspired scientists across multidisciplinary fields to reproduce its architectural structure, using biomimetic materials. A key approach to mimicking the hierarchical structure of a cell is a multicompartmentalized assembly, where the inner compartments resemble the organelles and the outer compartment mimic the cell membrane. Once the architectural principle of cell mimicry has been achieved, these artificial cell mimics have a wide implication in several biomedical applications, particularly in "metabolism mimicry," which commonly is also referred to as encapsulated catalysis [235]. Certainly, the ability to conduct enzymatic cascade reactions in a spatially controlled manner, confined within multicompartments, are key functions of healthy biological cells. The combination of multicompartmentalization and encapsulated catalysis is targeted at the assembly of biomimetic cellular systems, with the aim of substituting for missing or lost cellular function towards therapeutic cell mimicry.

During recent years a variety of successful multicompartmentalized systems have emerged, including vesosomes [236, 237], liposomes within a liposome carrier, polymersomes within a polymersome [238–240], polymer capsule within a polymer capsule [241, 242], polymer hydrogel capsules containing cubosomes [243] or polymersomes [244], and capsosomes, which are liposomes embedded into a polymer carrier capsule. Although, within the scope of this chapter, attention is focused on liposome-based assemblies, it should be noted that polymersome-based systems in particular are equally promising, with reported complexities in terms of assembly and function [239]. The concept of capsosomes, which are micron-sized polymer capsules containing layers of cargo-loaded liposomes, was first demonstrated and reported by Städler and colleagues [32]. The polymer carrier capsules provide the mechanically stability, while their semi-permeable nature allows for interaction of the interior with the environment. The embedded liposomes separate the interior of the capsules into multiple compartments, and they can encapsulate both hydrophilic and hydrophobic therapeutic cargo while controlling access to the solutes from outside [245].

The assembly of capsosomes requires a stable and strong incorporation of liposomes into polymer capsules. This was achieved by introducing noncovalent anchoring linkages based on cholesterol-modified polymers that included cholesterol-modified poly(L-lysine) (PLL_c) [32, 246], poly(MA-*co*-CMA) [32, 246], and poly(*N*-vinyl pyrrolidone-*b*-cholesteryl acrylate) (poly(PVP-*co*-CHA)) [247]. An important finding from these investigations was that poly(MA-*co*-CMA) was more effective in anchoring liposomes than poly(methacrylic acid)-*co*-(oleyl methacrylate) (poly(MA-*co*-OMA)), most likely because cholesterol – as a natural lipid membrane component – is more effective in anchoring liposomes than the oleyl moieties [231]. The use of poly(MA-*co*-CMA), poly(MA-*co*-OMA), and PLL_c

all led to the formation of capsosomes with polymer capsule membrane-associated liposomal compartments. However, when poly(PVP-*co*-CHA) was used, a "free-floating" of the liposomal subunits in the void of the polymer capsule was promoted [97]. As poly(PVP-*co*-CHA) is an uncharged block copolymer consisting of a long PVP fragment and a short block of CHA, physical entanglement of the copolymer with the membrane of the carrier capsule was minimized, and this resulted in a detachment of the liposomes from the polymer membrane into "free-floating" units within the void of the capsule at physiological pH. By using cholesterol-based polymers as anchors for liposomes, the same authors concluded that they could successfully assemble up to approximately 160 000 liposomal units within a 3 μm size polymer capsule – by far the highest level of subcompartmentalization described in the literature for cell mimics so far (Fig. 19a) [248].

The retention of cargo within the liposomal subcompartments is also highly important with regards to the use of capsosomes for conducting cascade reactions within confined environments for metabolic mimicry. Thus, several types of cargo were encapsulated within the subcompartments, and the activity of the cargo has been demonstrated in cell viability assays and enzymatic catalysis in confined environments [231, 246, 247, 249]. In particular, the functionality of the capsosomes was demonstrated by encapsulating enzymes – for example, β-lactamase [246, 249] or glutathione reductase [248] – within the liposomal subcompartments. The use of chemical and physical stimuli to trigger the enzymatic catalytic reactions confirmed the encapsulation and the activity of the enzymes.

To further address the complexity of therapeutic cell mimicry, a two-step enzymatic catalytic reaction in which production of the enzymatic conversion was used to trigger subsequent cargo release was demonstrated [248]. For this, the enzyme glutathione reductase was encapsulated within cholesterol-containing liposomal subunits, and the liposomes were then coencapsulated with peptide–polymer conjugates within the polymer capsules. Externally added oxidized glutathione (GSSG) was converted enzymatically into its reduced sulfhydryl form (GSH), leading to cleavage of the disulfide bond between the peptide and the polymer carrier, and subsequent release of the peptide.

While the aforementioned studies have demonstrated much success in encapsulated biocatalysis or drug delivery, their assembly relied on the labor-intensive sequential deposition of interacting polymer pairs. To overcome this shortcoming, Hosta-Rigau and coworkers used PDA to assemble the carrier capsules in one step [250]. The functionality of these capsosomes was demonstrated by performing a two-step enzymatic reaction simultaneously with a single-step enzymatic conversion within the same carrier. In the two-step enzymatic reaction, the enzyme uricase (UR) converts uric acid into hydrogen peroxide, CO_2 and allantoin, followed by the reaction of hydrogen peroxide with the reagent Amplex Ultra Red in the presence of the enzyme HRP to generate the fluorescent product, resorufin. In parallel, the enzyme ascorbate oxidase (AO) converts ascorbic acid into 2-L-dehydroascorbic acid (Fig. 19b, panel i). The authors compared the efficiency of the enzymatic reaction of capsosomes loaded with equal amounts of liposomes containing UR (L_{UR}), liposomes containing HRP (L_{HRP}), and empty liposomes (L)

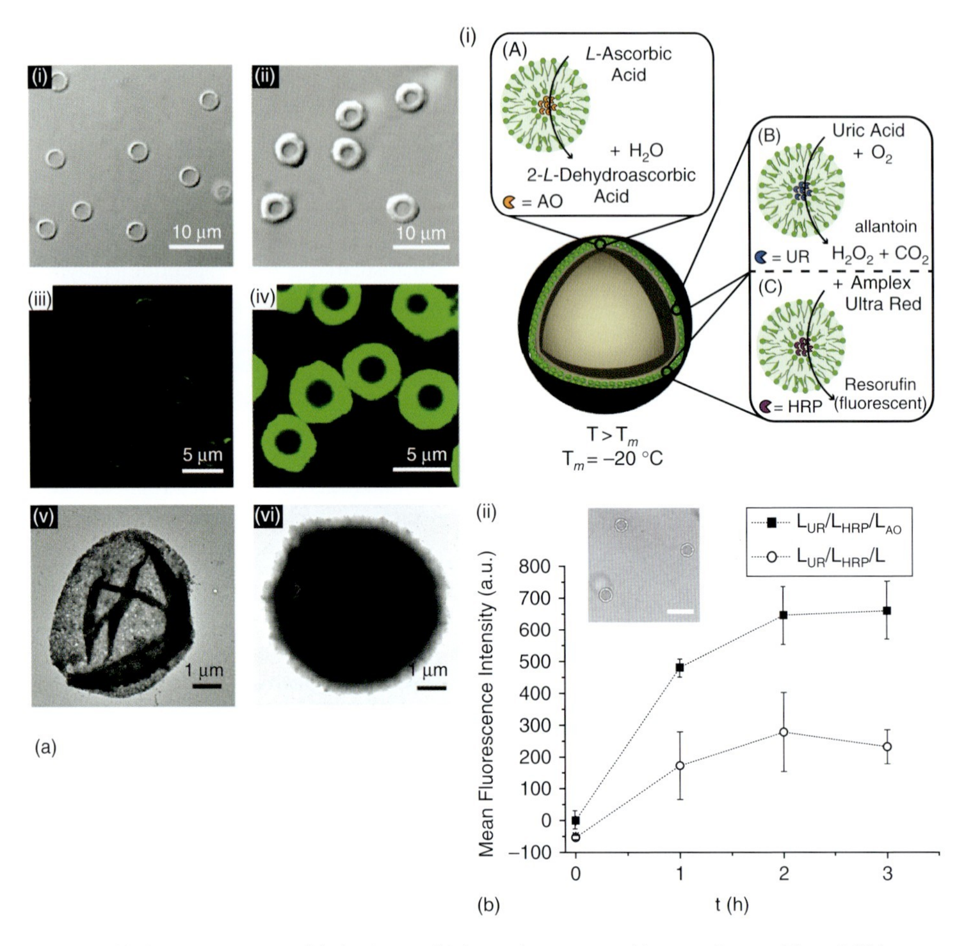

Fig. 19 (a) Capsosomes assembled using multiple liposome deposition steps. Images of capsosomes assembled using one (left) and eight (right) liposome deposition steps. Panels i and ii show differential interference contrast images, panels iii and iv show confocal laser scanning microscopy images, panels v and vi show transmission electron microscopy images. These different visualization techniques confirmed the higher loading of liposomal subunits for larger numbers of liposome deposition steps. Reprinted with permission from Ref. [249]; (b) Multiple enzymatic reactions. (i) Cartoon of the multiple enzymatic reactions performed within capsosomes. At $T > T_m$, ascorbic acid diffused out of the liposomes to be converted by ascorbate oxidase (AO) into the product 2-L-dehydroascorbic acid (panel A). In parallel, uric acid in the presence of O_2 diffused out of the liposomes and was converted by the enzyme uricase (UR) into H_2O_2, CO_2, and allantoin (panel B). H_2O_2 reacts with Amplex Ultra Red reagent in the presence of horseradish peroxidase (HRP) to form the fluorescent product resorufin (panel C); (ii) Reaction kinetics of capsosomes containing an equal mixture of liposomes loaded with UR (L_{UR}), HRP (L_{HRP}), and AO (L_{AO}) (■), compared to capsosomes loaded with an equal mixture of L_{UR}, L_{HRP}, and empty liposomes (L) (○). Reprinted with permission from Ref. [250].

versus capsosomes loaded with equal amounts of L_{UR}, L_{HRP}, and liposomes containing AO (L_{AO}). The results showed that capsosomes containing L_{AO} yielded a higher-fluorescence product than capsosomes containing L, thus illustrating the confined multiple enzymatic cascade reaction of the PDA-capsosomes (Fig. 19b, panel ii).

6 Conclusions

The importance of cholesterol for biomedical applications has been reviewed in this chapter. Cholesterol, a small biomolecule present in mammalian cells, was once described as "... the most highly decorated small molecule in biology" [251], and has since emerged as an indispensable building block in assemblies and devices in the field of nanomedicine. Cholesterol-containing liposomal therapeutic formulations were first approved for clinical use fifteen years ago by the FDA, when they were administered as cancer treatments. Their subsequent medical success proved the use of steroid additives to be correct, and has provided an explanation for the huge efforts invested during the past two decades that is reflected in the rising number of scientific publications in this field as well as the increasing numbers of these materials to be licensed for medical use. In contrast to former liposomal formulations, the benefits of using cholesterol as an integral building block in the lipid membranes of liposomes have been revealed in enhancements of membrane stability, yielding effective cargo loading, retention and release profiles. Moreover, cholesterol is currently being considered among the few attractive choices to connect liposomes and polymers. Cholesterol-modified PEG has been used to create "stealth" liposomal drug carriers, while cholesterol-modified polymers have been used to anchor liposomes to polymer films for SMDD or towards the assembly of artificial cells for therapeutic cell mimicry. Despite the field of therapeutic cell mimicry (especially considering multicompartment systems) still being in its infancy, cholesterol has already proven to be an important player in the assembly of these structures. Although cholesterol-modified liposomal formulations play the major role, polymeric biomaterials containing cholesterol moieties have recently become the focus of interest, a development which has been promoted by the advent of modern polymerization techniques. The synthesis of homopolymers with narrow molecular weight distributions, the preparation of random (or block) copolymers, tailoring of the length and nature of the spacer, or separating the functionality from the polymer backbone no longer represent serious challenges. In particular, block copolymers in which one block consists of cholesterol can self-assemble into micrometer-dimensioned particles, vesicles or even tubular structures, depending on the nature of the second block and/or the polymer architecture. These assemblies are currently being considered with regards to the controllable transport and release of entrapped cargoes. Although initial studies with such structures in mouse models were successful, further investigations – especially the design of biodegradable polymers with biocompatible degradation products – are necessary in order to fully assess the potential of these polymeric materials and, if possible, to translate them to the clinic.

Indeed, despite cholesterol being just a "small molecule," its versatility for biomedical applications is extremely impressive, and this is an aspect that is unlikely to change in the near future.

References

1. Berthelot, M. (1859) Sur plusieurs alcools nouveaux. *Ann. Chim. Phys.*, **56** 51–98.
2. Diels, O., Abderhalden, E. (1904) Information on cholesterol (II Announcement). *Ber. Dtsch Chem. Ges.*, **37** 3092–3103.
3. Windaus, A. (1932) Concerning the constitution of cholesterol and biliary acid. *Hoppe-Seyler's Z. Physiol. Chem.*, **213** 147–187.
4. Windaus, A., Neukirchen, K. (1919) The conversion of cholesterins in cholanic acid (28. Announcement concerning cholesterin). *Ber. Dtsch Chem. Ges.*, **52** 1915–1919.
5. Goluszko, P., Nowicki, B. (2005) Membrane cholesterol: a crucial molecule affecting interactions of microbial pathogens with mammalian cells. *Infect. Immun.*, **73** (12), 7791–7796.
6. Lingwood, D., Simons, K. (2010) Lipid rafts as a membrane-organizing principle. *Science*, **327** (5961), 46–50.
7. Grundy, S. M. (1983) Absorption and metabolism of dietary cholesterol. *Annu. Rev. Nutr.*, **3** 71–96.
8. Maxfield, F. R., Wustner, D. (2002) Intracellular cholesterol transport. *J. Clin. Invest.*, **110** (7), 891–898.
9. Ikonen, E. (2008) Cellular cholesterol trafficking and compartmentalization. *Nat. Rev. Mol. Cell Biol.*, **9** (2), 125–138.
10. Field, F. J., Born, E., Murthy, S., Mathur, S. N. (1998) Transport of cholesterol from the endoplasmic reticulum to the plasma membrane is constitutive in CaCo-2 cells and differs from the transport of plasma membrane cholesterol to the endoplasmic reticulum. *J. Lipid Res.*, **39** (2), 333–343.
11. Pucadyil, T. J., Chattopadhyay, A. (2006) Role of cholesterol in the function and organization of G-protein-coupled receptors. *Prog. Lipid Res.*, **45** (4), 295–333.
12. Bosch, M., Marí, M., Gross, S. P., Fernández-Checa, J. C., Pol, A. (2011) Mitochondrial cholesterol: a connection between caveolin, metabolism, and disease. *Traffic*, **12** (11), 1483–1489.
13. Farzadfar, F., Finucane, M. M., Danaei, G., Pelizzari, P. M., Cowan, M. J., Paciorek, C. J., *et al.* (2011) National, regional, and global trends in serum total cholesterol since 1980: systematic analysis of health examination surveys and epidemiological studies with 321 country-years and 3.0 million participants. *Lancet*, **377** (9765), 578–586.
14. Berry, J.D., Dyer, A., Cai, X., Garside, D.B., Ning, H., Thomas, A., *et al.* (2012) Lifetime risks of cardiovascular disease. *N. Engl. J. Med.*, **366** (4), 321–329.
15. Solomon, A., Kåreholt, I., Ngandu, T., Winblad, B., Nissinen, A., Tuomilehto, J., *et al.* (2007) Serum cholesterol changes after midlife and late-life cognition. *Neurology*, **68** (10), 751–756.
16. Moss, G.P. (1989) Nomenclature of steroids (Recommendations 1989). *Pure Appl. Chem.*, **61** (10), 1783–1822.
17. Gropper, S.S. and Smith, J.L. (2009) Advanced Nutrition and Human Metabolism, 5th edn, Wadsworth Publishing Co., Belmont, USA.
18. Lednicer, D. (2010) Steroid Chemistry at a Glance, John Wiley & Sons, Ltd, West Sussex.
19. Morzycki, J. W. (2014) Recent advances in cholesterol chemistry. *Steroids*, **83**, 62–79.
20. Achalkumar, A. S., Bushby, R. J., Evans, S. D. (2010) Cholesterol-based anchors and tethers for phospholipid bilayers and for model biological membranes. *Soft Matter*, **6** (24), 6036–6051.
21. Bladon, P. (1958), Cholesterol (ed. Cook, R. P.), Academic Press, New York, pp. 15–115.
22. Bijsterbosch MK, Rump ET, De Vrueh RL, Dorland R, van Veghel R, Tivel K, *et al.* (2000) Modulation of plasma protein binding and in vivo liver cell uptake of phosphorothioate oligodeoxynucleotides by cholesterol conjugation. *Nucleic Acids Res.*, **28** (14), 2717–2725.
23. Han, S.-O., Mahato, R. I., Kim, S. W. (2001) Water-soluble lipopolymer for gene delivery. *Bioconjugate Chem.*, **12** (3), 337–345.
24. Shannon, P. J. (1984) Photopolymerization in cholesteric mesophases. *Macromolecules*, **17** (9), 1873–1876.
25. Palakollu, V., Kanvah, S. (2014) Diphenylpolyene-cholesterol conjugates as fluorescent probes for microheterogeneous media. *J. Photochem. Photobiol., A*, **281**, 18–26.
26. Ahn, S.-K., Nguyen Le, L. T., Kasi, R. M. (2009) Synthesis and characterization of

side-chain liquid crystalline polymers bearing cholesterol mesogen. *J. Polym. Sci., Part A: Polym. Chem.*, **47** (10), 2690–2701.

27. Yusa, S.-I., Kamachi, M., Morishima, Y. (2000) Self-association of cholesterol-end-capped poly(sodium 2-(acrylamido)-2-methylpropanesulfonate) in aqueous solution. *Macromolecules*, **33** (4), 1224–1231.
28. De Visser, A. C., De Groot, K., Feyen, J., Bantjes, A. (1971) Thermal bulk polymerization of cholesteryl acrylate. *J. Polym. Sci., Part A: Polym. Chem.*, **9** (7), 1893–1899.
29. He, S.-J., Zhang, Y., Cui, Z.-H., Tao, Y.-Z., Zhang, B.-L. (2009) Controlled radical polymerization of cholesteryl acrylate and its block copolymer with styrene via the RAFT process. *Eur. Polym. J.*, **45** (8), 2395–2401.
30. Sevimli, S., Sagnella, S., Kavallaris, M., Bulmus, V., Davis, T. P. (2012) Synthesis, self-assembly and stimuli responsive properties of cholesterol conjugated polymers. *Polym. Chem.*, **3** (8), 2057–2069.
31. Sivakumar, P. A., Panduranga Rao, K. (2001) Stable polymerized cholesteryl methacrylate liposomes for vincristine delivery. *Biomed. Microdevices*, **3** (2), 143–148.
32. Städler, B., Chandrawati, R., Price, A. D., Chong, S.-F., Breheney, K., Postma, A., *et al.* (2009) A microreactor with thousands of subcompartments: enzyme-loaded liposomes within polymer capsules. *Angew. Chem. Int. Ed.*, **48** (24), 4359–4362.
33. Liu, J., Setijadi, E., Liu, Y., Whittaker, M. R., Boyer, C., Davis, T. P. (2010) PEGylated gold nanoparticles functionalized with β-cyclodextrin inclusion complexes: towards metal nanoparticle–polymer–carbohydrate cluster biohybrid materials. *Aust. J. Chem.*, **63** (8), 1245–1250.
34. Xu, J., Tao, L., Boyer, C., Lowe, A. B., Davis, T. P. (2011) Facile access to polymeric vesicular nanostructures: remarkable ω-end group effects in cholesterol and pyrene functional (co)polymers. *Macromolecules*, **44** (2), 299–312.
35. Zhu, Y., Zhou, Y., Chen, Z., Lin, R., Wang, X. (2012) Photoresponsive diblock copolymers bearing strong push–pull azo chromophores and cholesteryl groups. *Polymer*, **53** (16), 3566–3576.
36. Treiser, M., Abramson, S., Langer, R., Kohn, J. (2013) Degradable and resorbable Bio materials, in Biomaterials Science, 3rd edn, (eds Ratner, B. D., Hoffman, A. S., Schoen, F. J., Lemons, J. E.), Academic Press, Oxford, pp. 179–195.
37. Smith, D., Clark, S. H., Bertin, P. A., Mirkin, B. L., Nguyen, S. T. (2009) Synthesis and in vitro activity of ROMP-based polymer nanoparticles. *J. Mater. Chem.*, **19** (15), 2159–2165.
38. Liu, Y., Wang, Y., Zhuang, D., Yang, J., Yang, J. (2012) Bionanoparticles of amphiphilic copolymers polyacrylate bearing cholesterol and ascorbate for drug delivery. *J. Colloid Interface Sci.*, **377** (1), 197–206.
39. Yang, J., Li, Q., Li, Y., Jia, L., Fang, Q., Cao, A. (2006) Chemical preparation and characterization of new biodegradable aliphatic polyesters end-capped with diverse steroidal moieties. *J. Polym. Sci., Part A: Polym. Chem.*, **44** (6), 2045–2058.
40. Chabala, J. C., Shen, T. Y. (1978) The preparation of 3-cholesteryl 6-(thioglycosyl)hexyl ethers and their incorporation into liposomes. *Carbohydr. Res.*, **67** (1), 55–63.
41. Stübs, G., Fingerle, V., Wilske, B., Göbel, U. B., Zähringer, U., Schumann, R. R., *et al.* (2009) Acylated cholesteryl galactosides are specific antigens of *Borrelia* causing Lyme disease and frequently induce antibodies in late stages of disease. *J. Biol. Chem.*, **284** (20), 13326–13334.
42. Stübs, G., Rupp, B., Schumann, R. R., Schröder, N. W. J., Rademann, J. (2010) Chemoenzymatic synthesis of a glycolipid library and elucidation of the antigenic epitope for construction of a vaccine against Lyme disease. *Chem. Eur. J.*, **16** (11), 3536–3544.
43. Wang, B., Du, H., Zhang, J. (2011) Synthesis and characterisation of new types of side chain cholesteryl polymers. *Steroids*, **76** (1-2), 204–209.
44. Yu, Y.-L., Bai, J.-W., Zhang, J.-H. (2012) Synthesis and characterization of side-chain cholesterol derivatives based on double bond. *J. Mol. Struct.*, **1019**, 1–6.
45. Yu, Y.-L., Du, H.-Y., Zhang, J.-H. (2011) Self-assembly of novel cholesterol derivative based on hydrogen bond. *J. Mol. Struct.*, **1005** (1-3), 107–112.
46. Moad, G., Rizzardo, E., Thang, S. H. (2008) Toward living radical polymerization. *Acc. Chem. Res.*, **41** (9), 1133–1142.
47. Matyjaszewski, K. (2012) Atom transfer radical polymerization (ATRP): current status

and future perspectives. *Macromolecules*, **45** (10), 4015–4039.
48. Goto, A., Fukuda, T. (2004) Kinetics of living radical polymerization. *Prog. Polym. Sci.*, **29** (4), 329–385.
49. Jenkins, A. D., Jones, R. G., Moad, G. (2009) Terminology for reversible-deactivation radical polymerization previously called "controlled" radical or "living" radical polymerization (IUPAC Recommendations 2010). *Pure Appl. Chem.*, **82** (2), 483–491.
50. Fleet, R., McLeary, J. B., Grumel, V., Weber, W. G., Matahwa, H., Sanderson, R. D. (2007) Preparation of new multiarmed RAFT agents for the mediation of vinyl acetate polymerization. *Macromol. Symp.*, **255** (1), 8–19.
51. Postma, A., Davis, T. P., Li, G., Moad, G., O'Shea, M. S. (2006) RAFT polymerization with phthalimidomethyl trithiocarbonates or xanthates. On the origin of bimodal molecular weight distributions in living radical polymerization. *Macromolecules*, **39** (16), 5307–5318.
52. Chiefari, J., Chong, Y. K., Ercole, F., Krstina, J., Jeffery, J., Le, T. P. T., *et al.* (1998) Living free-radical polymerization by reversible addition–fragmentation chain transfer: the RAFT process. *Macromolecules*, **31** (16), 5559–5562.
53. Chong, Y. K., Krstina, J., Le, T. P. T., Moad, G., Postma, A., Rizzardo, E., *et al.* (2003) Thiocarbonylthio compounds [SC(Ph)S–R] in free radical polymerization with reversible addition-fragmentation chain transfer (RAFT Polymerization). Role of the free-radical leaving group (R). *Macromolecules*, **36** (7), 2256–2272.
54. Moad, G., Rizzardo, E., Thang, S. H. (2012) Living radical polymerization by the RAFT process – a third update. *Aust. J. Chem.*, **65** (8), 985–1076.
55. Semsarilar, M., Perrier, S. (2010) 'Green' reversible addition-fragmentation chain-transfer (RAFT) polymerization. *Nat. Chem.*, **2** (10), 811–820.
56. Matyjaszewski, K., Tsarevsky, N. V. (2014) Macromolecular engineering by atom transfer radical polymerization. *J. Am. Chem. Soc.*, **136** (18):6513–6533.
57. Matyjaszewski, K., Xia, J. (2001) Atom transfer radical polymerization. *Chem. Rev.*, **101** (9), 2921–2990.
58. Boyer, C., Bulmus, V., Davis, T. P., Ladmiral, V., Liu, J., Perrier, S. (2009) Bioapplications of RAFT polymerization. *Chem. Rev.*, **109** (11), 5402–5436.
59. Jochum, F. D., L. zur Borg, Roth, P. J., Theato, P. (2009) Thermo- and light-responsive polymers containing photoswitchable azobenzene end groups. *Macromolecules*, **42** (20), 7854–7862.
60. Muggli, M. W., Ward, T. C., Tchatchoua, C., Ji, Q., McGrath, J. E. (2003) End-group effect on physical aging and polymer properties for poly(ether sulfones). *J. Polym. Sci., Part B: Polym. Phys.*, **41** (22), 2850–2860.
61. Coessens, V., Pintauer, T., Matyjaszewski, K. (2001) Functional polymers by atom transfer radical polymerization. *Prog. Polym. Sci.*, **26** (3), 337–377.
62. Tasdelen, M. A., Kahveci, M. U., Yagci, Y. (2011) Telechelic polymers by living and controlled/living polymerization methods. *Prog. Polym. Sci.*, **36** (4), 455–567.
63. Henríquez, C., Bueno, C., Lissi, E. A., Encinas, M. V. (2003) Thiols as chain transfer agents in free radical polymerization in aqueous solution. *Polymer*, **44** (19), 5559–5561.
64. Sugiyama, K., Hanamura, R., Sugiyama, M. (2000) Assembly of poly[*N*-(2-hydroxypropyl)methacrylamide] having cholesteryl moiety as terminal groups. *J. Polym. Sci., Part A: Polym. Chem.*, **38** (18), 3369–3377.
65. Roth, P. J., Davis, T. P., Lowe, A. B. (2014) Novel α,α-bischolesteryl functional (co)polymers: RAFT radical polymerization synthesis and preliminary solution characterization. *Macromol. Rapid Commun.*, **35** (8), 813–820.
66. Haddleton, D. M., Edmonds, R., Heming, A. M., Kelly, E. J., Kukulj, D. (1999) Atom transfer polymerisation with glucose and cholesterol derived initiators. *New J. Chem.*, **23** (5), 477–479.
67. Lutz, J.-F., Pfeifer, S., Zarafshani, Z. (2007) In situ functionalization of thermoresponsive polymeric micelles using the "click" cycloaddition of azides and alkynes. *QSAR Comb. Sci.*, **26** (11-12), 1151–1158.
68. Xu, J. P., Ji, J., Chen, W. D., Shen, J. C. (2005) Novel biomimetic polymersomes as polymer therapeutics for drug delivery. *J. Controlled Release*, **107** (3), 502–512.

69. Xu, J. P., Ji, J., Chen, W. D., Shen, J. C. (2005) Novel biomimetic surfactant: synthesis and micellar characteristics. *Macromol. Biosci.*, **5** (2), 164–171.
70. Klok, H.-A., Hwang, J. J., Iyer, S. N., Stupp, S. I. (2002) Cholesteryl-(l-Lactic Acid)$_n$ building blocks for self-assembling biomaterials. *Macromolecules*, **35** (3), 746–759.
71. Zou, T., Cheng, S.-X., Zhuo, R.-X. (2005) Synthesis and enzymatic degradation of end-functionalized biodegradable polyesters. *Colloid Polym. Sci.*, **283** (10), 1091–1099.
72. Zou, T., Li, F., Cheng, S.-X., Zhuo, R.-X. (2006) Synthesis and characterization of end-capped biodegradable oligo/poly(trimethylene carbonate)s. *J. Biomater. Sci., Polym. Ed.*, **17** (10), 1093–1106.
73. Wan, T., Zou, T., Cheng, S.-X., Zhuo, R.-X. (2004) Synthesis and characterization of biodegradable cholesteryl end-capped polycarbonates. *Biomacromolecules*, **6** (1), 524–529.
74. Zhang, L., Wang, Q.-R., Jiang, X.-S., Cheng, S.-X., Zhuo, R.-X. (2005) Studies on functionalization of poly(ε-caprolactone) by a cholesteryl moiety. *J. Biomater. Sci., Polym. Ed.*, **16** (9), 1095–1108.
75. Wang, Y., Wang, H., Liu, G., Liu, X., Jin, Q., Ji, J. (2013) Self-assembly of near-monodisperse redox-sensitive micelles from cholesterol-conjugated biomimetic copolymers. *Macromol. Biosci.*, **13** (8), 1084–1091.
76. Roth, P. J., Kessler, D., Zentel, R., Theato, P. (2009) Versatile ω-end group functionalization of RAFT polymers using functional methane thiosulfonates. *J. Polym. Sci., Part A: Polym. Chem.*, **47** (12), 3118–3130.
77. Segui, F., Qiu, X.-P., Winnik, F. M. (2008) An efficient synthesis of telechelic poly (*N*-isopropylacrylamides) and its application to the preparation of α,ω-dicholesteryl and α,ω-dipyrenyl polymers. *J. Polym. Sci., Part A: Polym. Chem.*, **46** (1), 314–326.
78. Rasmussen, K. F., Smith, A. A., Ruiz-Sanchis, P., Edlund, K., Zelikin, A. N. (2014) Cholesterol modification of (bio)polymers using UV-Vis traceable chemistry in aqueous solutions. *Macromol. Biosci.*, **14** (1), 33–44.
79. Liu, J., Liu, H., Bulmus, V., Tao, L., Boyer, C., Davis, T. P. (2010) A simple methodology for the synthesis of heterotelechelic protein–polymer–biomolecule conjugates. *J. Polym. Sci., Part A: Polym. Chem.*, **48** (6), 1399–1405.
80. A. C. de Visser, Feyen, J., K. de Groot, Bantjes, A. (1970) Bulk polymerization of cholesteryl acrylate. *J. Polym. Sci., Part B: Polym. Lett.*, **8** (11), 805–808.
81. Mizuta, K., Katashima, M., Koga, T., Yamabuki, K., Onimura, K., Oishi, T. (2012) Synthesis of chiral side-chain liquid crystalline polyacetylenes bearing succinic acid spacer. *Polym. Bull.*, **68** (3), 623–634.
82. Odian, G. (2004), in Principles of Polymerization, John Wiley & Sons, Inc., New Jersey, pp. 464–543.
83. Young, L. J. (1961) Copolymerization parameters. *J. Polym. Sci.*, **54** (160), 411–455.
84. Hongbo, Z., Yaobang, L., Haoyu, Z., Xiaogong, W. (2004) Study on thermosensitive amphiphilic copolymers from *N*-isopropylacrylamide and cholesterol acrylate. *Acta Polym. Sin.*, **1** (3), 327–332.
85. Delbecq, F., Kawakami, K. (2014) Preparation of polyoligo(ethyleneglycol) methacrylate decorated with pendant cholesterol moieties: hydrogel and mesoglobule preparation and their use for entrapping lipophilic nanomaterials. *Colloids Surf., A*, **444**, 173–179.
86. Kaneko, T., Nagasawa, H., Gong, J. P., Osada, Y. (2003) Liquid crystalline hydrogels: mesomorphic behavior of amphiphilic polyacrylates bearing cholesterol mesogen. *Macromolecules*, **37** (1), 187–191.
87. Weidner, S., Wolff, D., Springer, J. (1996) Influence of chirality on the phase behaviour of copolymers containing cholesterol as mesogenic moiety. *Macromol. Chem. Phys.*, **197** (4), 1337–1348.
88. Filippov, S. K., Chytil, P., Konarev, P. V., Dyakonova, M., Papadakis, C., Zhigunov, A., *et al.* (2012) Macromolecular HPMA-based nanoparticles with cholesterol for solid-tumor targeting: detailed study of the inner structure of a highly efficient drug delivery system. *Biomacromolecules*, **13** (8), 2594–2604.
89. Sevimli, S., Inci, F., Zareie, H. M., Bulmus, V. (2012) Well-defined cholesterol polymers with pH-controlled membrane switching activity. *Biomacromolecules*, **13** (10), 3064–3075.
90. Iwasaki, Y., Akiyoshi, K. (2006) Synthesis and characterization of amphiphilic polyphosphates with hydrophilic graft chains

and cholesteryl groups as nanocarriers. *Biomacromolecules*, **7** (5), 1433–1438.

91. Ahn, S.-K., Deshmukh, P., Gopinadhan, M., Osuji, C. O., Kasi, R. M. (2011) Side-chain liquid crystalline polymer networks: exploiting nanoscale smectic polymorphism to design shape-memory polymers. *ACS Nano*, **5** (4), 3085–3095.
92. Ahn, S.-K., Kasi, R. M. (2011) Exploiting microphase-separated morphologies of side-chain liquid crystalline polymer networks for triple shape memory properties. *Adv. Funct. Mater.*, **21** (23), 4543–4549.
93. Städler, B., Price, A. D., Chandrawati, R., Hosta-Rigau, L., Zelikin, A. N., Caruso, F. (2009) Polymer hydrogel capsules: en route toward synthetic cellular systems. *Nanoscale*, **1** 68–73.
94. Ranucci, E., Suardi, M. A., Annunziata, R., Ferruti, P., Chiellini, F., Bartoli, C. (2008) Poly(amidoamine) conjugates with disulfide-linked cholesterol pendants self-assembling into redox-sensitive nanoparticles. *Biomacromolecules*, **9** (10), 2693–2704.
95. Yang, L., Zhang, B., Wen, L., Liang, Q., Zhang, L.-M. (2007) Amphiphilic cholesteryl grafted sodium alginate derivative: synthesis and self-assembly in aqueous solution. *Carbohydr. Polym.*, **68** (2), 218–225.
96. Jia, L., Liu, M., Di Cicco, A., Albouy, P.-A., Brissault, B., Penelle, J., *et al.* (2012) Self-assembly of amphiphilic liquid crystal polymers obtained from a cyclopropane-1,1-dicarboxylate bearing a cholesteryl mesogen. *Langmuir*, **28** (30), 11215–11224.
97. Hosta-Rigau, L., Chung, S. F., Postma, A., Chandrawati, R., Städler, B., Caruso, F. (2011) Capsosomes with "free-floating" liposomal subcompartments. *Adv. Mater.*, **23** (35), 4082–4087.
98. Zhou, Y., Kasi, R. M. (2008) Synthesis and characterization of polycholesteryl methacrylate–polyhydroxyethyl methyacrylate block copolymers. *J. Polym. Sci., Part A: Polym. Chem.*, **46** (20), 6801–6809.
99. Nguyen, C. T., Tran, T. H., Lu, X., Kasi, R. M. (2014) Self-assembled nanoparticles from thiol functionalized liquid crystalline brush block copolymers for dual encapsulation of doxorubicin and gold nanoparticles. *Polym. Chem.*, **5** (8), 2774–2783.
100. Zhou, Y., Ahn, S.-K., Lakhman, R. K., Gopinadhan, M., Osuji, C. O., Kasi, R. M. (2011) Tailoring crystallization behavior of PEO-based liquid crystalline block copolymers through variation in liquid crystalline content. *Macromolecules*, **44** (10), 3924–3934.
101. Shibaev, V. P., Ivanov, M. G., Boiko, N. I., Chernikova, E. V. (2009) A new approach to the synthesis of liquid crystalline triblock copolymers with a cholesteric structure. *Dokl. Chem.*, **427** (2), 183–185.
102. Hosta-Rigau, L., Jensen, B. E. B., Fjeldsø, K. S., Postma, A., Li, G., Goldie, K. N., *et al.* (2012) Surface-adhered composite poly(vinyl alcohol) physical hydrogels: polymersome-aided delivery of therapeutic small molecules. *Adv. Healthcare Mater.*, **1** (6), 791–795.
103. Zhang, X., Boisse, S., Bui, C., Albouy, P.-A., Brulet, A., Li, M.-H., *et al.* (2012) Amphiphilic liquid-crystal block copolymer nanofibers via RAFT-mediated dispersion polymerization. *Soft Matter*, **8** (4), 1130–1141.
104. Konikoff, F. M., Chung, D. S., Donovan, J. M., Small, D. M., Carey, M. C. (1992) Filamentous, helical, and tubular microstructures during cholesterol crystallization from bile. Evidence that cholesterol does not nucleate classic monohydrate plates. *J. Clin. Invest.*, **90** (3), 1155–1160.
105. Chung, D. S., Benedek, G. B., Konikoff, F. M., Donovan, J. M. (1993) Elastic free energy of anisotropic helical ribbons as metastable intermediates in the crystallization of cholesterol. *Proc. Natl Acad. Sci. USA*, **90** (23), 11341–11345.
106. Smith, B., Zastavker, Y. V., Benedek, G. B. (2001) Tension-induced straightening transition of self-assembled helical ribbons. *Phys. Rev. Lett.*, **87** (27), 278101.
107. Khaykovich, B., Kozlova, N., Choi, W., Lomakin, A., Hossain, C., Sung, Y., *et al.* (2009) Thickness–radius relationship and spring constants of cholesterol helical ribbons. *Proc. Natl Acad. Sci. USA*, **106** (37), 15663–15666.
108. Goodby, J. W. (1998) Liquid crystals and life. *Liq. Cryst.*, **24** (1), 25–38.
109. Hosta-Rigau, L., Zhang, Y., Teo, B. M., Postma, A., Stadler, B. (2013) Cholesterol – a biological compound as a building block in bionanotechnology. *Nanoscale*, **5** (1), 89–109.

110. Weiss, R. G. (1988) Thermotropic liquid crystals as reaction media for mechanistic investigations1. *Tetrahedron*, **44** (12), 3413–3475.
111. Zhou, Y., Briand, V. A., Sharma, N., Ahn, S.-K., Kasi, R. M. (2009) Polymers comprising cholesterol: synthesis, self-assembly, and applications. *Materials*, **2** (2), 636–660.
112. Akiyama, E., Morimoto, N., Kujawa, P., Ozawa, Y., Winnik, F. M., Akiyoshi, K. (2007) Self-assembled nanogels of cholesteryl-modified polysaccharides: effect of the polysaccharide structure on their association characteristics in the dilute and semidilute regimes. *Biomacromolecules*, **8** (8), 2366–2373.
113. Gu, X.-G., Schmitt, M., Hiasa, A., Nagata, Y., Ikeda, H., Sasaki, Y., *et al.* (1998) A novel hydrophobized polysaccharide/oncoprotein complex vaccine induces in vitro and in vivo cellular and humoral immune responses against HER2-expressing murine sarcomas. *Cancer Res.*, **58** (15), 3385–3390.
114. Yamane, S., Sugawara, A., Watanabe, A., Akiyoshi, K. (2009) Hybrid nanoapatite by polysaccharide nanogel-templated mineralization. *J. Bioact. Compat. Polym.*, **24** (2), 151–168.
115. Akiyoshi, K., Nishikawa, T., Mitsui, Y., Miyata, T., Kodama, M., Sunamoto, J. (1996) Self-assembly of polymer amphiphiles: thermodynamics of complexation between bovine serum albumin and self-aggregate of cholesterol-bearing pullulan. *Colloids Surf., A*, **112** (2-3), 91–95.
116. Kuroda, K., Fujimoto, K., Sunamoto, J., Akiyoshi, K. (2002) Hierarchical self-assembly of hydrophobically modified pullulan in water: gelation by networks of nanoparticles. *Langmuir*, **18** (10), 3780–3786.
117. Kumar, M. N. V. R., Muzzarelli, R. A. A., Muzzarelli, C., Sashiwa, H., Domb, A. J. (2004) Chitosan chemistry and pharmaceutical perspectives. *Chem. Rev.*, **104** (12), 6017–6084.
118. Yu, J.-M., Li, Y.-J., Qiu, L.-Y., Jin, Y. (2009) Polymeric nanoparticles of cholesterol-modified glycol chitosan for doxorubicin delivery: preparation and in-vitro and in-vivo characterization. *J. Pharm. Pharmacol.*, **61** (6), 713–719.
119. Li, X., Chen, M., Yang, W., Zhou, Z., Liu, L., Zhang, Q. (2012) Interaction of bovine serum albumin with self-assembled nanoparticles of 6-*O*-cholesterol modified chitosan. *Colloids Surf., B*, **92**, 136–141.
120. Wang, Y., Jiang, Q., Liu, L. R., Zhang, Q. (2007) The interaction between bovine serum albumin and the self-aggregated nanoparticles of cholesterol-modified *O*-carboxymethyl chitosan. *Polymer*, **48** (14), 4135–4142.
121. Wang, Y.-S., Liu, L.-R., Jiang, Q., Zhang, Q.-Q. (2007) Self-aggregated nanoparticles of cholesterol-modified chitosan conjugate as a novel carrier of epirubicin. *Eur. Polym. J.*, **43** (1), 43–51.
122. Yinsong, W., Lingrong, L., Jian, W., Zhang, Q. (2007) Preparation and characterization of self-aggregated nanoparticles of cholesterol-modified *O*-carboxymethyl chitosan conjugates. *Carbohydr. Polym.*, **69** (3), 597–606.
123. Wang, Y.-S., Jiang, Q., Li, R.-S., Liu, L.-L., Zhang, Q.-Q., Wang, Y.-M., *et al.* (2008) Self-assembled nanoparticles of cholesterol-modified *O*-carboxymethyl chitosan as a novel carrier for paclitaxel. *Nanotechnology*, **19** (14), 145101.
124. Long, L.-X., Yuan, X.-B., Chang, J., Zhang, Z.-H., Gu, M.-Q., Song, T.-T., *et al.* (2012) Self-assembly of polylactic acid and cholesterol-modified dextran into hollow nanocapsules. *Carbohydr. Polym.*, **87** (4), 2630–2637.
125. Abu Lila, A. S., Nawata, K., Shimizu, T., Ishida, T., Kiwada, H. (2013) Use of polyglycerol (PG), instead of polyethylene glycol (PEG), prevents induction of the accelerated blood clearance phenomenon against long-circulating liposomes upon repeated administration. *Int. J. Pharm.*, **456** (1), 235–242.
126. Allen, T. M., Hansen, C., Martin, F., Redemann, C., Yau-Young, A. (1991) Liposomes containing synthetic lipid derivatives of poly(ethylene glycol) show prolonged circulation half-lives in vivo. *Biochim. Biophys. Acta*, **1066** (1), 29–36.
127. Knop, K., Hoogenboom, R., Fischer, D., Schubert, U. S. (2010) Poly(ethylene glycol) in drug delivery: pros and cons as well as potential alternatives. *Angew. Chem. Int. Ed.*, **49** (36), 6288–6308.
128. Chern, C.-S., Chiu, H.-C., Chuang, Y.-C. (2004) Synthesis and characterization of amphiphilic graft copolymers with

poly(ethylene glycol) and cholesterol side chains. *Polym. Int.*, **53** (4), 420–429.
129. Jia, L., Albouy, P.-A., Di Cicco, A., Cao, A., Li, M.-H. (2011) Self-assembly of amphiphilic liquid crystal block copolymers containing a cholesteryl mesogen: effects of block ratio and solvent. *Polymer*, **52** (12), 2565–2575.
130. Nagahama, K., Ouchi, T., Ohya, Y. (2008) Temperature-induced hydrogels through self-assembly of cholesterol-substituted star PEG-b-PLLA copolymers: an injectable scaffold for tissue engineering. *Adv. Funct. Mater.*, **18** (8), 1220–1231.
131. Ooya, T., Huh, K. M., Saitoh, M., Tamiya, E., Park, K. (2005) Self-assembly of cholesterol-hydrotropic dendrimer conjugates into micelle-like structure: preparation and hydrotropic solubilization of paclitaxel. *Sci. Technol. Adv. Mater.*, **6** (5), 452–456.
132. Thompson, C., Ding, C., Qu, X., Yang, Z., Uchegbu, I., Tetley, L., *et al.* (2008) The effect of polymer architecture on the nano self-assemblies based on novel comb-shaped amphiphilic poly(allylamine). *Colloid Polym. Sci.*, **286** (13), 1511–1526.
133. F. van de Manakker, Vermonden, T., N. el Morabit, C. F. van Nostrum, Hennink, W. E. (2008) Rheological behavior of self-assembling PEG-β-cyclodextrin/PEG-cholesterol hydrogels. *Langmuir*, **24** (21), 12559–12567.
134. Wang, Y., Ke, C.-Y., Weijie Beh, C., Liu, S.-Q., Goh, S.-H., Yang, Y.-Y. (2007) The self-assembly of biodegradable cationic polymer micelles as vectors for gene transfection. *Biomaterials*, **28** (35), 5358–5368.
135. Yang, D.-B., Zhu, J.-B., Huang, Z.-J., Ren, H.-X., Zheng, Z.-J. (2008) Synthesis and application of poly(ethylene glycol)–cholesterol (Chol–PEGm) conjugates in physicochemical characterization of nonionic surfactant vesicles. *Colloids Surf., B*, **63** (2), 192–199.
136. Pescador, P., Brodersen, N., Scheidt, H. A., Loew, M., Holland, G., Bannert, N., *et al.* (2010) Microtubes self-assembled from a cholesterol-modified nucleoside. *Chem. Commun.*, **46** (29), 5358–5360.
137. Jesorka, A., Orwar, O. (2008) Liposomes: technologies and analytical applications. *Annu. Rev. Anal. Chem.*, **1**, 801–832.
138. Park, B.-W., Yoon, D.-Y., Kim, D.-S. (2010) Recent progress in bio-sensing techniques with encapsulated enzymes. *Biosens. Bioelectron.*, **26** (1), 1–10.
139. Ohvo-Rekilä, H. (2002) Cholesterol interactions with phospholipids in membranes. *Prog. Lipid Res.*, **41**, 66–97.
140. Kohsaku, K., Yoshitaka, N., Koichiro, H. (1999) Rigidity of lipid membranes detected by capillary electrophoresis. *Langmuir*, **15**, 1893–1895.
141. Pál, J., Mihaly, M. (2003) Effect of cholesterol on the properties of phospholipid membranes. 2. Free energy profile of small molecules. *J. Phys. Chem. B*, **107**, 5322–5332.
142. Waldeck, A. R., Nouri-Sorkhabi, M. H., David, R. S., Philip, W. K. (1995) Effects of cholesterol on transmembrane water diffusion in human erythrocytes measured using pulsed field gradient NMR. *Biophys. Chem.*, **55**, 197–208.
143. Kusumi, A., Tsuda, M., Akino, T., Ohnishi, S., Terayama, Y. (1983) Protein-phospholipid-cholesterol interaction in the photolysis of invertebrate rhodopsin. *Biochemistry*, **22** (5), 1165–1170.
144. Mouritsen, O., Jørgensen, K. (1994) Dynamical order and disorder in lipid bilayers. *Chem. Phys. Lipids*, **73** (1-2), 3–25.
145. Rog, T., Pasenkiewicz-Gierula, M., Vattulainen, I., Karttunen, M. (2009) Ordering effects of cholesterol and its analogues. *Biochim. Biophys. Acta*, **1788** (1), 97–121.
146. Ludewig, B., Barchiesi, F., Pericin, M., Zinkernagel, R., Hengartner, H., Schwendener, R. (2000) In vivo antigen loading and activation of dendritic cells via a liposomal peptide vaccine mediates protective antiviral and anti-tumour immunity. *Vaccine*, **19** (1), 23–32.
147. Hallaj-Nezhadi, S., Hassan, M. (2013) Nanoliposome-based antibacterial drug delivery. *Drug Delivery*, 1–9.
148. Hyodo, K., Yamamoto, E., Suzuki, T., Kikuchi, H., Asano, M., Ishihara, H. (2013) Development of liposomal anticancer drugs. *Biol. Pharm. Bull.*, **36** (5), 703–707.
149. Gaspar, M., Perez-Soler, R., Cruz, M. (1996) Biological characterization of L-asparaginase liposomal formulations. *Cancer Chemother. Pharmacol.*, **38** (4), 373–377.
150. Kuang, Y., Liu, J., Liu, Z., Zhuo, R. (2012) Cholesterol-based anionic long-circulating cisplatin liposomes with reduced renal toxicity. *Biomaterials*, **33** (5), 1596–1606.

151. Čeh, B., Winterhalter, M., Frederik, P. M., Vallner, J. J., and Lasic, D. D. (1997) Stealth® liposomes: from theory to product. *Adv. Drug Delivery Rev.*, **24** (2–3), 165–177.
152. Barenholz, Y. (2012) Doxil® – the first FDA-approved nano-drug: lessons learned. *J. Controlled Release*, **160** (2), 117–134.
153. Fassas, A., Anagnostopoulos, A. (2005) The use of liposomal daunorubicin (DaunoXome) in acute myeloid leukemia. *Leuk. Lymphoma*, **46** (6), 795–802.
154. Phuphanich, S., Maria, B., Braeckman, R., Chamberlain, M. (2007) A pharmacokinetic study of intra-CSF administered encapsulated cytarabine (DepoCyt®) for the treatment of neoplastic meningitis in patients with leukemia, lymphoma, or solid tumors as part of a phase III study. *J. Neuro-Oncol.*, **81** (2), 201–208.
155. Hartrick, C. T., Hartrick, K. A. (2008) Extended-release epidural morphine (DepoDur™): review and safety analysis. *Expert Rev. Neurother.*, **8** (11), 1641–1648.
156. Azanza Perea, J. R., Barberán, J. (2012) Anfotericina B forma liposómica: un perfil farmacocinético exclusivo. Una historia inacabada. *Rev. Esp. Quimioterapia*, **25** (1), 17–24.
157. Boswell, G., Buell, D., Bekersky, I. (1998) AmBisome (liposomal amphotericin B): a comparative review. *J. Clin. Pharmacol.*, **38** (7), 583–592.
158. Clemons, K., Stevens, D. (2004) Comparative efficacies of four amphotericin B formulations – Fungizone, amphotec (Amphocil), AmBisome, and Abelcet – against systemic murine aspergillosis. *Antimicrob. Agents Chemother.*, **48** (3), 1047–1050.
159. Stathopoulos, G., Boulikas, T. (2012) Lipoplatin formulation review article. *J. Drug Delivery*, **2012**, Article ID 581363.
160. Thomas, D., Kantarjian, H., Stock, W., Heffner, L., Faderl, S., Garcia-Manero, G., *et al.* (2009) Phase 1 multicenter study of vincristine sulfate liposomes injection and dexamethasone in adults with relapsed or refractory acute lymphoblastic leukemia. *Cancer*, **115** (23), 5490–5498.
161. Venegas, B., Zhu, W. W., Haloupek, N. B., Lee, J., Zellhart, E., Sugar, I. P., *et al.* (2012) Cholesterol superlattice modulates CA4P release from liposomes and CA4P cytotoxicity on mammary cancer cells. *Biophys. J.*, **102** (9), 2086–2094.
162. Yang, S. Y., Zheng, Y., Chen, J. Y., Zhang, Q. Y., Zhao, D., Han, D. E., *et al.* (2013) Comprehensive study of cationic liposomes composed of DC-Chol and cholesterol with different mole ratios for gene transfection. *Colloids Surf., B*, **101** 6–13.
163. Alhajlan, M., Alhariri, M., Omri, A. (2013) Efficacy and safety of liposomal clarithromycin and its effect on *Pseudomonas aeruginosa* virulence factors. *Antimicrob. Agents Chemother.*, **57** (6), 2694–2704.
164. Gómez-Hens, A., Fernández-Romero, J.M. (2005) The role of liposomes in analytical processes. *Trends Anal. Chem.*, **24**, 9–19.
165. Lundahl P, Yang Q. (1991) Liposome chromatography: liposomes immobilized in gel beads as a stationary phase for aqueous column chromatography. *J. Chromatogr. A*, **544** (1-2):283–304.
166. Lundahl P, Beigi F.(1997) Immobilized liposome chromatography of drugs for model analysis of drug-membrane interactions. *Adv. Drug Delivery Rev.*, **23**(1), 221–227.
167. Wiedmer, S.K., Jussila, M.S., Riekkola, M.L. (2004) Phospholipids and liposomes in liquid chromatographic and capillary electromigration techniques. *Trends Anal. Chem.*, **23**(8), 562–582.
168. Chapelle, S., Gillesbaillien, M. (1983) Phospholipids and cholesterol in brush-border and basolateral membranes from rat intestinal-mucosa. *Biochim. Biophys. Acta*, **753** (2), 269–271.
169. Geiger, M., Hogerton, A. L., Bowser, M. T. (2011) Capillary electrophoresis. *Anal. Chem.*, **84** (2), 577–596.
170. Cunliffe J.M., Baryla N.E., Lucy C.A. (2002) Phospholipid bilayer coatings for the separation of proteins in capillary electrophoresis. *Anal. Chem.*, **74**(4), 776–783.
171. Zhang, Y., Zhang, R., Hjertén, S., Lundahl, P. (1995) Liposome capillary electrophoresis for analysis of interactions between lipid bilayers and solutes. *Electrophoresis*, **16** (8), 1519–1523.
172. Lindén, M. V., Holopainen, J. M., Laukkanen, A., Riekkola, M.-L., Wiedmer, S. K. (2006) Cholesterol-rich membrane coatings for interaction studies in capillary electrophoresis: application to red blood cell lipid extracts. *Electrophoresis*, **27** (20), 3988–3998.
173. Tiala, H., Riekkola, M. L., Wiedmer, S. K. (2013) Study on capillaries covalently

bound with phospholipid vesicles for open-tubular CEC and application to on-line open-tubular CEC-MS. *Electrophoresis*, **34** (22-23), 3180–3188.
174. Li, B., Lv, X. F., Geng, L. N., Qing, H., Deng, Y. L. (2012) Proteoliposome-based capillary electrophoresis for screening membrane protein inhibitors. *J. Chromatogr. Sci.*, **50** (7), 569–573.
175. Liu, Q., Boyd, B. (2013) Liposomes in biosensors. *Analyst*, **138** (2), 391–409.
176. Ceccoli, J., Rosales, N., Tsimis, J., Yarosh, D. (1989) Encapsulation of the UV-DNA repair enzyme T4 endonuclease V in liposomes and delivery to human cells. *J. Invest. Dermatol.*, **93** (2), 190–194.
177. Orcutt, K. M., Wells, M. L. (2007) A liposome-based nanodevice for sequestering siderophore-bound Fe. *J. Membr. Sci.*, **288**, 247–254.
178. Filion, M.C., Philip, N. C. (1998) Major limitations in the use of cationic liposomes for DNA delivery. *Int. J. Pharm.*, **162**, 159–170.
179. Zhou X, Huang L (1992) Targeted delivery of DNA by liposomes and polymers. *J. Controlled Release*, **19**(1-3), 269–274.
180. Baeumner AJ, Cohen RN, Miksic V, Min J. (2003) RNA biosensor for the rapid detection of viable Escherichia coli in drinking water. *Biosens. Bioelectron.*, **18** (4), 405–413.
181. Chumbimuni-Torres, K., Wu, J., Clawson, C., Galik, M., Walter, A., Flechsig, G.-U., *et al.* (2010) Amplified potentiometric transduction of DNA hybridization using ion-loaded liposomes. *Analyst*, **135** (7), 1618–1623.
182. Shukla, S., Leem, H., Kim, M. (2011) Development of a liposome-based immunochromatographic strip assay for the detection of *Salmonella*. *Anal. Bioanal. Chem.*, **401** (8), 2581–2590.
183. Pakavadee, R., Akarin, I., Kavi, R. (2010) Luminol encapsulated liposome as a signal generator for the detection of specific antigen-antibody reactions and nucleotide hybridization. *Anal. Sci.*, **26**, (7), 767–772.
184. Nakane, Y., Ito, M., Kubo, I.(2008) Novel detection method of endocrine disrupting chemicals utilizing liposomes as cell membrane model. *Anal. Lett.*, **41** (16), 2923–2932.
185. Voccia, D., Bettazzi, F., Palchetti, I. (2014) Electrochemical liposome-based biosensors for nucleic acid detection. *Sens. Microsyst.*, **268**, 179–182.
186. Nana-Sinkam, S., Croce, C. (2013) Clinical applications for microRNAs in cancer. *Clin. Pharmacol. Ther.*, **93** (1), 98–104.
187. Guan, H., Liu, X., Wang, W. (2013) Encapsulation of tyrosinase within liposome bioreactors for developing an amperometric phenolic compounds biosensor. *J. Solid State Electrochem.*, **17** (11), 2887–2893.
188. Ruktanonchai, U., Nuchuchua, O., Charlermroj, R., Pattarakankul, T., Karoonuthaisiri, N. (2012) Signal amplification of microarray-based immunoassay by optimization of nanoliposome formulations. *Anal. Biochem.*, **429** (2), 142–147.
189. Román-Pizarro, V., Fernández-Romero, J. M., Gómez-Hens, A. (2014) Fluorometric determination of alkaline phosphatase activity in food using magnetoliposomes as on-flow microcontainer devices. *J. Agric. Food Chem.*, **62** (8), 1819–1825.
190. Connelly, J., Kondapalli, S., Skoupi, M., Parker, J. L., Kirby, B., Baeumner, A. (2012) Micro-total analysis system for virus detection: microfluidic pre-concentration coupled to liposome-based detection. *Anal. Bioanal. Chem.*, **402** (1), 315–323.
191. Rongen, H. A. H., Bult, A., W. P. van Bennekom (1997) Liposomes and immunoassays. *J. Immunol. Methods*, **204**, 105–133.
192. Edwards, K. A., Baeumner, A. J. (2014) Enhancement of heterogeneous assays using fluorescent magnetic liposomes. *Anal. Chem.*, **86**(13):6610–6616.
193. Shukla, S., Bang, J., Heu, S., Kim, M. (2012) Development of immunoliposome-based assay for the detection of *Salmonella typhimurium*. *Eur. Food Res. Technol.*, **234** (1), 53–59.
194. Durant, J. A., Young, C. R., Nisbet, D. J., Stanker, L. H., Ricke, S. C. (1997) Detection and quantification of poultry probiotic bacteria in mixed culture using monoclonal antibodies in an enzyme-linked immunosorbent assay. *Int. J. Food Microbiol.*, **38** (2-3), 181–189.
195. Blume, G., Cevc, G. (1990) Liposomes for the sustained drug release in vivo. *Biochim. Biophys. Acta*, **1029** (1), 91–97.
196. Nag, O. K., Awasthi, V. (2013) Surface engineering of liposomes for stealth behavior. *Pharmaceutics*, **5** (4), 542–569.

197. Sawant, R., Torchilin, V. (2012) Challenges in development of targeted liposomal therapeutics. *AAPS J.*, **14** (2), 303–315.
198. Torchilin, V. P. (2005) Recent advances with liposomes as pharmaceutical carriers. *Nat. Rev. Drug Discovery*, **4** (2), 145–160.
199. Lasic, D. D., Papahadjopoulos, D. (1995) Liposomes revisited. *Science*, **267** (5202), 1275–1276.
200. Lee, R. J., Low, P. S. (1994) Delivery of liposomes into cultured KB cells via folate receptor-mediated endocytosis. *J. Biol. Chem.*, **269** (5), 3198–3204.
201. Lee, R. J., Low, P. S. (1995) Folate-mediated tumor-cell targeting of liposome-entrapped doxorubicin in-vitro. *Biochim. Biophys. Acta*, **1233** (2), 134–144.
202. Zhao, X. B., Muthusamy, N., Byrd, J. C., Lee, R. J. (2007) Cholesterol as a bilayer anchor for PEGylation and targeting ligand in folate-receptor-targeted liposomes. *J. Pharm. Sci.*, **96** (9), 2424–2435.
203. Chiu, S. J., Marcucci, G., Lee, R. J. (2006) Efficient delivery of an antisense oligodeoxyribonucleotide formulated in folate receptor-targeted liposomes. *Anticancer Res.*, **26** (2A), 1049–1056.
204. Brynestad, K., Babbitt, B., Huang, L., Rouse, B. T. (1990) Influence of peptide acylation, liposome incorporation, and synthetic immunomodulators on the immunogenicity of a 1-23 peptide of glycoprotein-D of herpes-simplex virus - implications for subunit vaccines. *J. Virol.*, **64** (2), 680–685.
205. Mastrobattista, E., Koning, G. A., L. van Bloois, Filipe, A. C. S., Jiskoot, W., Storm, G. (2002) Functional characterization of an endosome-disruptive peptide and its application in cytosolic delivery of immunoliposome-entrapped proteins. *J. Biol. Chem.*, **277** (30), 27135–27143.
206. Harata, M., Soda, Y., Tani, K., Ooi, J., Takizawa, T., Chen, M. H., *et al.* (2004) CD19-targeting liposomes containing imatinib efficiently kill Philadelphia chromosome-positive acute lymphoblastic leukemia cells. *Blood*, **104** (5), 1442–1449.
207. Pinnaduwage, P., Huang, L. (1992) Stable target-sensitive immunoliposomes. *Biochemistry*, **31** (11), 2850–2855.
208. Beugin, S., Edwards, K., Karlsson, G., Ollivon, M., Lesieur, S. (1998) New sterically stabilized vesicles based on nonionic surfactant, cholesterol, and poly(ethylene glycol)-cholesterol conjugates. *Biophys. J.*, **74** (6), 3198–3210.
209. Beugin-Deroo, S., Ollivon, M., Lesieur, S. (1998) Bilayer stability and impermeability of nonionic surfactant vesicles sterically stabilized by PEG-cholesterol conjugates. *J. Colloid Interface Sci.*, **202** (2), 324–333.
210. Tang, J., Zhang, L., Fu, H., Kuang, Q., Gao, H., Zhang, Z., *et al.* (2014) A detachable coating of cholesterol-anchored PEG improves tumor targeting of cell-penetrating peptide-modified liposomes. *Acta Pharm. Sin. B*, **4** (1), 67–73.
211. Yuda, T., Maruyama, K., Iwatsuru, M. (1996) Prolongation of liposome circulation time by various derivatives of polyethyleneglycols. *Biol. Pharm. Bull.*, **19** (10), 1347–1351.
212. Ho, R. J., Huang, L. (1985) Interactions of antigen-sensitized liposomes with immobilized antibody: a homogeneous solid-phase immunoliposome assay. *J. Immunol.*, **134** (6), 4035–4040.
213. Rakthong, P., Intaramat, A., Ratanabanangkoon, K. (2010) Luminol encapsulated liposome as a signal generator for the detection of specific antigen-antibody reactions and nucleotide hybridization. *Anal. Sci.*, **26** (7), 767–772.
214. Bally, M., Bailey, K., Sugihara, K., Grieshaber, D., Voros, J., Stadler, B. (2010) Liposome and lipid bilayer arrays towards biosensing applications. *Small*, **6** (22), 2481–2497.
215. Bally, M., Voros, J. (2009) Nanoscale labels: nanoparticles and liposomes in the development of high-performance biosensors. *Nanomedicine*, **4** (4), 447–467.
216. Barenholz, Y. (2001) Liposome application: problems and prospects. *Curr. Opin. Colloid Interface Sci.*, **6** (1), 66–77.
217. Jung, L. S., Shumaker-Parry, J. S., Campbell, C. T., Yee, S. S., Gelb, M. H. (2000) Quantification of tight binding to surface-immobilized phospholipid vesicles using surface plasmon resonance: binding constant of phospholipase A(2). *J. Am. Chem. Soc.*. **122**(17), 4177–4184.
218. Dusseiller, M. R., Niederberger, B., Stadler, B., Falconnet, D., Textor, M., Voros, J. (2005) A novel crossed microfluidic device for the precise positioning of proteins and vesicles. *Lab Chip*, **5** (12), 1387–1392.
219. Stora, T., Dienes, Z., Vogel, H., Duschl, C. (2000) Histidine-tagged amphiphiles for

the reversible formation of lipid bilayer aggregates on chelator-functionalized gold surfaces. *Langmuir*, **16** (12), 5471–5478.

220. Niemeyer, C. M., Sano, T., Smith, C. L., Cantor, C. R. (1994) Oligonucleotide-directed self-assembly of proteins: semisynthetic DNA–streptavidin hybrid molecules as connectors for the generation of macroscopic arrays and the construction of supramolecular bioconjugates. *Nucleic Acids Res.*, **22** (25), 5530–5539.
221. Benkoski, J. J., Höök, F. (2005) Lateral mobility of tethered vesicle–DNA assemblies. *J. Phys. Chem. B*, **109** (19), 9773–9779.
222. Städler, B., Falconnet, D., Pfeiffer, I., Höök, F., Vörös, J. (2004) Micropatterning of DNA-tagged vesicles. *Langmuir*, **20** (26), 11348–11354.
223. Pfeiffer, I., Höök, F. (2004) Bivalent cholesterol-based coupling of oligonucleotides to lipid membrane assemblies. *J. Am. Chem. Soc.*, **126** (33), 10224–10225.
224. Städler, B., Bally, M., Grieshaber, D., Vörös, J., Brisson, A., Grandin, H. M. (2006) Creation of a functional heterogeneous vesicle array via DNA controlled surface sorting onto a spotted microarray. *Biointerphases*, **1** (4), 142–145.
225. Bailey, K., Bally, M., Leifert, W., Vörös, J., McMurchie, T. (2009) G-protein coupled receptor array technologies: site directed immobilisation of liposomes containing the H1-histamine or M2-muscarinic receptors. *Proteomics*, **9** (8), 2052–2063.
226. Gribova, V., Auzely-Velty, R., Picart, C. (2012) Polyelectrolyte multilayer assemblies on materials surfaces: from cell adhesion to tissue engineering. *Chem. Mater.*, **24** (5), 854–869.
227. Zelikin, A. N. (2010) Drug releasing polymer thin films: new era of surface-mediated drug delivery. *ACS Nano*, **4** (5), 2494–2509.
228. Teo, B. M., Hosta-Rigau, L., Lynge, M. E., Stadler, B. (2014) Liposome-containing polymer films and colloidal assemblies towards biomedical applications. *Nanoscale*, **6** (12), 6426–6433.
229. Stadler, B., Price, A. D., Chandrawati, R., Hosta-Rigau, L., Zelikin, A. N., Caruso, F. (2009) Polymer hydrogel capsules: en route toward synthetic cellular systems. *Nanoscale*, **1** (1), 68–73.
230. Graf, N., Thomasson, E., Tanno, A., Vörös, J., Zambelli, T. (2011) Spontaneous formation of a vesicle multilayer on top of an exponentially growing polyelectrolyte multilayer mediated by diffusing poly-l-lysine. *J. Phys. Chem. B*, **115** (43), 12386–12391.
231. Hosta-Rigau, L., Chandrawati, R., Saveriades, E., Odermatt, P. D., Postma, A., Ercole, F., *et al.* (2010) Noncovalent liposome linkage and miniaturization of capsosomes for drug delivery. *Biomacromolecules*, **11** (12), 3548–3555.
232. Lynge, M. E., Baekgaard Laursen, M., Hosta-Rigau, L., Jensen, B. E. B., Ogaki, R., Smith, A. A. A., *et al.* (2013) Liposomes as drug deposits in multilayered polymer films. *ACS Appl. Mater. Interfaces*, **5** (8), 2967–2975.
233. Lynge, M. E., Teo, B. M., Laursen, M. B., Zhang, Y., Stadler, B. (2013) Cargo delivery to adhering myoblast cells from liposome-containing poly(dopamine) composite coatings. *Biomater. Sci.*, **1** (11), 1181–1192.
234. Jensen, B. E. B., Hosta-Rigau, L., Spycher, P. R., Reimhult, E., Stadler, B., Zelikin, A. N. (2013) Lipogels: surface-adherent composite hydrogels assembled from poly(vinyl alcohol) and liposomes. *Nanoscale*, **5** (15), 6758–6766.
235. Zhang, Y., Ruder, W. C., LeDuc, P. R. (2008) Artificial cells: building bioinspired systems using small-scale biology. *Trends Biotechnol.*, **26** (1), 14–20.
236. Kisak, E. T., Coldren, B., Evans, C. A., Boyer, C., Zasadzinski, J. A. (2004) The vesosome – a multicompartment drug delivery vehicle. *Curr. Med. Chem.*, **11** (2), 199–219.
237. Paleos, C. M., Tsiourvas, D., Sideratou, Z., Pantos, A. (2013) Formation of artificial multicompartment vesosome and dendrosome as prospected drug and gene delivery carriers. *J. Controlled Release*, **170** (1), 141–152.
238. Chiu, H.-C., Lin, Y.-W., Huang, Y.-F., Chuang, C.-K., Chern, C.-S. (2008) Polymer vesicles containing small vesicles within interior aqueous compartments and pH-responsive transmembrane channels. *Angew. Chem. Int. Ed.*, **47** (10), 1875–1878.
239. Peters, R. J. R. W., Marguet, M., Marais, S., Fraaije, M. W., J. C. M. van Hest, Lecommandoux, S. (2014) Cascade reactions

in multicompartmentalized polymersomes. *Angew. Chem. Int. Ed.*, **53** (1), 146–150.

240. Siti, W., H.-P. M. de Hoog, Fischer, O., Shan, W. Y., Tomczak, N., Nallani, M., *et al.* (2014) An intercompartmental enzymatic cascade reaction in channel-equipped polymersome-in-polymersome architectures. *J. Mater. Chem. B*, **2** (18), 2733–2737.
241. De Geest, B. G., S. De Koker, Immesoete, K., Demeester, J., S. C. De Smedt, Hennink, W. E. (2008) Self-exploding beads releasing microcarriers. *Adv. Mater.*, **20** (19), 3687–3691.
242. Kulygin, O., Price, A. D., Chong, S.-F., Städler, B., Zelikin, A. N., Caruso, F. (2010) Subcompartmentalized polymer hydrogel capsules with selectively degradable carriers and subunits. *Small*, **6** (14), 1558–1564.
243. Driever, C. D., Mulet, X., Johnston, A. P. R., Waddington, L. J., Thissen, H., Caruso, F., *et al.* (2011) Converging layer-by-layer polyelectrolyte microcapsule and cubic lyotropic liquid crystalline nanoparticle approaches for molecular encapsulation. *Soft Matter*, **7** (9), 4257–4266.
244. Lomas, H., Johnston, A. P. R., Such, G. K., Zhu, Z., Liang, K., Van Koeverden, M. P., *et al.* (2011) Polymersome-loaded capsules for controlled release of DNA. *Small*, **7** (14), 2109–2119.
245. Hosta-Rigau, L., Städler, B. (2013) Subcompartmentalized Systems Towards Therapeutic Cell Mimicry, in: Regenerative Medicine, Artificial Cells and Nanomedicine (ed. Chang, T.M.S.), World Scientific Publishing Co. Pte, Singapore, Chapter 12.
246. Chandrawati, R., Staedler, B., Postma, A., Connal, L. A., Chong, S.-F., Zelikin, A. N., *et al.* (2009) Cholesterol-mediated anchoring of enzyme-loaded liposomes within disulfide-stabilized polymer carrier capsules. *Biomaterials*, **30** (30), 5988–5998.
247. Hosta-Rigau, L., Städler, B., Yan, Y., Nice, E. C., Heath, J. K., Albericio, F., *et al.* (2010) Capsosomes with multilayered subcompartments: assembly and loading with hydrophobic cargo. *Adv. Funct. Mater.*, **20** (1), 59–66.
248. Chandrawati, R., Odermatt, P. D., Chong, S.-F., Price, A.D., Städler, B., Caruso, F. (2011) Triggered cargo release by encapsulated enzymatic catalysis in capsosomes. *Nano Lett.*, **11** (11), 4958–4963.
249. Chandrawati, R., Hosta-Rigau, L., Vanderstraaten, D., Lokuliyana, S.A., Städler, B., Albericio, F., Caruso, F. (2010) Engineering advanced capsosomes: maximizing the number of subcompartments, cargo retention, and temperature-triggered reaction. *ACS Nano*, **4** (3), 1351–1361.
250. Hosta-Rigau, L., Zhang, Y., Goldie, K.N., Städler, B. (2014) Confined multiple enzymatic (Cascade) reactions within poly(dopamine)-based capsosomes. *ACS Appl. Mater. Interfaces*, **6** (15), 12771–12779.
251. Brown, M.S., Goldstein, J.L. (1986) A receptor-mediated pathway for cholesterol homeostasis. *Science*, **232** (4746), 34–47.

9
Nanoparticle Conjugates for Small Interfering RNA Delivery

Timothy L. Sita and Alexander H. Stegh
Northwestern University Ken and Ruth Davee Department of Neurology, The Northwestern Brain Tumor Institute, The Robert H. Lurie Comprehensive Cancer Center, The International Institute for Nanotechnology, 303 East Superior, Chicago, IL 60611, USA

Translational Nanomedicine. First Edition. Edited by Robert A. Meyers.

Keywords

Enhanced permeation and retention (EPR) effect
The phenomenon by which nanoparticles accumulate in tumor tissue due to both increased porosity of tumor vasculature and decreased lymphatic drainage

Mononuclear phagocyte system (MPS)
Previously known as the reticuloendothelial system (RES), which consists of monocytes and macrophages in the liver, spleen, bone marrow, blood, and lymph nodes, and is required for the body's defense against microorganisms and foreign materials

Proton sponge effect (PSE)
The phenomenon by which siRNA carriers become protonated in the acidic endosomal environment, leading to an influx of additional protons and chloride ions, and resulting in an osmotic imbalance that draws water into the endosome, causing it to burst and release its contents into the cytosol

RNA-induced silencing complex (RISC)
A protein complex mediating the binding and unwinding of double-stranded siRNA within the cytoplasm and the subsequent sequence-specific recognition and degradation of target mRNA

RNA interference (RNAi)
A fundamental pathway in eukaryotic cells, by which RNA oligonucleotide sequences induce the degradation of mRNA containing a complementary sequence

Small interfering RNA (siRNA)
RNA fragments approximately 21–23 nucleotides long that are capable of inducing sequence-specific degradation of complementary mRNA

Spherical nucleic acids (SNAs)
Nanoparticles for siRNA delivery that consist of a 13 nm gold nanoparticle core functionalized with a corona of thiolated siRNA duplexes and polyethylene glycol

Stable nucleic acid–lipid particles (SNALPs)
Nanoparticles for siRNA delivery, which have a lipid bilayer composed of cationic lipids and neutral fusogenic lipids, and which are coated with polyethylene glycol to provide a neutral surface

1

As a therapeutic strategy, small interfering RNA (siRNA) has a remarkable potential to treat genetic diseases driven by mutated or aberrantly expressed genes, including cancer, inflammatory conditions, neurodegenerative disorders, and viral infections. Since the discovery of the RNA interference (RNAi) pathway in 1998, a multitude of siRNA delivery methods have been designed and tested. However, siRNA-based therapeutics are currently limited by the inability to safely and robustly deliver nucleic acids to target cells and tissues. With the development of nanocarriers composed of organic (lipids, liposomes, conjugated polymers) and inorganic (iron oxide, gold) materials, nanotechnology has emerged as a discipline that offers one of the most powerful solutions to enable the stable and safe delivery of siRNA oligonucleotides. Many of these nanocarriers demonstrate acceptable safety profiles, enhance the *in-vivo* stability of siRNA, promote robust tissue penetration, and can be modified with a targeting ligand to allow for tissue-specific uptake. In this chapter, the challenges of delivering siRNA are reviewed, and the most important advances in the development of nanoparticle-based siRNA delivery systems currently in preclinical and clinical development will be highlighted.

1
The RNA Interference Pathway

In 1998, Fire and Mello demonstrated siRNA-based gene silencing in *Caenorhabditis elegans*, a discovery that earned them the Nobel Prize in Physiology or Medicine [1]. Tuschl and colleagues then confirmed the RNA interference (RNAi) pathway in mammalian cells [2], and the idea of harnessing RNAi-mediated gene silencing for therapeutic purposes has garnered enormous attention from scientists ever since [3–5]. The ability to knockdown any gene of interest opened up an entirely new way to combat diseases that were minimally responsive to traditional therapeutics, including (but not limited to) cancer, hereditary disorders, heart disease, inflammatory conditions, and viral infections [6–9]. These diseases are often driven by genetic aberrations, most of which are considered "undruggable" by conventional pharmacological strategies, foremost small molecules and biotherapeutic antibodies. RNAi, however, holds promise as a novel tool for precision medicine, as it is able to efficiently and persistently silence deregulated or mutated genes through the use of customizable siRNA oligonucleotides [10, 11].

The RNAi pathway inhibits gene expression through the selective cleavage of mRNA (Fig. 1) [12–14]. The pathway is activated by the presence of double-stranded RNA (dsRNA) in the cytoplasm. The cytoplasmic enzyme Dicer cleaves longer dsRNA into smaller segments, typically 21–23 nucleotides in length. These siRNA segments are then recognized and loaded into the RNA-induced silencing complex (RISC), activating the complex. Once activated, antisense strands bind complementary mRNA. Argonaute 2, an RNA endonuclease resident in the activated RISC complex, subsequently cleaves mRNA, thereby inhibiting its translation. Because the RISC complex is recycled and cleaves mRNA continuously, knockdown is persistent, lasting between three and seven days in dividing cells, and up to three to four weeks in nondividing cells [15].

However, a number of barriers to effective delivery are encountered when naked siRNA is systemically injected, limiting its therapeutic potential *in-vivo*. Unmodified

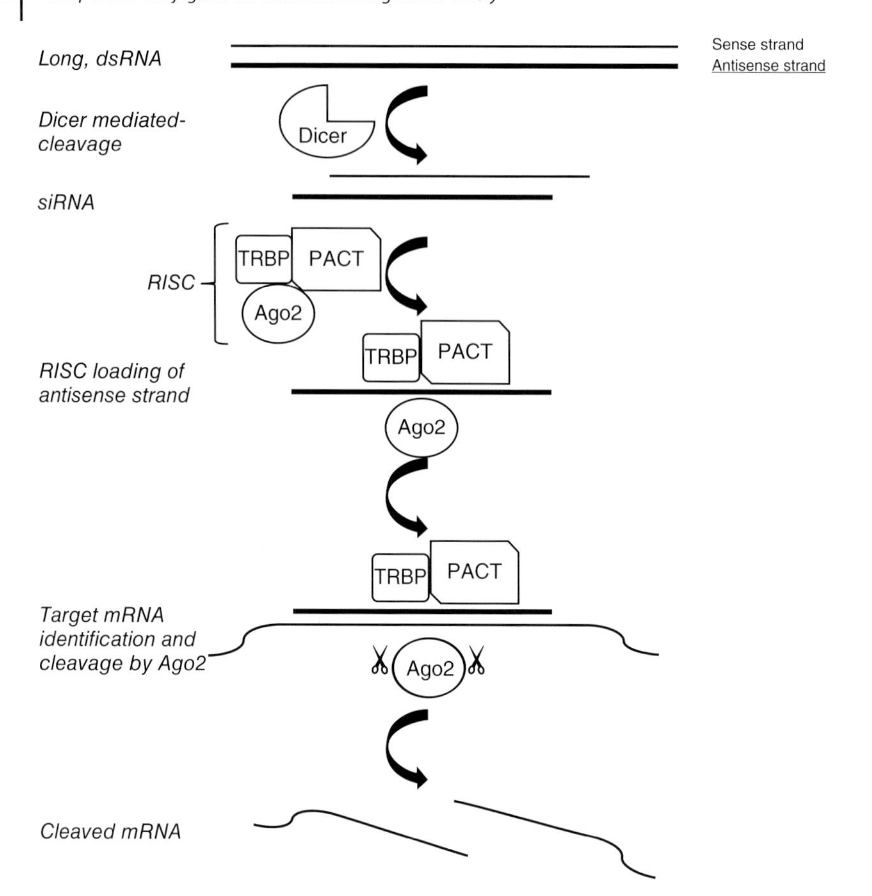

Fig. 1 The RNA interference pathway. Long, dsRNA is cleaved by the enzyme Dicer into 21- to 23-nucleotide siRNA segments. These segments are loaded into the RISC complex, which includes Protein ACTivator of the interferon-induced protein kinase (PACT), transactivation response RNA binding protein (TRBP) and Argonaute 2 (Ago2). The sense siRNA strand is degraded and the activated RISC complex recognizes mRNA complementary to the antisense siRNA strand. Target mRNA is then cleaved by Ago2, prohibiting its translation.

siRNA is subject to rapid serum cleavage, renal clearance, and distribution to nontarget organ systems. Furthermore, unmodified siRNA shows poor intracellular uptake and has been shown to trigger a cellular immune response [16–18]. The *in-vivo* obstacles to siRNA delivery, as well as the potential solutions afforded by siRNA nanocarriers currently in development, are discussed in the following sections.

2 Limitations of Unmodified siRNA Delivery and Carrier Design Considerations

2.1 Serum Stability

The phosphodiester backbone of unmodified siRNA makes it sensitive to rapid hydrolysis by serum RNases. On average, intravenously administered siRNA has an approximate serum half-life of 15 min, with some sequences hydrolyzed and rendered nonfunctional in as little as 1 min or less [19, 20]. Such rapid degradation renders therapeutic benefit of systemically delivered unmodified siRNA unlikely in humans. However, chemical modifications have demonstrated efficacy in prolonging the *in-vivo* stability of intravenously administered siRNA without affecting RISC complexation. Examples include modifications with 2′-*O*-methyl, 2′-fluoro, and 2′-*O*-methoxyethyl groups at the 2′ hydroxyl in the sugar ring, as well as at the phosphate backbone with phosphorothioate and boranophosphate (Fig. 2) [21–23]. These modifications are capable of disguising binding motifs recognized by RNases and significantly slowing the rate of siRNA degradation.

2.2 Immunogenicity

Numerous studies have demonstrated activation of the innate immune system following the systemic administration of siRNA [24]. Although this immunostimulatory potential may be advantageous in circumstances where a proinflammatory environment is desired (i.e., recent viral inoculation) [25], it is usually an unwanted outcome. Some of these immunostimulatory effects are sequence-specific, as siRNA sequences with uridine-, guanosine-, and uridine-rich regions have been shown to more robustly activate the immune system [26]. Other immunogenic responses are independent of the siRNA sequence. In particular, interferon-α induction via activation of toll-like receptor (TLR) 3 occurs independently of siRNA sequence [27].

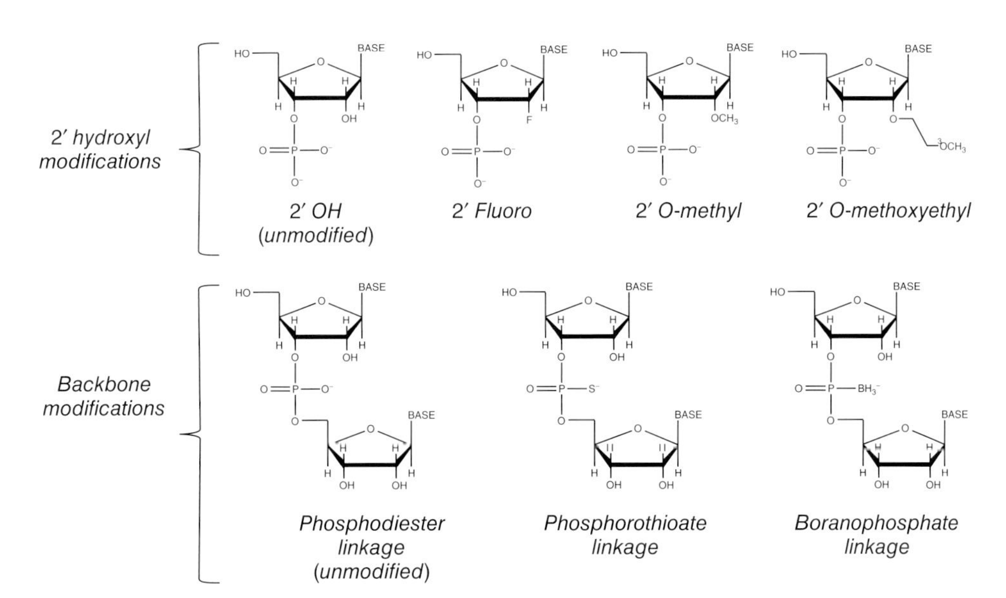

Fig. 2 Chemical modifications of siRNA. Common sites for modification include the 2′ hydroxyl in the sugar ring and the phosphate backbone. These modifications disguise binding motifs recognized by RNases.

Some of the chemical modifications intended to prolong *in-vivo* stability also reduce immune responses triggered by the intravenous administration of siRNA. The incorporation of as few as two 2′-*O*-methyl groups into a siRNA duplex can be sufficient to prevent TLR7 activation, as these groups serve as competitive TLR7 inhibitors [28, 29]. Other sugar ring modifications, such 2′-fluoro groups, have also been shown to confer protection against immune activation by abrogating the interaction between the siRNA and TLR7 [22]. Hence, most siRNA therapeutics that have reached clinical development utilize chemically modified siRNA to increase stability and reduce immunogenicity.

2.3 Renal Clearance

While chemically modified siRNA increases serum stability and reduces immunogenicity, it does not effectively prevent renal clearance, which is largely a function of molecular size. Physical filtration of the blood occurs at the basement membrane of the renal glomerulus through pores roughly 8 nm wide, allowing the passage of water and other small molecules into the urine for subsequent excretion [30]. This includes naked siRNA, which has been observed to pass freely through the glomerular basement membrane [31]. With the exception of urinary tract targets, renally cleared siRNA is not delivered to its target tissue and fails to engage the RNAi pathway at the desired organ site. When designing delivery systems for siRNA, a minimum size of 10–20 nm is typically required to avoid renal clearance [32, 33]. Molecular weight is an additional consideration, with polymers of >40 kDa more likely to be retained in circulation rather than be passed through the glomerular filtration barrier into urine [20].

2.4 Biodistribution

In addition to evading renal clearance, systemically delivered siRNA has to selectively reach its target site while avoiding off-target effects in nontarget tissues. Both passive and active targeting mechanisms influence the biodistribution of systemically administered siRNA. To reach organ sites, siRNA must passively leave the bloodstream by traversing fenestrations in the endothelium. Typical endothelial fenestrations are 10–50 nm in diameter, which prohibits the extravasation of larger molecules [34]. However, extravasation occurs more readily in organs such as the liver and spleen, as well as many solid tumors, which are characterized by more discontinuous endothelia [35]. During circulation and following extravasation from the bloodstream, molecules – particularly those more than 100 nm in size – are subject to engulfment by macrophages of the mononuclear phagocytic system (MPS) in the blood, liver, and spleen [36]. Thus, a suggested maximal size of 100 nm for siRNA carrier systems has been proposed to minimize engulfment by the MPS [37].

In solid tumor tissue, passive targeting can be increased with nanocarriers via the enhanced permeation and retention (EPR) effect – a phenomenon by which circulating nanoparticles preferentially accumulate in tumor tissue [38]. Uncontrolled tumor angiogenesis results in poorly formed and leaky tumor vasculature. Over time, nanoparticles accumulate due to this leaky vasculature and are retained, as the tumor microenvironment is characterized by decreased lymphatic drainage.

To achieve optimal permeation with concomitant retention in tumor tissue, molecules less than 200 nm in diameter are ideal, further supporting the notion that a particle size of 100 nm is ideal for tissue delivery [36].

In addition to size-based considerations, both carrier charge and shape influence the passive biodistribution of siRNA. For example, increasing the carrier charge through the introduction of 1,2-dioleoyl-3-tri-methylammonium-propane (DOTAP) has been shown to decrease siRNA accumulation in the spleen and normal vasculature, while increasing accumulation in the liver and tumor vasculature [39]. Furthermore, multiple studies have suggested that nonspherical particles – including filomicelles, discoidal, and rod-shaped particles – promote a greater tumor accumulation and reduced uptake via the MPS compared to spherical particles, which suggests that nonspherical carriers have a physiologic advantage for siRNA delivery [40–42].

For active targeting, siRNA and siRNA carriers may be conjugated with a variety of ligands to enable tissue-specific uptake. For example, the galactose derivative *N*-acetylgalactosamine (GalNAc) has been studied extensively in hepatocyte targeting. GalNAc has a strong affinity for the asialoglycoprotein receptor (ASPGR), which is highly expressed in hepatocytes [43]. Additionally, ligands with the capacity to bind the transferrin receptor (TfR) are frequently conjugated to nanoparticles for targeting purposes. While TfR is ubiquitously expressed at low levels in most human tissues, it is overexpressed on most cancer cell surfaces, and its expression on malignant cells has been shown to correlate with tumor grade and rate of progression [44]. Furthermore, TfR is overexpressed on brain capillary endothelial cells and can facilitate increased blood–brain barrier (BBB) penetration via receptor-mediated transcytosis [45]. Moreover, an abundance of other tissue-specific receptors, small molecules, antibodies, proteins, peptides and aptamers have been employed to increase target tissue uptake. These include small molecules such as folic acid, clinically relevant monoclonal antibodies (e.g., rituximab), and aptamers that are capable of specifically binding prostate-specific membrane antigen (PSMA). These targeting moieties are discussed elsewhere in detail by Sanna *et al.* [46].

2.5 Intracellular Uptake

siRNA must first cross the plasma membrane in order to engage the RNAi machinery in the cytosol. Naked, linear siRNA cannot freely diffuse across this membrane due to its negative charge and relatively large size; hence, it must be either coupled with a delivery agent able to cross or fuse with the membrane, or be presented in a spherical configuration that is recognized by scavenger receptors and internalized via receptor-mediated endocytosis (see Sect. 3.3) [47]. Common classes of delivery agents conjugated to siRNA include lipid moieties, cell-penetrating peptides, and small molecules [17]. The vast majority of these materials mediate intracellular uptake via clathrin- or lipid raft/caveolin-mediated endocytosis [48].

2.6 Endosomal Escape

As receptor-mediated endocytosis is the most common route by which siRNA can enter the cytoplasm, it must first escape the endosomal pathway to gain access to

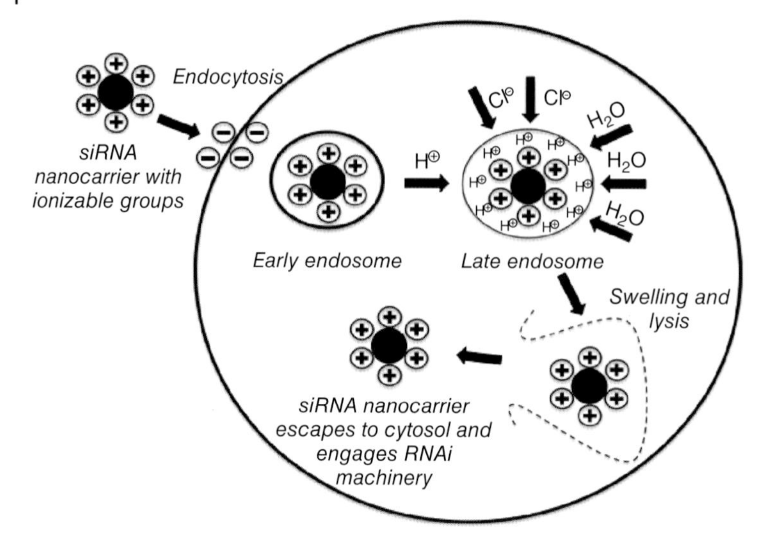

Fig. 3 The proton sponge effect. siRNA nanocarriers with ionizable groups are protonated in the acidic environment of the endosome, leading to an influx of additional protons and chloride ions. This results in an osmotic imbalance that draws water into the endosome, causing swelling and lysis of the endosome and allowing siRNA to escape to the cytosol, where it can engage the RNAi pathway.

the RNAi machinery. In the endosomal pathway, extracellular material is entrapped in membrane-bound vesicles, which then fuse with early endosomes, mature into increasingly acidic late endosomes, and finally fuse with lysosomes. If siRNA is unable to escape the endosomal pathway, it is degraded by lysosomal RNases. Certain types of siRNA carrier can promote endosomal escape, although the precise mechanism of escape is often elusive. One theory, based on the "proton sponge effect" (PSE), suggests that siRNA carriers harboring amine groups with pK_a values ranging from 5 to 7, such as those found on polyethylenimine (PEI) and β-amino esters, are protonated as the endosome becomes increasingly acidic [49, 50]. This protonation is followed by an influx of additional protons and chloride ions, resulting in an osmotic imbalance that draws water into the endosome, causing it to burst and release its contents into the cytosol (Fig. 3) [51]. Additionally, pH-sensitive groups on lipid carriers, such as the citraconic anhydride-modified phospholipid 1,2-dioleoyl-3-phosphatidylethanolamine (C-DOPE), are designed to degrade under acidic conditions and induce structural transformations that promote the endosomal escape of siRNA [17, 52].

3 siRNA Nanocarriers in Clinical Development

Currently, there are 24 different siRNA-based therapeutics in 43 different clinical trials (Table 1). The first siRNA therapeutics to enter clinical trials were based on the local delivery of naked siRNA or chemically modified siRNA [10]. These included intravitreal delivery for age-related macular degeneration and intranasal delivery to treat respiratory syncytial virus (RSV) [53–59]. Subsequent generations of siRNA therapeutics were designed for systemic delivery, and are based on a variety of nanoparticle carriers capable of overcoming the delivery challenges described above.

3.1
Conjugate Delivery Systems

Conjugate delivery systems (CDSs) are the simplest of all siRNA delivery materials in design, consisting of siRNA directly conjugated to either polymers, peptides, antibodies, aptamers, and small molecules [60]. CDSs that are composed of siRNA attached to cholesterol and other lipid moieties have demonstrated efficacy *in-vivo*. One particular CDS system, termed Dynamic PolyConjugates (DPCs) technology, consists of siRNA conjugated to an amphipathic, membrane-active polymer, a shielding polyethylene glycol (PEG) that can mask the polymer until it has reached the endosome, and a GalNAc hepatocyte targeting ligand [61]. DPCs were shown to achieve a functional knockdown of two different siRNA targets in the liver – apolipoprotein B (ApoB) and peroxisome proliferator-activated receptor alpha (PPARα) [61]. The quantification of target mRNA levels, using quantitative real-time reverse transcriptase polymerase chain reaction (qRT-PCR), in murine liver tissue indicated that DPC silenced *ApoB* expression up to 74% and *PPARα* expression up to 61%, as compared to mice treated with control siRNA. Furthermore, serum liver enzymes and cytokine levels were not statistically different from saline-treated mice, indicating that DPCs were well tolerated. These constructs are currently undergoing Phase I and II trials for hepatitis B infection (Table 1).

3.2
Gold Nanoparticles

Gold nanoparticles (AuNPs) serve as excellent siRNA vectors due to their biocompatibility, tunable size, and ease of functionalization. Although AuNPs have not yet been approved by the United States Food and Drug Administration (FDA) for use in humans, gold salts have been used for many years in the clinic as treatments for arthritis; indeed, recent preclinical studies of AuNPs conjugated with siRNA have indicated little to no toxicity in mice [62–67]. Additionally, monodisperse AuNPs ranging from 1 to 150 nm in diameter can be reproducibly generated by reducing gold salts in the presence of stabilizing agents that prevent the AuNPs from aggregating. For particles ranging from ~1.5 to 6 nm in diameter, the one-pot protocol developed by Schiffrin and colleagues can be utilized, in which tetrachloroaurate ($AuCl_4^-$) is reduced with sodium borohydride ($NaBH_4$) in the presence of an alkanethiol [68]. For particles with core sizes ranging from ~10 to 150 nm, chloroauric acid ($HAuCl_4$) can be reduced with sodium citrate ($Na_3C_6H_5O_7$) [69–73].

AuNPs can be easily modified to carry siRNA through both covalent and noncovalent interactions. The strong metal–ligand interaction between gold and sulfur (S-Au bond) allows for the covalent attachment of almost any thiolated biomolecule (including siRNA) to the surface of AuNPs [74–76]. In the past, investigators have taken full advantage of this gold–thiol chemistry to develop multimodal drug delivery constructs, using the S-Au bond as a scaffold for the attachment of targeting ligands (e.g., folic acid, anti-HER2 antibody), PEG moieties for increased colloidal stability and blood circulation times, and

Tab. 1 siRNA-based compounds in clinical trials.

Compound	*Target*	*Delivery system*	*Condition*	*Phase*	*Status*	*Sponsor*	*ClinicalTrials.gov identifier*
AGN211745 (previously known as Sirna-027)	VEGFR1	Naked siRNA	Age-related macular degeneration, choroidal neovascularization	II	Terminated	Allergan	NCT00395057
ALN-RSV01	RSV nucleocapsid	Naked siRNA	Respiratory syncytial virus infections	II	Completed	Alnylam Pharmaceuticals	NCT00496821
ALN-RSV01	RSV nucleocapsid	Naked siRNA	Respiratory syncytial virus infections	II	Completed	Alnylam Pharmaceuticals	NCT00658086
ALN-RSV01	RSV nucleocapsid	Naked siRNA	Respiratory syncytial virus infections	II	Completed	Alnylam Pharmaceuticals	NCT01065935
Bevasiranib	VEGF	Naked siRNA	Wet age related-macular degeneration	II	Completed	Opko Health, Inc.	NCT00259753
Bevasiranib	VEGF	Naked siRNA	Diabetic macular edema	II	Completed	Opko Health, Inc.	NCT00306904
I5NP	p53	Naked siRNA	Kidney injury, acute renal failure	I	Completed	Quark Pharmaceuticals	NCT00554359
I5NP	p53	Naked siRNA	Delayed graft function, kidney transplant complications	I	Active, not recruiting	Quark Pharmaceuticals	NCT00802347
PF-04523655	RTP801 (proprietary target)	Naked siRNA	Choroidal neovascularization, diabetic retinopathy, diabeter macular edema	II	Active, not recruiting	Quark Pharmaceuticals	NCT01445899
QPI-1007	CASP2	Naked siRNA	Optic atrophy, non-arteritic anterior ischemic optic neuropathy	I	Completed	Quark Pharmaceuticals	NCT01064505
SYL040012	ADRB2	Naked siRNA	Ocular hypertension, glaucoma	I, II	Completed	Sylentis, S.A.	NCT01227291
SYL040012	ADRB2	Naked siRNA	Ocular hypertension, open-angle glaucoma	II	Completed	Sylentis, S.A.	NCT01739244
SYL1001	TRPV1	Naked siRNA	Ocular pain, dry-eye syndrome (in healthy volunteers)	I	Completed	Sylentis, S.A.	NCT01438281
SYL1001	TRPV1	Naked siRNA	Ocular pain, dry-eye syndrome	I, II	Recruiting	Sylentis, S.A.	NCT01776658
TD101	K6a	Naked siRNA	Pachyonychia congenita	I	Completed	Pachyonychia Congenita Project	NCT00716014

ALN-AT3SC	Antithrombin	siRNA-GalNAc conjugate	Hemophilia A, Hemophilia B	I	Recruiting	Alnylam Pharmaceuticals	NCT02035605
ALN-TTRsc	Transthyretin	siRNA-GalNAc conjugate	Transthyretin-mediated amyloidosis	I	Recruiting	Alnylam Pharmaceuticals	NCT01814839
ALN-TTRsc	Transthyretin	siRNA-GalNAc conjugate	Transthyretin-mediated amyloidosis	II	Recruiting	Alnylam Pharmaceuticals	NCT01981837
ARC-520	Conserved regions of HBV	Dynamic polyconjugates	Chronic hepatitis B	I	Recruiting	Arrowhead Research Corporationq	NCT01872065
ARC-520	Conserved regions of HBV	Dynamic polyconjugates	Chronic hepatitis B	II	Recruiting	Arrowhead Research Corporationq	NCT02065336
RXi-109	CTGF	Self-delivering RNAi compound	Cicatrix, scar prevention	I	Active, not recruiting	RXi Pharmaceuticals, Corp.	NCT01640912
RXi-109	CTGF	Self-delivering RNAi compound	Cicatrix, scar prevention	I	Active, not recruiting	RXi Pharmaceuticals, Corp.	NCT01780077
RXi-109	CTGF	Self-delivering RNAi compound	Hypertrophic scar	II	Recruiting	RXi Pharmaceuticals, Corp.	NCT02030275
RXi-109	CTGF	Self-delivering RNAi compound	Keloid	II	Recruiting	RXi Pharmaceuticals, Corp.	NCT02079168
siG12D LODER	KRAS	LODER polymer	Pancreatic cancer	I	Active, not recruiting	Silenseed Ltd.	NCT01188785
siG12D LODER	KRAS	LODER polymer	Unresectable locally advanced pancreatic cancer	II	Not yet recruiting	Silenseed Ltd.	NCT01676259
CALAA-01	RRM2	CDP NP	Solid tumors	I	Terminated	Calando Pharmaceuticals	NCT00689065
ALN-PCS02	PCSK9	Lipid nanoparticle	Hypercholesterolemia	I	Completed	Alnylam Pharmaceuticals	NCT01437059
ALN-VSP02	KSP/VEGF	Lipid nanoparticle	Solid tumors	I	Completed	Alnylam Pharmaceuticals	NCT00882180
ALN-VSP02	KSP/VEGF	Lipid nanoparticle	Solid tumors	I	Completed	Alnylam Pharmaceuticals	NCT01158079

(continued overleaf)

Tab. 1 (*Continued.*)

Compound	*Target*	*Delivery system*	*Condition*	*Phase*	*Status*	*Sponsor*	*ClinicalTrials.gov identifier*
ND-L02-s0201	HSP47	Lipid nanoparticle	Healthy subjects	I	Completed	Nitto Denko Corporation	NCT01858935
Patisiran (ALN-TTR02)	Transthyretin	Lipid nanoparticle	Transthyretin-mediated amyloidosis	I	Completed	Alnylam Pharmaceuticals	NCT01559077
Patisiran (ALN-TTR02)	Transthyretin	Lipid nanoparticle	Transthyretin-mediated amyloidosis	II	Active, not recruiting	Alnylam Pharmaceuticals	NCT01617967
Patisiran (ALN-TTR02)	Transthyretin	Lipid nanoparticle	Transthyretin-mediated amyloidosis	III	Recruiting	Alnylam Pharmaceuticals	NCT01960348
Patisiran (ALN-TTR02)	Transthyretin	Lipid nanoparticle	Transthyretin-mediated amyloidosis	II	Recruiting	Alnylam Pharmaceuticals	NCT01961921
Patisiran (ALN-TTR02)	Transthyretin	Lipid nanoparticle	Transthyretin-mediated amyloidosis	I	Recruiting	Alnylam Pharmaceuticals	NCT02053454
Atu027	PKN3	Cationic liposome	Advanced solid tumors	I	Completed	Silence Therapeutics AG	NCT00938574
Atu027	PKN3	Cationic liposome	Advanced or metastatic pancreatic cancer	I, II	Recruiting	Silence Therapeutics AG	NCT01808638
siRNA-EphA2-DOPC	EphA2	Neutral liposome	Advanced cancers	I	Not yet recruiting	M.D. Anderson Cancer Center	NCT01591356
PRO-040201	ApoB	SNALP	Hypercholesterolemia	I	Terminated	Tekmira Pharmaceuticals Corporation	NCT00927459
TKM-080301	PLK1	SNALP	Neuroendocrine tumors, adrenocortical carcinoma	I, II	Recruiting	Tekmira Pharmaceuticals Corporation	NCT01262235
TKM-080301	PLK1	SNALP	Primary or secondary liver cancer	I	Completed	National Cancer Institute	NCT01437007
TKM-100802	VP24, VP35, Zaire Ebola L-polymerase	SNALP	Ebola virus infection	I	Recruiting	Tekmira Pharmaceuticals Corporation	NCT02041715

LODER, LOcal Drug EluterR.

gadolinium for magnetic resonance imaging (MRI) [77–80]. The addition of thiolated hydrophobic drugs (e.g., paclitaxel) to AuNPs has even been used as a method to increase drug solubility in blood and thus to enhance drug efficacy [81, 82]. Noncovalent AuNP–siRNA nanoconjugates are generated by electrostatically complexing negatively charged siRNA with cationic AuNPs and/or cationic polymers. The layer-by-layer assembly of positively charged PEI with siRNA on AuNPs is another technique that has been used extensively, resulting in particles that exhibit both efficient siRNA transfection and high rates of endosomal escape [83, 84]. Other groups have demonstrated the successful delivery of siRNA using positively charged biomolecules, complexing protamine, dendrons, and cysteamine to siRNA on AuNPs [62, 64, 66].

AuNPs also benefit from a distinct localized surface plasmon resonance (LSPR), which is the collective oscillation of electrons in a solid or liquid stimulated by incident light. Resonance occurs when the frequency of incident light photons matches the natural frequency of surface electrons oscillating against the restoring force of positive nuclei. Depending on the size and shape of the AuNPs, LSPR occurs in the visible and near-infrared (NIR) range of the spectrum [85]. This property opens up a multitude of applications in imaging, diagnosis, and photothermal therapy [73, 79, 86–89]. For example, Eghtedari *et al.* used NIR light to detect gold nanorods (50 nm × 15 nm) in mouse tissue (4 cm depth) using an optoacoustic method [90]. Likewise, Shim *et al.* took advantage of the LSPR of AuNPs in conjunction with siRNA delivery, by combining gene therapy and stimuli-responsive optical imaging with 15 nm AuNPs conjugated to acid-degradable ketalized linear polyethylenimine (KL-PEI) electrostatically complexed with siRNA [91]. In this construct, *N*-succinimidyl 3-(2-pyridyldithio)-propionate (SPDP) was directly conjugated to the AuNP surface via pyridyldithio groups, while the amine-reactive *N*-succinimidyl (NHS) groups on SPDP reacted with KL-PEI/siRNA polyplexes. In this way, Shim and colleagues showed that in a mildly acidic environment (e.g., tumor microenvironment), the acid-sensitive PEI/siRNA complexes would disassociate from the AuNPs, resulting in both gene knockdown and changes in optical signals, including diminished scattering intensity, increased variance of Doppler frequency, and blue-shifted ultraviolet absorbance.

Given their straightforward synthesis and conjugation protocols, biologically inert nature, and multimodal imaging and gene delivery capabilities, the preclinical outlook on AuNP–siRNA constructs is particularly encouraging [92]. Although clinical trials with AuNPs for siRNA delivery have not yet commenced, AuNPs conjugated to PEG and tumor necrosis factor alpha (TNFα) have already been investigated in Phase I trials for the treatment of cancer. It is very likely, therefore, that AuNP–siRNA constructs will shortly be enrolled in clinical trials [93, 94].

3.3 Spherical Nucleic Acids

In 1996, Mirkin and colleagues pioneered the spherical nucleic acid (SNA) platform, which originally consisted of a 13-nm citrate-capped AuNP core functionalized with thiolated DNA [95]. Over time, many variations of the original structure have been explored (e.g., adding fluorophores, modifying oligonucleotide sequence, and length), and these constructs have proven to be incredibly useful in a vast amount of applications, including (but not limited to) the *in-vitro* biodetection of mRNAs (more than 1500 different RNA detection assays are currently marketed by Millipore as SmartFlare™ technology), DNA-based materials synthesis and engineering [96, 97], and RNAi-mediated gene regulation [67, 98, 99]. Subsequent generations of SNAs intended for *in-vivo* gene regulation exchanged thiolated DNA for thiolated siRNA duplexes, and backfilled the nanocarriers with PEG molecules for increased colloidal stability and prolonged *in-vivo* circulation times [67, 98, 100].

Unique properties arise that address siRNA delivery challenges when siRNA duplexes are arranged in this spherical fashion, including intracellular delivery into almost all cell types, increased nuclease resistance, little to no immune activation *in-vitro* and *in-vivo*, and the persistent knockdown of target genes [95, 101–105]. Importantly, these properties are retained even when the gold core is dissolved or changed to a lipid-based core, crediting the spherical three-dimensional (3D) presentation of nucleic acid for the majority of these distinctive properties [106]. Despite the highly negative surface charge (zeta potential = −34 mV) associated with presenting oligonucleotides in a spherical architecture, SNAs manage to rapidly cross almost all negatively charged cell membranes [107]. This seemingly atypical intracellular uptake of SNAs was found to be facilitated by scavenger receptor A-mediated endocytosis, and allows for the accumulation of

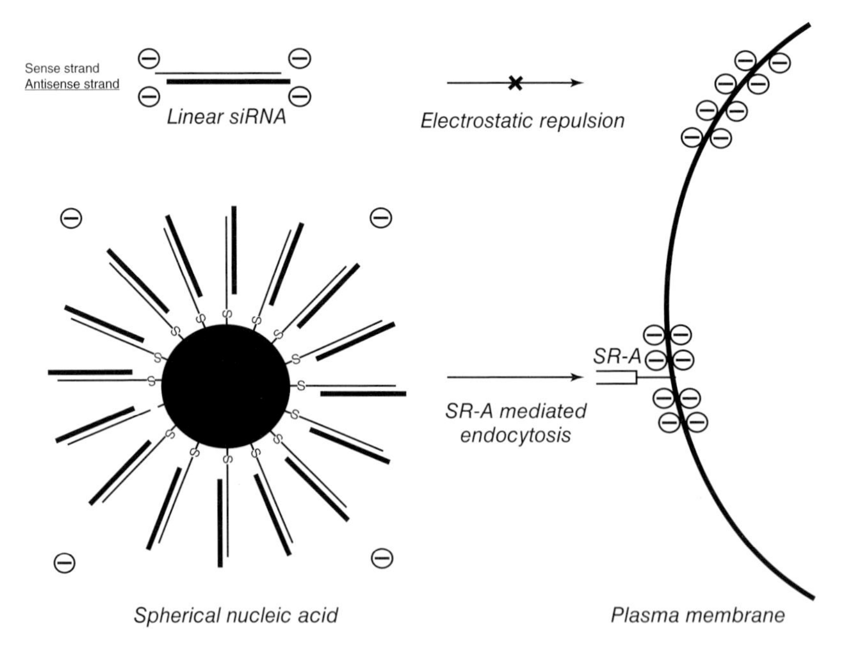

Fig. 4 Mechanism for intracellular entry of spherical nucleic acids. Negatively charged, linear siRNA duplexes are unable to independently cross the negatively charged plasma membrane. However, the spherical presentation of siRNA on SNAs allows for rapid intracellular uptake via scavenger receptor-A (SR-A) -mediated endocytosis.

hundreds of thousands of AuNPs per cell (Fig. 4) [47].

The SNA structure has been leveraged for the delivery of siRNA, microRNA and antisense DNA target sequences, and had its applicability expanded by cofunctionalization with targeting antibodies and chemotherapeutics for translational applications, including skin diseases, breast cancer, prostate cancer, and high-grade gliomas [67, 78, 82, 98, 108, 109]. To assess the potential of topically applied SNAs to treat skin diseases, Zhang *et al.* utilized SNAs with a 13 nm AuNP core conjugated to epidermal growth factor receptor (EGFR)-targeting thiolated siRNA duplexes [98]. The results showed that even without the assistance of auxiliary transfection agents, the EGFR-targeting SNAs could enter 100% of human keratinocytes (hKCs) within 2 h *in-vitro* (this was an especially impressive feature given that hKCs are traditionally a difficult cell line to transfect) [110]. The EGFR-targeting SNAs were then used to treat, topically, the skin of hairless mice three times weekly for three weeks. Relative to a control SNA treatment, the EGFR-targeting SNAs reduced Ki-67 (a proliferation marker) staining of keratinocytes in the basal layer by 40%, and reduced epidermal thickness by 40%. Following EGFR-targeting SNA treatment, EGFR expression was almost abolished, while the downstream phosphorylation of extracellular-signal-regulated kinase (ERK) 1/2 was decreased by 74% relative to control SNA treatment, as assessed by Western blotting. Importantly, the treated skin did not show any clinical or histological evidence of toxicity, including a lack of cytokine activation in mouse blood and tissue samples. In addition, after the three-week treatment, SNAs were almost undetectable in the internal organs, with 0.0003% and 0.00015% of the injected AuNP dose detected in the liver and spleen, respectively, as assessed by inductively coupled plasma mass spectrometry (ICP-MS). These data suggest that SNA conjugates may be safe and efficacious for the treatment of cutaneous tumors, skin inflammatory conditions, and dominant-negative genetic skin disorders.

In a glioblastoma multiforme (GBM) model, Jensen *et al.* demonstrated the ability of SNAs to cross the BBB and blood–tumor barrier (BTB) and to pervasively penetrate glioma tissue, both *in-vitro* and *in-vivo* [67]. Using an *in-vitro* coculture model of the human BBB, which consisted of human primary brain microvascular endothelial cells (huBMECs) and human astrocytes, Jensen and colleagues showed that SNAs were able to undergo transcytosis through the huBMEC layer and enter human astrocytes [111–113]. This BBB-penetrating capacity was abolished when polyinosinic acid (Poly I) was added prior to SNA treatment, which blocks SR-A (scavenger receptor-A) uptake and likely mediates transcytosis. The penetration of both BBB and glioma tissue was then investigated *in-vivo* in both healthy and glioma-bearing mice. In addition to Cy-5-labeled SNAs, gadolinium (Gd(III)) was conjugated to SNAs to visualize and quantify tissue penetration. SNA distribution was evaluated via ICP-MS, MRI, and confocal fluorescence microscopy. Following local administration, as assessed by the 3D reconstruction of MRI images and confocal fluorescence, SNAs exhibited extensive intratumoral dissemination. ICP-MS further substantiated these results with a 10-fold higher accumulation of SNAs in tumor versus nontumor brain regions, possibly due to the EPR effect. Analogous results were obtained when Cy5-labeled SNAs were injected into the tail vein of mice and fluorescence monitored with an *in-vivo* imaging system (IVIS). Both, IVIS imaging and the quantification of radiant intensities showed a 1.8-fold higher accumulation of SNAs in GBM xenograft-bearing mice compared to sham GBM-inoculated mice. Notably, SNA accumulation in the brain tissue of healthy mice was also extensive, with approximately 10^{10} SNA particles per gram of tissue, as determined with ICP-MS.

Jensen and colleagues then developed SNAs which consisted of a 13-nm AuNP core conjugated to thiolated siRNA duplexes targeting the GBM oncogene *Bcl2L12,* an effector caspase and p53 inhibitor overexpressed in the vast majority (>90%) of GBM patients. Systemically delivered Bcl2L12-targeting SNAs neutralized *Bcl2L12* expression, increased intratumoral apoptosis, reduced tumor burden, and augmented the survival of GBM-xenografted mice. Furthermore, rodent toxicity studies did not reveal any adverse side effects or signs of toxicity, as the systemically administered SNAs did not induce inflammatory cytokines and did not cause any changes in blood chemistry, complete blood counts, or histopathology compared to saline or control SNAs. With no evidence to date of toxicity, and encouraging *in-vivo* results, SNAs represent a promising construct for siRNA delivery that will shortly enter clinical testing.

3.4
Iron Oxide Nanoparticles

Iron oxide nanoparticles (IONs) possess superparamagnetic properties that enable them to exhibit magnetic interaction only in the presence of an external magnetic field. Thus, IONs can serve as contrast agents for MRI applications and also as therapeutic constructs for magnetic hyperthermia treatment [114, 115]. Dextran-coated IONs are biocompatible, biodegradable, and have been approved by the FDA for certain imaging and treatment applications, such as the imaging of liver lesions (Feridex I.V.®) and the treatment of iron-deficiency anemia in patients with chronic kidney disease (Feraheme®) [116]. For siRNA delivery purposes, IONs are commonly functionalized with oligonucleotides by first fabricating them with a polycationic layer such as PEI, polyarginine, polylysine, or cationic lipids, after which siRNA can be electrostatically adsorbed to the cationic nanoparticle surface [117]. Specifically, monodisperse IONs can be synthesized via a high-temperature organic phase decomposition of an iron precursor such as iron(III) acetylacetonate with 1,2-hexadecanediol, oleic acid, and oleylamine [118]. The oleic acid layer on these hydrophobic IONs can then be directly exchanged with a cationic molecule that has an affinity for the ION surface (e.g., PEI, which has a strong affinity for IONs due to the amine coordination of iron) through a ligand-exchange reaction to form cationic layer-coated, water-soluble IONs [119]. siRNA or other oligonucleotides may then be electrostatically tethered to the cationic nanoparticle surface for delivery purposes.

Several ION systems have successfully delivered siRNA to cells and tissues [89, 117, 119–123]. Notably, Liu *et al.* showed that PEI-coated IONs complexed to luciferase-targeting siRNA (lucsiRNA) were capable of increasing the serum stability of lucsiRNA, could efficiently release lucsiRNA in the presence of heparin, and could knockdown luciferase expression *in-vitro* and, via local intratumoral injection *in-vivo*, in a 4T1 breast tumor model [117]. Other ION systems, upon the application of an external magnetic field, can increase target site accumulation and siRNA transfection efficiency [119, 121–123]. For example, Anderson and colleagues developed an epoxide-derived lipidoid-coated ION that was capable of complexing siRNA and DNA for transfection [121]. These authors showed that green fluorescent protein (GFP)-targeting siRNA could be transfected into HeLa cells using a lipidoid–ION construct, and could silence GFP expression by ~80% at siRNA concentrations as low as 1.5 nM with external magnetic field application, compared to ~50% reduction in GFP expression at the same dose without an external magnetic field. Furthermore, when the lipoid–ION construct was complexed to plasmid DNA encoding GFP, an approximate 70% transfection efficiency of DNA was achieved at a DNA concentration as low as 0.05 nM with the application of an external magnetic field, compared to ~20% transfection efficiency without the external magnetic field. Although preliminary *in-vitro* results are encouraging, it is important to stress that there is currently no experimental evidence to support the claim that magnetic-based delivery *in-vivo* is superior to EPR-based passive delivery [89]. Despite the enticing combination of *in-vivo* imaging and siRNA delivery potential afforded by ION systems, the biodegradability and biocompatibility profiles of siRNA–ION constructs are currently undefined and more preclinical studies are required to establish their safety and efficacy *in-vivo* prior to human trials.

3.5 Cyclodextrin Polymer Nanoparticles

Cyclodextrin is a natural polymer produced during the bacterial digestion of cellulose [124]. Cyclodextrins have been incorporated into many pharmaceutical formulations as they do not activate the immune system, display low toxicity, and are not enzymatically degraded in humans [124]. Cyclodextrin polymer (CDP) nanoparticles were the first siRNA nanoparticle-based delivery system to enter clinical trials for cancer [125]. CDPs are composed of polycationic oligomers with amidine functional groups; the positively charged amidine groups serve to complex siRNA. CDP NPs can also be end-capped with imidazole groups, which can become protonated to enable endosomal escape [126–128]. For *in-vivo* targeting and stability purposes, CDP NPs can be functionalized with PEG and transferrin molecules through conjugation with adamantane, a hydrophobic molecule that stably interacts with the cyclic core of CDP NPs [15, 126, 129, 130]. CDP–PEG–transferrin nanoparticles loaded with siRNA targeting the EWS-FL11 fusion gene successfully inhibited tumor growth in a mouse model of Ewing's sarcoma, without inducing any toxicity or immune activation [126]. Another type of CDP–PEG–transferrin nanoparticle, CALAA-01, was designed with siRNA specific for the M2 subunit of ribonucleotide reductase (RRM2) and demonstrated efficacy in mouse models [131]. In cynomolgus monkeys, CALAA-01 was found to be active at doses as low as 0.6 $mg\,kg^{-1}$ while being tolerated at doses as high as 27 $mg\,kg^{-1}$ [132]. Phase I clinical trials with CALAA-01 are currently under way for the treatment of solid tumors shown to be refractory to standard-care therapies, and thus far nanoparticles have been detected in tumor biopsies taken from melanoma patients treated with CALAA-01. Immunohistochemical staining and qRT-PCR of tumor tissue have demonstrated the knockdown of RRM2, proving that RNAi can be achieved in humans by systemically administering a nanoparticle-based siRNA delivery system [133].

3.6 Liposomes and Lipid-Based Materials

Liposomes are the most prevalent nanocarrier for siRNA, with five different siRNA–liposomal formulations currently being investigated in clinical trials. In addition to their ability to effectively deliver siRNA into cells, there is significant past clinical experience with liposomal drug formulations. In fact, seven liposomal drug formulations have approved by the FDA, dating back to 1995 with the introduction of liposomal doxorubicin (Doxil®) [20, 33].

Analogous to the cell membrane, liposomes consist of a phospholipid bilayer. The polar, hydrophilic head groups of the phospholipids face the exterior of the liposome and the internal core of the liposome, while the nonpolar, hydrophobic phospholipid tail groups interact with each other to form the bilayer. This structure allows the entrapment of a variety of molecules in the core of the liposome, including siRNA, DNA, proteins, chemotherapeutics, and other therapeutic payloads. Additionally, liposome size, lipid composition and charge can be tailored to optimize target site accumulation. Similar to metal-based nanoconjugates, PEG molecules can also be conjugated to liposomes to prevent aggregation during the fabrication process and reduce clearance by the MPS [134, 135].

Liposomes are typically categorized by their charge, and thus can be subdivided into cationic, neutral, and anionic liposomes. As anionic liposomal formulations have only shown a limited success in delivering RNAi-active compounds due to impaired fusion with the negatively charged cell membrane, only cationic and neutral liposomes will be discussed here. In addition to cationic and neutral liposomes, stable nucleic acid–lipid particles (SNALPs; a hybrid lipid nanoparticle containing a cationic liposomal core but exhibiting a neutral exterior due to PEG conjugation) and synthetic exosomes (a mimic of naturally occurring liposomes that can be either cationic or neutral in surface charge) will be described.

3.6.1 Cationic Liposomes

Cationic liposomes composed of lipids such as DOTAP and DOTMA (*N*-[1-(2,3-dioleoyloxy)propyl]-*N*,*N*,*N*-trimethylammonium methyl sulfate) can form complexes with negatively charged siRNA via electrostatic interactions, thus facilitating delivery and transfection [20, 136]. However, due to the high-level intracellular stability of cationic liposomes, siRNA often remains entrapped in the core and does not reach the cytoplasm [137, 138], resulting in only a moderate target gene knockdown. In addition, unmodified cationic liposomes are toxic on systemic administration. Possibly due to TLR4 activation, dose-dependent hepatotoxicity, pulmonary inflammation, the induction of reactive oxygen species (ROS) production and a systemic interferon type I response have been observed in mice treated with cationic liposomes [137–139]. However, liposomes containing a mixture of cationic lipids, neutral fusogenic lipids, and PEG-modified lipids have been developed that are less toxic and release their siRNA contents more effectively. One such siRNA-based liposomal carrier that targets protein kinase N3, Atu027, inhibited lymph node metastasis in prostate and pancreatic cancer mouse models, as well as pulmonary metastasis in a variety of mouse models [140, 141], and has been enrolled into clinical trials. In Phase I trials, doses of up to 0.180 mg kg^{-1} Atu027 were well tolerated by patients, though dose escalation studies are expected to be performed in the near future [142].

3.6.2 Neutral Liposomes

Given the significant toxicity associated with cationic lipids, liposomes composed of neutral lipids have been investigated. While neutral liposomes show a greatly increased biocompatibility compared to cationic liposomes, the efficiency with which neutral liposomes can load siRNA tends to be much lower due to a poor interaction between neutral lipids and siRNA [143, 144]. One neutral liposomal formulation is composed of 1,2-dioleoyl-*sn*-glycero-3-phosphatidylcholine (DOPC) and involves a synthesis method that increases the loading efficiency of siRNA via conjugation of the RNA with DOPC in the presence of excess *tert*-butanol and Tween 20 [145]. These DOPC-based liposomes have been loaded with a variety of different siRNA sequences, including those targeting EphA2 [145], focal adhesion kinase (FAK) [146], neuropilin-2 [147], interleukin (IL)-8 [148], transmembrane protease, serine 2 (TMPRRS2)/ETS-related gene (ERG) [149], and Bcl-2 [150], and have been used successfully in preclinical models of ovarian, breast, pancreatic, prostate cancer, and melanoma [20, 151–153]. The constructs did not induce any adverse side effects after four weeks of repeated intravenous injection in mice and nonhuman primates [145–148, 150]. Consequently, DOPC-based liposomes packed with EphA2-targeting siRNA entered Phase I clinical trials in 2012 for patients with advanced, recurrent solid tumors.

3.6.3 Stable Nucleic Acid–Lipid Particles

SNALPs contain a lipid bilayer composed of cationic lipids and neutral fusogenic lipids that are coated with PEG-lipids to provide a neutral surface (Fig. 5).

SNALPs have been shown to facilitate the cellular uptake and endosomal escape of siRNA, and to effectively regulate gene expression *in-vivo*. SNALPs made with siRNA targeting hepatitis B virus (HBV) RNA were able to reduce serum HBV levels more than 10-fold, with effects persisting for up to one week [19]. Additionally, Zimmerman and coworkers were able to achieve up to 90% knockdown of *ApoB* mRNA in nonhuman primate livers; SNALPs containing ApoB-targeting siRNA were injected once intravenously and demonstrated RNAi-mediated gene silencing for up to 11 days [154]. Based on the results of preclinical trials, multiple SNALP-based siRNA constructs are currently undergoing Phase I clinical trials for the treatment of hypercholesterolemia, lymphomas, advanced tumors with liver involvement, and ebola-virus infection [155].

3.6.4 Exosome-Based siRNA Carriers

Exosomes are biological vesicles that have a diameter of 50–90 nm and contain a lipid bilayer composed of glycerophospholipids, sphingomyelins, and cholesterol [156, 157]. Exosomes are naturally occurring liposomes, and are secreted from cells with payloads of protein and/or nucleic acids (including mRNA and miRNA); they are important for intercellular communication [158]. As these vesicles are nonimmunogenic and serve as the body's natural carrier for RNAi, the aim to design synthetic exosomes that mimic naturally occurring exosomes is an exciting new opportunity for RNAi-based therapy development [159]. Alvarez-Erviti and colleagues genetically engineered dendritic cells to produce targeted exosomes containing neuron-specific rabies virus glycoprotein

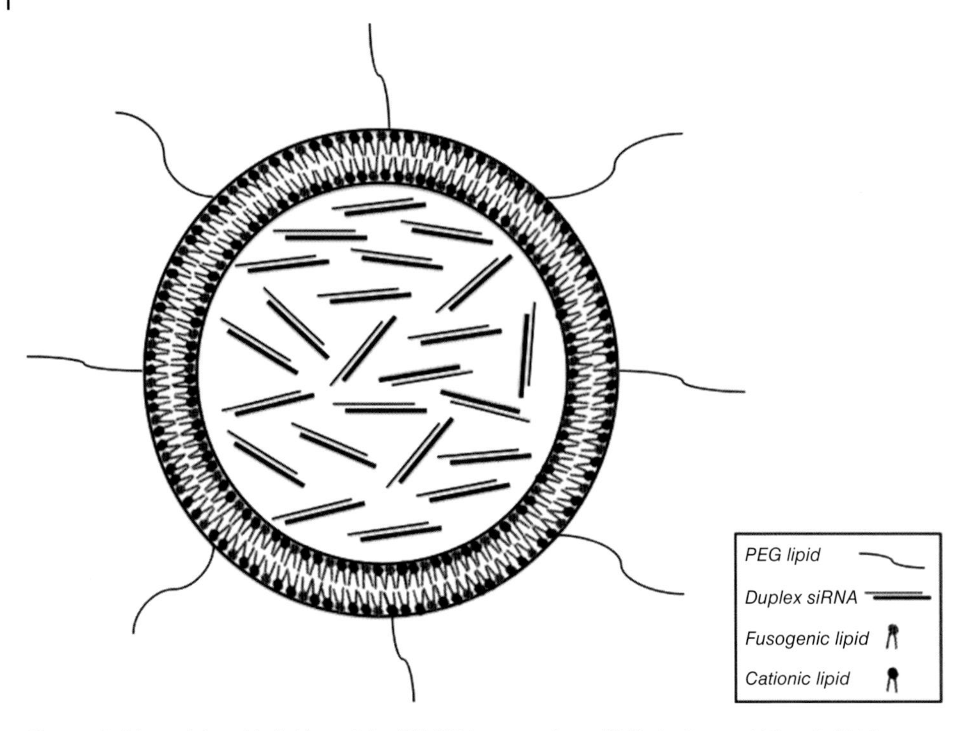

Fig. 5 Stable nucleic acid–lipid particles (SNALPs) encapsulate siRNA duplexes within a lipid bilayer composed of cationic lipids and neutral fusogenic lipids. The surface is coated with polyethylene glycol (PEG) lipids to provide a neutral exterior.

(RVG) peptides; the exosomes were then loaded with glyceraldehyde 3-phosphate dehydrogenase (GAPDH)-targeting siRNA via electroporation [160]. These vesicles downregulated GAPDH expression in the central nervous system (CNS), but not in the liver or other organs. Additionally, Wahlgren *et al.* purified human exosomes, used them to encapsulate siRNA targeted to mitogen-activated protein kinase (MAPK)-1 via electroporation, and then demonstrated gene silencing in human monocytes and lymphocytes [161]. These promising initial studies have laid the groundwork for the future development of exosome-based siRNA carriers. Additional *in-vitro* and *in-vivo* validation studies are required, however, prior to the advancement of these conjugates into human trials.

4 The Future of Nanoparticle-Based siRNA Delivery

Nanoparticle delivery systems for siRNAs have emerged as a powerful gene regulation platform, with over 24 conjugates undergoing clinical trials just 13 years after the first demonstration of the RNAi pathway in mammalian cells. The major challenges in siRNA delivery can be addressed with nanovectors, though these systems carry their own problems that are related to scaling-up syntheses, achieving organ site-selective or site-specific delivery, an incomplete understanding of immune stimulation by these constructs, and a lack of long-term safety/toxicity profiles. However, as these different classes of constructs continue to be tested in pre-clinical and clinical trials, more is being learned regarding their synthesis, physico-chemical characterization, biodistribution, toxicity, immunogenicity, and *in-vivo* efficacy. Design heuristics, such as the recommended size, charge, and shape of nanoparticles for optimal *in-vivo* delivery, are also beginning to emerge as investigators study the pharmacokinetics and pharmacodynamics of nanoparticle-based constructs. These design considerations will accelerate the pace of siRNA therapeutic development and decrease the time for bench-to-bedside translation.

With multiple classes of nanoparticle-based delivery systems undergoing Phase II clinical trials, the implementation of these conjugates into clinical practice is becoming close, and will bring humankind one giant step closer to realizing the concept of precision medicine to target virtually any disease-causing genetic element. If the genetic driver of a specific patient's disease can be identified, then siRNA against this sequence can be rapidly designed and loaded into one of these carriers. This adaptability becomes even more important when considering that the genetic base of many diseases is constantly changing, such as multidrug-resistant bacteria, continuously evolving viral genomes, and the acquisition of therapy-resistant phenotypes in cancer. Yet, given the speed by which advancements have been made in RNAi therapy thus far, perhaps this technology will be available sooner than expected.

Acknowledgments

This research was supported by the Center for Cancer Nanotechnology Excellence (CCNE) initiative of the National Institutes of Health (NIH) (U54 CA151880), the Dixon Translational Research Grants Initiative of the Northwestern Memorial Foundation, the James S. MacDonnell 21st Century Initiative, the Coffman Charitable Trust, the John McNicholas foundation, the American Cancer Society, and the Association for Cancer Gene Therapy. T.L.S. is supported by the NIH Ruth L. Kirschstein National Research Service Award for Predoctoral MD/PhD Fellows (F30CA174058) awarded by the National Cancer Institute (NCI) and the Northwestern Ryan Fellowship.

References

1. Fire, A., Xu, S., Montgomery, M.K., Kostas, S.A. et al. (1998) *Nature*, **391**, 806–811.
2. Elbashir, S.M., Harborth, J., Lendeckel, W., Yalcin, A. et al. (2001) *Nature*, **411**, 494–498.
3. Novina, C.D., Sharp, P.A. (2004) *Nature*, **430**, 161–164.
4. Hannon, G.J., Rossi, J.J. (2004) *Nature*, **431**, 371–378.
5. Castanotto, D., Rossi, J.J. (2009) *Nature*, **457**, 426–433.
6. Kanasty, R., Dorkin, J.R., Vegas, A., Anderson, D. (2013) *Nat. Mater.*, **12**, 967–977.
7. Burnett, J.C., Rossi, J.J. (2012) *Chem. Biol.*, **19** (1), 60–71.
8. Davidson, B.L., McCray, P.B. (2011) *Nat. Rev. Genet.*, **12**, 329–340.
9. Kim, D.H., Rossi, J.J. (2007) *Nat. Rev. Genet.*, **8**, 173–184.
10. Burnett, J.C., Rossi, J.J., Tiemann, K. (2011) *Biotechnol. J.*, **6**, 1130–1146.
11. Stegh, A.H. (2013) *Integr. Biol. (Camb.)*, **5**, 48–65.
12. Zamore, P.D., Tuschl, T., Sharp, P.A., Bartel, D.P. (2000) *Cell*, **101**, 25–33.
13. Hannon, G.J. (2002) *Nature*, **418**, 244–251.

14. McManus, M.T., Sharp, P.A. (2002) *Nat. Rev. Genet.*, **3**, 737–747.
15. Bartlett, D.W., Davis, M.E. (2006) *Nucleic Acids Res.*, **34**, 322–333.
16. Whitehead, K.A., Langer, R., Anderson, D.G. (2010) *Nat. Rev. Drug Discovery*, **9**, 412.
17. Schroeder, A., Levins, C.G., Cortez, C., Langer, R., Anderson, D.G. (2010) *J. Intern. Med.*, **267**, 9–21.
18. Kanasty, R.L., Whitehead, K.A., Vegas, A.J., Anderson, D.G. (2012) *Mol. Ther.*, **20**, 513–524.
19. Morrissey, D.V., Lockridge, J.A., Shaw, L., Blanchard, K. *et al.* (2005) *Nat. Biotechnol.*, **23**, 1002–1007.
20. Ozpolat, B., Sood, A.K., Lopez-Berestein, G. (2014) *Adv. Drug Delivery Rev.*, **66**, 110–116.
21. Layzer, J.M., McCaffrey, A.P., Tanner, A.K., Huang, Z. *et al.* (2004) *RNA*, **10** (5), 766–771.
22. Behlke, M.A. (2008) *Oligonucleotides*, **18**, 305–320.
23. Jackson, A.L., Burchard, J., Leake, D., Reynolds, A. *et al.* (2006) *RNA*, **12**, 1197–1205.
24. Robbins, M., Judge, A., MacLachlan, I. (2009) *Oligonucleotides*, **19**, 89–102.
25. Gantier, M.P., Tong, S., Behlke, M.A., Irving, A.T. *et al.* (2010) *Mol. Ther.*, **18**, 785–795.
26. Judge, A.D., Sood, V., Shaw, J.R., Fang, D. *et al.* (2005) *Nat. Biotechnol.*, **23**, 457–462.
27. Kleinman, M.E., Yamada, K., Takeda, A., Chandrasekaran, V. *et al.* (2008) *Nature*, **452**, 591–597.
28. Judge, A., Bola, G., Lee, A., Maclachlan, I. (2006) *Mol. Ther.*, **13**, 494–505.
29. Robbins, M., Judge, A., Liang, L., McClintock, K. *et al.* (2007) *Mol. Ther.*, **15**, 1663–1669.
30. Wartiovaara, J., Öfverstedt, L.-G., Khoshnoodi, J., Zhang, J. *et al.* (2004) *J. Clin. Invest.*, **114**, 1475–1483.
31. Huang, Y., Hong, J., Zheng, S., Ding, Y. *et al.* (2011) *Mol. Ther.*, **19** (2), 381–385.
32. Choi, H.S., Liu, W., Liu, F., Nasr, K. *et al.* (2009) *Nat. Nanotechnol.*, **5**, 42–47.
33. Petros, R.A., DeSimone, J.M. (2010) *Nat. Rev. Drug Discovery*, **9**, 615–627.
34. Aird, W.C. (2007) *Circ. Res.*, **100** (2), 174–190.
35. Braet, F., Wisse, E., Bomans, P., Frederik, P. *et al.* (2007) *Microsc. Res. Tech.*, **70**, 230–242.
36. Alexis, F., Pridgen, E., Molnar, L.K., Farokhzad, O.C. (2008) *Mol. Pharm.*, **5**, 505–515.
37. Li, W., Szoka, F.C. Jr. (2007) *Pharm. Res.*, **24**, 438–449.
38. Maeda, H. (2010) *Bioconjugate Chem.*, **21**, 797–802.
39. Campbell, R.B., Fukumura, D., Brown, E.B., Mazzola, L.M. *et al.* (2002) *Cancer Res.*, **62** (23), 6831–6836.
40. Champion, J.A., Mitragotri, S. (2006) *Proc. Natl Acad. Sci. USA*, **103** (13), 4930–4934.
41. Liu, Z., Cai, W., He, L., Nakayama, N. *et al.* (2006) *Nat. Nanotechnol.*, **2**, 47–52.
42. Park, J.-H., von Maltzahn, G., Zhang, L., Schwartz, M. P. *et al.* (2008) *Adv. Mater.*, **20**, 1630–1635.
43. Wu, J. (2002) *Front. Biosci.*, **7**, d717–d725.
44. Daniels, T.R., Bernabeu, E., Rodríguez, J.A., Patel, S. (2012) *Biochim. Biophys. Acta*, **1820** (3), 291–317.
45. Abbott, N.J., Patabendige, A.A.K., Dolman, D.E.M., Yusof, S.R., Begley, D.J. (2010) *Neurobiol. Dis.*, **37**, 13–25.
46. Sanna, V., Pala, N., Sechi, M. (2014) *Int. J. Nanomed.*, **9**, 467–483.
47. Choi, C.H.J., Hao, L., Narayan, S.P., Auyeung, E., Mirkin, C.A. (2013) *Proc. Natl Acad. Sci. USA*, **110**, 7625–7630.
48. Lu, J.J., Langer, R., Chen, J. (2009) *Mol. Pharm.*, **6**, 763–771.
49. Reddy, J.A., Low, P.S. (2000) *J. Controlled Release*, **64** (1-3), 27–37.
50. Kichler, A., Leborgne, C., Coeytaux, E., Danos, O. (2001) *J. Gene Med.*, **3**, 135–144.
51. Sonawane, N.D., Szoka, F.C., Verkman, A.S. (2003) *J. Biol. Chem.*, **278**, 44826–44831.
52. Drummond, D.C., Daleke, D.L. (1995) *Chem. Phys. Lipids*, **75** (1), 27–41.
53. Shen, J., Samul, R., Silva, R.L., Akiyama, H. *et al.* (2005) *Gene Ther.*, **13**, 225–234.
54. Dejneka, N.S., Wan, S., Bond, O.S., Kornbrust, D.J., Reich, S.J. (2008) *Mol. Vision*, **14**, 997–1005.
55. Garba, A.O., Mousa, S.A. (2010) *Ophthalmol. Eye Dis.*, **2**, 75–83.
56. Martínez, T., Wright, N., López-Fraga, M., Jimenez, A.I., Pañeda, C. (2013) *Hum. Genet.*, **132** (5), 481–493.

57. Alvarez, R., Elbashir, S., Borland, T., Toudjarska, I. *et al.* (2009) *Antimicrob. Agents Chemother.*, **53**, 3952–3962.
58. DeVincenzo, J., Lambkin-Williams, R., Wilkinson, T., Cehelsky, J. *et al.* (2010) *Proc. Natl Acad. Sci. USA*, **107**, 8800–8805.
59. Zamora, M.R., Budev, M., Rolfe, M., Gottlieb, J. *et al.* (2011) *Am. J. Respir. Crit. Care Med.*, **183**, 531–538.
60. Jeong, J.H., Mok, H., Oh, Y.K., Park, T.G. (2008) *Bioconjugate Chem.*, **20** (1), 5–14.
61. Rozema, D.B., Lewis, D.L., Wakefield, D.H., Wong, S.C. *et al.* (2007) *Proc. Natl Acad. Sci. USA*, **104**, 12982–12987.
62. DeLong, R.K., Akhtar, U., Sallee, M., Parker, B. *et al.* (2009) *Biomaterials*, **30**, 6451–6459.
63. Sashin, D., Spanbock, J., Kling, D.H. (1939) *J. Bone Joint Surg. Am.*, **21**, 723–734.
64. Kim, S.T., Chompoosor, A., Yeh, Y.-C., Agasti, S.S. *et al.* (2012) *Small*, **8**, 3253–3256.
65. Fraser, T.N. (1945) *Ann. Rheum. Dis.*, **4**, 71–75.
66. Lee, S.H., Bae, K.H., Kim, S.H., Lee, K.R., Park, T.G. (2008) *Int. J. Pharm.*, **364**, 94–101.
67. Jensen, S.A., Day, E.S., Ko, C.H., Hurley, L.A. *et al.* (2013) *Sci. Transl. Med.*, **5**, 209ra152.
68. Brust, M., Walker, M., Bethell, D., Schiffrin, D.J., Whyman, R. (1994) *J. Chem. Soc. Chem. Commun.*, >801–802.
69. Frens, G. (1973) *Nat. Phys. Sci.*, **241**, 20–22.
70. Sutherland, W.S., Winefordner, J.D. (1992) *J. Colloid Interface Sci.*, **148**, 129–141.
71. Grabar, K.C., Freeman, R.G., Hommer, M.B., Natan, M.J. (1995) *Anal. Chem.*, **67**, 735–743.
72. Templeton, A.C., Wuelfing, W.P., Murray, R.W. (2000) *Acc. Chem. Res.*, **33**, 27–36.
73. Jiang, X.-M., Wang, L.-M., Wang, J., Chen, C.-Y. (2012) *Appl. Biochem. Biotechnol.*, **166**, 1533–1551.
74. Daniel, M.C., Astruc, D. (2004) *Chem. Rev.*, **104** (1), 293–346.
75. Pakiari, A.H., Jamshidi, Z. (2010) *J. Phys. Chem. A*, **114**, 9212–9221.
76. Ding, Y., Jiang, Z., Saha, K., Kim, C.S. *et al.* (2014) *Mol. Ther.*, **22** (6), 1075–1083.
77. Dixit, V., Van den Bossche, J., Sherman, D.M., Thompson, D.H., Andres, R.P. (2006) *Bioconjugate Chem.*, **17** (1), 603–609.
78. Zhang, K., Hao, L., Hurst, S.J., Mirkin, C.A. (2012) *J. Am. Chem. Soc.*, **134**, 16488–16491.
79. Ghosh, P., Han, G., De, M., Kim, C.K., Rotello, V.M. (2008) *Adv. Drug Delivery Rev.*, **60** (11), 1307–1315.
80. Debouttière, P.J., Roux, S., Vocanson, F., Billotey, C. *et al.* (2006) *Adv. Funct. Mater.*, **16**, 2330–2339.
81. Gibson, J.D., Khanal, B.P., Zubarev, E.R. (2007) *J. Am. Chem. Soc.*, **129**, 11653–11661.
82. Zhang, X.-Q., Xu, X., Lam, R., Giljohann, D. *et al.* (2011) *ACS Nano*, **5**, 6962–6970.
83. Elbakry, A., Zaky, A., Liebl, R., Rachel, R., Goepferich, A. (2009) *Nano Lett.*, **9** (5), 2059–2206.
84. Guo, S., Huang, Y., Jiang, Q., Sun, Y. *et al.* (2010) *ACS Nano*, **4** (9), 5505–5511.
85. Kelly, K.L., Coronado, E., Zhao, L.L., Schatz, G.C. (2003) *J. Phys. Chem. B*, **107**, 668–677.
86. Alkilany, A.M., Murphy, C.J. (2010) *J. Nanopart. Res.*, **12**, 2313–2333.
87. Zhang, Z., Wang, J., Chen, C. (2013) *Theranostics*, **3**, 223–238.
88. Shim, M.S., Kwon, Y.J. (2012) *Adv. Drug Delivery Rev.*, **64**, 1046–1059.
89. Wang, Z., Liu, G., Zheng, H., Chen, X. (2014) Biotechnol. *Adv.*, **32** (4), 831–843.
90. Eghtedari, M., Oraevsky, A., Copland, J.A., Kotov, N.A. *et al.* (2007) *Nano Lett.*, **7**, 1914–1918.
91. Shim, M.S., Kim, C.S., Ahn, Y.-C., Chen, Z., Kwon, Y.J. (2010) *J. Am. Chem. Soc.*, **132**, 8316–8324.
92. Weintraub, K. (2013) *Nature*, **495**, S14–S16.
93. Libutti, S.K., Paciotti, G.F., Byrnes, A.A., Alexander, H.R. *et al.* (2010) *Clin. Cancer Res.*, **16**, 6139–6149.
94. Haynes, R., Gannon, W., Walker, M. (2009) *J. Clin. Oncol.*, **27**, 3586.
95. Mirkin, C.A., Letsinger, R.L., Mucic, R.C., Storhoff, J.J. (1996) *Nature*, **382**, 607–609.
96. Park, S.Y., Lytton-Jean, A.K.R., Lee, B., Weigand, S. *et al.* (2008) *Nature*, **451**, 553–556.
97. Nykypanchuk, D., Maye, M.M., van der Lelie, D., Gang, O. (2008) *Nature*, **451**, 549–552.
98. Zheng, D., Giljohann, D.A., Chen, D.L., Massich, M.D. *et al.* (2012) *Proc. Natl Acad. Sci. USA*, **109**, 11975–11980.

99. Cutler, J.I., Auyeung, E., Mirkin, C.A. (2012) *J. Am. Chem. Soc.*, **134**, 1376–1391.
100. Giljohann, D.A., Seferos, D.S., Prigodich, A.E., Patel, P.C., Mirkin, C.A. (2009) *J. Am. Chem. Soc.*, **131**, 2072–2073.
101. Seferos, D.S., Prigodich, A.E., Giljohann, D.A., Patel, P.C., Mirkin, C.A. (2009) *Nano Lett.*, **9**, 308–311.
102. Massich, M.D., Giljohann, D.A., Seferos, D.S., Ludlow, L.E. *et al.* (2009) *Mol. Pharm.*, **6**, 1934–1940.
103. Massich, M.D., Giljohann, D.A., Schmucker, A.L., Patel, P.C., Mirkin, C.A. (2010) *ACS Nano*, **4**, 5641–5646.
104. Williams, S.C.P. (2013) *Proc. Natl Acad. Sci. USA*, **110**, 13231–13233.
105. Mirkin, C.A., Stegh, A.H. (2013) *Oncotarget*, **5** (1), 9–10.
106. Cutler, J.I., Zhang, K., Zheng, D., Auyeung, E. *et al.* (2011) *J. Am. Chem. Soc.*, **133**, 9254–9257.
107. Rosi, N.L. (2006) *Science*, **312**, 1027–1030.
108. Alhasan, A.H., Patel, P.C., Choi, C.H.J., Mirkin, C.A. (2014) *Small*, **10**, 186–192.
109. Alhasan, A.H., Kim, D.Y., Daniel, W.L., Watson, E. *et al.* (2012) *Anal. Chem.*, **84**, 4153–4160.
110. Dickens, S., Van den Berge, S., Hendrickx, B., Verdonck, K. *et al.* (2010) Tissue Eng. *Part C Methods*, **16**, 1601–1608.
111. Boveri, M., Berezowski, V., Price, A., Slupek, S. *et al.* (2005) *Glia*, **51**, 187–198.
112. Cecchelli, R., Dehouck, B., Descamps, L., Fenart, L. *et al.* (1999) *Adv. Drug Delivery Rev.*, **36**, 165–178.
113. Culot, M., Lundquist, S., Vanuxeem, D., Nion, S. *et al.* (2008) *Toxicol. in Vitro*, **22**, 799–811.
114. Hergt, R., Dutz, S., Müller, R., Zeisberger, M. (2006) *J. Phys.: Condens. Matter*, **18**, S2919–S2934.
115. Mahmoudi, M., Sant, S., Wang, B., Laurent, S., Sen, T. (2011) *Adv. Drug Delivery Rev.*, **63**, 24–46.
116. Tassa, C., Shaw, S.Y., Weissleder, R. (2011) *Acc. Chem. Res.*, **44**, 842–852.
117. Liu, G., Xie, J., Zhang, F., Wang, Z. *et al.* (2011) *Small*, **7**, 2742–2749.
118. Sun, S., Zeng, H., Robinson, D.B., Raoux, S. *et al.* (2004) *J. Am. Chem. Soc.*, **126**, 273–279.
119. Zhang, H., Lee, M.Y., Hogg, M.G., Dordick, J.S. (2010) *ACS Nano*, **4** (8), 4733–4743.
120. Liu, G., Wang, Z., Lu, J., Xia, C. *et al.* (2011) *Biomaterials*, **32**, 528–537.
121. Jiang, S., Eltoukhy, A.A., Love, K.T., Langer, R., Anderson, D.G. (2013) *Nano Lett.*, **13**, 1059–1064.
122. Shubayev, V.I., Pisanic, T.R. II, Jin, S. (2009) *Adv. Drug Delivery Rev.*, **61**, 467–477.
123. Del Pino, P., Munoz-Javier, A., Vlaskou, D., Gil, P.R. (2010) *Nano Lett.*, **10** (10), 3914–3921.
124. Davis, M.E., Brewster, M.E. (2004) *Nat. Rev. Drug Discovery*, **3**, 1023–1035.
125. Davis, M.E. (2009) *Mol. Pharm.*, **6**, 659–668.
126. Hu-Lieskovan, S. (2005) *Cancer Res.*, **65**, 8984–8992.
127. Mishra, S., Heidel, J.D., Webster, P., Davis, M.E. (2006) *J. Controlled Release*, **116** (2), 179–191.
128. Bartlett, D.W., Davis, M.E. (2007) *Bioconjugate Chem.*, **18**, 456–468.
129. Bellocq, N.C., Pun, S.H., Jensen, G.S., Davis, M.E. (2003) *Bioconjugate Chem.*, **14**, 1122–1132.
130. Bartlett, D.W., Su, H., Hildebrandt, I.J., Weber, W.A., Davis, M.E. (2007) *Proc. Natl Acad. Sci. USA*, **104**, 15549–15554.
131. Bartlett, D.W., Davis, M.E. (2008) *Biotechnol. Bioeng.*, **99**, 975–985.
132. Heidel, J.D., Yu, Z., Liu, J.Y.C., Rele, S.M. *et al.* (2007) *Proc. Natl Acad. Sci. USA*, **104**, 5715–5721.
133. Davis, M.E., Zuckerman, J.E., Choi, C., Seligson, D. (2010) *Nature*, **464** (7291), 1067–1070.
134. Klibanov, A.L., Maruyama, K., Torchilin, V.P., Huang, L. (1990) *FEBS Lett.*, **268** (1), 235–237.
135. Ishida, T., Harashima, H., Kiwada, H. (2002) *Biosci. Rep.*, **22**, 197–224.
136. Miller, C.R., Bondurant, B., McLean, S.D., McGovern, K.A., O'Brien, D.F. (1998) *Biochemistry*, **37**, 12875–12883.
137. Dokka, S., Toledo, D., Shi, X., Castranova, V., Rojanasakul, Y. (2000) *Pharm. Res.*, **17**, 521–525.
138. Spagnou, S., Miller, A.D., Keller, M. (2004) *Biochemistry*, **43**, 13348–13356.
139. Lv, H., Zhang, S., Wang, B., Cui, S., Yan, J. (2006) *J. Controlled Release*, **114**, 100–109.
140. Aleku, M., Schulz, P., Keil, O., Santel, A. *et al.* (2008) *Cancer Res.*, **68**, 9788–9798.
141. Santel, A., Aleku, M., Röder, N., Möpert, K. *et al.* (2010) *Cancer Res.*, **16**, 5469–5480.

142. Strumberg, D., Schultheis, B., Traugott, U., Vank, C. (2012) Int. J. *Clin. Pharmacol. Ther.*, **50** (1), 76–78.
143. Lee, J.-M., Yoon, T.-J., Cho, Y.-S. (2013) Biomed. *Res. Int.*, **2013**, 782041.
144. Wu, S.Y., McMillan, N.A.J. (2009) *AAPS J.*, **11**, 639–652.
145. Landen, C.N., Chavez-Reyes, A., Bucana, C., Schmandt, R. *et al.* (2005) *Cancer Res.*, **65** (15), 6910–6918.
146. Halder, J., Kamat, A.A., Landen, C.N., Han, L.Y. *et al.* (2006) *Clin. Cancer Res.*, **12**, 4916–4924.
147. Gray, M.J., Van Buren, G., Dallas, N.A., Xia, L. *et al.* (2008) *J. Natl Cancer Inst.*, **100** (2), 109–120.
148. Merritt, W.M., Lin, Y.G., Spannuth, W.A., Fletcher, M.S. *et al.* (2008) *J. Natl Cancer Inst.*, **100**, 359–372.
149. Shao, L., Tekedereli, I., Wang, J., Yuca, E. *et al.* (2012) *Clin. Cancer Res.*, **18**, 6648–6657.
150. Tekedereli, I., Alpay, S.N., Akar, U., Yuca, E. *et al.* (2013) *Mol. Ther. Nucleic Acids*, **2**, e121.
151. Nick, A.M., Stone, R.L., Armaiz-Pena, G., Ozpolat, B. *et al.* (2011) *J. Natl Cancer Inst.*, **103**, 1596–1612.
152. Pan, X., Arumugam, T., Yamamoto, T., Levin, P.A. *et al.* (2008) *Clin. Cancer Res.*, **14** (24), 8143–8151.
153. Villares, G.J., Zigler, M., Wang, H., Melnikova, V.O. *et al.* (2008) *Cancer Res.*, **68**, 9078–9086.
154. Zimmermann, T.S., Lee, A.C.H., Akinc, A., Bramlage, B. *et al.* (2006) *Nature*, **441**, 111–114.
155. Barros, S.A., Gollob, J.A. (2012) *Adv. Drug Delivery Rev.*, **64** (15), 1730–1737.
156. Vlassov, A.V., Magdaleno, S., Setterquist, R. (2012) Biochim. *Biophys. Acta*, **1820** (7), 940–948.
157. Subra, C., Laulagnier, K., Perret, B., Record, M. (2007) *Biochimie*, **89**, 205–212.
158. Valadi, H., Ekström, K., Bossios, A., Sjöstrand, M., Lee, J.J. (2007) *Nat. Cell Biol.*, **9** (6), 654–659.
159. Kooijmans, S.A., Vader, P., van Dommelen, S.M., van Solinge, W.W., Schiffelers, R.M. (2012) *Int. J. Nanomed.*, **7**, 1525–1541.
160. Alvarez-Erviti, L., Seow, Y., Yin, H.F., Betts, C. (2011) *Nat. Biotechnol.*, **29** (4), 341–345.
161. Wahlgren, J., Karlson, T.D.L., Brisslert, M., Vaziri Sani, F. *et al.* (2012) *Nucleic Acids Res.*, **40**, e130.

10
Role of Scavenger Receptors in Immune Recognition and Targeting of Nanoparticles

Guankui Wang and Dmitri Simberg
University of Colorado Anschutz Medical Campus, The Skaggs School of Pharmacy and Pharmaceutical Sciences, 12850 East Montview Blvd, Aurora, CO 80045, USA

Keywords

Nanotechnology
The field of science that deals with objects on the nanometer scale.

Nanomedicine
The division of nanotechnology that deals with the medicinal application of nanometer-sized materials.

Nanoparticles
Particles with sizes between 1 and 1000 nm.

Scavenger receptors
A group of receptors responsible for the recognition of molecular patterns.

Translational Nanomedicine. First Edition. Edited by Robert A. Meyers.

Clearance
The process of removing molecules, particles, and pathogens from the body.

Macrophages
Cells that specialize in neutralizing and ingesting foreign substances that penetrate the body.

With the recent rapid advances in nanotechnology, nanomedicine has become an emerging field that aims to utilize various nanoparticles (NPs) with unique properties to improve disease diagnostics and drug delivery. Notwithstanding the advances made in NP engineering, most of the currently available NPs are not ready for clinical trials. The first major barrier that NPs encounter after entering the body is the innate immune system. Before reaching the target sites, NPs are readily cleared by macrophages in the liver and spleen. To overcome the rapid body clearance and minimize the immune recognition of NPs using different chemical or physical surface modifications, the mechanisms of NP interaction with the immune system must be understood. In this chapter, an overview and discussion are provided of the role of scavenger receptors (SRs) as an important part of the innate immune response. The use of SRs as markers for selective targeting to various cells is also discussed. A better understanding of the innate immunity response and immune recognition will be crucial for improving NP design for medical applications.

1 Introduction

Nanotechnology is a rapidly evolving field in which objects are studied on a nanometer scale. Nanomedicine is the extension of nanotechnology that aims to use nanoparticles (NPs), which are tens to hundreds of nanometers in size, for medicinal uses. Currently, the most prominent examples of NPs used in clinical nanomedicine are liposomes as carriers of drugs, and iron oxide NPs as molecular imaging diagnostic agents. By definition, nanomedicines are supposed to interact with the body milieu in order to perform the designed task, for example, to deliver a drug/imaging agent safely and precisely to a desired location. However, chemically synthesized and engineered NPs may encounter significant barriers inside the body before they reach the desired target tissue/receptor. No matter how carefully they are masked, the nanosized particles are relatively quickly recognized and picked up by the immune cells, mostly by resident macrophages (i.e., resident in the tissues), blood monocytes and neutrophils. Macrophages are the most important component of the immune system, as they are responsible for the recognition and uptake of foreign pathogens and particulates [1].

Immune clearance is an evolutionarily evolved mechanism that ensures survival of the host via the rapid removal of pathogens. Depending on the application, the uptake of NPs by macrophages and other immune cells could be either desirable or undesirable. On the one hand, a rapid clearance of NPs (often within minutes after injection) will affect critical pharmacokinetic parameters, such as the area-under-curve (AUC) – that is, the integrated levels of NPs in blood over time. A decrease in the AUC can reduce the ability of NPs to bind to the tissue/receptor of interest [2]. On the other hand, macrophages play a critical role in disease, and account for the majority of uptake of therapeutic and imaging NPs in diseased tissues (e.g., tumor, atherosclerosis). While such uptake can serve the purpose of a vaccine delivery, therapeutic targeting and imaging of macrophages, it may limit drug availability and result in off-target toxicities in applications where NP-transported agents are intended for delivery into non-immune cells. It must also be remembered that the uptake of NPs could lead to undesirable toxicity to macrophages, and may impair the body's immune function.

The non-self property of NPs is the primary reason for their immune recognition and clearance. Many studies have been conducted to address the role of the physico-chemical properties of NPs, such as size, shape and surface chemistry [3, 4], and the results obtained and conclusions drawn are highly informative [3–5]. The aim of the present chapter, however, is to address the problem from biological and immunological perspectives rather than from a chemical/engineering perspective, and to focus on the scavenger receptors (SRs) as a group of pattern recognition receptors (PRRs) implicated in NP clearance. Within the chapter, attempts will be made to draw a parallel between conventional targets of the immune system (pathogens, apoptotic cells, and modified proteins/lipoproteins) and NPs. The finding that, due to their broad specificity, SRs are able to recognize NPs as efficiently as other immunological targets, is described in the following sections.

2 Overview of the Mechanisms of Endocytosis

The cells of the body have developed numerous pathways by which to ingest the nutrients, biomolecules and particulates present in the extracellular space (for a summary, see Table 1). Some of these pathways are intrinsic only to "professional" phagocytic cells such as macrophages and neutrophils, whereas other pathways are shared among many cell types. The relative efficiencies of uptake via different pathways are different.

The simplest form of endocytosis is pinocytosis, which is used to capture biomolecules, fluids and solutes. Some NPs can enter the cells via pinocytosis, but the uptake is relatively inefficient. In a classical study [6] on the size range of particles, by using rat peritoneal macrophages it was suggested that pinocytosis is restricted to macromolecules and smaller NPs, and that the rate of uptake is much lower than via phagocytosis. Moreover, pinocytosis is nonspecific (it does not require a receptor recognition) as in effect it simply "grabs" the fluid by means of invaginations in the membrane [7].

Receptor-mediated endocytosis (RME) is a more efficient uptake mechanism, but there is an associated size selectivity and smaller NPs (up to 50 nm) have been shown to enter cells more efficiently than larger NPs [8]. In addition to pinocytosis and RME, professional phagocytes – as opposed to semi-professional and non-professional phagocytes – have developed highly efficient internalization mechanisms that are not limited by the size of the cargo and are able to internalize whole apoptotic cells and damaged erythrocytes. Phagocytosis leads to the formation of a phagosome and to a downstream activation of the inflammatory response. Most knowledge

Tab. 1 Different types of endocytosis.

Mechanism	*Micropinocytosis*	*Macropinocytosis*	*Clathrin-dependent endocytosis*	*Caveolin-dependent endocytosis*	*Clathrin- and caveolin independent endocytosis*	*Phagocytosis*
Substrate	Fluid, solutes, macromolecules, particles	Macromolecules, larger particles	Macromolecules, viruses, small particles, proteins, toxins	Proteins, toxins, some viruses, small particles	GPI-anchored proteins, cytokines, toxins, viruses, modified proteins	Particulates, pathogens, apoptotic cells
Mechanism	Clathrin-coated pits	Actin polymerization	Clathrin-coated pits	Caveolin	Flotillin, CLIC/GEEC, other adaptors	Actin polymerization
Specificity for the cargo	Non-specific (no receptors)		Specific cell-surface receptors are usually involved			Range of phagocytic receptors with narrow or broad specificity
Size cutoff based on the size of endocytic vesicles	<30 nm	0.5–2 μm	100–150 nm	50–80 nm	<100 nm	<20 μm

on phagocytosis has been acquired via studies of pathogen clearance. Yet, with the increased interest in nanomedicine, much more attention has been focused on the mechanisms of clearance of NPs by phagocytes. The phagocytosis event starts with a contact between the pathogen surface and the cell membrane; this is termed the phagocytic synapse [9]. While phagocytic receptors trigger the initial binding event, the results of proteomic analyses have suggested that there are hundreds of proteins present in the synapse, and that these are responsible for phagosome and phagolysosome formation, downstream signal transduction, and regulation of the inflammatory response [9, 10].

3 Pattern Recognition Receptors

Innate immunity is an ancient and universal form of host defense against invading pathogens, and plays a crucial role in the early recognition and subsequent triggering of a proinflammatory response. The innate immune response relies on the recognition of evolutionarily conserved structures on pathogens, referred to by Janeway [11] as pathogen-associated molecular patterns (PAMPs), through a limited number of PRRs, of which Toll-like receptors (TLRs) are the most prominent [12]. PAMPs are characterized by being invariant among entire classes of pathogens, which is essential for survival of the pathogen, and distinguishable from "self." However, it is being increasingly appreciated that some host factors, when they are present in aberrant locations or abnormal molecular complexes, or are chemically modified, could be also recognized by PRRs as "danger" signals. Hence, the term danger-associated molecular pattern (DAMP) has been introduced to distinguish this situation from PAMP. Both, PAMPs and DAMPs are defined by the presence, type and spatial arrangement of chemical groups such as sugars, charged moieties, proteins, and nucleic acids in the combinations, conformations, and in a context not known to the host. Following interaction between the foreign pathogen or altered "self" and the PRRs, either one or both of the following events take place: (i) phagocytosis and clearance by macrophages, leukocytes, and dendritic cells; and (ii) activation of intracellular signaling pathways leading to pro-inflammatory or anti-inflammatory responses. These events are critical in order to mount the early host response to infection and, if necessary, to elicit an effective adaptive immune response. Albeit both processes are often interrelated and mutually dependent [13], attention here will be focused principally on the role of PRRs in the clearance of NPs. As TLRs are not strictly phagocytic receptors and influence phagocytosis indirectly via downstream interaction with professional phagocytic receptors and by inducing an inflammatory response, the TLR role will not be discussed here. Rather, attention will be mostly focused on the extracellular recognition mechanisms, for which the reader is referred to in-depth reviews on the intracellular pathways activated by PRRs [9, 13].

The main pathways responsible for the recognition and clearance of PAMPs and DAMPs are described in Fig. 1 and reviewed extensively elsewhere [1, 13, 14]. As can be seen from this scheme, two main types of innate immunity-mediated clearance are apparent: (i) direct recognition via cell surface PRR; and (ii) indirect recognition via plasma opsonins. Two main groups of PRRs have been identified, namely lectins that mediate sugar

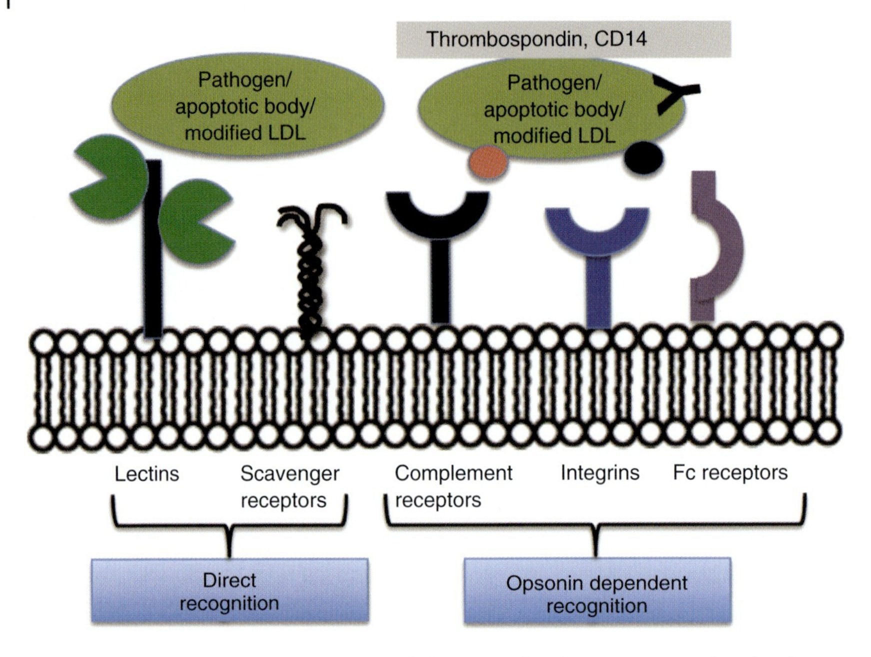

Fig. 1 General overview of mechanisms of clearance of pathogen-associated molecular patterns (PAMPs) and danger-associated molecular patterns (DAMPs).

recognition, and SRs. The separation between these two groups is artificial, as many lectins (e.g., collectins) also have SR function, and many SRs mediate ligand recognition via lectin domains (see below). Opsonin-dependent phagocytosis is based on antibody–Fc receptor, RGD–integrin, and complement factors–complement receptor (mannose-binding lectin receptor, CR3/4, C1qR) interactions. Soluble pattern recognition proteins such as complement, thrombospondin, CD14 and vitronectin are extremely important for innate immunity in the fluid phase.

4 Role of SRs in the Clearance of Modified Lipoproteins, Pathogens, and Apoptotic Cells

Brown and Goldstein discovered SRs in 1979 [15] as receptors responsible for the binding and uptake of oxidized low-density lipoprotein (LDL). Since then, multiple members of the SR family have been identified and described in the seminal studies of Krieger, Gordon, and others [16]. The SRs are structurally very heterogeneous, being subdivided into classes in which, although the members of each class share structural features there is little or no homology among classes [17]. The cartoon representations of the main classes of SRs are shown in Fig. 2, and the main types are summarized in Table 2. The range of binding activities and functions performed by SRs are overwhelming. The SRs "recognition portfolio" is extremely diverse and includes unmodified endogenous proteins and lipoproteins, bacterial lipopolysaccharide (LPS) and lipoteichoic acid (LTA), apoptotic cells, phosphatidylserine (PS), nucleic acids, bacteria, and viruses [17]. Therefore, the definition of an SR includes

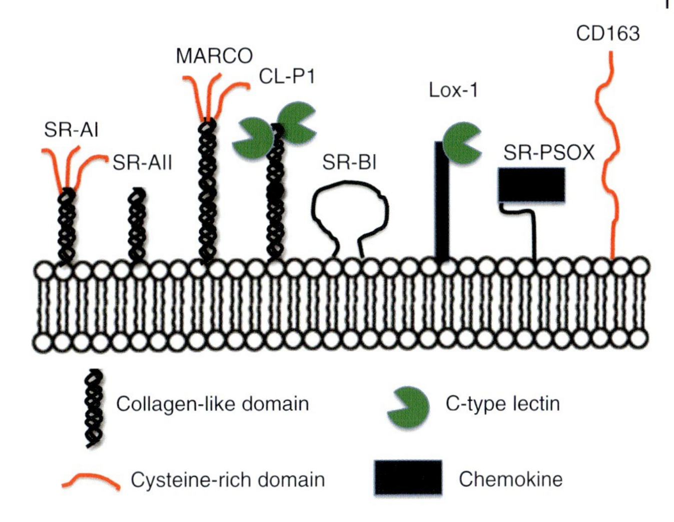

Fig. 2 Structures of the main scavenger receptors (SRs).

not only the recognition of DAMPs but also the recognition of several PAMPs. As such, SRs are considered to be a subclass of the membrane-bound PRRs.

Perhaps the most universal feature of SRs is that their ligands usually possess anionic charge. Thus, PS, polynucleotides, polyanions, malonylated bovine serum albumin (BSA), acetylated low-density lipoprotein (ac-LDL) and oxidized low-density lipoprotein (ox-LDL) have overall negative charge and have been shown to bind to SRs. That being said, the interactions between SRs and their ligands are not based solely on electrostatics, and there are definitely "rules of recognition." Thus, being polyanionic is insufficient to confer receptor binding; for example, poly-inosinic and poly-glutamic acids are efficient inhibitors of scavenger receptor type I/II (SR-AI/II), whereas poly-A and double-stranded DNA are not [16]. The differential inhibition of polynucleotides was explained by formation of tetraplexes that favor the binding [18]. The charge density also seems to play a role, since dextran sulfate and fucoidan are more efficient inhibitors of SRs than are heparan sulfate and chondroitin sulfate [16].

Another perplexing feature of SRs is that even receptors which are structurally similar use different mechanisms for ligand recognition. Thus, class A SR-AI/II and macrophage receptor with collagenous structure (MARCO), despite being structurally similar (both possess a collagen-like domain and a cysteine-rich domain), recognize their ligands differently. SR-AI recognizes polyanionic ligands via its collagen-like domain, whereas MARCO can recognize ligands via its cysteine-rich domain [19, 20]. The mechanisms by which structurally unrelated receptors recognize anionic ligands are still not clear, but are likely related to the presence of charged domains or patches in the receptor structure. Thus, SR-AI and SR-AII recognize the anionic ligands via a positively charged collagen sequence in the upper (C-terminal) part of the receptor [21]. MARCO, CD36, scavenger receptor for phosphatidylserine and oxidized lipoprotein (SR-PSOX) and lectin-like oxidized LDL receptor-1 (LOX-1) appear to recognize anionic ligands

Tab. 2 Types of scavenger receptors.

Scavenger receptors				
Class A	SR-AI/II	Macrophages, endothelial cells, epithelial cells, astrocytes, subpopulations of dendritic cells (DCs), mast cells, and smooth muscle cells in atherosclerotic plaques	acLDL, oxLDL, malondialdehyde LDL, maleylated LDL, lysophosphatidylcholine, phosphatidic acid, cholesterol, Apo A-I, Apo E, glycated type IV collagen, modified collagen type I, III, and IV, biglycan, decorin, modified albumin, AGE-BSA, β-amyloid fibrils, calreticulin, gp96, HSP70 family members, LPS, LTA, CpG DNA, calciprotein particles, *Neisseria meningitidis* surface proteins, Gram-positive and Gram-negative bacteria, C reactive protein, hepatitis C virus NS3 protein, and Tamm–Horsfall protein	Lipid metabolism, clearance of modified host components, clearance of pathogens, clearance of apoptotic cells, B cell–macrophage interactions, antigen presentation, binding of macro phages to extracellular matrix components, and intracellular signaling
	MARCO	Macrophages, endothelial cells, and astrocytes	acLDL, oxLDL, LPS, *N. meningitidis* surface proteins, Gram-positive and Gram-negative bacteria, TiO_2, Fe_2O_3, latex beads, and $CSiO_2$	Clearance of pathogens, B cell–macrophage interactions, clearance of apoptotic cells, clearance of unopsonized particles, and DC and microglial cell maturation
	CSR12	Fibroblasts and leukocytes	No data	Clearance of pathogens, B cell–macrophage interactions, clearance of apoptotic cells, clearance of unopsonized particles, and DC and microglial cell maturation
	Collectin placenta	Endothelial cells, alveolar macrophages, stromal cells, astrocytes, and microglia	oxLDL, yeast, Gram-positive and Gram-negative bacteria, β-amyloid, desialylated Lewis X-containing glycoproteins, lacto-ferrin, and matrix metalloproteinases 8 and 9	Adherence of Lewis X-positive cells to vascular endothelium, clearance of desialylated glycoproteins, and clearance of β-amyloid

(*continued overleaf*)

Tab. 2 (*Continued.*)

Scavenger receptors				
	SCARA5	Epithelial cells	Gram-positive and Gram-negative bacteria, L-ferritin, and haptoglobin-hemoglobin	Tumor suppression and iron trafficking
Class B	SR-BI	Macrophages, dendritic cells, hepatocytes, and steroidogenic tissue	AcLDL, oxLDL, native LDL, native HDL, very low-density lipoprotein VLDL, LPS, Gram-negative bacteria and Gram-positive bacteria, apoptotic cells, *Mycobacterium fortuitum*, Hepatitis C virus, *Plasmodium falciparum*, and Tamm–Horsfall protein	Lipid transfer activity, clearance of bacterial pathogens, and vitamin E transport
	CD36	Macrophages, platelets, adipocytes, and some endothelial and epithelial cells, T cells	AcLDL, oxLDL, LTA and a diacylated lipopeptides, Gram-negative bacteria, phosphatidylserine, β-glucan, *Cryptococcus neoformans*, and *Plasmodium falciparum*	Lipid transfer activity, clearance of apoptotic cells, inhibition of angiogenesis, propagation of proapoptotic signals, induction of pro-inflammatory response, clearance of *Plasmodium falciparum*-infected erythrocytes, and fat-soluble vitamin uptake
	LIMP-2	Expressed ubiquitously	Enterovirus 71, Coxsackie virus A7, A14, and A16, β-glucocerebrosidase, and *Listeria monocytogenes*	Trafficking of β-glucocerebrosidase to lysosomes
Class D	CD68/macrosialin	Macrophages, dendritic cells, Langerhans cells, osteoclasts, fibroblasts, and endothelial cells	OxLDL, phosphatidylserine-rich liposomes, and ICAM-L (*Leishmania* surface protein)	Exact function unknown

(*continued overleaf*)

Tab. 2 (Continued.)

Scavenger receptors				
Class E	LOX-1	Endothelial cells, macrophages, smooth muscle cells, platelets, and adipocytes	OxLDL, acLDL, hypochlorite-modified LDL, remnant-like lipoprotein particle, electronegative LDL, carbamylated LDL, (CRP), fibronectin, HSP60, HSP70, phosphatidylserine, phosphatidylinositol, phosphatidic acid, cardiolipin, phosphatidylglycerol, heparin, dextran sulfate, poly(I), Advanced glycation end products (AGEs), *trans*-2-nonenal, 4-oxo-2(*E*)-nonenal, Gram- negative, and Gram- positive bacteria, and pancreatic bile salt-dependent lipase	Induction of apoptosis of endothelial cells, monocyte adhesion to endothelial cells, release of proinflammatory cytokines and increase in reactive oxygen species (ROS) production by endothelial cells, and antigen presentation
Class F	SREC-1	Endothelial cells and macrophages	Gram-positive and Gram-negative bacteria, hepatitis C virus, acLDL, oxLDL, carbamylated LDL, calreticulin, HSP70, HSP90, HSP 110, glucose-regulated protein 170, fungal pathogens, zymogen granule protein 2, and Tamm–Horsfall protein	Clearance of modified host components, antigen clearance and cross-presentation, and cell morphology changes
Class G	FEEL-1/stabilin-1/CLEVER-1	Macrophages, mono-nuclear cells, hematopoietic stem cells, and endothelial cells	acLDL, Hsp70, AGE, secreted protein acidic and rich in cysteine (SPARC), stabilin-I chitinase-like protein (SICLP), Gram-negative and Gram-positive bacteria, phosphatidylserine, placental lactogen, heparin, and growth differentiation factor 15 (GDF-15)	Lymphocyte adhesion and transmigration, clearance of modified self, angiogenesis, apoptotic cell clearance, and intracellular trafficking

(*continued overleaf*)

Tab. 2 (*Continued.*)

Scavenger receptors				
	FEEL-2/stabilin-2/HARE	Sinusoidal endothelial cells	acLDL, AGE, Gram-negative and Gram-positive bacteria, pro-collagen, hyaluronic acid, phosphatidylserine, heparin, and GDF-15	Clearance of modified self, lymphocyte adhesion, and necrotic/apoptotic cell clearance
Non classified	CD163	Macrophages/monocytes	Hb:Hp, TWEAK, Gram-positive and Gram-negative bacteria	Clearance of Hb:Hp complexes, erythroblast adhesion to macrophages, TWEAK sequestration and degradation, and limit inflammation
	CD5	B-1 cells and T cells	CD5, CD72, gp35-40, gp150, framework region of IgVH, and fungal pathogens	Modulation of T-cell receptor (TCR) and B-cell receptor (BCR) signaling
	CD6	B-1 cells, T cells, and natural killer (NK) cells	Gram-negative and Gram-positive bacteria and CD166/ALCAM	Modulation of TCR and BCR signaling, clearance of bacterial pathogens, contributes to NK-derived cytokine and chemokine secretion
	KIM-1/TIM-1/4	Epithelial cells, T cells, B cells, and dendritic cells	OxLDL, phosphatidylserine, hepatitis A virus (HAV), Ebola virus, Marburg virus, Tim-4, and CD300b	Clearance of apoptotic cells and necrotic cell debris, activation of invariant natural killer T cells (iNKT cells), and regulation of T-cell activation and differentiation
	P2X7	Monocytes, macrophages, dendritic cells, and T cells	Latex beads, bacteria, and apoptotic cells	Clearance of apoptotic cells and clearance of bacterial pathogens

Modified from Ref. [17].

through arginine- and lysine-rich patches in their extracellular domains [20, 22–24].

Below, a brief overview will be provided of the role of SRs, based on studies that used receptor-deficient cells, knockout mice, and receptor inhibitors. Comprehensive reviews on the role of SRs in the uptake of apoptotic cells, lipoproteins, and pathogens are available elsewhere [13, 16, 17, 25–28]. The results of pioneering studies performed by Brown and Goldstein suggested that SR-A efficiently takes up ac-LDL, although SR-AI/II is much less efficient regarding the uptake of ox-LDL [26]. The existence of several receptors with different specificities of ac-LDL and ox-LDL has been suggested in a classical study performed by Steinberg and coworkers [29]. In a remarkable follow-up study, the uptake of ac-LDL was 70% inhibited in SR-AI/II deficient cells, but only 30% of ox-LDL uptake was inhibited in those cells. The residual activity was inhibited completely by poly-inosinic acid [30]. In subsequent investigations, attempts were made to find bypass pathways of ox-LDL uptake in the absence of SR-AI/II. A subsequent search for the ox-LDL receptor led to the finding that CD36 mediates the uptake of ox-LDL by macrophages [31]. In addition, SR-AI/II and CD36 were both shown to be important for ox-LDL uptake. At the same time, the reduction in the binding and degradation of ox-LDL in SR-AI/II/CD36-deficient macrophages was no more than 50–70% [32]. The available data suggest that there may be additional, as yet unidentified, SRs that bind to ox-LDL. It is possible that such receptors are responsible for some of the actions of ox-LDL that involve the activation of cell signaling pathways or the induction of gene expression. Thus, knockout of SR-AI/II and CD36 did not prevent the formation of foam macrophages in a mouse model of atherosclerosis [33]. Recently, another SR-PSOX was found on atherogenic macrophages [34]; however, the disruption of SR-AI/II and SR-PSOX only accelerated the atherosclerosis [35], suggesting that alternative lipid uptake mechanisms may contribute to macrophage cholesterol ester accumulation *in vivo.*

Another example of particulate cargo that is phagocytosed via SRs is pathogens (bacteria, viruses). The role of SRs in the pathogen uptake has been extensively studied. For example, bacterial ligands for SR-AI/II include LPS from the Gram-negative bacterium *Neisseria meningitidis* [36, 37], LTA from Gram-positive bacteria [38], bacterial CpG DNA [39], as well as bacterial surface proteins [40]. The role of SR-AI/II in the uptake of *N. meningitidis* has been shown *in vivo*, where the survival of SR-AI/II-deficient mice was significantly lower than for wild-type mice [40]. SR-AI/II-deficient mice have also been shown to be more susceptible to experimental *Listeria monocytogenes* and *Staphylococcus aureus* infections as a result of deficient bacterial clearance from the spleen and liver [41]. *Streptococcus pyogenes* and group B streptococci (GBS; e.g., *Streptococcus agalactiae*) have evolved mechanisms to evade recognition by macrophages via SR-AI/II. Despite the fact that SR-AI/II participates in the uptake of pathogens, the inhibition of pathogen uptake in SR-AI/II KO mice is only partial. Another class A SR, MARCO has been shown to be involved in host defense against *S. pneumonia* [42] and *N. meningitidis* [43]. The downregulation of MARCO on alveolar macrophages (AMs) has also been linked to a possible enhanced susceptibility to secondary pneumococcal infection subsequent to an influenza infection [44].

The significant involvement of the innate immune system in the recognition of pathogens and dying cells has led to the suggestion that apoptotic cells (ACs), like pathogens, carry conserved molecular patterns, termed apoptotic cell-associated molecular patterns (ACAMPs). The best-characterized of the specific surface changes relates to the exposure of the anionic phospholipid PS, and its presence in the outer leaflet of the plasma membrane generates a negative charge at the cell surface, and can mediate the clearance of ACs [45]. In addition, other lipid modifications that generate a negative charge, such as oxidized PS and other phospholipids, were proposed to play a role in AC recognition [25]. SR-AI/II has been shown to promote the uptake of ACs, but this is not the only receptor responsible for the clearance as its deficiency in mice does not impair the clearance of ACs [46, 47]. Many other PS receptors expressed on the macrophage subtypes have been identified so far. Thus, SR-PSOX [39], T-cell immunoglobulin mucin (TIM)-1/4 [48], LOX-1, CD36 and CD14 have each been shown to promote the internalization of ACs [49–51]. CD36 is very interesting in that regard as it requires an adaptor molecule, thrombospondin, and cooperation with other receptors such as integrins [52], for AC clearance. At the same time, additional evidence suggests that the presence of other, not yet completely characterized, receptors for PS [25, 48, 53, 54]. It has also been shown that soluble opsonins, including mannose-binding lectin and C1q, participate in AC clearance [25].

5
Role of SRs in the Immune Recognition of NPs

The role of SRs in the clearance of NPs has only recently begun to attract attention, due to the importance of the clearance process in the biodistribution, pharmacokinetics and toxicity of NPs [55]. Numerous reports have been made that polyanionic SR inhibitors could inhibit the uptake of NPs, but only a few studies have focused on the molecular mechanisms of NP recognition and the validation of a specific receptor type involved in the uptake. Some of the first studies on the role of SRs in the clearance by macrophages were performed using hazardous and environmental particles. Exposure of the lungs to silica can cause fibrosis, with the initial uptake of silica by human AMs being mediated via SRs so as to trigger apoptosis [56]. Five years later, SR-AI was reported as being essential for caspase activation and apoptosis in murine AMs, as well as necrosis induced by silica [57]. Orr *et al.* showed that the uptake of engineered silica NPs (100 nm) by RAW 264.7 cells (a murine macrophage cell line) was mediated via SR-AI [58]. In this study, silica NPs were incubated in a serum-free medium [58], thereby excluding any effects of serum protein absorption on cellular uptake mechanisms. The SR-AI/II expression level was high in RAW 264.7 cells, but MARCO and SR-BI (a major class B SR) were not expressed in this cell line [58]. As opposed to RAW 264.7 cells, SR-AI/II is not the only SR that mediates the uptake of silica NPs in AMs. Thus, MARCO has been shown to mediate the uptake of silica NPs by primary AMs from C57BL/6 mice [59]. MARCO was also identified to clear other unopsonized environmental particles, including TiO_2(1.3 μm), Fe_2O_3(1.3 μm), and latex beads (1.0 μm), by AMs [60]. Thus, MARCO has been reported to partially mediate the uptake of polystyrene particles (1 μm, 200 nm, and 20 nm) in MARCO-transfected COS-7 cells (a monkey kidney cell line) [61]. Kobzik and colleagues, who pioneered studies on the role of MARCO in the uptake of environmental particles, generated immortalized MARCO/SR-AI/II-deficient AMs and demonstrated a significantly decreased uptake of unopsonized TiO_2, fluorescent latex beads, and bacteria (*Staphylococcus aureus*) compared to the primary AMs from wild-type C57BL/6 mice. At the same time, as the inhibition of uptake was not complete, it was concluded that other receptors are involved in the uptake [62].

SRs play a role in endocytosis and phagocytosis of other types of inorganic and organic NPs. The uptake of Au NP great than 100 nm by RAW 264.7 murine macrophages was mediated by SR-A, while smaller Au NPs (30 nm) were internalized via multiple endocytic mechanisms, including clathrin- and caveolin-mediated pathways [63]. These study results indicated that the inhibition or interruption of one particular pathway would be unlikely to prevent NP uptake by macrophages that utilize multiple mechanisms to recognize and internalize NPs. Mirkin and colleagues showed that the uptake of Au NP decorated with nucleic acids (spherical nucleic acids) by HeLa cells could be inhibited by polyinosinic acid [64]. Although HeLa cells are cervical carcinoma cells, similar mechanisms of DNA–Au NP uptake might be utilized by immune cells with a much more abundant expression of SRs. In addition, the interaction of the spherical nucleic acid and SRs was dependent on the density of DNA on the NP surface, with a higher density promoting a more efficient uptake [64]. This should not be surprising, considering the fact that DNA-coated particles are negatively charged, despite DNA not being a very efficient SR ligand [16]. By using RNA interference knockdown, the same group later showed that the uptake of spherical nucleic acids by C166 mouse endothelial cell line is SR-I/II-dependent, but SR-BI-independent [65]. Silver NPs (44 nm) were inhibited by dextran sulfate in the presence of cell medium [66]. Another study showed an inhibition of 50 nm silver NPs by polyinosinic acid in J774 macrophages, whereby the silver NPs were negatively charged due to a modification with carboxyl-tethered poly(ethylene glycol) (PEG), and had a zeta-potential of −28.4 mV [67], which could explain their affinity for SRs. Unfortunately, the type of SR responsible for the uptake was not reported in this study. The cellular uptake of quantum dots (QDs, 6–12 nm in size, which have great potential in biological and medical applications) by human epidermal keratinocytes was regulated by G protein-coupled receptors and SR-BI [68]. These data suggest that SRs could mediate the uptake of NPs, but the involvement of a specific SR still needs to be determined.

Liposomes are lipid-based NPs which have been investigated in terms of their interactions with SRs. Nishikawa *et al.* were the first to report that liposomes containing various phospholipids and cholesterol were taken up by mouse peritoneal macrophages via SRs [69]. By using SR-binding ligands such as ac-LDL, dextran sulfate or fucoidan, the liposome uptake was decreased by 60% while the ox-LDL competition reduced the liposome uptake by up to 90% [69]. By using ligand-binding assays and SR-transfected cell lines, Rigotti *et al.* were the first to report that SR-BI and CD36 were the receptors which recognized and internalized anionic liposomes containing PS or phosphatidylinositol (PI) [70]. SR-BI was later reported to recognize not only negatively charged liposomes but also apoptotic cells that have PS in their outer plasma membrane [71]. Poly(acrylic acid) (PAA)-conjugated liposomes were internalized by macrophages via RME, including SRs [72]. These studies showed that the uptake of PAA liposomes was inhibited by dextran sulfate and maleylated-BSA [72], both of which are known to bind SRs [73]. Poly(C) and BSA, which are not SR-binding ligands, did not compete the uptake of PAA liposomes [72].

Mechanisms of SR-mediated recognition were studied in detail using dextran-coated superparamagnetic iron oxide (SPIO) NPs, which serve as an effective magnetic resonance imaging contrast agent. SPIO consists of iron oxide (Fe_3O_4 and $\gamma - Fe_2O_3$)

crystals embedded in a meshwork of the polysaccharide (dextran, carboxydextran). The SPIO surface represents a classical molecular pattern due to the presence of periodical anionic charges of iron oxide crystals. Raynal *et al.* showed that the uptake of SPIO (120–180 nm) and ultrasmall superparamagnetic iron oxide (USPIO; 15–30 nm) by mouse peritoneal macrophages was SR-dependent [74]. By using an overexpression assay in human embryonic kidney (HEK) 293T cells, Chao *et al.* showed that SPIO is taken up by SR-AI, but not by the main carbohydrate receptors [75]. The results of a more recent study showed that the overexpression of SR-AI/II, MARCO, endothelial receptor collectin-12, and chemokine SR-PSOX promoted the cellular uptake of SPIO NPs by HEK 293T cells, while the overexpression of receptor SR-BI did not increase SPIO uptake [76]. When Chao *et al.* continued to investigate the binding of SR-AI to SPIO with different surface coatings [77], their results showed that the uptake of SPIO could be blocked by conjugating 20 kDa dextran on the NP surface. Computer modeling has revealed that the direct recognition of an iron oxide crystalline core by the positively charged collagen-like domain of SR-AI [77] could explain the efficient recognition of NPs when not coated by dextran chains, and that 20 kDa dextran chains sterically interfere with the uptake. This is one of the few studies that has attempted to explain the mechanisms of interaction between SRs and engineered NPs.

The effect of serum proteins and protein corona on the recognition and uptake by SRs is not completely understood, and existing reports are conflicting. Some groups have shown an inhibitory effect of serum, while others have shown an enhancing effect of serum on the uptake by SRs. Thus, serum protein binding inhibited the cellular uptake of polyvalent oligonucleotide-functionalized Au NPs mediated via the SR [64]. On the other hand, Fleischer *et al.* found that blood serum protein binding to cationic polystyrene NPs enhanced cell binding, while the serum protein binding to anionic polystyrene NPs inhibited cell binding [78]. Furthermore, the protein–NP complex formed from cationic NPs was recognized by SRs to initialize the cellular uptake, while the protein–NP complex formed from anionic NPs binds to native protein receptors, which were not identified in this study [78]. Serum protein binding also inhibited the cellular uptake of polyvalent oligonucleotide-functionalized Au NPs mediated via the SRs. [64] Gottstein *et al.* showed that 50 nm polystyrene particles could be inhibited by poly-inosinic acid, and the inhibition effect persisted even after coating particles with 2% BSA or 30% fetal bovine serum (FBS) [79]. P2X7, which is a purinergic channel and also a SR, was shown to mediate the uptake of carboxylate beads by monocytes under serum-free conditions, but the uptake was strongly inhibited by serum. This inhibitory effect was explained by the presence of glycoproteins in the serum [80]. Several groups have investigated the role of absorption of albumin and the resultant changes in protein secondary structure on the recognition of particles by SR-AI [81]. Briefly, when albumin becomes adsorbed onto the particles there is a partial denaturation (as confirmed by circular dichroism). As the modified albumin is a classical SR ligand [16], this hypothesis is very plausible. In another study, the effect of the protein corona was different for monocytes and macrophages; macrophage uptake was mediated by denatured albumin on the surface of the particles, whereas monocyte uptake was inhibited by the protein corona [82]. The uptake of single-walled carbon nanotubes

(SWCNTs) by RAW264.7 macrophages was also mediated by SRs, and shown to depend on albumin absorption. Thus, SWCNTs inhibited the induction of cyclooxygenase-2 (Cox-2) by LPS, and this anti-inflammatory response was inhibited by fucoidan (a SR polyanionic antagonist). Fucoidan also reduced the uptake of fluorescently labeled SWCNTs [83]. This effect was reproduced under serum-free conditions (cell medium without FBS) by precoating the SWCNTs with serum albumin. In another study, it was shown that rat lung epithelial and rat aortic endothelial cells could internalize silver (Ag) NPs via SR-BI [84]. This uptake was independent of coating by human serum albumin, BSA and high-density lipoprotein (HDL), possibly due to a negative charge on the Ag NPs. The same group showed that the uptake of Ag NPs and subsequent mast cell degranulation, as evaluated by monitoring the release of β-hexosaminidase, was SR-BI dependent. Interestingly, the uptake of citrate- and polyvinyl pyrrolidone-coated 20-nm NPs was much more dependent on SR-BI than that of larger, 550-nm NPs [85]. Collectively, these data suggest that there could be a strong effect of individual serum proteins on SR-mediated uptake, but that the directional effect (either enhancing or inhibiting) was dependent on the physico-chemical properties of the NPs.

Several reports have addressed the role of SRs in NP uptake within the context of the protein corona *in vivo*. In one reports, a rat liver *ex-vivo* perfusion model was used to analyze the deposition of carboxylated latex particles in Kupffer cells in the liver [86]. The most important observation here was that the opsonin-independent deposition was SR-dependent because it was inhibited by poly-inosinic acid. Moreover, the deposition was increased in the presence of serum, and it was inhibitable by SRs. This suggests that either protein-coated particles are recognized by SRs, or that proteins did not mask the sites for SRs while blocking other uptake pathways. The follow-up study identified fetuin as one of the serum proteins that could mediate the recognition via adsorption onto the NPs [87]. Unfortunately, no studies in fetuin-deficient mice were reported, although *in-vivo* studies could shed more light on the mechanisms of clearance and the role of the protein corona.

The mechanisms of endocytosis of NPs involving SRs appear to differ depending on the cell type and the NP used. Thus, Lunov *et al.* showed that the uptake of 100 nm carboxydextran iron oxide NPs is mediated by SR-AI/II and also clathrin-mediated endocytosis. The uptake of engineered silica NPs (100 nm) by RAW 264.7 cells was mediated via a clathrin-dependent pathway, and single or small clusters of silica NPs during intracellular trafficking within SR-A containing vesicles were detected using confocal fluorescence microscopy [58]. In contrast, the uptake of spherical nucleic acids by a C166 cell line (non-macrophage cell line) was via lipid-raft-dependent, caveolae-mediated endocytosis [65]. Silver NPs can be internalized by the adherent mouse macrophage cell line J774A.1 by SR via actin- and clathrin-dependent endocytosis [67]. Again, in this case it appears that the cell type, NP type and the SRs involved dictate the mechanisms of uptake.

6
Targeting of NPs to SRs for Therapy and Imaging Applications

Scavenger receptors are involved in various pathologies, including Alzheimer's disease, atherosclerosis, diabetes mellitus, and cancer [17]. This situation is partly and directly related to the ability of SRs to transport fatty acids [88], ox-LDL [89], and beta-amyloid [90]. On the other hand, SRs including CD163 and SR-AI/II are over-expressed in disease-associated IL-10^{high} anti-inflammatory macrophages (so-called M2 type), and therefore are considered valid markers of several inflammatory conditions such as atherosclerosis and cancer [17]. A significant effort was spent on designing NPs that would inhibit the uptake of ox-LDL by macrophages. For example, Moghe and colleagues developed anionic lipopolymer particles termed nano-lipoblockers, whereby macromolecules containing mucic acid, lauroyl chloride and PEG would self-assemble into nano-lipoblockers (15–20 nm) that were capable of binding to SR-AI and CD36 (and possibly to other receptors) in order to inhibit the uptake of ox-LDLs by macrophages [91] (up to 80% inhibition was achieved *in vitro* [92, 93]). The blocking efficiency was dependent on the presence of negative charges on the nanolipoblockers, although increasing the number of charges did not necessarily improve the blocking ability [92, 93].

Novel peptide–phospholipid nanocarriers (NCs) for drug delivery were designed to mimic the HDL structure that can interact with SR-BI [94]. Upon recognition of the biomimetic NCs by SR-BI, it was shown in a double-tumor-bearing mouse that the NCs would accumulate in SR-BI-positive tumors at a 3.8-fold higher level than in SR-BI-negative tumors [94]. Based on the fact that the HDL-mimicking peptide-phospholipid scaffold (HPPS) NCs could be internalized into the cytosol directly, bypassing the endolysosomal degradation, the same research group also loaded siRNA onto HPPS NCs and demonstrated an efficient cytosolic delivery of siRNA mediated by SR-BI, and a more effective gene knockdown compared to siRNA delivery using the cationic transfection agent Lipofectamine® [95].

Another research group established an efficient siRNA delivery system by incorporating siRNA into reconstituted high-density lipoprotein (rHDL) NPs that can target to tumors via the SR-BI [96]. Subsequent *in-vivo* gene silencing using those siRNA-rHDL NPs of two critical cancer targets (STAT3: signal transducer and activator of transcription 3; and FAK: focal adhesion kinase) in ovarian and colorectal cancer models resulted in a significantly greater apoptosis ($P < 0.05$) after a combination treatment with docetaxel compared to docetaxel monotherapy. As SR-BI is overexpressed in human nasopharyngeal carcinoma (NPC) cell lines, and in 75% of NPC biopsies, SR-BI is a potential biomarker for targeting to NPC [97]. HPPS NCs were also utilized for targeting NPC as a therapeutic for this study. The HPPS NCs significantly suppressed NPC growth in nude mice, though neither necrosis nor apoptosis was induced [97]. Yang *et al.* also described HDL-coated Au NPs as a novel therapeutic which targeted B-cell lymphoma mediated by SR-BI [98]. In another interesting study, CD163 was also targeted by anti-CD163 antibody-linked, PEGylated liposomes to deliver drugs into M2 macrophages [99]. Doxorubicin, when loaded into CD163-targeted liposomes, showed strong cytotoxic effects on CD163-expressing human monocytes [99]. Seger *et al.* conjugated a peptidic SR-AI ligand with USPIO to target SR-AI in macrophages. In a mouse model of atherosclerosis, SR-AI-targeted USPIO NPs showed a significantly greater accumulation in plaque macrophages, as well as an enhanced MRI contrast [100]. Most recently, PEG-coated USPIO NPs labeled with monoclonal anti-mouse LOX-1 antibodies were used to target LOX-1, class E

SR, in mouse macrophages for the detection of carotid atherosclerotic lesions [101] and inflammatory renal lesions in early diabetic nephropathy [102]. LOX-1-targeted USPIO NPs showed a significantly greater uptake in RAW264.7 macrophages [101, 102]. These promising data suggest that, with an appropriate design, NPs can be targeted to specific receptor types, although the nonspecific passive uptake by immune cells could still be an issue in these applications. One interesting application of a passive targeting to SRs was described recently, where carboxylated NPs were taken predominantly by inflammatory Ly6C^{high} monocytes via MARCO, thereby inhibiting monocyte migration to inflammatory tissues and significantly alleviating pathological conditions in inflammatory disease models in mice [103].

7 Conclusions and Future Directions

As evidenced by the results of the above-described studies, SRs play an important role in the clearance NPs by various immune cells and endothelial cells. SRs recognize negatively charged ligands; in the case of NPs, this recognition is due to the intrinsic negative surface charge or to the absorption of serum proteins that confer a negative charge. At the same time, the structural requirements for SR-mediated recognition of NPs are more complex than

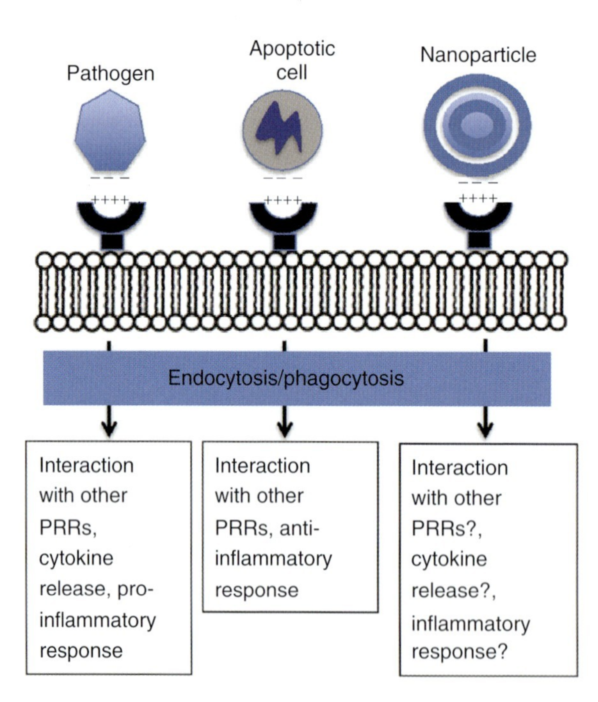

Fig. 3 The recognition of nanoparticles by scavenger receptors is in many regards similar to the recognition of other immune targets. However, many mechanistic details of nanoparticle uptake, such as interactions with other pattern recognition receptor (PRRs) and downstream signaling events, need to be elucidated.

the mere presence of a negative charge. Similar to pathogens and lipoproteins, the SR-mediated uptake of NPs is redundant and single receptor knockouts – in particular SR-AI/II – are unlikely to lead to the prevention of NP clearance. It is plausible to suggest that, in order to inhibit the clearance of NPs, multiple receptors and pathways – both opsonin-dependent (e.g., complement) and opsonin-independent – need to be inhibited at the same time. The pathways leading to NP uptake cannot be generalized, and it is possible that each NP type exhibits its own clearance mechanism; it is entirely possible that more SRs remain to be discovered. Unlike the immune recognition of lipoproteins, pathogens and apoptotic cells, the clearance of NPs has not been studied to the same level of detail (understandably so, since nanomedicine is a relatively new field). Moreover, the downstream pro- and anti-inflammatory pathways following the uptake of NPs via SRs and interaction with TLRs and other intracellular PRRs (Fig. 3) are only just beginning to unravel. The ability to modulate or inhibit these pathways will have a critical effect on the development of nanomedicine, and can help to improve NP design.

Abbreviations

SR	scavenger receptors
PRR	pattern recognition receptor
NP	nanoparticle
RME	receptor-mediated endocytosis
PAMP	pathogen-associated molecular pattern
DAMP	danger-associated molecular pattern
LDL	low-density lipoproteins
TLR	toll-like receptor
MARCO	macrophage receptor with collagenous structure
SR-PSOX	scavenger receptor for phosphatidylserine and oxidized lipoprotein
SPIO	superparamagnetic iron oxide

References

1. Taylor, P.R., Martinez-Pomares, L., Stacey, M., Lin, H.H. *et al.* (2005) Macrophage receptors and immune recognition. *Annu. Rev. Immunol.*, **23**, 901–944.
2. Moghimi, S.M., Hunter, A.C., Murray, J.C. (2001) Long-circulating and target-specific nanoparticles: theory to practice. *Pharmacol. Rev.*, **53** (2), 283–318.
3. Oh, N., Park, J.H. (2014) Endocytosis and exocytosis of nanoparticles in mammalian cells. *Int. J. Nanomedicine*, **9** (Suppl. 1), 51–63.
4. Albanese, A., Tang, P.S., Chan, W.C. (2012) The effect of nanoparticle size, shape, and surface chemistry on biological systems. *Annu. Rev. Biomed. Eng.*, **14**, 1–16.
5. Verma, A., Stellacci, F. (2010) Effect of surface properties on nanoparticle–cell interactions. *Small*, **6** (1), 12–21.
6. Pratten, M.K., Lloyd, J.B. (1986) Pinocytosis and phagocytosis: the effect of size of a particulate substrate on its mode of capture by rat peritoneal macrophages cultured in vitro. *Biochim. Biophys. Acta*, **881** (3), 307–313.
7. Doherty, G.J., McMahon, H.T. (2009) Mechanisms of endocytosis. *Annu. Rev. Biochem.*, **78**, 857–902.
8. Rejman, J., Oberle, V., Zuhorn, I.S., Hoekstra, D. (2004) Size-dependent internalization of particles via the pathways of clathrin- and caveolae-mediated endocytosis. *Biochem. J.*, **377** (Pt. 1), 159–169.
9. Stuart, L.M., Ezekowitz, R.A. (2005) Phagocytosis: elegant complexity. *Immunity*, **22** (5), 539–550.
10. Garin, J., Diez, R., Kieffer, S., Dermine, J.F. *et al.* (2001) The phagosome proteome: insight into phagosome functions. *J. Cell Biol.*, **152** (1), 165–180.
11. Janeway, C.A. Jr., (1989) Approaching the asymptote? Evolution and revolution in immunology. *Cold Spring Harb. Symp. Quant. Biol.*, **54** (Pt. 1), 1–13.

12. O'Neill, L.A., Golenbock, D., Bowie, A.G. (2013) The history of Toll-like receptors – redefining innate immunity. *Nat. Rev. Immunol.*, **13** (6), 453–460.
13. Kumar, H., Kawai, T., Akira, S. (2011) Pathogen recognition by the innate immune system. *Int. Rev. Immunol.*, **30** (1), 16–34.
14. Pluddemann, A., Mukhopadhyay, S., Gordon, S. (2011) Innate immunity to intracellular pathogens: macrophage receptors and responses to microbial entry. *Immunol. Rev.*, **240** (1), 11–24.
15. Goldstein, J.L., Ho, Y.K., Basu, S.K., Brown, M.S. (1979) Binding site on macrophages that mediates uptake and degradation of acetylated low density lipoprotein, producing massive cholesterol deposition. *Proc. Natl Acad. Sci. USA*, **76** (1), 333–337.
16. Platt, N., Gordon, S. (1998) Scavenger receptors: diverse activities and promiscuous binding of polyanionic ligands. *Chem. Biol.*, **5** (8), R193–R203.
17. Canton, J., Neculai, D., Grinstein, S. (2013) Scavenger receptors in homeostasis and immunity. *Nat. Rev. Immunol.*, **13** (9), 621–634.
18. Pearson, A.M., Rich, A., Krieger, M. (1993) Polynucleotide binding to macrophage scavenger receptors depends on the formation of base-quartet-stabilized four-stranded helices. *J. Biol. Chem.*, **268** (5), 3546–3554.
19. Jozefowski, S., Arredouani, M., Sulahian, T., Kobzik, L. (2005) Disparate regulation and function of the class A scavenger receptors SR-AI/II and MARCO. *J. Immunol.*, **175** (12), 8032–8041.
20. Ojala, J.R., Pikkarainen, T., Tuuttila, A., Sandalova, T. *et al.* (2007) Crystal structure of the cysteine-rich domain of scavenger receptor MARCO reveals the presence of a basic and an acidic cluster that both contribute to ligand recognition. *J. Biol. Chem.*, **282** (22), 16654–16666.
21. Doi, T., Higashino, K., Kurihara, Y., Wada, Y. *et al.* (1993) Charged collagen structure mediates the recognition of negatively charged macromolecules by macrophage scavenger receptors. *J. Biol. Chem.*, **268** (3), 2126–2133.
22. Kar, N.S., Ashraf, M.Z., Valiyaveettil, M., Podrez, E.A. (2008) Mapping and characterization of the binding site for specific oxidized phospholipids and oxidized low density lipoprotein of scavenger receptor CD36. *J. Biol. Chem.*, **283** (13), 8765–8771.
23. Ohki, I., Ishigaki, T., Oyama, T., Matsunaga, S. *et al.* (2005) Crystal structure of human lectin-like, oxidized low-density lipoprotein receptor 1 ligand binding domain and its ligand recognition mode to OxLDL. *Structure*, **13** (6), 905–917.
24. Shimaoka, T., Nakayama, T., Hieshima, K., Kume, N. *et al.* (2004) Chemokines generally exhibit scavenger receptor activity through their receptor-binding domain. *J. Biol. Chem.*, **279** (26), 26807–26810.
25. Erwig, L.P., Henson, P.M. (2008) Clearance of apoptotic cells by phagocytes. *Cell Death Differ.*, **15** (2), 243–250.
26. Krieger, M., Herz, J. (1994) Structures and functions of multiligand lipoprotein receptors: macrophage scavenger receptors and LDL receptor-related protein (LRP). *Annu. Rev. Biochem.*, **63**, 601–637.
27. Witztum, J.L., Lichtman, A.H. (2014) The influence of innate and adaptive immune responses on atherosclerosis. *Annu. Rev. Pathol.*, **9**, 73–102.
28. Greaves, D.R., Gordon, S. (2009) The macrophage scavenger receptor at 30 years of age: current knowledge and future challenges. *J. Lipid Res.*, **50** (Supp. l), S282–S286.
29. Sparrow, C.P., Parthasarathy, S., Steinberg, D. (1989) A macrophage receptor that recognizes oxidized low density lipoprotein but not acetylated low density lipoprotein. *J. Biol. Chem.*, **264** (5), 2599–2604.
30. Lougheed, M., Lum, C.M., Ling, W., Suzuki, H. *et al.* (1997) High- affinity saturable uptake of oxidized low density lipoprotein by macrophages from mice lacking the scavenger receptor class A type I/II. *J. Biol. Chem.*, **272** (20), 12938–12944.
31. Endemann, G., Stanton, L.W., Madden, K.S., Bryant, C.M. *et al.* (1993) CD36 is a receptor for oxidized low-density lipoprotein. *J. Biol. Chem.*, **268** (16), 11811–11816.
32. Kunjathoor, V.V., Febbraio, M., Podrez, E.A., Moore, K.J. *et al.* (2002) Scavenger receptors class A-I/II and CD36 are the principal receptors responsible for the uptake of modified low-density lipoprotein leading to lipid loading in macrophages. *J. Biol. Chem.*, **277** (51), 49982–49988.
33. Moore, K.J., Kunjathoor, V.V., Koehn, S.L., Manning, J.J. *et al.* (2005) Loss of

receptor-mediated lipid uptake via scavenger receptor A or CD36 pathways does not ameliorate atherosclerosis in hyperlipidemic mice. *J. Clin. Invest.*, **115** (8), 2192–2201.
34. Shimaoka, T., Kume, N., Minami, M., Hayashida, K. *et al.* (2000) Molecular cloning of a novel scavenger receptor for oxidized low-density lipoprotein, SR-PSOX, on macrophages. *J. Biol. Chem.*, **275** (52), 40663–40666.
35. Aslanian, A.M., Charo, I.F. (2006) Targeted disruption of the scavenger receptor and chemokine CXCL16 accelerates atherosclerosis. *Circulation*, **114** (6), 583–590.
36. Arredouani, M.S., Yang, Z.P., Imrich, A., Ning, Y.Y. *et al.* (2006) The macrophage scavenger receptor SR-AI/II and lung defense against pneumococci and particles. *Am. J. Respir. Cell Mol. Biol.*, **35** (4), 474–478.
37. Peiser, L., de Winther, M.P.J., Makepeace, K., Hollinshead, M. *et al.* (2002) The class A macrophage scavenger receptor is a major pattern recognition receptor for *Neisseria meningitidis* which is independent of lipopolysaccharide and not required for secretory responses. *Infect. Immun.*, **70** (10), 5346–5354.
38. Dunne, D.W., Resnick, D., Greenberg, J., Krieger, M. *et al.* (1994) The type-I macrophage scavenger receptor binds to Gram-positive bacteria and recognizes lipoteichoic acid. *Proc. Natl Acad. Sci. USA*, **91** (5), 1863–1867.
39. Zhu, F.G., Reich, C.F., Pisetsky, D.S. (2001) The role of the macrophage scavenger receptor in immune stimulation by bacterial DNA and synthetic oligonucleotides. *Immunology*, **103** (2), 226–234.
40. Peiser, L., Makepeace, K., Plüddemann, A., Savino, S. *et al.* (2006) Identification of *Neisseria meningitidis* nonlipopolysaccharide ligands for class A macrophage scavenger receptor by using a novel assay. *Infect. Immun.*, **74** (9), 5191–5199.
41. Thomas, C.A., Li, Y., Kodama, T., Suzuki, H. *et al.* (2000) Protection from lethal Gram-positive infection by macrophage scavenger receptor-dependent phagocytosis. *J. Exp. Med.*, **191** (1), 147–156.
42. Arredouani, M., Yang, Z., Ning, Y., Qin, G. *et al.* (2004) The scavenger receptor MARCO is required for lung defense against pneumococcal pneumonia and inhaled particles. *J. Exp. Med.*, **200** (2), 267–272.
43. Mukhopadhyay, S., Chen, Y., Sankala, M., Peiser, L. *et al.* (2006) MARCO, an innate activation marker of macrophages, is a class A scavenger receptor for *Neisseria meningitidis*. *Eur. J. Immunol.*, **36** (4), 940–949.
44. Sun, K., Metzger, D.W. (2008) Inhibition of pulmonary antibacterial defense by interferon-gamma during recovery from influenza infection. *Nat. Med.*, **14** (5), 558–564.
45. Fadok, V.A., Voelker, D.R., Campbell, P.A., Cohen, J.J. *et al.* (1992) Exposure of phosphatidylserine on the surface of apoptotic lymphocytes triggers specific recognition and removal by macrophages. *J. Immunol.*, **148** (7), 2207–2216.
46. Terpstra, V., Kondratenko, N., Steinberg, D. (1997) Macrophages lacking scavenger receptor A show a decrease in binding and uptake of acetylated low-density lipoprotein and of apoptotic thymocytes, but not of oxidatively damaged red blood cells. *Proc. Natl Acad. Sci. USA*, **94** (15), 8127–8131.
47. Platt, N., Suzuki, H., Kodama, T., Gordon, S. (2000) Apoptotic thymocyte clearance in scavenger receptor class A-deficient mice is apparently normal. *J. Immunol.*, **164** (9), 4861–4867.
48. Miyanishi, M., Tada, K., Koike, M., Uchiyama, Y. *et al.* (2007) Identification of Tim4 as a phosphatidylserine receptor. *Nature*, **450** (7168), 435–439.
49. Oka, K., Sawamura, T., Kikuta, K., Itokawa, S. *et al.* (1998) Lectin-like oxidized low-density lipoprotein receptor 1 mediates phagocytosis of aged/apoptotic cells in endothelial cells. *Proc. Natl Acad. Sci. USA*, **95** (16), 9535–9540.
50. Devitt, A., Moffatt, O.D., Raykundalia, C., Capra, J.D. *et al.* (1998) Human CD14 mediates recognition and phagocytosis of apoptotic cells. *Nature*, **392** (6675), 505–509.
51. Sambrano, G.R., Steinberg, D. (1995) Recognition of oxidatively damaged and apoptotic cells by an oxidized low-density lipoprotein receptor on mouse peritoneal macrophages: role of membrane phosphatidylserine. *Proc. Natl Acad. Sci. USA*, **92** (5), 1396–1400.

52. Savill, J., Hogg, N., Ren, Y., Haslett, C. (1992) Thrombospondin cooperates with CD36 and the vitronectin receptor in macrophage recognition of neutrophils undergoing apoptosis. *J. Clin. Invest.*, **90** (4), 1513–1522.
53. Fadok, V.A., Bratton, D.L., Rose, D.M., Pearson, A. *et al.* (2000) A receptor for phosphatidylserine-specific clearance of apoptotic cells. *Nature*, **405** (6782), 85–90.
54. Bose, J., Gruber, A.D., Helming, L., Schiebe, S. *et al.* (2004) The phosphatidylserine receptor has essential functions during embryogenesis but not in apoptotic cell removal. *J. Biol.*, **3** (4), 15.
55. Li, S.D., Huang, L. (2008) Pharmacokinetics and biodistribution of nanoparticles. *Mol. Pharm.*, **5** (4), 496–504.
56. Iyer, R., Hamilton, R.F., Li, L., Holian, A. (1996) Silica-induced apoptosis mediated via scavenger receptor in human alveolar macrophages. *Toxicol. Appl. Pharmacol.*, **141** (1), 84–92.
57. Chao, S.K., Hamilton, R.F., Pfau, J.C., Holian, A. (2001) Cell surface regulation of silica-induced apoptosis by the SR-A scavenger receptor in a murine lung macrophage cell line (MH-S). *Toxicol. Appl. Pharmacol.*, **174** (1), 10–16.
58. Orr, G.A., Chrisler, W.B., Cassens, K.J., Tan, R. *et al.* (2011) Cellular recognition and trafficking of amorphous silica nanoparticles by macrophage scavenger receptor A. *Nanotoxicology*, **5** (3), 296–311.
59. Hamilton, R.F. Jr., Thakur, S.A., Mayfair, J.K., Holian, A. (2006) MARCO mediates silica uptake and toxicity in alveolar macrophages from C57BL/6 mice. *J. Biol. Chem.*, **281** (45), 34218–34226.
60. Palecanda, A., Paulauskis, J., Al-Mutairi, E., Imrich, A. *et al.* (1999) Role of the scavenger receptor MARCO in alveolar macrophage binding of unopsonized environmental particles. *J. Exp. Med.*, **189** (9), 1497–1506.
61. Kanno, S., Furuyama, A., Hirano, S. (2007) A murine scavenger receptor MARCO recognizes polystyrene nanoparticles. *Toxicol. Sci.*, **97** (2), 398–406.
62. Zhou, H., Imrich, A., Kobzik, L. (2008) Characterization of immortalized MARCO and SR-AI/II-deficient murine alveolar macrophage cell lines. *Part. Fibre Toxicol.*, **5**, 7.
63. Franca, A., Aggarwal, P., Barsov, E.V., Kozlov, S.V. *et al.* (2011) Macrophage scavenger receptor A mediates the uptake of gold colloids by macrophages *in vitro*. *Nanomedicine*, **6** (7), 1175–1188.
64. Patel, P.C., Giljohann, D.A., Daniel, W.L., Zheng, D. *et al.* (2010) Scavenger receptors mediate cellular uptake of polyvalent oligonucleotide-functionalized gold nanoparticles. *Bioconjug. Chem.*, **21** (12), 2250–2256.
65. Choi, C.H., Hao, L., Narayan, S.P., Auyeung, E. *et al.* (2013) Mechanism for the endocytosis of spherical nucleic acid nanoparticle conjugates. *Proc. Natl Acad. Sci. USA*, **110** (19), 7625–7630.
66. Singh, R.P., Ramarao, P. (2012) Cellular uptake, intracellular trafficking and cytotoxicity of silver nanoparticles. *Toxicol. Lett.*, **213** (2), 249–259.
67. Wang, H., Wu, L., Reinhard, B.M. (2012) Scavenger receptor-mediated endocytosis of silver nanoparticles into J774A.1 macrophages is heterogeneous. *ACS Nano*, **6** (8), 7122–7132.
68. Zhang, L.W., Monteiro-Riviere, N.A. (2009) Mechanisms of quantum dot nanoparticle cellular uptake. *Toxicol. Sci.*, **110** (1), 138–155.
69. Nishikawa, K., Arai, H., Inoue, K. (1990) Scavenger receptor-mediated uptake and metabolism of lipid vesicles containing acidic phospholipids by mouse peritoneal macrophages. *J. Biol. Chem.*, **265** (9), 5226–5231.
70. Rigotti, A., Acton, S.L., Krieger, M. (1995) The class B scavenger receptors SR-BI and CD36 are receptors for anionic phospholipids. *J. Biol. Chem.*, **270** (27), 16221–16224.
71. Fukasawa, M., Adachi, H., Hirota, K., Tsujimoto, M. *et al.* (1996) SRB1, a class B scavenger receptor, recognizes both negatively charged liposomes and apoptotic cells. *Exp. Cell Res.*, **222** (1), 246–250.
72. Fujiwara, M., Baldeschwieler, J.D., Grubbs, R.H. (1996) Receptor-mediated endocytosis of poly(acrylic acid)-conjugated liposomes by macrophages. *Biochim. Biophys. Acta*, **1278** (1), 59–67.
73. Krieger, M. (1992) Molecular flypaper and atherosclerosis – structure of the macrophage scavenger receptor. *Trends Biochem. Sci.*, **17** (4), 141–146.

74. Raynal, I., Prigent, P., Peyramaure, S., Najid, A. *et al.* (2004) Macrophage endocytosis of superparamagnetic iron oxide nanoparticles: mechanisms and comparison of ferumoxides and ferumoxtran-10. *Invest. Radiol.*, **39** (1), 56–63.
75. Chao, Y., Karmali, P.P., Simberg, D. (2012) Role of carbohydrate receptors in the macrophage uptake of dextran-coated iron oxide nanoparticles. *Adv. Exp. Med. Biol.*, **733**, 115–123.
76. Chao, Y., Makale, M., Karmali, P.P., Sharikov, Y. *et al.* (2012) Recognition of dextran-superparamagnetic iron oxide nanoparticle conjugates (Feridex) via macrophage scavenger receptor charged domains. *Bioconjug. Chem.*, **23** (5), 1003–1009.
77. Chao, Y., Karmali, P.P., Mukthavaram, R., Kesari, S. *et al.* (2013) Direct recognition of superparamagnetic nanocrystals by macrophage scavenger receptor SR-AI. *ACS Nano*, **7** (5), 4289–4298.
78. Fleischer, C.C., Payne, C.K. (2012) Nanoparticle surface charge mediates the cellular receptors used by protein–nanoparticle complexes. *J. Phys. Chem. B*, **116** (30), 8901–8907.
79. Gottstein, C., Wu, G., Wong, B.J., Zasadzinski, J.A. (2013) Precise quantification of nanoparticle internalization. *ACS Nano*, **7** (6), 4933–4945.
80. Gu, B.J., Duce, J.A., Valova, V.A., Wong, B. *et al.* (2012) P2X7 receptor-mediated scavenger activity of mononuclear phagocytes toward non-opsonized particles and apoptotic cells is inhibited by serum glycoproteins but remains active in cerebrospinal fluid. *J. Biol. Chem.*, **287** (21), 17318–17330.
81. Mortimer, G.M., Butcher, N.J., Musumeci, A.W., Deng, Z.J. *et al.* (2014) Cryptic epitopes of albumin determine mononuclear phagocyte system clearance of nanomaterials. *ACS Nano*, **8** (4), 3357–3366.
82. Yan, Y., Gause, K.T., Kamphuis, M.M., Ang, C.S. *et al.* (2013) Differential roles of the protein corona in the cellular uptake of nanoporous polymer particles by monocyte and macrophage cell lines. *ACS Nano*, **7** (12), 10960–10970.
83. Dutta, D., Sundaram, S.K., Teeguarden, J.G., Riley, B.J. *et al.* (2007) Adsorbed proteins influence the biological activity and molecular targeting of nanomaterials. *Toxicol. Sci.*, **100** (1), 303–315.
84. Shannahan, J.H., Podila, R., Aldossari, A.A., Emerson, H. *et al.* (2015) Formation of a protein corona on silver nanoparticles mediates cellular toxicity via scavenger receptors. *Toxicol. Sci.*, **143** (1), 136–146.
85. Aldossari, A.A., Shannahan, J.H., Podila, R., Brown, J.M. (2014) Influence of physicochemical properties of silver nanoparticles on mast cell activation and degranulation. *Toxicol. In Vitro*, **29** (1), 195–203.
86. Furumoto, K., Nagayama, S., Ogawara, K., Takakura, Y. *et al.* (2004) Hepatic uptake of negatively charged particles in rats: possible involvement of serum proteins in recognition by scavenger receptor. *J. Control. Release*, **97** (1), 133–141.
87. Nagayama, S., Ogawara, K., Minato, K., Fukuoka, Y. *et al.* (2007) Fetuin mediates hepatic uptake of negatively charged nanoparticles via scavenger receptor. *Int. J. Pharm.*, **329** (1-2), 192–198.
88. Coort, S.L., Willems, J., Coumans, W.A., van der Vusse, G.J. *et al.* (2002) Sulfo-*N*-succinimidyl esters of long-chain fatty acids specifically inhibit fatty acid translocase (FAT/CD36)-mediated cellular fatty acid uptake. *Mol. Cell. Biochem.*, **239** (1-2), 213–219.
89. Miller, Y.I., Chang, M.K., Binder, C.J., Shaw, P.X. *et al.* (2003) Oxidized low-density lipoprotein and innate immune receptors. *Curr. Opin. Lipidol.*, **14** (5), 437–445.
90. Santiago-Garcia, J., Mas-Oliva, J., Innerarity, T.L., Pitas, R.E. (2001) Secreted forms of the amyloid-beta precursor protein are ligands for the class A scavenger receptor. *J. Biol. Chem.*, **276** (33), 30655–30661.
91. Chnari, E., Nikitczuk, J.S., Wang, J.Z., Uhrich, K.E. *et al.* (2006) Engineered polymeric nanoparticles for receptor-targeted blockage of oxidized low- density lipoprotein uptake and atherogenesis in macrophages. *Biomacromolecules*, **7** (6), 1796–1805.
92. Iverson, N.M., Sparks, S.M., Demirdirek, B., Uhrich, K.E. *et al.* (2010) Controllable inhibition of cellular uptake of oxidized low-density lipoprotein: structure–function relationships for nanoscale amphiphilic polymers. *Acta Biomater.*, **6** (8), 3081–3091.

93. Hehir, S., Plourde, N.M., Gu, L., Poree, D.E. *et al.* (2012) Carbohydrate composition of amphiphilic macromolecules influences physicochemical properties and binding to atherogenic scavenger receptor A. *Acta Biomater.*, **8** (11), 3956–3962.
94. Zhang, Z., Cao, W., Jin, H., Lovell, J.F. *et al.* (2009) Biomimetic nanocarrier for direct cytosolic drug delivery. *Angew. Chem.*, **48** (48), 9171–9175.
95. Yang, M., Jin, H., Chen, J., Ding, L. *et al.* (2011) Efficient cytosolic delivery of siRNA using HDL-mimicking nanoparticles. *Small*, **7** (5), 568–573.
96. Shahzad, M.M., Mangala, L.S., Han, H.D., Lu, C. *et al.* (2011) Targeted delivery of small interfering RNA using reconstituted high-density lipoprotein nanoparticles. *Neoplasia*, **13** (4), 309–319.
97. Zheng, Y., Liu, Y., Jin, H., Pan, S. *et al.* (2013) Scavenger receptor B1 is a potential biomarker of human nasopharyngeal carcinoma and its growth is inhibited by HDL-mimetic nanoparticles. *Theranostics*, **3** (7), 477–486.
98. Yang, S., Damiano, M.G., Zhang, H., Tripathy, S.*et al.* (2013) Biomimetic, synthetic HDL nanostructures for lymphoma. *Proc. Natl Acad. Sci. USA*, **110** (7), 2511–2516.
99. Etzerodt, A., Maniecki, M.B., Graversen, J.H., Moller, H.J. *et al.* (2012) Efficient intracellular drug-targeting of macrophages using stealth liposomes directed to the hemoglobin scavenger receptor CD163. *J. Control. Release*, **160** (1), 72–80.
100. Segers, F.M., den Adel, B., Bot, I., van der Graaf, L.M. *et al.* (2013) Scavenger receptor-AI-targeted iron oxide nanoparticles for *in vivo* MRI detection of atherosclerotic lesions. *Arterioscler. Thromb. Vasc. Biol.*, **33** (8), 1812–1819.
101. Wen, S., Liu, D.F., Cui, Y., Harris, S.S. *et al.* (2014) *In vivo* MRI detection of carotid atherosclerotic lesions and kidney inflammation in ApoE-deficient mice by using LOX-1 targeted iron nanoparticles. *Nanomed.: Nanotechnol., Biol. Med.*, **10** (3), 639–649.
102. Luo, B., Wen, S., Chen, Y.C., Cui, Y. *et al.* (2015) LOX-1-targeted iron oxide nanoparticles detect early diabetic nephropathy in db/db mice. *Mol. Imaging Biol.* (in press).
103. Getts, D.R., Terry, R.L., Getts, M.T., Deffrasnes, C. *et al.* (2014) Therapeutic inflammatory monocyte modulation using immune-modifying microparticles. *Sci. Transl. Med.*, **6** (219), 219ra7.

Part IV
Cancer

Translational Nanomedicine. First Edition. Edited by Robert A. Meyers.

11
Gold and Iron Oxide Nanoparticles with Antibody Guides to Find and Destroy Cancer Cells

Stephanie A. Parker[1], *Isabel A. Soto*[1], *Dickson K. Kirui*[3], *Cameron L. Bardliving*[2], *and Carl A. Batt*[2]
[1] *Cornell University, Department of Biomedical Engineering, 101 Weill Hall, Ithaca, New York 14853, USA*
[2] *Cornell University, Department of Food Science, 351 Stocking Hall, Ithaca, New York 14853, USA*
[3] *Houston Methodist Research Institute, Department of NanoMedicine, 6670 Bertner Ave, Houston, Texas 77030, USA*

Translational Nanomedicine. First Edition. Edited by Robert A. Meyers.

Keywords

Bioconjugation
A chemical strategy that forms a stable covalent link between two biomolecules.

Theranostics
A combination of diagnostic and therapeutic capabilities into a single agent, in this instance multimodal nanoparticles used for cancer imaging and treatment.

Gold nanoparticles (AuNPs)
Gold particles with diameters between 1 to 100 nm in size.

Superparamagnetic iron oxide nanoparticles (SPIONs)
Iron oxide particles that have a core diameter of less than 20 nm and exhibit superparamagnetic properties at room temperature.

Iron oxide nanoparticles (IONPs)
Iron oxide particles with diameters of between 1 and 100 nm.

Gold and iron oxide hybrid nanoparticles (HNPs)
Metal particles bearing multiple components that are structurally very close to the actual catalysts, with diameters of between 1 and 100 nm. Here, gold and iron oxide composites nanoparticles are reviewed.

Nanoparticles represent an increasingly important field of theranostic agents which can not only sense but also treat disease. Gold nanoparticles (AuNPs) and iron oxide nanoparticles (IONPs) functionalized with antibodies (Abs) specific to tumor antigens can detect and treat cancer at the cellular level. These particles are being investigated for clinical applications, including imaging contrast, thermal tumor therapy, drug delivery, and radiosensitizing agents. In this chapter, a review is provided of the unique properties of these particles which, when combined with targeting moieties, make them suitable for cancer imaging and therapeutics, as evidenced by the recent preclinical, *in-vivo* and *in-vitro* studies, conducted with AuNP and IONP Ab conjugates. Results from these preclinical studies provide evidence for the potential clinical applications, which will advance the diagnosis and treatment of cancer.

1 Introduction

Despite significant advances in cancer research, treatment and diagnosis over past two decades, cancer remains a major health problem in the United States and many other parts of the world. Cancer is the second leading cause of death in the US, accounting for 23% of all deaths in 2009, and is the leading cause of death for both men and women aged 40–79 years [1, 2]. The current standard treatment strategies include surgical resection, chemotherapy and radiotherapy, but these are indiscriminate and unnecessarily lead to the damage of healthy tissues. The limitations of treatments such as chemotherapy and radiotherapy can be attributed to resistance to chemotherapeutics [3], physiological barriers, and poor bioavailability [3, 4]. Advances in nanomedicine have led to the development of immunotargeted nanomaterials that have the potential to address the limitations of conventional cancer therapeutics by targeting cancer at the molecular level. Gold nanoparticles (AuNPs) and iron oxide nanoparticles (IONPs), when conjugated to antibodies, are of particular interest for cancer theranostic applications (Fig. 1). Both, AuNPs and IONPs possess unique optical and magnetic properties that can be exploited for simultaneous diagnostic imaging and therapy, including thermal and controlled drug release for cancer therapy [5] (Fig. 2), which can lead to improved treatments. Recent *in-vitro* and *in-vivo* studies conducted with immunotargeted nanoparticles have shown that these conjugates exhibit a targeted specificity, improved pharmacokinetics, and can even deliver drugs via a leaky vasculature [6].

The synthesis and composition of AuNPs, IONPs and hybrid gold and IO nanoparticles (HNPs), in addition to the clinically relevant properties of these nanomaterials, will first be reviewed. An overview will then be provided of the functionalization strategies used to couple nanoparticles with cancer-targeting antibodies, and some of the targeting moieties currently available will be outlined. Preclinical studies focused on the development of antibody-conjugated AuNPs, IONPs and HNPs for cancer imaging and therapeutics will then be examined, including hyperthermia, targeted drug delivery, and radiation therapy.

2 Composition

During recent years, AuNPs, IONPs and HNPs have been extensively studied and synthesized using several methods that may be applied for biomedical purposes. The unique characteristics of nanoparticles that have spurred interest in them for clinical applications include: (i) their large surface-to-volume ratio; (ii) the physical and chemical properties that can be tuned depending on their size, composition, and shape; (iii) their enhanced penetration and preferential accumulation in tumor tissues; (iv) their high absorption band in the near-infrared (NIR) region [7]; and (v) their ease of surface functionalization with biomolecules [8]. The nanoscale dimensions of nanoparticles also make them suited for targeting cancer at the cellular level [9].

2.1 Gold Nanoparticles

The most common method used for AuNP synthesis is the chemical or electrochemical reduction of a gold(III) precursor [10].

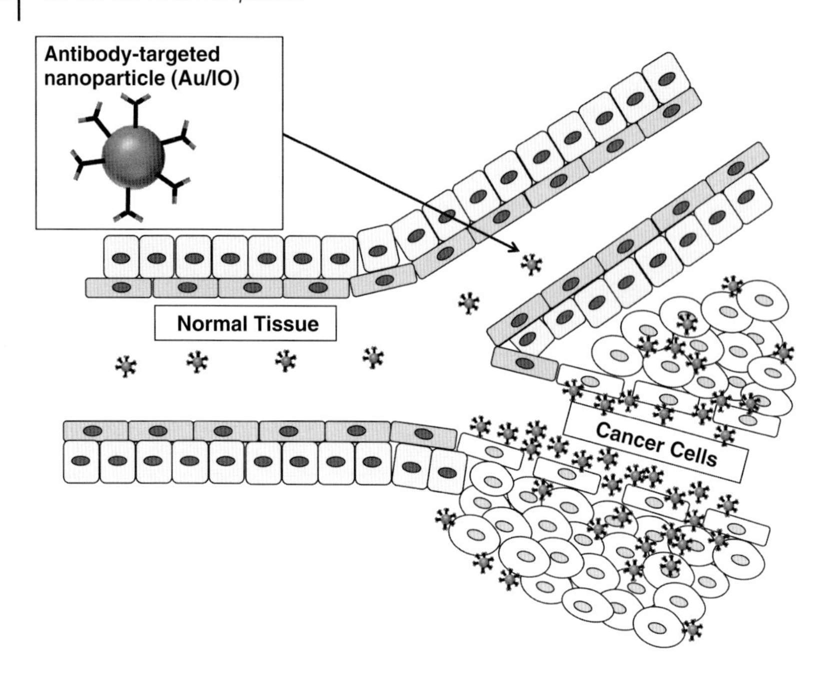

Fig. 1 Antibody-targeted gold (Au)/iron oxide (IO)/hybrid nanoparticles moving through the vascular system of the body to selectively search for and destroy cancerous cells. Normal cells are portrayed as rectangles, while cancer cells are portrayed as oval. Antibody-targeted nanoparticles are able to penetrate the epithelial layer, and antibodies on the particles bind to tumor-specific antigens on the cancer cells, resulting in an accumulation of nanoparticles at the site of disease.

A variety of shapes and structures can be synthesized, with the most prevalent AuNP structures including nanorods, nanospheres and nanoshells, as well as composite particles where gold is combined with other elements such as silica and iron (see Sect. 2.3) [10]. The geometry and composition of AuNPs both affect the nanoparticles' optical properties, including absorption wavelength [11], as well as their interactions with biological molecules, including enzyme catalysis [12]. Hence, AuNPs can be tuned for specific imaging and thermal therapy applications; the varying properties of the nanoparticles are described more fully in Refs [11, 13–15].

AuNPs possess several distinct properties that make them attractive for clinical applications, including an easily modified surface chemistry and size-dependent electrical and optical properties [16–18]. AuNPs possess intense surface plasmon resonance (SPR) scattering, which results in an enhanced electromagnetic field (EMF) at the nanoparticle surface [19]. Scattered light from AuNPs produces an array of colors under dark-field and confocal microscopy imaging applications [20–22]. Due to their high atomic mass AuNPs also absorb significantly more X-ray radiation than do soft tissue cells, and consequently they can be used to enhance cancer radiation therapy and provide an increased contrast in diagnostic computed tomography (CT) scanning [23].

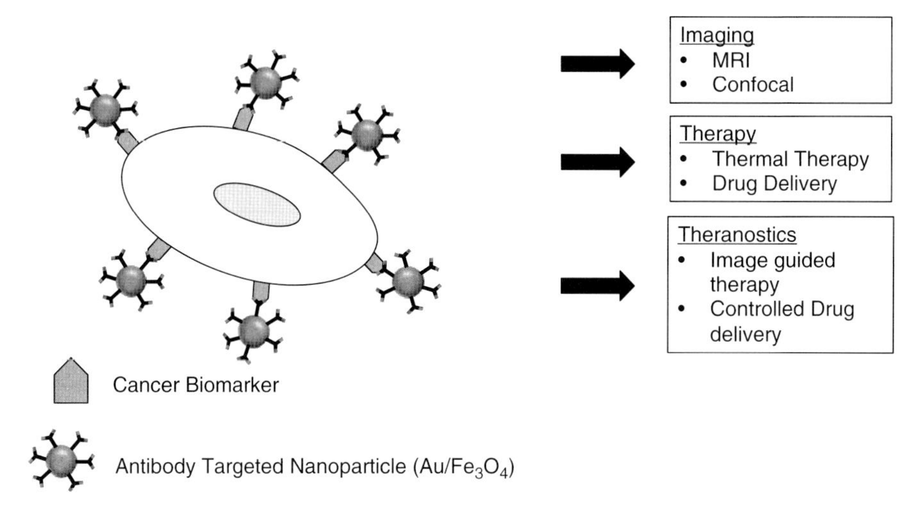

Fig. 2 Antibody-targeted nanoparticles gold (Au), iron oxide (IO), hybrid (Au and IO) nanoparticles binding selectively to cancer biomarkers located on the surface of a cancer cell. This selective and specific binding and the unique optical and magnetic properties of the nanoparticles allows for the imaging and therapy of cancer cells.

2.2 Iron Oxide Nanoparticles

IONPs can be synthesized with a precise size, morphology, surface treatment and magnetic properties [24, 25]. The common methods of synthesis include coprecipitation, microemulsions, and the high-temperature decomposition of organic precursors [25]. Maghemite (Fe_2O_3) and magnetite (Fe_3O_4) iron oxide can each be prepared as monodispersed surface-derivatized nanoparticles [24, 26], and ferric magnetic materials such as iron oxide also possess unique paramagnetic properties. When the size domain of the nanoparticle is reduced below that of a single magnetic domain [27] the magnetic properties are altered, and this results in superparamagnetic behavior [25]. IONPs with a core diameter <20 nm exhibit superparamagnetic properties at room temperature, and are termed superparamagnetic iron oxide nanoparticles (SPIONs).

SPIONs have a significant ability to reduce the magnetic resonance imaging (MRI) signal, thus acting as negative contrast agents and causing a darkening of the MR image [28]. Superparamagnetic iron oxide (SPIO) agents, such as ferumoxides (Feridex IV; Berlex Laboratories. Inc. in the USA and Endorem, Guerbet S.A in Europe), have been approved for clinical applications in the USA [29]. Feridex SPIONs have been approved for intravenous administration and are used as contrast agents to identify liver tumors [30]. These SPIONs serve as attractive platforms for targeted drug delivery, magnetic hyperthermia, immunotherapy, tumor visualization, and the detection of circulating tumor cells [31]. Preclinical studies in animal models have shown that antibody-conjugated SPIONs retain their specificity for antigen-expressing tissues and are also effective contrast-enhancing agents for MRI.

2.3 Hybrid Nanoparticles

The potential benefits of integrating nanomaterials with disparate properties (i.e., magnetism and energy absorption) into HNPs have spurred great interest in the development of various synthetic strategies. HNPs retain the optical and magnetic properties of AuNPs and IONPs, respectively [32–36], and can be fabricated by the aqueous-phase [37] and high-temperature organic phase thermal decomposition of their respective precursors [37–40]. The addition of a gold coating to IONPs has been reported to enhance their chemical stability by protecting the core against oxidation and corrosion [41, 42]. HNPs with gold cores encapsulated within hollow iron oxide shells and gold decorated with iron oxide have been developed for dual-imaging agents, MRI and X-ray detection [43], and multifunctional imaging and thermal applications, respectively [42, 44, 45].

3 Nanoparticle Surface Functionalization

Following their synthesis, several strategies have been employed to functionalize AuNPs and IONPs by allowing the addition of targeting moieties – that is, cancer-specific antibodies. A successful functionalization strategy will not alter the desired properties of the nanoparticles, and will preserve the targeting ability of the immobilized antibodies. First, however, the nanoparticles must be adequately coated, as bare or surfactant-stabilized AuNPs and IONPs are unstable and cytotoxic under physiological conditions [46–48]. These unmodified nanoparticles are also prone to aggregation, and would be rapidly cleared from the body via the reticuloendothelial system (RES) [49]. The resultant shortened blood half-life reduces the efficacy of AuNP- and IONP-based theranostic platforms, where longer a circulation time will increase the probability that the particles can find, attach to and accumulate in, targeted cancer cells.

Fortunately, facile chemistries provide methods for stabilizing the nanoparticles and conjugating antibodies to their surfaces [50]. The most common methods of nanoparticle functionalization include ligand exchange [51–64], coating with amphiphilic polymers [65–78], coating with silica [79–83], polymer coating, and encapsulation (Fig. 3) [84–88]. Following surface coating, AuNPs and IONPs have been shown to retain the desired relativities, biocompatibility properties, native optical properties, and can easily be additionally functionalized with biomolecules, notably monoclonal antibodies (mAbs) and single-chain variable fragments (ScFvs) [89].

3.1 Antibody Immobilization and Orientation on Nanoparticles

Following stabilization coating, nanoparticles can then be functionalized with antibodies for targeted cancer therapeutic applications. Five classes of antibodies have been identified, with immunoglobulin G (IgG) being the most abundant in human serum, and thus the most widely used for biomedical applications [88]. mAbs consist of four polypeptide chains with a molecular weight of 180 kDa, which complicates their conjugation to nanoparticles. mAb and antibody fragments including ScFvs and

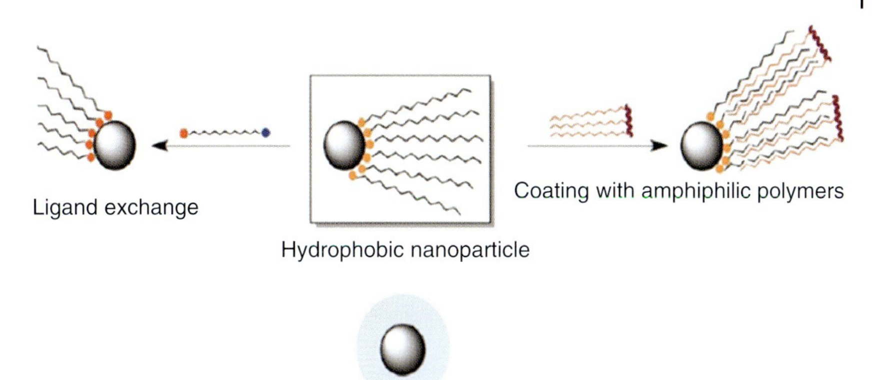

Fig. 3 Primary coating strategies for engineering the surfaces of nanoparticles. The diagram shows ligand exchange, hydrophobic interactions, coating with amphiphilic polymers, and silica coating. Reprinted with permission from Ref. [88]; © 2014, American Chemical Society.

fragment antigen-binding (Fab) antibodies have been investigated in preclinical studies for tumor-targeting nanoparticle treatment applications. The strategies used to produce functional nanoparticles with antibodies include: (i) physical adsorption [90]; (ii) covalent binding via amine groups or carbohydrate groups on antibodies [91]; (iii) the use of adapter biomolecules (streptavidin–biotin, Protein G); and (iv) ionic adsorption. However, a drawback to streptavidin–biotin is that streptavidin is a bacterial protein without a mammalian analog and may cause an immunogenic response in patients [92, 93]. The choice of coupling method depends on the type and amount of functional groups present on the surface-coating layer of the nanoparticles.

It is also important to note that the immobilization of biologically active molecules requires a correct steric availability of the active binding sites in order to avoid reducing or completely eliminating the binding capacity. The orientation of antibodies on the nanoparticle surface must be such that antigen-binding sites are well exposed, without compromising the function of the protein. The steric hindrance of antigen-recognition sites due to randomly oriented antibodies can result in a 100-fold decrease in binding affinity [94]. End-on or flat-on orientations have been found to yield higher antigen-binding capacities compared to random (head-on and sideways) orientations [88]. Typically, to retain full biological activity of the immobilized antibody, the recognition sites – the Fab regions – should be oriented away from surface of the nanoparticle. A directed immobilization of antibody fragments, Fabs and ScFvs, has been achieved via the addition of functionalities to the terminal regions of ScFvs [95] and Fabs [96]. Modification of the internal linker regions of ScFvs has also been reported for site-specific immobilization on the surface of nanoparticles [97].

4 Cancer-Targeting Antibodies

Antibodies are an example of targeting molecules that can allow the nanoparticles

Tab. 1 FDA-approved unmodified monoclonal antibodies for solid tumor and hematological malignancies.

Therapeutic antibody	*Target*	*FDA-approved indication*	*Mechanism of action*
Solid tumor malignancies			
Trastuzumab (Herceptin; Genentech); humanized IgG1	HER2/neu	HER2-overexpressing breast cancer	Inhibition of HER2/neu signaling and ADCC
		HER2-overexpressing metastatic gastric or gastroesophageal junction adenocarcinoma	
Cetuximab (Erbitux; Bristol-Myers Squibb); chimeric human-murine IgG1	EGFR	Colorectal cancer that has metastasized	Inhibition of EGFR signaling and ADCC
		Squamous cell carcinoma of the head and neck	
Bevacizumab (Avastin; Genetech/Roche); humanized IgG1	VEGF	Colorectal cancer that has metastasized	Inhibition of VEGF signaling
		Glioblastoma in patients not responding to other treatments	
		Non-small-cell lung that is locally advance, cannot be surgically removed, has metastasized, or has recurred	
		Renal cell cancer that has metastasized	
Panitumumab (Vectibix; Amgen); human IgG2	EGFR	Colorectal cancer (certain types) in patients whose disease has not responded with other chemotherapy and has metastasized	Inhibition of EGFR signaling
Ipilimumab (Yervoy; Bristol-Myers Squibb); IgG1	CTLA4	Melanoma that cannot be surgically removed or has metastasized	Inhibition of CTLA4 signaling
Pertuzumab (Perjeta; Genentech); humanized IgG1	HER2/neu	Neoadjuvant treatment of HER2-positive breast cancer in combination with trastuzumab and docetaxel (chemotherapeutic)	Inhibition of HER2/neu signaling
		HER2-overexpressing metastatic breast cancer	
Hematological malignancies			
Rituximab (Mabthera; Roche); chimeric human-murine IgG1	CD20	For treatment of CD20-positive B-cell non-Hodgkin's lymphoma CLL	ADCC, induction of apoptosis and CDC
Alemtuzumab (Campath; Genzyme); humanized IgG1	CD52	Approved for single-agent treatment of B-cell CLL.	Induction of apoptosis and CDC
Ofatumumab (Arzerra; Genmab); human IgG1	CD20	For treatment of patients with CLL no longer controlled by other forms of chemotherapy	ADCC and CDC
Obinutuzumab (Gazya; Genetech/Roche); humanized IgG1	CD20	For patients with previously untreated CLL	ADCC and CDC

ADCC, antibody-dependent cellular cytotoxicity; CDC, complement-dependent cytotoxicity; CLL, chronic lymphocytic leukemia; EGFR, epidermal growth factor receptor.

to locate cancerous cells and identify them at a cellular level. On contacting the cancerous cell, the NPs can be used to provide treatment on a highly region-specific basis. Tumor stromal and vascular cells express specific antigens that distinguish them from normal tissue [98], and the identification of these antigens has led to the clinical development of therapeutic mAbs that recognize cancer cells as compared with normal tissues [99, 100]. These cancer antigen-specific antibodies are among the most successful targeted cancer therapeutic strategies for patients with hematological malignancies and solid tumors [98, 101, 102]. Currently, 15 unmodified antibodies have been approved by the Food and Drug Administration for cancer therapy [103, 104]. Tumor-associated antigens recognized by therapeutic antibodies are cell-surface differentiation antigens. Hematopoietic differentiation antigens are generally associated with antibodies that target cluster of differentiation (CD) groups, including CD20, CD20, CD33, and CD52. Cell-surface antigens are glycoproteins and carbohydrates found on the surface of both normal and tumor cells, but they are overexpressed in tumor cell populations. The unmodified mAbs currently approved for solid tumor and hematological malignancies are detailed in Table 1.

ScFv, Ab fragments, have also been explored for use as targeting ligands for cancer therapeutics [105, 106]. ScFvs are recombinant proteins generated from mAbs, and their fragments contain the highly specific variable domains of mAbs, though they are smaller in size. Cancer-targeting ScFvs conjugated to nanoparticles have been used for site-specific therapeutic delivery and imaging [107]. This reduction in size also facilitates a direct and dense immobilization of the antibodies on the surface of nanoparticles [108].

5
AuNPs and IONPs with Antibody Conjugates for Cancer Imaging and Detection

The first step in the process of treating cancer with an antibody-conjugated AuNP and/or IONP is to identify the cancer. Immunotargeted AuNPs and IONPs are of particular interest as these nanoparticles, when selectively targeted to diseased sites, increase the signal-to-noise ratio of traditional imaging technologies, thus making it possible to visualize antigen-expressing cancer cells [109]. The ability to image cancer at the cellular level would provide a significant improvement over traditional methods by helping to diagnose cancers earlier than is currently feasible, and with unaided imaging techniques.

The current standard optical technologies for cancer detection are optical coherence tomography (OCT) and reflectance confocal microscopy (RCM). These technologies image the micro-anatomic features of diseased tissue, and cannot distinguish among cancers based specifically on molecular biomarkers associated with carcinogenesis [110]. The image is generated via reflected light from the endogenous chromophores present in the tissue [111]. The resulting optical signal is weak, and results in poor intracellular contrast and subtle spectral differences between malignant and benign cells. To overcome these limitations, exogenous chemicals have been explored to function as contrast agents; unfortunately, however, organic dyes are subject to rapid photobleaching, biological and chemical degradation, and their signal quickly decreases upon exposure to an excitation light source [112, 113].

AuNPs, IONPs and HNPs would be an improvement over dyes as they are photostable [114, 115] and can carry fluorescent dyes for NIR imaging [113, 116]. Preclinical studies have employed several imaging techniques to examine the capacity of antibody-conjugated AuNPs, IONPs and HNPs for cancer imaging, including NIR, confocal reflectance [117], dark-field, optoacoustic microscopy [118], X-ray microtomography (micro-CT) [119], OCT [120], and MRI [76, 121] contrast agents [122]. Some recent antibody-targeted imaging studies conducted with AuNPs, IONPs and HNPs are outlined in Table 2. The objective of preclinical studies has been to demonstrate the selectivity and quantify the improved imaging as a result of using antibody-conjugated nanoparticles, thereby demonstrating their feasibility for clinical studies. Highlights from recent *in-vivo* imaging studies with AuNPs and IONPs conjugated with cancer-targeting antibodies are provided in the following section. The variety of targeting agents and the subsequent imaging modalities demonstrate the robust nature of these approaches for the detection of a variety of cancers.

5.1 Imaging Studies

Specific examples of imaging studies using antibody-directed nanoparticles demonstrate the utility of the approach. For example, in a recent study by Karmani *et al.* [126], AuNPs conjugated with the anti-CD105 mAb, specific to tumor angiogenesis and tumor progression, were investigated for theranostic, dual-imaging and therapeutic applications in an *in-vivo* mouse model The antibodies were radiolabeled with ^{89}Zr to facilitate positron emission tomography (PET) imaging before their conjugation to 5 nm AuNPs. Male mice bearing melanoma xenografts were intravenously injected with the radiolabeled, antibody-conjugated AuNPs, as well as radiolabeled CD-105 antibody alone and nontargeted AuNPs for comparison. Whole-body PET images and an inductively coupled plasma mass spectrometry (ICP-MS) analysis of the blood and tumors were used to assess the *in-vivo* distribution of the nanoparticles. A high tumor contrast and selective

Tab. 2 Cancer imaging studies conducted with antibody-functionalized gold and iron oxide nanoparticles.

Nanoparticle type	*Cancer type(s)*	*Antibody conjugate*	*Imaging technique (Reference)*
AuNP, GNR	Lung cancer, solid tumor malignancies	Anti-EGFR	Confocal [123], dark-field microscopy [124]
GNR	Breast cancer	Herceptin (anti-HER2/neu)	Micro CT [119]
GNR	Prostate cancer	Anti-PSMA	Optoacoustic imaging [118]
AuNP	Colorectal cancer	Anti-β-catenin and anti-E-cadherin	Confocal [111]
AuNP, HNP	Solid tumor malignancies	Anti-CD105	PET imaging [18], MRI [76]
IONP	Breast cancer	Anti-HER2/neu	MRI [19, 125]

AuNP, gold nanoparticle; GNR, gold nanorod; HNP, hybrid nanoparticle; IONP, iron oxide nanoparticle; MRI, magnetic resonance imaging; Micro CT, X-ray microtomography; PET, positron emission tomography.

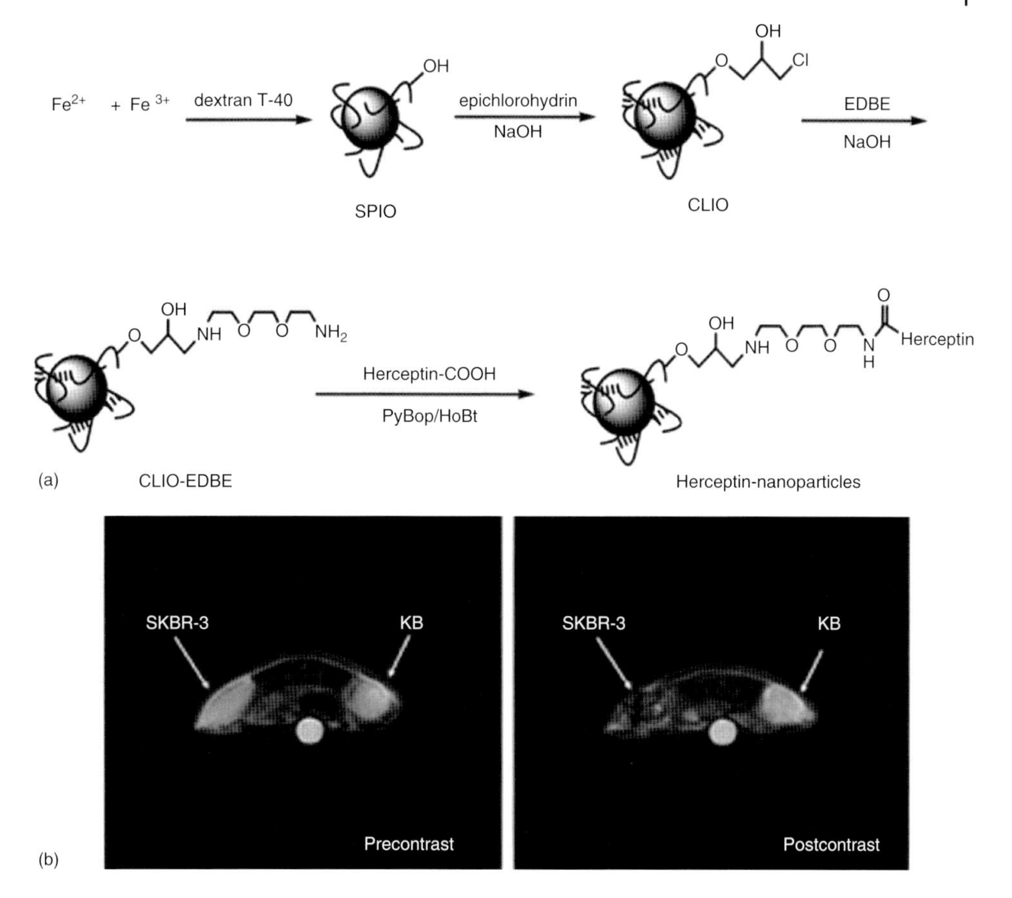

Fig. 4 SPION conjugation scheme and enhanced MRI images. (a) The synthetic scheme of herceptin–nanoparticles. SPIO, superparamagnetic iron oxide; CLIO, cross-linked iron oxide, 2,2′-(ethylenedioxy)bisethylamine; PyBop, (benzotriazol-1-yloxy) tripyrrolidinophosphonium hexafluorophosphate; HoBt, 1-hydroxybenzotriazole; (b) T_2-weighted MRI before (left) and after (right) injection of herceptin–iron oxide nanoparticles. The cell lines used here are SKBR-3 HER2/neu-expressing breast cancer cells and KB non-HER2/neu-expressing epidermal carcinoma cells. Reprinted with permission from Ref. [124]; © 2009, John Wiley & Sons, Ltd.

tumor targeting were observed with the immunotargeted AuNPs, as compared to the antibody–target radioisotope alone or the nontargeted AuNPs. Yet, the major finding of this study was that conjugation of the anti-CD105 antibody with AuNPs did not inhibit tumor uptake, nor the efficacy of tumor-targeting of the antibody.

In another recent study, nanoparticles targeting HER2/neu demonstrated an effective MRI imaging of HER2-expressing cells in xenograft mice [127] (Fig. 4). In this case, SPIONs were first surface-modified with dextran, followed by the addition of primary amine functional groups, and then functionalized with Herceptin in the presence of coupling agents. The targeting of Herceptin–SPIONs to HER2/neu receptor-positive breast cancer cell lines was detected via *in-vitro* and *in-vivo* MRI.

The *in-vitro* MRI signal enhancement increased proportionally as a function of the relative HER2/neu expression level. For *in-vivo* studies, the Herceptin–SPIONs were administered intravenously to tumor-bearing mice prepared by the subcutaneous injection of KB cells (human nasopharyngeal epidermal carcinoma cells lacking HER2/neu receptors) and SKBR-3 (HER2/neu-expressing breast cancer) cells. The tumor site was detected via T_2-weighted MRI, and the administration of Herceptin–SPIONs produced a 45% signal enhancement in the HER2/neu-positive tumor compared to controls, indicating a high level of accumulation within the tumor (Fig. 4). The results of this study demonstrated the potential of SPIONs to be used as MRI contrast agents for the detection of HER2/neu-expressing breast cancer.

6 Antibody-Conjugated AuNPs and IONPs for Cancer Therapeutics

Nanoparticles can serve as theranostic agents, allowing the clinician to initially image and diagnose the cancer and then to follow-up with a target-specific therapy. The advantage is the simplicity of the coupled approach and the ability to initiate treatment only when the diagnostic indicates a need for treatment. Therapeutic strategies for the treatment of cancers employing antibody-conjugated AuNPs and IONPs include hyperthermia, targeted drug delivery, and the enhancement of radiation therapies. These therapeutic methods can be combined with imaging modalities for theranostics, including image-guided hyperthermia and drug delivery. The advantage of the theranostic approach would be that, before the administration of therapeutics, a physician would be able to track the particles and direct treatment at the site of disease. This technology presents a powerful tool that could be used to discover cancer cells and provide treatment immediately, in a potential combined screening and treatment strategy. A brief overview of recent developments in the area of AuNPs and IONPs for cancer therapeutics is provided in the following section.

6.1 Targeted Hyperthermia

One example of a targeted therapeutic approach that utilizes nanoparticles is the exogenous application of energy which is then concentrated by the nanoparticles, resulting in a region-specific increase in temperature. Hyperthermia is the elevation of temperature up to 42–46 °C, usually by photon flux of 5–48 W cm^{-2} [128], for an extended period of time to damage and kill cancer cells by DNA denaturation [129–131]. Hyperthermia has been widely used for cancer treatment [132] to enhance cancer treatment by directly causing cell death, sensitizing cells to chemotherapy [128] and radiotherapy [133], and by promoting tumor reoxygenation [133, 135]. The major technical impedances of traditional hyperthermia treatment are a difficulty of heating the local tumor region without damaging normal tissue [138], and an inability to create hyperthermia uniformly throughout the tumor, leaving cancerous cells and resulting in regrowth of tumor [137].

Targeted AuNP- and IONP-mediated hyperthermia has been explored to enhance treatment specificity and increase heating uniformity [138], and can be initiated via magnetic or light induction. Magnetic hyperthermia is induced by exposing magnetic nanoparticles to an alternating magnetic field (AMF) [31, 139]. Light-induced hyperthermia, photothermal therapy (PTT), is the use of electromagnetic radiation (infrared wavelengths) to convert NIR light energy into heat [140, 141]. The objective of preclinical studies in this area has been to determine the feasibility of administering antibody-conjugated AuNPs and IONPs directed to the disease site by tumor antigen-targeting antibodies, followed by the initiation of thermal activation by either magnetic or light induction. Overviews of some recent *in-vitro* and *in-vivo* studies in this area are detailed in the following sections.

6.1.1 *In-Vitro* Hyperthermia Studies

Nanoparticle-mediated hyperthermia with bioconjugated antigen-targeted photothermal therapeutic agents have been of particular interest for breast cancer theranostic applications [142]. Loo *et al.* reported dual gold nanoshells conjugated with anti-HER2 mAb via attachment with a poly(ethylene glycol) (PEG) linker for imaging and PTT of breast cancer cells. The gold nanoshells were incubated with SKBR3 breast cancer cells for binding studies. The results of the study showed a significantly increased scatter based on optical contrast due to nanoshell selective binding to breast cancer cells as compared to control cell groups [143]. The anti-HER2 nanoshells also exhibited a targeted photothermal effect, with cell death occurring as a result of treatment with a NIR laser after exposure to anti-HER2 nanoshells. The effect was not seen with nanoshells labeled with a nonspecific control antibody and unconjugated nanoshells. The study demonstrated the bioimaging and cancer therapeutic potential of mAb-conjugated gold nanoshells with a clinically relevant biomarker, HER2.

HNPs have also been investigated for use in MRI-guided PTT. Melancon *et al.* reported HNPs for use in MRI-guided laser ablation of head and neck cancer as a minimally invasive alternative for patients who are not candidates for surgery [34]. In this study, a SPIO core covered in amorphous silica was coated with a gold nanoshell (SPIO@Au). The particles were then conjugated with anti-epidermal growth factor receptor (EGFR) Cetuximab (C225) and tested for their capacity to selectively ablate human squamous carcinoma cell lines overexpressing EGFR *in vitro.* The results of the study showed that the C225–SPIO@Au conjugates could produce a temperature in excess of 65 °C, and cell death was only apparent with cells incubated with C225–SPIO@Au nanoshells after exposure to NIR laser light for 3 min. MRI detection was carried out using cells inoculated with the HNP conjugates and then suspended in agar. Cells inoculated with C225–SPIO@Au nanoshells showed darker contrast than the nontargeted control samples and the antibody-blocked control. The biodistribution of these nanoshell conjugates showed a selective uptake into tumors in comparison to PEG-coated SPIO@Au. However, a majority of the conjugates localized to the liver, spleen and kidneys. In addition, the authors did not demonstrate any *in-vivo* ablation of tumors using the C225 functionalized nanoshells.

6.1.2 *In-Vivo* Hyperthermia Studies

Preclinical *in-vivo* studies for targeted hyperthermia have investigated the potential for antibody-conjugated nanoparticles to improve hyperthermia treatment. In one study, multifunctional gold–iron oxide HNPs targeted with humanized single-chain antibody conjugates (A33ScFv) were studied in SW1222 cells (antigen-expressing

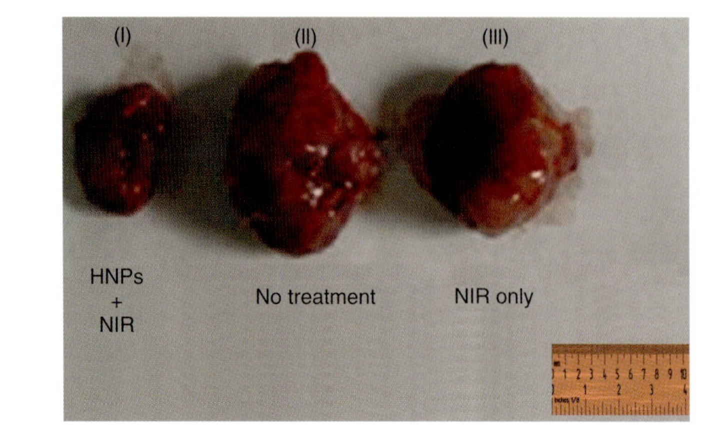

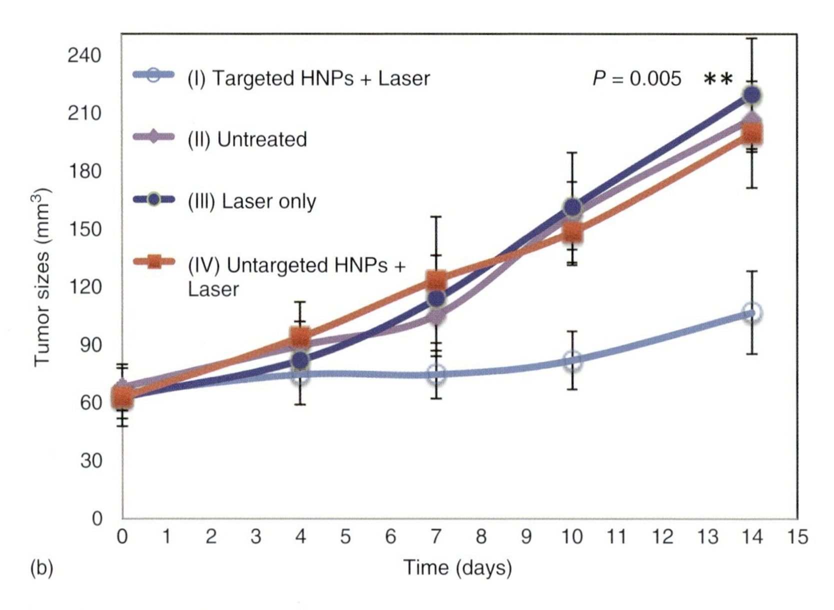

Fig. 5 Photothermal therapy (PTT) response on subcutaneous colorectal tumors. (a) Tumor masses harvested after treatment with HNPs and NIR laser (group I), untreated control (group II), or PBS plus NIR laser (group III); (b) Tumor growth curves after PTT. Tumor size calculated as $a \times b^2/2$ (a = length, b = width), measured 24 h after irradiation and the duration of therapy, (n = 5 mice per group). Reprinted with permission from Ref. [144]; © 2013, Elsevier.

colorectal cancer cell line) for use in laser-assisted PTT using NIR irradiation and MRI contrast agents *in vivo*. The HNPs in this study were effective MRI contrast agents, reducing the post-contrast phase T_2 value by 50%. The HNPs showed localization to SW1222 tumor xenografts within 12 h of systemic injection. The efficacy of photothermal treatment was confirmed by a significant reduction in tumor volume in samples treated with target HNPs exposed to a NIR laser (Fig. 5). Histological analysis further confirmed degradation of the extracellular matrix, cell necrosis, and nucleus damage/shrinkage in the treatment group. PTT resulted in necrosis in approximately 65% of tumor tissues [144].

6.2 Targeted Drug Delivery

Current cancer drug therapies are limited by inefficient delivery due to insolubility, toxicity, and non-specificity [145], but using nanoparticles as the delivery platform for many of these therapies has the potential to overcome these limitations. Nanoparticles provide a large surface area-to-volume ratio, which allows them to bind, adsorb, and carry a large variety of chemical compounds [147]. Coupling chemical compounds with nanoparticles has been shown to increase the aqueous stability of many insoluble drugs and provide specificity, which leads to lower dosage requirements, thereby reducing harmful side effects [145, 147].

The use of AuNPs and IONPs as a drug-delivery system has been shown to increase the effectiveness of several anticancer drugs. Several mechanisms are attributed to this increase in efficiency, including an increased solubility of the active therapy, nanoparticle-mediated cellular uptake, and prolonged circulation and receptor wrapping time [148]. For example, AuNPs have been shown to increase the efficacy of the chemotherapy drugs doxorubicin (DOX) and gemcitabine (Gemzar) [149, 150]. The stability in solution and cytotoxicity of the originally hydrophobic chemotherapy drug, paclitaxel, was also significantly increased after its conjugation to AuNPs [145, 146]. Optimization of this system using PEGylated AuNPs with biotin receptors has further enhanced the cytotoxic effects of paclitaxel [146]. IONPs loaded with chemotherapeutics have been shown to increase the cellular uptake – and thus the potency – of the drugs, including in multidrug-resistant cancer cells [151, 152]. Methotrexate, an antimetabolite used in the treatment of various cancers has been successfully conjugated with IONPs, resulting in an increased cytotoxicity towards target cells due to an improvement in cellular uptake [147].

AuNPs and IONPs can not only carry drugs to target locations, but can also be used to produce a controlled release of the drugs. Triggered drug release from AuNPs and IONPs has been achieved with various activators, including X-rays, glutathione and changes in the environmental pH [67, 148, 153–155]. Notably, the local irradiation of gold nanorods bound to a therapeutic agent via a thiol group has been shown to result in a detachment of the drug in the local area, thus reducing the nonspecific dispersion of stand-alone drugs, which can in turn lead to toxicity in non-targeted organs [158]. Preliminary studies have also shown that radiofrequency electromagnetic field (RF-EMF) activation can also be an effective trigger for the independent release of multiple drugs on a single nanoparticle, using nucleic acid tethers [157]. Some details of recent *in-vitro* and *in-vivo* preclinical studies in this area are outlined in the following sections.

6.2.1 *In-Vitro* Targeted Drug-Delivery Studies

Bisker *et al.* conducted a controlled-release study using AuNPs conjugated with rituximab, an anti-CD20 mAb, for the treatment of leukemia (Fig. 6). [158]. AuNPs of 20 nm diameter were conjugated with PEGylated anti-CD20 via thiol chemistry, and the functionalized AuNPs were then incubated with a human CD-20-positive B-cell line for controlled-release studies. The controlled release of anti-CD20 was activated upon irradiation by a femtosecond pulse train, the wavelength of which was tuned to match

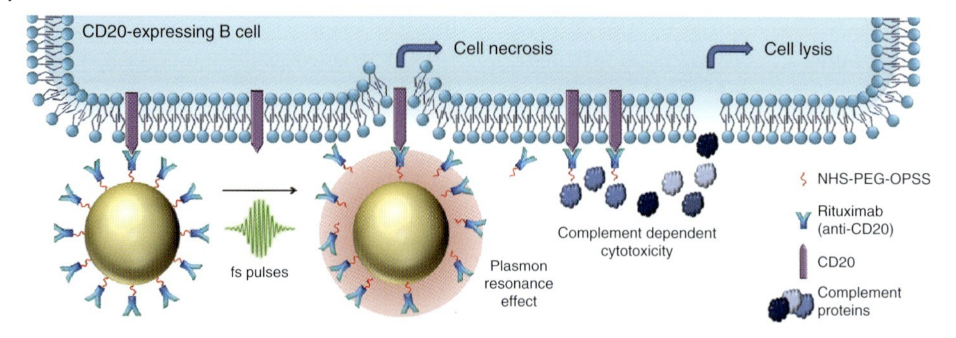

Fig. 6 Controlled release of rituximab from gold nanoparticles. Targeted nanoparticles bind to CD20-expressing cells and are laser-irradiated to release rituximab, which recruits complement proteins to induce complement-dependent cytotoxicity (CDC)-mediated cell death. Reprinted with permission from Ref. [158]; © 2012, Elsevier.

the plasmon resonance of the AuNPs. The released anti-CD20 retained its functionality, inducing a complement-dependent cytotoxicity, where the complement proteins in blood serum acted to induce necrosis of cells marked by rituximab [158]. In addition, plasmonic shock waves emanating from the AuNPs, as a result of irradiation, initiated cell death. The advantages of this system included a high specificity and efficiency, low toxicity, and high levels of repeatability and consistency due to the morphological stability of the nanoparticles.

Dilnawaz *et al.* conducted an *in-vitro* study with HER2 antibody-conjugated glycerol monooleate (GMO) -coated IONPs loaded with different anticancer chemotherapeutic drugs (paclitaxel, rapamycin either alone or combination) and for use as active cancer therapeutics (Fig. 7) [159]. Approximately 95% drug entrapment

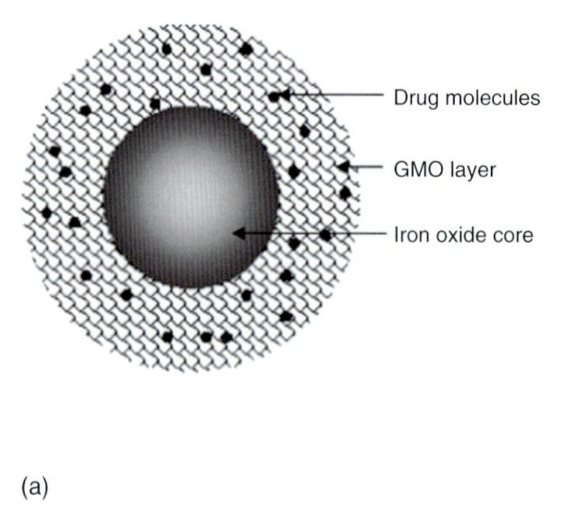

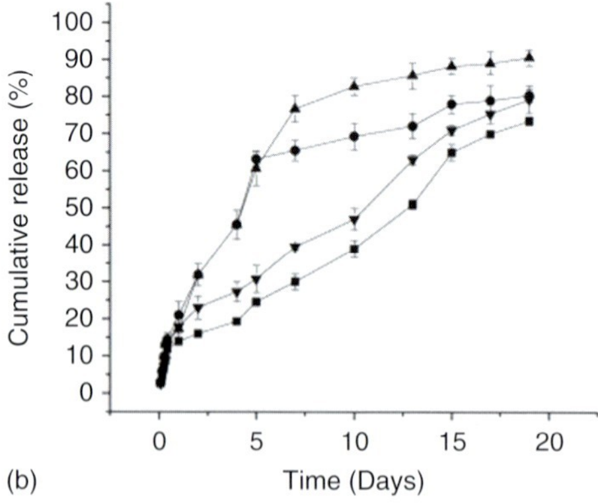

Fig. 7 Drug adsorption and drug release of GMO-coated drug-loaded magnetic iron oxide nanoparticles (IONPs). (a) Schematic representation of drug adsorption in the GMO coating surrounding the IONP core; (b) Release of rapamycin from rapa-GMO-IONPs (•), paclitaxel (*) from pac-GMO-IONPs, rapamycin (▲) from combo-GMO-IONPs, and paclitaxel (▼) from combo-GMO-IONPs under *in-vitro* conditions. Reprinted with permission from Ref. [159]; © 2010, Elsevier.

was achieved via absorption in the GMO coating, and sustained release occurred over a two-week period. Conjugation to the HER2 antibody increased the cellular uptake of GMO–IONPs by human breast cancer carcinoma cell lines (MCF-7) cells threefold when compared to non-targeted GMO–IONPs. The targeted, drug-loaded GMO–IONPs were 55-fold more effective than the native drug, and sevenfold more effective than the unconjugated NPs, which indicated the potential antiproliferation effect in MCF-7 [159]. These findings demonstrated the potential of drug-loaded, antibody-conjugated IONPs as potential drug carriers for improved drug efficiency, while providing selectivity that would reduce the adverse side effects caused by damage to normal healthy tissues.

6.2.2 *In-Vivo* Imaging and Targeted Drug Delivery

Multifunctional IONPs have been developed for combined targeted imaging and drug delivery. Of particular interest are IONPs as nanocarriers for MRI imaging and drug delivery, conjugated with anti-HER2 trastuzumab and DOX [125]. In a recent *in-vivo* study conducted by Zolata *et al.*, polymer-coated IONPs were ^{11}In-labeled and conjugated with trastuzumab–DOX for tumor targeting, drug delivery, controlled release, and dual-mode tumor imaging [160]. The theranostic effects of the immunotargeted and drug-loaded IONPs were evaluated in an HER2-positive breast tumor model bearing BALB/c mice, using biodistribution, molecular imaging, and tumor morphology measurements. Antibody-conjugated nanoparticles from this study accumulated at very high concentrations in the tumors due to their targeting specificity, when compared to non-targeted NP controls. Results obtained using single photon emission computed tomography (SPECT) and MRI showed the tumor region to be clearly contrasted by the antibody-conjugated IONPs. Additionally, the tumor volumes were measured at 12, 24, and 48 h after treatment and were 50%, 25%, and 23% lower, respectively, for treated animals compared to the control group, indicating a significant therapeutic effect. These findings supported the use of multifunctional modified SPIONs for cancer theranostics, namely imaging and drug-delivery applications.

6.3 AuNPs for the Enhancement of Externally Delivered Radiation Therapy

The two major limitations to the efficacy of radiation therapy are the dose limitations that are required to spare healthy surrounding tissues, and the hypoxia-associated reduction in the radiosensitivity of solid tumor cells [161, 162]. AuNPs have been used to address these limitations as radiosensitizers, as they possess a significantly high X-ray absorption coefficient [163, 164]. Both, *in-vitro* and *in-vivo* models have provided evidence that pretreatment with AuNPs has the potential to improve the imaging and radiosenstitization of a tumor, while minimizing the dose to normal tissues [165–171]. AuNPs are expected to increase the effects of radiation treatment due to lethal DNA double-strand breaks (DSBs), when the tumor cells are exposed to radiation [172–174]. Preclinical studies in this area have relied heavily on the leaky tumor microenvironment and intratumoral injection to localize AuNPs to tumor tissues [175]. However, the introduction of targeting moieties on AuNP-based radiosensitizers has the potential to improve tumor specificity and internalization [173, 174, 176].

Thus, antibody-conjugated AuNPs are among the next generation of targeted AuNP radiosensitizers [177]. In a recent study evaluating the effect of Herceptin (trastuzumab)-conjugated AuNPs (Au-T) on HER2-overexpressing MDA-MB-361 on a human breast cancer cell culture and a tumor xenograft mouse model showed that the clonogenic survival of cells exposed to Au-T and X-rays was significantly lower than that of cells exposed to radiation alone, resulting in a dose enhancement of 1.6. *In vivo*, the combination of Au-T and X-irradiation resulted in a regression of MDA-MB-361 tumors by 46% compared to X-irradiation treatment alone [172]. This study was one of the first to demonstrate the potential for nanoparticles conjugated with cancer-targeting antibodies in enhancing externally delivered radiation therapy.

7 Conclusions

AuNPs, IONPs and HNPs functionalized with cancer-specific antibodies provide attractive platforms for potent next-generation cancer theranostics, which couple diagnostics with therapies. Preclinical studies with immunotargeted nanoparticles have demonstrated more specific therapies with the potential to result in reduced adverse side effects in comparison to convectional treatment strategies. Further, treatment at the nanoscale allows for cellular level cancer detection, providing a means to eliminate cancer at the cellular level and thus reduce cancer recurrence. Thus, AuNPs and IONPs, when conjugated with antibodies, enhance the optical signals and provide a sensitive strategy for the detection and destruction of cancer cells which could, potentially, lead to a paradigm shift in cancer treatment strategies.

References

1. Siegel, R., Naishadham, D., Jemal, A. (2013) Cancer statistics, 2013. CA. *Cancer J. Clin.*, **63** (1), 11–30.
2. Kochanek, K.D., Xu, J., Murphy, S.L., Minino, A.M., Kung, H.C. (2011) Deaths: final data for 2009. *Natl. Vital Stat. Rep.*, **60**, 1–117.
3. Holohan, C., Van Schaeybroeck, S., Longley, D.B., Johnston, P.G. (2013) Cancer drug resistance: an evolving paradigm. *Nat. Rev. Cancer*, **13** (10), 714–726.
4. Hull, L.C., Farrell, D., Grodzinski, P. (2014) Highlights of recent developments and trends in cancer nanotechnology research – view from NCI Alliance for Nanotechnology in Cancer. *Biotechnol. Adv.*, **32** (4), 666–678.
5. Curry, T., Kopelman, R., Shilo, M., Popovtzer, R. (2014) Multifunctional theranostic gold nanoparticles for targeted CT imaging and photothermal therapy. *Contrast Media Mol. Imaging*, **9** (1), 53–61.
6. Julien, D.C., Behnke, S., Wang, G., Murdoch, G.K. *et al.* (2011) Utilization of monoclonal antibody-targeted nanomaterials in the treatment of cancer. *MAbs*, **3** (5), 467–478.
7. He, X., Wang, K., Cheng, Z. (2010) In vivo near-infrared fluorescence imaging of cancer with nanoparticle-based probes. *Wiley Interdiscip. Rev. Nanomed. Nanobiotechnol.*, **2** (4), 349–366.
8. Chow, E.K.-H., D. Ho (2013) Cancer nanomedicine: from drug delivery to imaging. *Sci. Transl. Med.*, **5** (216), 1–12.
9. Alivisatos, P. (2004) The use of nanocrystals in biological detection. *Nat. Biotechnol.*, **22** (1), 47–52.
10. Thakor, A.S., Jokerst, J., Zavaleta, C., Massoud, T.F. *et al.* (2011) Gold nanoparticles: a revival in precious metal administration to patients. *Nano Lett.*, **11** (10), 4029–4036.
11. Cai, W., Gao, T., Hong, H., Sun, J. (2008) Applications of gold nanoparticles in cancer nanotechnology. *Nanotechnol. Sci. Appl.*, **1**, 17–32.

12. Heddle, J. (2013) Gold nanoparticle–biological molecule interactions and catalysis. *Catalysts*, **3** (3), 683–708.
13. Huang, X.H., Jain, P.K., El-Sayed, I.H., El-Sayed, M.A. (2007) Gold nanoparticles: interesting optical properties and recent applications in cancer diagnostic and therapy. *Nanomedicine*, **2** (5), 681–693.
14. Huang, X., El-Sayed, M.A. (2010) Gold nanoparticles: optical properties and implementations in cancer diagnosis and photothermal therapy. *J. Adv. Res.*, **1** (1), 13–28.
15. Ku, K.H., Yang, H., Shin, J.M., Kim, B.J. (2015) Aspect ratio effect of nanorod surfactants on the shape and internal morphology of block copolymer particles. *J. Polym. Sci. Part A - Polym. Chem.*, **53** (2), 188–192.
16. Cao-Milan, R., Liz-Marzan, L.M. (2014) Gold nanoparticle conjugates: recent advances toward clinical applications. Expert Opin. *Drug Delivery*, **11** (5), 741–752.
17. Zhou, J., Ralston, J., Sedev, R., Beattie, D.A. (2009) Functionalized gold nanoparticles: synthesis, structure and colloid stability. *J. Colloid Interface Sci.*, **331** (2), 251–262.
18. Fratoddi, I., Venditti, I., Cametti, C., Russo, M.V. (2014) Gold nanoparticles and gold nanoparticle-conjugates for delivery of therapeutic molecules. Progress and challenges. J. Mater. *Chem. B*, **2** (27), 4204–4220.
19. Dykman, L., Khlebtsov, N. (2012) Gold nanoparticles in biomedical applications: recent advances and perspectives. *Chem. Soc. Rev.*, **41** (6), 2256–2282.
20. Klein, S., S. Petersen, U. Taylor, D. Rath. Quantitative visualization of colloidal and intracellular gold nanoparticles by confocal microscopy. *J. Biomed. Optics*, 2010. **15** (3), 036015–036015-11.
21. van Dijk, M.A., Tchebotareva, A.L., Orrit, M., Lippitz, M. (2006) Absorption and scattering microscopy of single metal nanoparticles. *Phys. Chem. Chem. Phys.*, **8** (30), 3486–3495.
22. Tcherniak, A., Ha, J.W., Dominguez-Medina, S., Slaughter, L.S. *et al.* (2010) Probing a century old prediction one plasmonic particle at a time. *Nano Lett.*, **10** (4), 1398–1404.
23. Dreaden, E.C., Alkilany, A.M., Huang, X.H., Murphy, C.J. *et al.* (2012) The golden age: gold nanoparticles for biomedicine. *Chem. Soc. Rev.*, **41** (7), 2740–2779.
24. Cho, S.J., Idrobo, J.C., Olamit, J., Liu, K. *et al.* (2005) Growth mechanisms and oxidation resistance of gold-coated iron nanoparticles. *Chem. Mater.*, **17** (12), 3181–3186.
25. Teja, A.S., Koh, P.-Y. (2009) Synthesis, properties, and applications of magnetic iron oxide nanoparticles. *Prog. Cryst. Growth Charact. Mater.*, **55** (1-2), 22–45.
26. Wang, D., He, J., Rosenzweig, N., Rosenzweig, Z. (2004) Superparamagnetic Fe_2O_3 beads–CdSe/ZnS quantum dots core–shell nanocomposite particles for cell separation. *Nano Lett.*, **4** (3), 409–413.
27. Rosen, J.E., Chan, L., Shieh, D.-B., Gu, F.X. (2012) Iron oxide nanoparticles for targeted cancer imaging and diagnostics. *Nanomed. Nanotechnol. Biol. Med.*, **8** (3), 275–290.
28. Wang, Y.-X., Hussain, S., Krestin, G. (2001) Superparamagnetic iron oxide contrast agents: physicochemical characteristics and applications in MR imaging. *Eur. Radiol.*, **11** (11), 2319–2331.
29. Wang, Y.X. (2011) Superparamagnetic iron oxide based MRI contrast agents: current status of clinical application. *Quant. Imaging Med. Surg.*, **1** (1), 35–40.
30. Clement, O., Frija, G., Chambon, C., Schouman-Clayes, E. *et al.* (1991) Liver tumors in cirrhosis: experimental study with SPIO-enhanced MR imaging. *Radiology*, **180** (1), 31–36.
31. Deryl, L.T., Stefan, H.B. (2011) *Utilization of Magnetic Nanoparticles for Cancer Therapy, in Nanomedicine and Cancer*, Science Publishers, pp. 84–104.
32. Kim, J., Park, S., Lee, J.E., Jin, S.M. *et al.* (2006) Designed fabrication of multifunctional magnetic gold nanoshells and their application to magnetic resonance imaging and photothermal therapy. *Angew. Chem.*, **118** (46), 7918–7922.
33. Wang, L.Y., Luo, J., Fan, Q., Suzuki, M. *et al.* (2005) Monodispersed core–shell Fe_3O_4@Au nanoparticles. *J. Phys. Chem. B*, **109** (46), 21593–21601.
34. Melancon, M.P., Lu, W., Zhong, M., Zhou, M. *et al.* (2011) Targeted multifunctional gold-based nanoshells for magnetic resonance-guided laser ablation of head and neck cancer. *Biomaterials*, **32** (30), 7600–7608.

35. Li, L., Mak, K.Y., Leung, C.W., Leung, C.H. *et al.* (2014) Synthesis and morphology control of gold/iron oxide magnetic nanocomposites via a simple aqueous method. *IEEE Trans. Magn.*, **50** (1), 1–5.
36. Li, L., Du, Y.M., Mak, K.Y., Leung, C.W. *et al.* (2014) Novel hybrid Au/Fe_3O_4 magnetic octahedron-like nanoparticles with tunable size. *IEEE Trans. Magn.*, **50** (1), 1–5.
37. Xu, Z., Hou, Y., Sun, S. (2007) Magnetic core/shell Fe_3O_4/Au and Fe_3O_4/Au/Ag nanoparticles with tunable plasmonic properties. *J. Am. Chem. Soc.*, **129** (28), 8698–8699.
38. Lyon, J.L., Fleming, D.A., Stone, M.B., Schiffer, P. *et al.* (2004) Synthesis of Fe oxide core/Au shell nanoparticles by iterative hydroxylamine seeding. *Nano Lett.*, **4** (4), 719–723.
39. Yu, H., Chen, M., Rice, P.M., Wang, S.X. *et al.* (2005) Dumbbell-like bifunctional Au–Fe_3O_4 nanoparticles. *Nano Lett.*, **5** (2), 379–382.
40. Kirui, D.K., Rey, D.A., Batt, C.A. (2010) Gold hybrid nanoparticles for targeted phototherapy and cancer imaging. *Nanotechnology*, **21** (10), 105105.
41. Barnett, C.M., Gueorguieva, M., Lees, M.R., McGarvey, D.J. *et al.* (2013) Physical stability, biocompatibility and potential use of hybrid iron oxide–gold nanoparticles as drug carriers. *J. Nanopart. Res.*, **15** (6), 1689.
42. Mezni, A., Balti, I., Mlayah, A., Jouini, N. *et al.* (2013) Hybrid Au–Fe_3O_4 nanoparticles: plasmonic, surface enhanced Raman scattering, and phase transition properties. *J. Phys. Chem. C*, **117** (31), 16166–16174.
43. Shevchenko, E.V., Bodnarchuk, M.I., Kovalenko, M.V., Talapin, D.V. *et al.* (2008) Gold/iron oxide core/hollow–shell nanoparticles. *Adv. Mater.*, **20** (22), 4323–4329.
44. Park, H.Y., Schadt, M.J., Wang, L., Lim, I.I.S. *et al.* (2007) Fabrication of magnetic core–shell Fe oxide@Au nanoparticles for interfacial bioactivity and bio-separation. *Langmuir*, **23** (17), 9050–9056.
45. Gautier, J., Allard-Vannier, E., Munnier, E., Souce, M. *et al.* (2013) Recent advances in theranostic nanocarriers of doxorubicin based on iron oxide and gold nanoparticles. *J. Controlled Release*, **169** (1-2), 48–61.
46. Zavisova, V., Koneracka, M., Kovac, J., Kubovcikova, M. *et al.* (2015) The cytotoxicity of iron oxide nanoparticles with different modifications evaluated in vitro. *J. Magn. Magn. Mater.*, **380**, 85–89.
47. Tiwari, P.M., Eroglu, E., Bawage, S.S., Vig, K. *et al.* (2014) Enhanced intracellular translocation and biodistribution of gold nanoparticles functionalized with a cell-penetrating peptide (VG-21) from vesicular stomatitis virus. *Biomaterials*, **35** (35), 9484–9494.
48. Bogdanov, A.A., Gupta, S., Koshkina, N., Corr, S.J. *et al.* (2015) Gold nanoparticles stabilized with MPEG-grafted poly(L-lysine): in vitro and in vivo evaluation of a potential theranostic agent. *Bioconjugate Chem.*, **26** (1), 39–50.
49. Choi, Y.-W., Lee, H., Song, Y., Sohn, D. (2015) Colloidal stability of iron oxide nanoparticles with multivalent polymer surfactants. *J. Colloid Interface Sci.*, **443**, 8–12.
50. Maldonado, C.R., Salassa, L., Gomez-Blanco, N., Mareque-Rivas, J.C. (2013) Nano-functionalization of metal complexes for molecular imaging and anticancer therapy. *Coord. Chem. Rev.*, **257** (19–20), 2668–2688.
51. Aryal, S., Remant, B.K.C., Dharmaraj, N., Bhattarai, N. *et al.* (2006) Spectroscopic identification of S–Au interaction in cysteine-capped gold nanoparticles. Spectrochim. *Acta, Part A*, **63** (1), 160–163.
52. Gupta, R.K., Srinivasan, M.P., Dharmarajan, R. (2011) Synthesis of short chain thiol capped gold nanoparticles, their stabilization and immobilization on silicon surface. *Colloids Surf., A*, **390** (1–3), 149–156.
53. Gupta, R.K., Srinivasan, M.P., Dharmarajan, R. (2012) Synthesis of 16-mercaptohexadecanoic acid-capped gold nanoparticles and their immobilization on a substrate. *Mater. Lett.*, **67** (1), 315–319.
54. Lavenn, C., Albrieux, F., Tuel, A., Demessence, A. (2014) Synthesis, characterization and optical properties of an amino-functionalized gold thiolate cluster: Au-10(SPh-pNH(2))(10). *J. Colloid Interface Sci.*, **418**, 234–239.
55. Lin, S.Y., Tsai, Y.T., Chen, C.C., Lin, C.M. *et al.* (2004) Two-step functionalization of neutral and positively charged thiols onto

citrate-stabilized Au nanoparticles. *J. Phys. Chem. B*, **108** (7), 2134–2139.
56. Zhou, J.F., Beattie, D.A., Ralston, J., Sedev, R. (2007) Colloid stability of thymine-functionalized gold nanoparticles. *Langmuir*, **23** (24), 12096–12103.
57. Tan, H., Zhan, T., Fan, W.Y. (2006) Direct functionalization of the hydroxyl group of the 6-mercapto-1-hexanol (MCH) ligand attached to gold nanoclusters. J. Phys. *Chem. B*, **110** (43), 21690–21693.
58. Yoo, C.I., Seo, D., Chung, B.H., Chung, I.S. *et al.* (2009) A facile one-pot synthesis of hydroxyl-functionalized gold polyhedrons by a surface regulating copolymer. *Chem. Mater.*, **21** (5), 939–944.
59. Goren, M., Galley, N., Lennox, R.B. (2006) Adsorption of alkylthiol-capped gold nanoparticles onto alkylthiol self-assembled monolayers: an SPR study. *Langmuir*, **22** (3), 1048–1054.
60. Hofmann, A., Schmiel, P., Stein, B., Graf, C. (2011) Controlled formation of gold nanoparticle dimers using multivalent thiol ligands. *Langmuir*, **27** (24), 15165–15175.
61. Shichibu, Y., Negishi, Y., Tsukuda, T., Teranishi, T. (2005) Large-scale synthesis of thiolated Au-25 clusters via ligand exchange reactions of phosphine-stabilized Au-11 clusters. *J. Am. Chem. Soc.*, **127** (39), 13464–13465.
62. Woehrle, G.H., Brown, L.O., Hutchison, J.E. (2005) Thiol-functionalized, 1.5-nm gold nanoparticles through ligand exchange reactions: scope and mechanism of ligand exchange. *J. Am. Chem. Soc.*, **127** (7), 2172–2183.
63. Davis, K., Qi, B., Witmer, M., Kitchens, C.L. *et al.* (2014) Quantitative measurement of ligand exchange on iron oxides via radiolabeled oleic acid. *Langmuir*, **30** (36), 10918–10925.
64. Hofmann, A., Thierbach, S., Semisch, A., Hartwig, A. *et al.* (2010) Highly monodisperse water-dispersable iron oxide nanoparticles for biomedical applications. *J. Mater. Chem.*, **20** (36), 7842–7853.
65. Shem, P.M., Sardar, R., Shumaker-Parry, J.S. (2009) One-step synthesis of phosphine-stabilized gold nanoparticles using the mild reducing agent 9-BBN. *Langmuir*, **25** (23), 13279–13283.
66. Scaravelli, R.C.B., Dazzi, R.L., Giacomelli, F.C., Machado, G. *et al.* (2013) Direct synthesis of coated gold nanoparticles mediated by polymers with amino groups. *J. Colloid Interface Sci.*, **397**, 114–121.
67. Wang, X.G., Kawanami, H., Islam, N.M., Chattergee, M. *et al.* (2008) Amphiphilic block copolymer-stabilized gold nanoparticles for aerobic oxidation of alcohols in aqueous solution. *Chem. Commun.*, **37**, 4442–4444.
68. Rey, D.A., Strickland, A.D., Kirui, D., Niamsiri, N. *et al.* (2010) In vitro self-assembly of gold nanoparticle-coated poly(3-hydroxybutyrate) granules exhibiting plasmon-induced thermo-optical enhancements. *ACS Appl. Mater. Interfaces*, **2** (7), 1804–1810.
69. Xu, S.Y., Tu, G.L., Peng, B., Han, X.Z. (2006) Self-assembling gold nanoparticles on thiol-functionalized poly(styrene-*co*-acrylic acid) nanospheres for fabrication of a mediatorless biosensor. *Anal. Chim. Acta*, **570** (2), 151–157.
70. Fuentes, M., Mateo, C., Guisán, J.M., Fernández-Lafuente, R. (2005) Preparation of inert magnetic nano-particles for the directed immobilization of antibodies. *Biosens. Bioelectron.*, **20** (7), 1380–1387.
71. Almeida, J.P.M., Figueroa, E.R., Drezek, R.A. (2014) Gold nanoparticle mediated cancer immunotherapy. Nanomed. Nanotechnol. *Biol. Med.*, **10** (3), 503–514.
72. Huang, G., Zhang, C., Li, S., Khemtong, C. *et al.* (2009) A novel strategy for surface modification of superparamagnetic iron oxide nanoparticles for lung cancer imaging. *J. Mater. Chem.*, **19**, 6367–6372.
73. Chen, H.W., Zou, H., Paholak, H.J., Ito, M. *et al.* (2014) Thiol-reactive amphiphilic block copolymer for coating gold nanoparticles with neutral and functionable surfaces. *Polym. Chem.*, **5** (8), 2768–2773.
74. Bloemen, M., Brullot, W., Luong, T., Geukens, N. *et al.* (2012) Improved functionalization of oleic acid-coated iron oxide nanoparticles for biomedical applications. *J. Nanopart. Res.*, **14** (9), 1–10.
75. von Maltzahn, G., Park, J.-H., Agrawal, A., Bandaru, N.K. *et al.* (2009) Computationally guided photothermal tumor therapy using long-circulating gold nanorod antennas. *Cancer Res.*, **69** (9), 3892–3900.
76. Zhang, S., Gong, M., Zhang, D., Yang, H. *et al.* (2014) Thiol-PEG-carboxyl-stabilized Fe_2O_3/Au nanoparticles targeted to CD105:

synthesis, characterization and application in MR imaging of tumor angiogenesis. *Eur. J. Radiol.*, **83** (7), 1190–1198.

77. Tassa, C., Shaw, S.Y., Weissleder, R. (2011) Dextran-coated iron oxide nanoparticles: a versatile platform for targeted molecular imaging, molecular diagnostics, and therapy. *Acc. Chem. Res.*, **44** (10), 842–852.
78. Li, S., Liu, H., He, N. (2010) Covalent binding of streptavidin on gold magnetic nanoparticles for bead array fabrication. *J. Nanosci. Nanotechnol.*, **10** (8), 4875–4882.
79. Giaume, D., Poggi, M., Casanova, D., Mialon, G. *et al.* (2008) Organic functionalization of luminescent oxide nanoparticles toward their application as biological probes. *Langmuir*, **24** (19), 11018–11026.
80. Erathodiyil, N., Ying, J.Y. (2011) Functionalization of inorganic nanoparticles for bioimaging applications. *Acc. Chem. Res.*, **44** (10), 925–935.
81. Graf, C., Vossen, D.L.J., Imhof, A., van Blaaderen, A. (2003) A general method to coat colloidal particles with silica. *Langmuir*, **19** (17), 6693–6700.
82. Liberman, A., Mendez, N., Trogler, W.C., Kummel, A.C. (2014) Synthesis and surface functionalization of silica nanoparticles for nanomedicine. *Surf. Sci. Rep.*, **69** (2-3), 132–158.
83. Guerrero-Martínez, A., Pérez-Juste, J., Liz-Marzán, L.M. (2010) Recent progress on silica coating of nanoparticles and related nanomaterials. *Adv. Mater.*, **22** (11), 1182–1195.
84. Ladj, R., Bitar, A., Eissa, M.M., Fessi, H. *et al.* (2013) Polymer encapsulation of inorganic nanoparticles for biomedical applications. *Int. J. Pharm.*, **458** (1), 230–241.
85. Gun, S., Edirisinghe, M., Stride, E. (2013) Encapsulation of superparamagnetic iron oxide nanoparticles in poly-(lactide-*co*-glycolic acid) microspheres for biomedical applications. Mater. *Sci. Eng., C*, **33** (6), 3129–3137.
86. Boisselier, E., Diallo, A.K., Salmon, L., Ornelas, C. *et al.* (2010) Encapsulation and stabilization of gold nanoparticles with "click" polyethyleneglycol dendrimers. *J. Am. Chem. Soc.*, **132** (8), 2729–2742.
87. Li, K., Liu, B. (2012) Polymer encapsulated conjugated polymer nanoparticles for fluorescence bioimaging. *J. Mater. Chem.*, **22** (4), 1257–1264.
88. Fratila, R.M., Mitchell, S.G., del Pino, P., Grazu, V. *et al.* (2014) Strategies for the biofunctionalization of gold and iron oxide nanoparticles. *Langmuir*, **30** (50), 15057–15071.
89. Sperling, R.A., Parak, W.J. (2010) Surface modification, functionalization and bioconjugation of colloidal inorganic nanoparticles. *Philos. Trans. R. Soc. A*, **368** (1915), 1333–1383.
90. Rayavarapu, R.G., Petersen, W., Ungureanu, C., Post, J.N. *et al.* (2007) Synthesis and bioconjugation of gold nanoparticles as potential molecular probes for light-based imaging techniques. *Int. J. Biomed. Imaging*, **2007**, 10.
91. Puertas, S., Batalla, P., Moros, M., Polo, E. *et al.* (2011) Taking advantage of unspecific interactions to produce highly active magnetic nanoparticle–antibody conjugates. *ACS Nano*, **5** (6), 4521–4528.
92. Chinol, M., Casalini, P., Maggiolo, M., Canevari, S. *et al.* (1998) Biochemical modifications of avidin improve pharmacokinetics and biodistribution, and reduce immunogenicity. *Br. J. Cancer*, **78** (2), 189–197.
93. Cheung, N.K., Modak, S., Lin, Y., Guo, H. *et al.* (2004) Single-chain Fv-streptavidin substantially improved therapeutic index in multistep targeting directed at disialoganglioside GD2. *J. Nucl. Med.*, **45** (5), 867–877.
94. Tajima, N., Takai, M., Ishihara, K. (2011) Significance of antibody orientation unraveled: well-oriented antibodies recorded high binding affinity. *Anal. Chem.*, **83** (6), 1969–1976.
95. Colombo, M., Sommaruga, S., Mazzucchelli, S., Polito, L. *et al.* (2012) Site-specific conjugation of ScFvs antibodies to nanoparticles by bioorthogonal strain-promoted alkyne–nitrone cycloaddition. *Angew. Chem. Int. Ed.*, **51** (2), 496–499.
96. Hutchins, B.M., Kazane, S.A., Staflin, K., Forsyth, J.S. *et al.* (2011) Site-specific coupling and sterically controlled formation of multimeric antibody fab fragments with unnatural amino acids. *J. Mol. Biol.*, **406** (4), 595–603.
97. Ackerson, C.J., Jadzinsky, P.D., Jensen, G.J., Kornberg, R.D. (2006) Rigid, specific,

and discrete gold nanoparticle/antibody conjugates. *J. Am. Chem. Soc.*, **128** (8), 2635–2640.

98. Scott, A.M. (2001) Specific targeting, biodistribution and lack of immunogenicity of chimeric anti-GD3 monoclonal antibody KM871 in patients with metastatic melanoma – results of a Phase I trial. *J. Clin. Oncol.*, **19**, 3976–3987.
99. Nelson, A.L. (2010) Development trends for human monoclonal antibody therapeutics. *Nat. Rev. Drug Discov.*, **9**, 767–774.
100. Reichert, J.M. (2008) Monoclonal antibodies as innovative therapeutics. *Curr. Pharm. Biotechnol.*, **9** (6), 423–430.
101. Weiner, L.M., Surana, R., Wang, S. (2010) Monoclonal antibodies: versatile platforms for cancer immunotherapy. *Nat. Rev. Immunol.*, **10**, 317–327.
102. Alderson, K.L., Sondel, P.M. (2011) Clinical cancer therapy by NK cells via antibody-dependent cell-mediated cytotoxicity. *J. Biomed. Biotechnol.*, doi:10.1155/2011/379123.
103. Bronte, G., Sortino, G., Passiglia, F., Rizzo, S. *et al.* (2015) Monoclonal antibodies for the treatment of non-haematological tumours: update of an expanding scenario. *Expert Opin. Biol. Ther.*, **15** (1), 45–59.
104. Ribatti, D. (2014) From the discovery of monoclonal antibodies to their therapeutic application: An historical reappraisal. *Immunol. Lett.*, **161** (1), 96–99.
105. Davis, M.E., Chen, Z., Shin, D.M. (2008) Nanoparticle therapeutics: an emerging treatment modality for cancer. *Nat. Rev. Drug Discov.*, **7** (9), 771–782.
106. Parker, S.A., Diaz, I.L., Anderson, K.A., Batt, C.A. (2013) Design, production, and characterization of a single-chain variable fragment (ScFv) derived from the prostate specific membrane antigen (PSMA) monoclonal antibody J591. *Protein Expr. Purif.*, **89** (2), 136–145.
107. Huang, X., Yi, C., Fan, Y., Zhang, Y. *et al.* (2014) Magnetic Fe_3O_4 nanoparticles grafted with single-chain antibody (scFv) and docetaxel-loaded β-cyclodextrin potential for ovarian cancer dual-targeting therapy. *Mater. Sci. Eng., C*, **42**, 325–332.
108. Liu, Y., Liu, Y., Mernaugh, R.L., Zeng, X. (2009) Single chain fragment variable recombinant antibody functionalized gold nanoparticles for a highly sensitive colorimetric immunoassay. *Biosens. Bioelectron.*, **24** (9), 2853–2857.
109. Lee, N., Kim, H., Choi, S.H., Park, M. *et al.* (2011) Magnetosome-like ferrimagnetic iron oxide nanocubes for highly sensitive MRI of single cells and transplanted pancreatic islets. *Proc. Natl Acad. Sci. USA*, **108** (7), 2662–2667.
110. Jain, P.K., El-Sayed, I.H., El-Sayed, M.A. (2007) Au nanoparticles target cancer. *Nano Today*, **2** (1), 18–29.
111. Lima, K.M.G., Junior, R.F.A., Araujo, A.A., Oliveira, A.L.C.S.L. *et al.* (2014) Environmentally compatible bioconjugated gold nanoparticles as efficient contrast agents for colorectal cancer cell imaging. *Sens. Actuators, B*, **196**, 306–313.
112. Luo, S., Zhang, E., Su, Y., Cheng, T. *et al.* (2011) A review of NIR dyes in cancer targeting and imaging. *Biomaterials*, **32** (29), 7127–7138.
113. Altinoglu, E.I., Adair, J.H. (2010) Near infrared imaging with nanoparticles. *Wiley Interdiscip. Rev. Nanomed. Nanobiotechnol.*, **2** (5), 461–477.
114. Murphy, C.J., Gole, A.M., Stone, J.W., Sisco, P.N. *et al.* (2008) Gold nanoparticles in biology: beyond toxicity to cellular imaging. *Acc. Chem. Res.*, **41** (12), 1721–1730.
115. Lee, C.N., Wang, Y.M., Lai, W.F., Chen, T.J. *et al.* (2012) Super-paramagnetic iron oxide nanoparticles for use in extrapulmonary tuberculosis diagnosis. *Clin. Microbiol. Infect.*, **18** (6), E149–E157.
116. Skaat, H., Corem-Slakmon, E., Grinberg, I., Last, D. *et al.* (2013) Antibody-conjugated, dual-modal, near-infrared fluorescent iron oxide nanoparticles for antiamyloidgenic activity and specific detection of amyloid-beta fibrils. *Int. J. Nanomed.*, **8**, 4063–4076.
117. Sokolov, K., Follen, M., Aaron, J., Pavlova, I. *et al.* (2003) Real-time vital optical imaging of precancer using anti-epidermal growth factor receptor antibodies conjugated to gold nanoparticles. *Cancer Res.*, **63** (9), 1999–2004.
118. Schol, D., Fleron, M., Greisch, J.F., Jaeger, M. *et al.* (2013) Anti-PSMA antibody-coupled gold nanorods detection by optical and electron microscopies. *Micron*, **50**, 68–74.
119. Hainfeld, J.F., O'Connor, M.J., Dilmanian, F.A., Slatkin, D.N. *et al.* (2011) Micro-CT enables microlocalisation and quantification

of Her2-targeted gold nanoparticles within tumour regions. *Br. J. Radiol.*, **84** (1002), 526–533.

120. Coughlin, A.J., Ananta, J.S., Deng, N., Larina, I.V. *et al.* (2014) Gadolinium-conjugated gold nanoshells for multimodal diagnostic imaging and photothermal cancer therapy. *Small*, **10** (3), 556–565.
121. Lee, N., Choi, Y., Lee, Y., Park, M. *et al.* (2012) Water-dispersible ferrimagnetic iron oxide nanocubes with extremely high r_2 relaxivity for highly sensitive in vivo MRI of tumors. *Nano Lett.*, **12** (6), 3127–3131.
122. Serda, R.E., Adolphi, N.L., Bisoffi, M., Sillerud, L.O. (2007) Targeting and cellular trafficking of magnetic nanoparticles for prostate cancer imaging. *Mol. Imaging*, **6** (4), 277–288.
123. Kao, H.W., Lin, Y.Y., Chen, C.C., Chi, K.H. *et al.* (2013) Evaluation of EGFR-targeted radioimmuno-gold-nanoparticles as a theranostic agent in a tumor animal model. *Bioorg. Med. Chem. Lett.*, **23** (11), 3180–3185.
124. Gong, T., Olivo, M., Dinish, U.S., Goh, D. *et al.* (2013) Engineering bioconjugated gold nanospheres and gold nanorods as label-free plasmon scattering probes for ultrasensitive multiplex dark-field imaging of cancer cells. *J. Biomed. Nanotechnol.*, **9** (6), 985–991.
125. Choi, W.I., Lee, J.H., Kim, J.Y., Heo, S.U. *et al.* (2015) Targeted antitumor efficacy and imaging via multifunctional nano-carrier conjugated with anti-HER2 trastuzumab. *Nanomedicine*, **11** (2), 359–368.
126. Karmani, L., Bouchat, V., Bouzin, C., Leveque, P. *et al.* (2014) Zr-89-labeled anti-endoglin antibody-targeted gold nanoparticles for imaging cancer: implications for future cancer therapy. *Nanomedicine*, **9** (13), 1923–1937.
127. Chen, T.J., Cheng, T.H., Chen, C.Y., Hsu, S.C. *et al.* (2009) Targeted herceptin-dextran iron oxide nanoparticles for noninvasive imaging of HER2/neu receptors using MRI. *J. Biol. Inorg. Chem.*, **14** (2), 253–260.
128. Vertrees, R.A., Das, G.C., Popov, V.L., Coscio, A.M. *et al.* (2005) Synergistic interaction of hyperthermia and gemcitabine in lung cancer. *Cancer Biol. Ther.*, **4** (10), 1144–1153.
129. Hildebrandt, B., Wust, P., Ahlers, O., Dieing, A. *et al.* (2002) The cellular and molecular basis of hyperthermia. *Crit. Rev. Oncol. Hematol.*, **43** (1), 33–56.
130. Ito, A., Honda, H., Kobayashi, T. (2006) Cancer immunotherapy based on intracellular hyperthermia using magnetite nanoparticles: a novel concept of 'heat-controlled necrosis' with heat shock protein expression. Cancer Immunol. *Immunother*, **55** (3), 320–328.
131. Chae, S.Y., Kim, Y.-S., Park, M.J., Yang, J. *et al.* (2014) High-intensity focused ultrasound-induced, localized mild hyperthermia to enhance anti-cancer efficacy of systemic doxorubicin: an experimental study. *Ultrasound Med. Biol.*, **40** (7), 1554–1563.
132. Overgaard, J., Gonzalez-Gonzalez, D., Hulshof, M.C., Arcangeli, G. *et al.* (1995) Randomised trial of hyperthermia as adjuvant to radiotherapy for recurrent or metastatic malignant melanoma. *European Society for Hyperthermic Oncology. Lancet*, **345** (8949), 540–543.
133. Roa, W., Zhang, X., Guo, L., Shaw, A. *et al.* (2009) Gold nanoparticle sensitize radiotherapy of prostate cancer cells by regulation of the cell cycle. *Nanotechnology*, **20** (37), 375101.
134. Atkinson, R.L., Zhang, M., Diagaradjane, P., Peddibhotla, S. *et al.* (2010) Thermal enhancement with optically activated gold nanoshells sensitizes breast cancer stem cells to radiation therapy. *Sci. Transl. Med.*, **2** (55), 3001447.
135. Kampinga, H.H. (2006) Cell biological effects of hyperthermia alone or combined with radiation or drugs: a short introduction to newcomers in the field. *Int. J. Hyperthermia*, **22** (3), 191–196.
136. Cui, Z.-G., Piao, J.-L., Kondo, T., Ogawa, R. *et al.* (2014) Molecular mechanisms of hyperthermia-induced apoptosis enhanced by docosahexaenoic acid: implication for cancer therapy. *Chem.- Biol. Interact.*, **215**, 46–53.
137. Saniei, N. (2009) Hyperthermia and cancer treatment. *Heat Transfer Eng.*, **30** (12), 915–917.
138. Sawdon, A., Weydemeyer, E., Peng, C.A. (2014) Antitumor therapy using nanomaterial-mediated thermolysis. *J. Biomed. Nanotechnol.*, **10** (9), 1894–1917.

139. Dennis, C.L., Jackson, A.J., Borchers, J.A., Hoopes, P.J. *et al.* (2009) Nearly complete regression of tumors via collective behavior of magnetic nanoparticles in hyperthermia. *Nanotechnology*, **20** (39), 395103.
140. Song, X., Gong, H., Yin, S., Cheng, L. *et al.* (2014) Ultra-small iron oxide doped polypyrrole nanoparticles for in vivo multimodal imaging guided photothermal therapy. *Adv. Funct. Mater.*, **24** (9), 1194–1201.
141. Huang, P., Lin, J., Wang, S., Zhou, Z. *et al.* (2013) Photosensitizer-conjugated silica-coated gold nanoclusters for fluorescence imaging-guided photodynamic therapy. *Biomaterials*, **34** (19), 4643–4654.
142. Lee, J., Chatterjee, D.K., Lee, M.H., Krishnan, S. (2014) Gold nanoparticles in breast cancer treatment: promise and potential pitfalls. *Cancer Lett.*, **347** (1), 46–53.
143. Loo, C., Lowery, A., Halas, N., West, J. *et al.* (2005) Immunotargeted nanoshells for integrated cancer imaging and therapy. *Nano Lett.*, **5** (4), 709–711.
144. Kirui, D.K., Khalidov, I., Wang, Y., Batt, C.A. (2013) Targeted near-IR hybrid magnetic nanoparticles for in vivo cancer therapy and imaging. *Nanomed. Nanotechnol. Biol. Med.*, **9** (5), 702–711.
145. Gibson, J.D., Khanal, B.P., Zubarev, E.R. (2007) Paclitaxel-functionalized gold nanoparticles. *J. Am. Chem. Soc.*, **129** (37), 11653–11661.
146. Ding, Y., Zhou, Y.-Y., Chen, H., Geng, D.-D. *et al.* (2013) The performance of thiol-terminated PEG-paclitaxel-conjugated gold nanoparticles. *Biomaterials*, **34** (38), 10217–10227.
147. Sun, C., C. Fang, Z. Stephen, O. Veiseh *et al.* (2008) Tumor-targeted drug delivery and MRI contrast enhancement by chlorotoxin-conjugated iron oxide nanoparticles. *Nanomed.* (London, Engl.), **3** (4), 495–505.
148. Wang, F., Wang, Y.C., Dou, S., Xiong, M.H. *et al.* (2011) Doxorubicin-tethered responsive gold nanoparticles facilitate intracellular drug delivery for overcoming multidrug resistance in cancer cells. *ACS Nano*, **5** (5), 3679–3692.
149. Venkatesan, R., Pichaimani, A., Hari, K., Balasubramanian, P.K. *et al.* (2013) Doxorubicin conjugated gold nanorods: a sustained drug delivery carrier for improved anti-cancer therapy. *J. Mater. Chem. B*, **1** (7), 1010–1018.
150. Patra, C.R., Bhattacharya, R., Mukhopadhyay, D., Mukherjee, P. (2010) Fabrication of gold nanoparticles for targeted therapy in pancreatic cancer. *Adv. Drug Delivery Rev.*, **62** (3), 346–361.
151. Laurent, S., Mahmoudi, M. (2011) Superparamagnetic iron oxide nanoparticles: promises for diagnosis and treatment of cancer. *Int. J. Mol. Epidemiol. Genet.*, **2** (4), 367–390.
152. Kapse-Mistry, S., Govender, T., Srivastava, R., Yergeri, M. (2014) Nanodrug delivery in reversing multidrug resistance in cancer cells. *Front. Pharmacol*, **5**, 159.
153. Starkewolf, Z.B., L. Miyachi, J. Wong, T. Guo. X-ray triggered release of doxorubicin from nanoparticle drug carriers for cancer therapy. *Chem. Commun.* (Cambridge, Engl.), 2013. **49** (25), 2545–2547.
154. Madhusudhan, A., Reddy, G.B., Venkatesham, M., Veerabhadram, G. *et al.* (2014) Efficient pH dependent drug delivery to target cancer cells by gold nanoparticles capped with carboxymethyl chitosan. *Int. J. Mol. Sci.*, **15** (5), 8216–8234.
155. Derfus, A.M., von Maltzahn, G., Harris, T.J., Duza, T. *et al.* (2007) Remotely triggered release from magnetic nanoparticles. *Adv. Mater.*, **19** (22), 3932–3936.
156. Zhang, X.-D., Wu, D., Shen, X., Chen, J. *et al.* (2012) Size-dependent radiosensitization of PEG-coated gold nanoparticles for cancer radiation therapy. *Biomaterials*, **33** (27), 6408–6419.
157. Dani, R.K., Schumann, C., Taratula, O. (2014) Temperature-tunable iron oxide nanoparticles for remote-controlled drug release. *AAPS Pharm. Sci. Technol.*, **15** (4), 963–972.
158. Bisker, G., Yeheskely-Hayon, D., Minai, L., Yelin, D. (2012) Controlled release of Rituximab from gold nanoparticles for phototherapy of malignant cells. *J. Control Release*, **162** (2), 303–309.
159. Dilnawaz, F., Singh, A., Mohanty, C., Sahoo, S.K. (2010) Dual drug loaded superparamagnetic iron oxide nanoparticles for targeted cancer therapy. *Biomaterials*, **31** (13), 3694–3706.
160. Zolata, H., Abbasi Davani, F., Afarideh, H. (2015) Synthesis, characterization and

theranostic evaluation of Indium-111 labeled multifunctional superparamagnetic iron oxide nanoparticles. *Nucl. Med. Biol.*, **42** (2), 164–170.
161. Lacombe, S., Sech, C.L. (2009) Advances in radiation biology: radiosensitization in DNA and living cells. *Surf. Sci.*, **603** (10-12), 1953–1960.
162. Harrison, L.B., Chadha, M., Hill, R.J., Hu, K. *et al.* (2002) Impact of tumor hypoxia and anemia on radiation therapy outcomes. *Oncologist*, **7** (6), 492–508.
163. Xi, D., Dong, S., Meng, X., Lu, Q. *et al.* (2012) Gold nanoparticles as computerized tomography (CT) contrast agents. *RSC Adv.*, **2** (33), 12515–12524.
164. Silvestri, A., Polito, L., Bellani, G., Zambelli, V. *et al.* (2015) Gold nanoparticles obtained by aqueous digestive ripening: their application as X-ray contrast agents. *J. Colloid Interface Sci.*, **439**, 28–33.
165. Chithrani, D.B., Jelveh, S., Jalali, F., van Prooijen, M. *et al.* (2010) Gold nanoparticles as radiation sensitizers in cancer therapy. *Radiat. Res.*, **173** (6), 719–728.
166. Geng, F., Xing, J.Z., Chen, J., Yang, R. *et al.* (2014) Pegylated glucose gold nanoparticles for improved in-vivo bio-distribution and enhanced radiotherapy on cervical cancer. *J. Biomed. Nanotechnol.*, **10** (7), 1205–1216.
167. Khoshgard, K., Hashemi, B., Arbabi, A., Rasaee, M.J. *et al.* (2014) Radiosensitization effect of folate-conjugated gold nanoparticles on HeLa cancer cells under orthovoltage superficial radiotherapy techniques. *Phys. Med. Biol.*, **59** (9), 2249.
168. Zhang, X.-D., Luo, Z., Chen, J., Song, S. *et al.* (2015) Ultrasmall glutathione-protected gold nanoclusters as next generation radiotherapy sensitizers with high tumor uptake and high renal clearance. *Sci. Rep.*, **5**, 8669.
169. Yasui, H., Takeuchi, R., Nagane, M., Meike, S. *et al.* (2014) Radiosensitization of tumor cells through endoplasmic reticulum stress induced by PEGylated nanogel containing gold nanoparticles. *Cancer Lett.*, **347** (1), 151–158.
170. Wolfe, T., Chatterjee, D., Lee, J., Grant, J.D. *et al.* (2015) Targeted gold nanoparticles enhance sensitization of prostate tumors to megavoltage radiation therapy in vivo. *Nanomedicine*, **11** (5), 1277–1283.
171. Joh, D.Y., Kao, G.D., Murty, S., Stangl, M. *et al.* (2013) Theranostic gold nanoparticles modified for durable systemic circulation effectively and safely enhance the radiation therapy of human sarcoma cells and tumors. *Transl. Oncol.*, **6** (6), 722–732.
172. Chattopadhyay, N., Cai, Z., Kwon, Y., Lechtman, E. *et al.* (2013) Molecularly targeted gold nanoparticles enhance the radiation response of breast cancer cells and tumor xenografts to X-radiation. *Breast Cancer Res. Treat.*, **137** (1), 81–91.
173. Herold, D.M., Das, I.J., Stobbe, C.C., Iyer, R.V. *et al.* (2000) Gold microspheres: a selective technique for producing biologically effective dose enhancement. *Int. J. Radiat. Biol.*, **76** (10), 1357–1364.
174. Hainfeld, J.F., Slatkin, D.N., Smilowitz, H.M. (2004) The use of gold nanoparticles to enhance radiotherapy in mice. *Phys. Med. Biol.*, **49** (18), N309–N315.
175. Hainfeld, J.F., Lin, L., Slatkin, D.N., Avraham Dilmanian, F. *et al.* (2014) Gold nanoparticle hyperthermia reduces radiotherapy dose. *Nanomed. Nanotechnol. Biol. Med.*, **10** (8), 1609–1617.
176. Burger, N., Biswas, A., Barzan, D., Kirchner, A. *et al.* (2014) A method for the efficient cellular uptake and retention of small modified gold nanoparticles for the radiosensitization of cells. *Nanomed. Nanotechnol. Biol. Med.*, **10** (6), 1365–1373.
177. Chattopadhyay, N., Cai, Z.L., Pignol, J.P., Keller, B. *et al.* (2010) Design and characterization of HER-2-targeted gold nanoparticles for enhanced X-radiation treatment of locally advanced breast cancer. *Mol. Pharmaceutics*, 7 (6), 2194–2206.

12
RNA Interference in Cancer Therapy

Barbara Pasculli[1,2] and George A. Calin[2,3,4]
[1] *University of Bari "Aldo Moro", Department of Biosciences, Biotechnology and Pharmacological Sciences, Via Orabona 4, Bari 70125, Italy*
[2] *The University of Texas MD Anderson Cancer Center, Department of Experimental Therapeutics, Unit 1950, 1881 East Road, Houston, TX 77054, USA*
[3] *The University of Texas MD Anderson Cancer, Department of Leukemia, 1515 Holcombe Blvd, Houston, TX 77030, USA*
[4] *The University of Texas MD Anderson Cancer Center, Center for RNA Interference and Non-Coding RNAs, Unit 1950, 1881 East Road, Houston, TX 77030, USA*

Translational Nanomedicine. First Edition. Edited by Robert A. Meyers.

Keywords

Cancer
Pathologic phenotype defined by abnormal and uncontrolled cell division, arising from the accumulation of epigenetic and genetic alterations affecting the expression profile of both protein-coding genes and noncoding RNAs.

RNA interference
A natural, sequence-specific mechanism of post-transcriptional gene silencing in animal and plants, triggered and mediated by small, double-stranded RNAs.

RISC (RNA-induced silencing complex)
A multicomponent ribonucleoprotein complex consisting of a member of the Argonaute (AGO) family proteins, and a mature miRNA/siRNA incorporated into AGO, together with a number of accessory factors, acting as the molecular effectors of translational repression/degradation of miRNA/siRNA -targeted transcripts.

microRNAs (miRNAs; miRs)
Small, endogenous, single-stranded, noncoding RNAs, 19–25 nucleotides in length, which enter the RNAi apparatus to mediate the post-transcriptional regulation of gene expression by specific but imperfect binding to target mRNAs.

siRNAs
Small-interfering, double-stranded RNAs, 21–23 nucleotides in length, which can be endogenously produced or exogenously provided (synthetic oligonucleotides) and, similar

to miRNAs, are able to associate with the RISC complex and mediate post-transcriptional gene silencing.

antimiRs
Antisense oligonucleotides which sequester, by complementary binding, the mature miRNA in competition with cellular target mRNAs, leading to a functional inhibition of the miRNA and derepression of target mRNA translation.

Personalized medicine
Healthcare approach which designs a therapeutic regimen and selects the patients to treat, according to the individual clinical, molecular, and environmental information.

RNAi-based therapy
The repertoire of strategies harnessing the RNAi endogenous apparatus by using small RNA molecules to target any gene and inhibit oncogenic pathways supporting the tumor.

Current research in cancer therapeutics is leading the way in exploiting the growing amounts of genetic and molecular information on human oncobiology to design new anticancer agents to target specific tumor-related signaling pathways, and to treat patients according to their clinical characteristics and molecular profile. In this context, RNA interference (RNAi), a ubiquitous cellular pathway of post-transcriptional gene regulation, provides an intriguing tool for an innovative rational cancer drug design. Among the endogenous mediators of RNAi, microRNAs (miRNAs) represent the most important class of small RNAs whose global dysregulation is a typical feature of human tumors. Harnessing of the RNAi machinery by using small, synthetic RNAs that target or mimic endogenous miRNAs offers the opportunity to reach virtually any gene and pathway relevant to tumor maintenance. However, along with assessments of the safety profile and effectiveness, strategies for assuring the specificity of these drugs, in terms of their effective delivery to cancer cells, must be devised before RNAi formulations can be made into ideal personalized therapeutics for the clinic.

1
Introduction: Cancer and RNAi

Cancer is a leading cause of death worldwide, second only to cardiovascular disorders. It is a genetic disease involving mutational and epigenetic alterations that cumulate over time, and cause activation of oncogenes and/or loss of function of tumor suppressor genes. More importantly, the past decade of research has revealed that the widespread deregulation of physiological mechanisms supporting the transformation of normal cells into malignant cells is due not only to alterations in protein-coding genes but also to global changes in noncodingRNAs (ncRNAs), such as the microRNAs (miRNAs) profile. miRNAs are involved in the regulation of most biological functions, including development, life span and metabolism, and their deregulation contributes to abnormal and uncontrolled cell division and survival, as well as the capability of migration and metastatization that characterizes the cancer phenotype. The mechanism through which miRNAs exert their regulatory control of gene expression in cells has been recognized as endogenous RNA interference (RNAi). Originally, this pathway was first described as a process triggered by double-stranded RNAs (dsRNAs), able to induce sequence-specific gene silencing [1]. Having arisen in the context of genetic manipulation experiments, the discovery of RNAi implied the existence of specific mechanisms within the cells, from worms to mammals, mediating the unwinding of dsRNAs. Hence, the search for complementary base-pairing partners, sequestration from the intracellular pool of nucleic acid sequences, and the activation of downstream processes culminated in translational repression. Since its discovery, RNAi and its related pathways have been exploited to identify gene functions and new putative targets for cancer therapy. Indeed, the endogenous RNAi machinery can be properly harnessed by providing exogenous triggers that enter the pathway and, similar to miRNAs, promote gene-silencing. If translated into a disease context such as cancer, RNAi is likely to provide an appealing tool that can be used to switch off the causative genes and/or restore deregulated expression patterns.

Whole-genome sequencing and integrated "omics" analyses have greatly speeded up the recent advances in dissecting the molecular basis of cancer, and have pointed out the marked heterogeneity of human tumors, as well as the need to design and develop new rational therapeutic strategies according to the patient's molecular profile. In this regard, RNA interference (RNAi) represents an emerging, powerful approach for personalized cancer medicine that can be used to tailor medical intervention to an individual's genetic and physiological background.

2 RNA Interference: Breakthrough of the Year 2002

RNA interference refers to an evolutionarily conserved cellular process that is mediated by dsRNAs. In worms, plants and flies, RNAi likely provides an innate defense to both endogenous parasitic and exogenous pathogenic nucleic acids (DNA or RNA) [1], whereas its primary role in mammals and human involves regulating the expression of protein-coding genes at the post-transcriptional level [1].

The finding that a dsRNA could activate a sequence-specific silencing, and the whole set of intracellular mediators and functional activities orchestrating the process, was such a high-impact discovery that it was hailed as the biggest "Breakthrough of the Year" in 2002 by the journal *Science* [2], and honored with the 2006 Nobel Prize in Physiology or Medicine to Andrew Z. Fire and Craig C. Mello.

Following on from their studies in the nematode *Caenorhabditis elegans* [3], several research groups have moved quickly to large mammals [4, 5], to start investigating the potential for harnessing RNAi to address biological questions and treat human diseases. Indeed, by enabling reversible gene silencing and modulating the expression of any gene, RNAi provides an appealing approach for loss-of-function gene analyses in vertebrate systems and, more importantly, for clinical purposes.

2.1 RNA Interference: a Brief Historical Digression

Retracing the history of RNAi discovery, the first hints were derived from experiments on engineering transgenic petunias, when Richard Jorgensen was attempting to alter pigmentation and obtain a more intensely purple flower strain [6]. Unexpectedly, the introduction of exogenous pigment-producing transgenes resulted in totally white or variegated pigmentation. As the endogenous loci and "foreign" transgenes appeared to be turning each other off, this phenomenon was called "co-suppression." Similar outcomes were subsequently derived from other plants, fungi (*Neurospora crassa*) [7], and metazoans (*Drosophila* and *C. elegans*) [8–10], where transmission of the silencing effect was also observed through several generations.

At that time, 1995, Guo and Kemphus were investigating the function of the *C. elegans par-1* gene by using an "antisense-mediated silencing" approach, in which a small, synthetic RNA oligonucleotide, complementary to the *par-1* sequence, once delivered to the cells, was expected to bind the messenger RNA (mRNA) and block its translation. Guo and Kemphus found that both the exogenous antisense and sense oligonucleotides (the latter being used as negative control) were able to produce the same lethal effect by inducing *par-1* mRNA silencing [11]. The turning point came when Fire, Mello and colleagues, during their experiments of manipulation of gene expression in *C. elegans*, found that the dsRNA mixture was 10-fold more potent in inducing translational repression than sense or antisense RNAs were alone [3]. This effect became known as RNA interference.

Shortly thereafter, this effect – which became variously known as post-transcriptional gene silencing (PTGS) [12], co-suppression [6], quelling [7], and RNAi – was demonstrated to be even more widespread, occurring in flies [8] as well as mammals [4, 5]. The role of RNAi pathways in the normal regulation of endogenous protein-coding genes was suggested by a flurry of many results. Small RNAs were found to be produced in plants undergoing PTGS [13, 14], while studies in *C. elegans* revealed that these small RNAs – which were now known as short interfering RNAs (siRNAs) were produced by the dsRNA-processing enzyme Dicer [15] and promoted gene silencing by triggering the assembly of a nuclease effector known as the RNA-induced silencing complex (RISC) [8]. Subsequently, a class of natural hairpin dsRNAs [16, 17], now called miRNAs, was shown to be processed by Dicer [18, 19]. Finally, the role of RNAi in chromatin regulation in yeast [20], and chromosomal rearrangement during development of the somatic macronucleus in *Tetrahymena* [21], led to a definite identification of the RNAi machinery as a natural developmental gene regulatory mechanism.

2.2 Short Regulatory RNAs: miRNA and siRNA Biogenesis

When attempting to manipulate and take advantage of RNAi for any experimental application, it is critical to have an overview of the molecular basics of this pathway.

Since its discovery, more insights have been attained about the genes and the molecules implicated in the endogenous RNAi machinery.

RNAi, also defined as a sequence-specific, PTGS process [1, 12] is mainly performed by small RNAs including endogenous miRNAs [22], short interfering RNAs (siRNAs) [4], and exogenous chemically synthetized short hairpin RNAs (shRNAs) [23].

Although similar in function, a distinction can be made between miRNAs and siRNAs based on their different biogenesis [24, 25] and mechanisms of recognition of the target RNAs [26].

miRNAs are initially transcribed in the nucleus by an RNA Polymerase II or III as long primary transcripts (pri-miRNAs), containing single or clustered double-stranded hairpins bearing single-stranded 5′- and 3′-terminal overhangs, and 10 nt distal loops [26]. The pri-miRNA is processed into a 70- to 100-nt, imperfectly base-paired hairpin precursor RNAs (pre-miRNAs) by the Microprocessor Complex, which consists of the RNase III enzyme Drosha and a protein called Pasha in *Drosophila* or DGCR8 (Di George Syndrome critical region gene 8) in mammals [27–29]. This initial cleavage is then followed by the Exportin-5/RanGTP-mediated pre-miRNA translocation to the cytoplasm for further processing into a 19- to 25-nt duplex by the RNase III endonuclease Dicer and TRBP (Human immunodeficiency virus-1 transactivating response RNA-binding protein) [30]. Final processing by Dicer is likely to culminate in incorporation of the miRNA into the RISC, which acts as effector of RNAi [31] (Fig. 1).

In contrast, siRNAs are produced from long, perfectly base-paired dsRNA precursors which can be produced endogenously or provided exogenously (Fig. 1).

Indeed, siRNAs are often referred to either genetic material deriving from virus infections or artificial inhibitory products. However, large-scale analyses in plants, fungi and animals have detected small RNAs that lack hairpin-loop precursors, and thus do not belong to miRNA species [32–34]. The larger part of these RNAs is encoded by repetitive elements of the genome, as trasponsons, especially located at heterochromatic regions such as centromeres and telomeres, and confirming that the RNAi machinery contributes to the establishment of a repressed chromatin state at these regions [20]. Other siRNAs are instead transcribed from the antisense strand of chromosome regions that often contain sequences for longer noncoding RNAs [32, 35]. Although their function is far from being completely understood, several organisms (including yeast, flies, plants, and protozoa) have been found to contain many of these siRNAs.

Nevertheless, once in the cytoplasm, the siRNA processing, as well as its assembly into the RISC complex, is mostly Dicer-dependent [36] (Fig. 1).

Thus, both for miRNA and siRNA precursors, the resulting dsRNA is a duplex of 19- to 25-nt strands bearing a 2-nt overhang at its 3′ terminus and a phosphate group at its 5′ terminus.

2.3 The RNAi Machinery

The key component of the RISC complex is an Argonaute (AGO) protein. AGO proteins have been consistently found in bacteria, archaea and eukaryotes [37]. In humans, among the eight AGO-classified proteins [38], AGO2 is the only member which exhibits endonuclease activity associated with its RNaseH-like domain [39] and accomplishes an miRNA/siRNA-induced silencing of target mRNAs. AGO2, together with Dicer and TRBP, represent the minimal RISC-loading complex (RLC) [26]. Additional proteins as Vasa intronic gene (VIG) protein, the Tudor-SN protein, Fragile X-related protein, the putative RNA helicases Dmp68, and Gemin3, as well as GW182 and TTP [40–43], have been found to associate with the RLC, although their functions are still not fully understood.

Once the diced dsRNAs interact with AGO2, one strand of the duplex – the

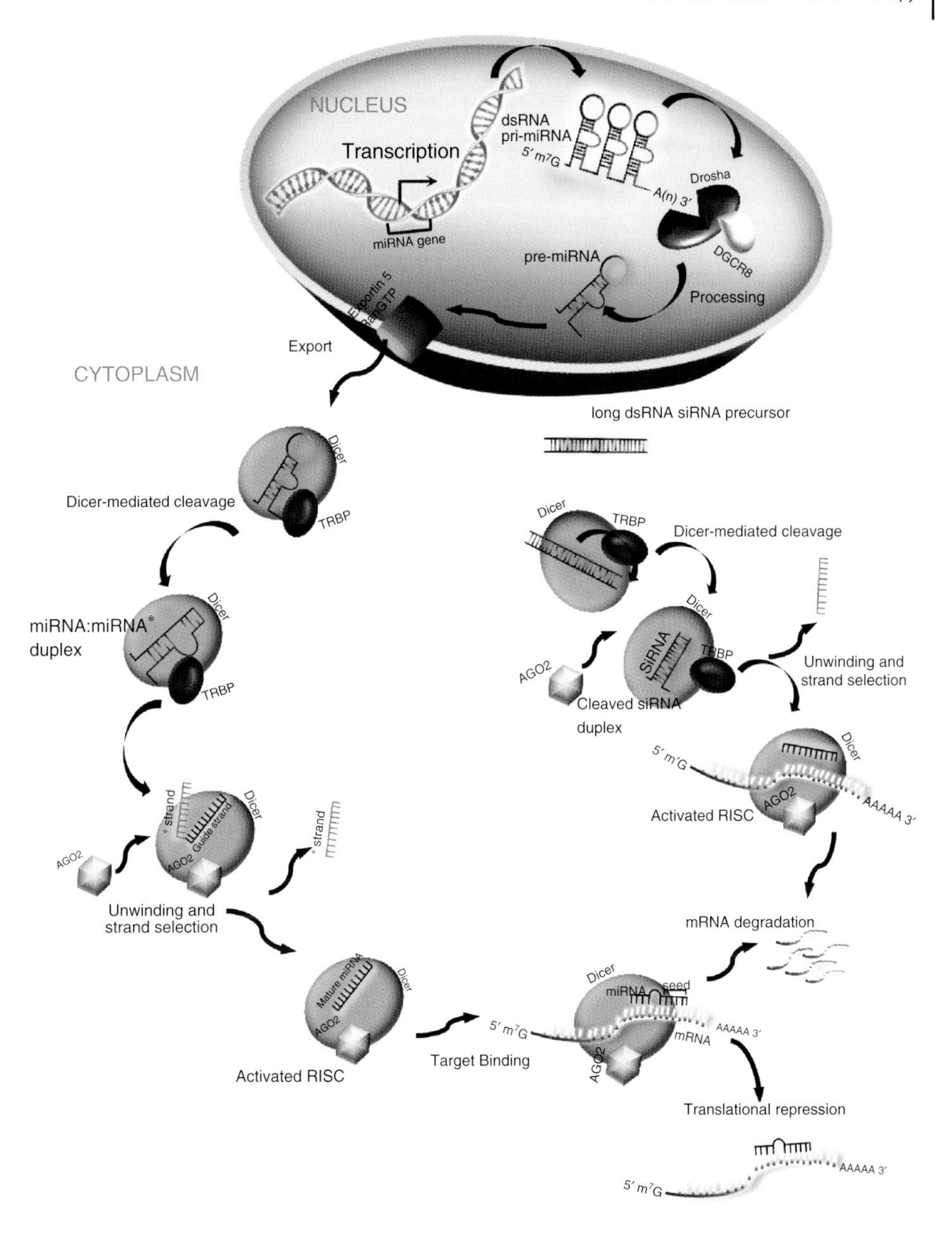

"guide" strand – is selected to be loaded into the RISC complex, while the other strand – the "passenger" strand (often indicated by *) – is discarded. Selection of the guide strand, which generally is based on thermodynamic properties, specifically addresses the silencing activity of the RISC complex to cytoplasmic single-stranded RNAs (ssRNAs), as endogenous mRNAs, through either perfect or imperfect

Fig. 1 The miRNA and siRNA pathways of RNAi. Primary miRNAs (pri-miRNAs) are transcribed in the nucleus by RNA Polymerase II as long, capped, and polyadenylated (A-(n)) precursors, and are trimmed by the microprocessor complex (including Drosha and DCGR8) into 70- to 100-nucleotide hairpin precursors, usually containing interspersed mismatches along the duplex, called pre-miRNAs. Pre-miRNAs associate with the Exportin/RanGTP complex and are translocated to the cytoplasm (left side of figure), where they are further processed into a 19- to 25-nucleotide duplex (miRNA:miRNA*) by the Dicer–TRBP complex. Finally, the duplex interacts with an Argonaute (AGO2) protein into the RISC (RNA-induced silencing complex) apparatus: one strand of the duplex (the passenger * strand) is removed, whereas the guide strand, the mature miRNA, remains stably associated within the RISC, and directs the complex to the target mRNA for the post-transcriptional gene silencing. The interactions between miRNA *seed* sequence and the 3′-UTR of the mRNA generally culminate in translation repression. However, cleavage and degradation of the mRNA have also been documented. Long, perfectly base-paired, dsRNA (right side of figure) are also processed by the Dicer–TRBP complex into small siRNAs. Once they enter the RISC complex, an Argonaute protein cleaves the passenger strand, and the guide strand is used to bind the complementary mRNA target. SiRNAs usually anneals to the mRNAs by a perfect complementarity, mainly leading to their degradation.

complementarity to the AGO2-bound guide strand. Indeed, the guide strand nucleotides 2–8 (counting from the 5′ end) constitute the "*seed* region," that is able to directly bind the target mRNAs. The degree of complementarity can be either perfect (as occurs mostly for siRNAs) or partial (as is usual for miRNAs), and determines subsequent AGO2-mediated target cleavage or, more frequently (if partial complementarity occurs) a non-endonucleolytic translational repression [26] (Fig. 1).

3
MicroRNAome: The Biological Toolkit for the Regulation of Gene Expression

A large number of miRNA genes (>1000) have been predicted to exist in the human genome, accounting for 1–5% of all predicted human genes. Altogether, miRNAs in human cells are estimated to regulate the expression of between 30% and 90% of human genes [44]. In particular, miRNAs mainly operate the PTGS by repressing translation or accelerating mRNA decay [45]. In both cases, after being loaded into the RISC complex, binding of the target mRNAs to the 3′ untranslated regions (UTRs) does not require perfect complementarity with the "*seed*" sequence. Thus, in contrast to siRNAs, a single miRNA may regulate multiple messenger RNAs and, in turn, each gene can be regulated by different miRNAs [45]. Moreover, recent studies have reported that miRNAs can also bind to the 5′ UTR or the open reading frame [46–49] and, even more surprisingly, they can upregulate translation upon growth arrest conditions [50]. It has also been noted that mature miRNAs may localize in the nucleus, where they may regulate pre-mRNA processing, act as chaperones modifying mRNA structures, or modulate mRNA–protein interactions [51]. The observations that miRNAs can be imported into the nucleus [52] or even secreted from the cells [53] suggest that the mechanistic details and the number of cellular functions of miRNAs are still far from being fully unraveled.

3.1 microRNA Deregulation is a Molecular Feature of Cancer

As data have accumulated, miRNAs have been found to be involved in a number of cellular processes such as proliferation, differentiation, survival, and apoptosis. Hence, it is not surprising that alterations affecting miRNA expression and function may promote the pathogenesis and progression of human tumors. Indeed, like classical genes, miRNAs can be either overexpressed or downregulated and act as oncogenes or tumor-suppressors, depending on their downstream targets [54]. Abnormalities in their expression or functions are associated with the typical hallmarks of cancer, including increased cell proliferation, abrogated apoptosis, enhanced cell motility and invasiveness, and neoangiogenesis [54]. For instance, miR-21 is highly upregulated in the majority of cancer tissues, and the repression of pro-apoptotic genes, such as *PTEN* (*phosphatase and tensin homolog*) or *PDCD4* (*programmed cell death 4*), stimulates proliferation and tumor initiation [55, 56]. Conversely, miR-15a and miR-16-1 were the first tumor-suppressor miRNAs to be described that were expressed from the fragile 13q14 region that is frequently deleted in patients with chronic lymphocytic leukemia (CLL) [57]. Their loss, by releasing the inhibition of tumor-promoting genes, such as *BCL2*, *BMI1*, *CCND2*, and *CCND1*, is able to promote cell growth and tumor progression [57–60].

Following the initial discovery of the miR-15a/16-1 cluster, a steadily growing number of reports have established that miRNA expression is globally deregulated in neoplastic cells compared to the corresponding normal tissue [61]. The progressive development of different high-throughput platforms for assessing miRNA expression in normal and diseased tissues (microarrays, bead-based flow cytometry, and deep sequencing) has allowed the miRNA expression profiling of several malignancies, including CLL [62], breast cancer [63], thyroid papillary carcinoma [64], lung cancer [65], glioblastoma [66], pancreatic tumors [67], hepatocellular carcinoma [68], prostate cancer [69], and gastric cancers [70]. This multi-combined approach has revealed the ability of miRNA signatures to not only differentiate between normal and cancerous tissues, but also to identify the tissue of origin. The latter point would be especially relevant when the tumor has already spread to distant metastatic sites, as it would allow the discrimination of different subtypes of a particular cancer even more successfully than previous mRNA panels [71]. Notably, the exciting discovery that miRNAs can be also detected at high levels in body fluids has spurred a rush to investigate miRNAs as potential diagnostic and prognostic biomarkers, as well as a novel class of drug targets, within the clinical setting of human malignancies [54].

3.2 Mechanisms of microRNA Deregulation in Cancer

Overall, large-scale miRNA profiling studies have revealed that the cancer phenotype is characterized by a globally reduced miRNA level relative to normal tissues [72], which suggests a role for miRNAs in the maintenance of a differentiated cell state and homeostasis.

The cause of the widespread differential expression of miRNA genes between neoplastic and normal cells can be linked to different mechanisms, including chromosomal rearrangements of miRNA genes,

DNA point mutations, epigenetic mechanisms, or alterations in the machinery responsible for miRNAs biogenesis [54, 73]. Along with the typical CLL breakpoint at the miR-15a-16-1 locus in the 13q14 chromosomal region, members of the let-7 family of tumor suppressor miRNAs map to fragile sites, as 3p2 (let-7g/let-7a-1), 9q22 (let-7f), 11q24 (let-7a-2), and 21q21 (let-7c), are frequently deleted in breast, lung, ovarian and cervical cancers [74]. Conversely, the oncogenic miR17~92 cluster in 13q31, encoding for six different miRNAs (miR-17, miR-18a, miR-19a, miR-20a, miR-19b-1, and miR-92-1), is amplified in several hematopoietic malignancies [54]. Interestingly, miR-17-5p and miR-20 are known to specifically repress the cell-cycle regulator E2F1, which mediates cell-cycle progression through G_1/S checkpoints, and whose expression is also regulated by the transcription factor c-MYC [75]. *In-vivo* studies in Eμ-Myc transgenic mice (a well-established mouse model of B-cell lymphoma) showed that the miR17~92 promoter also contains c-Myc E-box binding sites, and suggests that c-MYC is likely to fine-tune the cell cycle by regulating the expression of both mRNAs and miRNAs.

In addition to structural genetic alterations, miRNA expression in cancer can be also affected by epigenetic changes, such as altered DNA methylation [76]. Indeed, it is likely that half of the genomic sequences of miRNA genes contain CpG islands [77]. In this regard, the hypermethylation of miR-127 has been found to induce its downregulation, and to significantly increase the target proto-oncogene *BCL6* in bladder cancer cells [78]. Similarly, miR-9-1 [79] and the clustered miR-34b and miR-34c [80] were also found to be modulated by DNA methylation. Alternatively, defects in the epigenetic machinery, including DNA (cytosine-5)-methyltransferase 1 (DNMT1) and DNA (cytosine-5)-methyltransferase 3β (DNMT3β), also affect miR-124a expression, as found in colorectal cancer cells [81]. Conversely, the overexpression of putative oncogenic miRNAs in cancer can be due to DNA hypomethylation [82], and alterations affecting other chromatin-remodeling processes [83–85].

Single nucleotide polymorphisms (SNPs) in miRNAs, in either a heterozygous or homozygous configuration, represent another type of genetic variability that can affect gene and protein expression, and may be responsible for the interindividual variability of the cancer phenotype. Indeed, sequence variations in miRNA genes, including pri-miRNAs, pre-miRNAs and mature miRNAs, could also influence the processing and/or target selection of miRNAs.

A C→T germline alteration in the primary transcript of miR-15a/miR-16 reduces their expression levels, and has been found in patients with familial CLL [86]. miRNA precursor-related SNPs, such as pre-miR-196a2, pre-miR-499 and pre-miR-146a, have been associated with the risk of breast cancer [87] and papillary thyroid carcinoma [88]. Moreover, the SNP rs11614913 in pre-miR-196a2 has also shown prognostic significance correlating with survival in patients with non-small-cell lung cancer (NSCLC) [89].

SNPs in the miRNA *seed* region can impair – either by weakening/abolishing or, conversely, by strengthening – miRNA–mRNA interaction, and thus differentially affect the expression of miRNA targets [90, 91]. Despite the low (<1%) probability of SNP occurrence in a miRNA *seed* region, a polymorphism here would significantly impair a miRNA function, as compared to a polymorphism present in a 3′-mismatch-tolerant region (3′-MTR) binding region which, conversely, is very sensitive to mismatches. And last, but not least, miR-polymorphisms can alter the epigenetic regulation of a miRNA (either methylation or acetylation) and promote disease progression [79].

Several reports have also shown that changes in the miRNA biogenesis and processing machinery can be responsible for miRNA dysregulation in cancer [90]. For instance, a generalized downregulation of Dicer and/or Drosha has been significantly correlated to a poor clinical outcome in different tumors such as lung, ovarian, colon, prostate and breast cancer, nasopharyngeal carcinoma, and neuroblastoma [92–100]. In addition, the *DICER1* gene is commonly featured as a haploin-sufficient tumor suppressor because frequently it is single-copy-deleted in cancer [101]. Interestingly, a somatic hot-spot missense mutation in *DICER1* gene restricted to non-epithelial ovarian tumors has also been reported [102]. Inactivating mutations and a consequent loss of the *TARBP2* gene, encoding for TRBP, have been shown to destabilize *DICER1* and impair miRNA processing in cancers bearing microsatellite instability [103]. These tumors are also characterized by mutations of the *XPO5* gene, coding for the Expotin-5 transporter, which generate a truncated protein that is unable to associate with pre-miRNA molecules and exit the nucleus; this decreases the trafficking of pre-miRNAs in the cytoplasm and reduces the pool of diced mature miRNAs [104].

A dysregulation of AGO proteins also occurs in cancer. The loss of AGO1 (*EIF2C1*), AGO3 (*EIF2C3*) and AGO4 (*EIF2C4*) is a recurrent event in Wilms tumor of the kidney and in neuroectodermal tumors [37, 73]. However, the expression of AGO proteins, such as AGO2 (*EIF2C2*), is reported to be reduced in some cases, for example in melanoma samples (primary and metastatic) compared to normal epidermal melanocytes, but elevated in other tumors such as breast and colon cancer; this suggests that expression can be regulated in a cell context-dependent manner [105]. The RNA-binding proteins LIN28/LIN28b represent further potential oncogenic factors; they are normally highly expressed in hematopoietic stem cells (HSC), and have the unique ability to revert human somatic cells to pluripotent cells when coexpressed with the reprogramming factors OCT4, NANOG, and SOX2. The oncogenic roles of LIN28 and LIN28b have been suggested by elevated levels in cancer stem cells from various cancer types, including ovarian, breast and NSCLC [105].

Experimental findings have also documented how miRNA processing can be affected by other miRNAs, either directly, whereby the mouse miR-709 is able to bind a recognition element on pri-miR-15a/16-1 in the nucleus [106], or indirectly, whereby the miR-103-107 family targets Dicer and leads to a downstream downregulation of miRNAs as the miR-200 family, and to epithelial-to-mesenchymal transition (EMT) [100].

Finally, the deregulation of miRNA expression, as either increased or decreased transcription, can also result from an altered transcription factor activity. Indeed, miRNAs can be either positively or negatively regulated by transcription factors such as p53, activating miR-34a [107] and miR-205 [108], MYC, activating miR-17~92 cluster [75] but repressing let-7 [109] and miR-29 family members [110], and HIF-1α, regulating miR-210 [111] or ZEB1, and repressing the transcription of members of the miR-200 family, which in turn are able to directly target ZEB1 and ZEB2 [112].

3.3 Role of microRNAs in Cancer

As noted above, aberrant expressed miRNAs can act either as oncogenes (oncomiRs) or tumor suppressor genes (tumor suppressor miRNAs), and be crucially involved in the initiation and progression of the neoplastic disease in all human tumors.

Of note, their role is strictly dependent on the cellular context and tumor system: miR-221 and miR-222, for example, can function as tumor suppressor miRNAs in leukemia by targeting the *KIT* oncogene. [113]. Furthermore, they are also known for silencing tumor suppressors like *PTEN* or *TIMP3* (tissue inhibitor of metalloproteinases 3), playing the role of oncomiRs in several solid tumors including breast and lung cancer, hepatocellular carcinoma, or glioblastoma [114].

Here, a few examples of miRNAs classified as oncomiRs (Table 1) or tumor suppressor miRs (Table 2) are reported, according to the main role that they play in the majority of cancers.

3.3.1 OncomiRs

miR-155 miR-155, encoded by the B-cell integration cluster (BIC) ncRNA, exerts its role in regulating the response and adaptation of cells to the tumor microenvironment, in particular when hypoxic conditions occur. Indeed, miR-155 mediates the downregulation of FOXO3A upon HIF1α activation [115]. In this context, other targets of miR-155 are SOC1 and RhoA, which collectively act in the signaling pathways promoting EMT [116, 117].

A high expression of miR-155 has been reported in several solid tumors such as breast cancer, colon cancer, ovarian cancer, pancreatic ductal adenocarcinoma, thyroid carcinoma, and lung cancer [71], where it is also considered a marker of poor prognosis. Similarly, miR-155 levels are increased in various B-cell malignancies, including Hodgkin lymphoma, some subtypes of non-Hodgkin lymphoma, and acute myeloid leukemia (AML) [71]. Interestingly, alongside the contribution to cancer genesis and progression, *in-vitro* studies have shown that miR-155 overexpression can confer resistance to radiation therapy, and the administration of antisense oligonucleotides (ASOs) against miR-155 can overturn this effect, which suggests that miR-155 may serve as a predictive biomarker and putative therapeutic target for treatment-resistant tumors [118].

The miR-17∼92 Cluster and Paralogs The polycistronic miR-17∼92 cluster is located within approximately 1 kb of an intron of the C13orf25 locus on human chromosome 13q31.3, a region that is frequently amplified in several types of B-cell lymphoma and solid tumors [119]. The transcriptional regulation of the miR-17∼92 cluster is mostly mediated by the *MYC* oncogene [75], to inhibit the apoptotic activity of E2F1, and induce cell proliferation. Overexpression of the cluster is observed in a variety of human cancers, including small cell lung cancer (SCLC), colon and gastric cancer, retinoblastoma, neuroblastoma, medulloblastoma, and osteosarcoma [71]. The cluster, which is processed to produce mature miR-17, miR-18a, miR-19a, miR-20a, miR-19b-1, and miR-92a-1, can target several factors involved in cancer initiation (HIF1α), cell proliferation (PTEN, E2F1-3, TNFα, RAB14), survival (BIM, TGFBR2), and angiogenesis [TSP1, connective tissue growth factor (CTGF)] [54, 120, 121]. The paralog miR-106b-25 cluster, comprising highly conserved miR-106b, miR-93 and miR-25, has shown to be similarly regulated with its host gene *MCM7* by E2F1, building

Tab. 1 MicroRNAs which act mainly as oncomiRs and are upregulated in cancer.

microRNA	*Locus*	*Cancer type(s)*	*microRNA deregulation effects*	*Key targets*
miR-155	21q21.3	CLL, AML, B-lymphomas; lung, breast, colon, gastric, pancreatic, thyroid cancers	Resisting growth suppressors	FOXO3A, SOCS1, SHIP1
			Supporting inflammation	CEBPB, MEIS1, ETS1, C-MAF, CUTL1
			Mismatch repair impairment	hMSH2, hMSH6, hMLH1, WEE1
			Invasion and metastasis	RhoA
miR-17~92	13q31.3	Leukemias/lymphomas; lung, breast, colon, prostate, pancreatic, cancers; retinoblastoma, glioblastoma	Sustaining proliferation	PTEN, E2F1-3, TNF-α, RAB14, p63
			Resisting growth suppressors	BIM, TGFBR2
			Promoting angiogenesis	TSP1, CTGF, TGF-β, SMAD4
			Hypoxia response	HIF-1α
miR-221/222	Xp11.3	Lung, breast, prostate, papillary thyroid cancers; hepatocellular carcinoma; glioblastoma	Sustaining proliferation	PTEN, TIMP3, DICER
			Resisting growth suppressors	KIT, $p27^{Kip1}$, $p57^{Kip2}$, FOXO3A
			Evading apoptosis	PUMA
			EMT	TRSP1
miR-21	17q23.2	CLL, AML, myeloma; lung, breast, gastric, prostate pancreatic cancers; glioblastoma	Sustaining proliferation and ECM remodeling	PTEN, SPRY1, SPRY2
			Resisting growth suppressor	APAF1
			Evading apoptosis	PDCD4
			Metastasis	TPM-1, TPM-3, RECK, TIMP3

AML, acute myeloid leukemia; CLL, chronic lymphocytic leukemia; EMT, epithelial-to-mesenchymal transition.

a negative feedback loop where the cluster controls in turn the E2F1 intracellular level. An overexpression of miR-106b/25 has been reported in cancers of the prostate and pancreas, as well as in neuroblastoma, multiple myeloma [71], and is also likely to be responsible for the development of transforming growth factor-beta (TGF-β) resistance by suppressing p21 and BIM in gastric cancer [122].

Tab. 2 MicroRNAs which act mainly as tumor suppressor miRNAs and are downregulated in cancer.

microRNA	*Locus*	*Cancer type(s)*	*microRNA deregulation effects*	*Key targets*
miR-15a-16-1	13q14.2	CLL, multiple myeloma, lymphomas; lung, breast, colon prostate, pancreatic, ovarian cancers	Sustaining cell proliferation	CDC2, JUN, FGF-2, FGFR1 CCND-1, CHK1
			Evading apoptosis	BCL2, SIRT1
			Promoting angiogenesis	VEGF, VEGFR2, FGFR1
let-7 family	*LET7A1* 9q22.32	Lymphomas; lung, breast, colon, prostate, gastric, liver cancers	Sustaining cell proliferation	KRAS, NRAS, CDC25A, c-MYC
	LET7A2 11q24.1		Evading apoptosis	BCL-XL
	LET7A3,-7B 22q13.31		Metastasis	HMGA2, TWIST1
	LET7C 21q21.1			
	LET7D,-7F1 9q22.32			
	LET7E 19q13.41			
	LET7F2 Xp11.22			
miR-34a	1p36.22	Lymphoma; lung, breast, colon, renal, bladder pancreatic cancers; neuroblastoma, glioblastoma	Sustaining cell proliferation	CDC25A, CDK4, CDK6, c-MYC
			Evading apoptosis	BCL2, SIRT1
			EMT and metastasis	MET, SNAIL1
miR-200 family	*MIR200A*, *MIR200B*, *MIR429* 1p36.33	Lung, breast, endometrial, ovarian, nasopharyngeal, bladder cancers	EMT and metastasis	ZEB1, ZEB2, CTNNB1, BMI-1, PLCγ1, VEGFR1, FN1, LEPR, MALM2, MALM3
	MIR200C, *MIR141* 12p13.31			

FGF, fibroblast growth factor; LEPR, leptine receptor; NRAS, neuroblastoma rat sarcoma viral oncogene homolog; VEGFR1, vascular endothelial growth factor receptor 1.

The miR-222/221 Cluster Deregulation of the miR-222/221 cluster is mainly involved in promoting cell survival and the metastatic spread of cancer cells. miR-222/221 have been found highly upregulated in hepatocarcinoma, thyroid cancer, melanoma, estrogen receptor-negative breast cancer, and glioblastoma [123–127]. Cell-cycle control and anti-apoptotic activity are related to the repression of kit, p27Kip1, PTEN, and PUMA [128]. It has also been reported that miR-221 and miR-222 transcription is induced by FOSL1, leading to a repression of TRPS1 and an E-cadherin decrease, contributing to the invasive phenotype of basal-like breast cancer [129]. Recently, it was reported that miR-221 and miR-222, controlled by both MET and epidermal growth factor (EGF) receptors, may play a pivotal role in developing tyrosine kinase inhibitor resistance in NSCLC [130].

miR-21 The well-documented miR-21, transcribed from the 17q23.2 genomic locus, is typically highly expressed in virtually all human malignancies, including those derived from breast, colon, liver, brain, pancreas, and prostate.

The wide array of molecular targets, including the tumor suppressors maspin (*SERPINB5*), PDCD4, tropomyosin 1 (TPM1) and PTEN [131, 132], negative regulators of the RAS (rat sarcoma viral oncogene homolog) signaling axis, Spry1, Spry2 and Btg2 [133], or the metalloproteases RECK (reversion-inducing-cysteine-rich protein with kazal motifs) and TIMP3 (tissue inhibitor of metalloproteinase 3) [134] allocates miR-21 within the regulation of several oncogenic mechanisms supporting the aberrant growth and survival of cancer cells, as well as invasion and metastatic spread.

The central role of miR-21 is supported by a number of cell culture and animal model-based studies, which also indicate that the inhibition of miR-21 with ASOs represents a good strategy for reducing cell cancer proliferation and inducing apoptosis. For example, Medina and colleagues used Cre and Tet-off technologies to generate mice which, as a consequence of conditional expression of miR-21, developed a pre-B-cell malignant lymphoid-like phenotype, while subsequent miR-21 inactivation in the same model led to apoptosis and tumor regression [56]. This suggested that human cancers might be treated through the pharmacological inactivation of miRNAs such as miR-21.

The microarray-based profiling of tumor samples from breast cancer patients found that miR-21 was consistently upregulated in several subtypes of breast cancer, and suggested a role for this miRNA in mammary tumor initiation or progression [63]. Similarly, lung cancer profiling detected not only an elevated expression of miR-21 but also a direct correlation between its expression level and the presence of mutations in the epidermal growth factor receptor (*EGFR*) gene in patients [135]. Moreover, the blockade of miR-21 with ASO was able to enhance phosphorylation of epidermal growth factor receptor (p-EGFR) and restore apoptotic mechanisms in glioblastoma cell culture [136].

3.3.2 Tumor Suppressor miRs

The miR-15a/16-1 Cluster The link between miRNAs and cancer was first established with miR-15a and miR-16-1 [57], when it was discovered that they are often deleted and downregulated in CLL and, later, in prostate, lung cancer, and multiple myeloma [137–139], which suggested a main role in tumor suppression. Loss of their expression has been associated with the dysregulation of several pathways controlling cell proliferation, survival, migration and invasion, in which they regulate the levels of BCL-2, CDC2, ETS1, PDCD4, FGF-2, FGFR1 and other key cascade effectors. Moreover, miR-15a/16-1 was shown to regulate the response to cisplatin by targeting WEE1 and CHK1, which are commonly over-expressed in cisplatin-resistant cells, and also the restoration of normal miR-15/16 re-induced sensitivity to the drug [140], supporting the potential of a therapeutic solution to suppress aberrant growth in a variety of cancers.

let-7 Family The let-7 family includes 12 human homologs mapping to fragile sites associated with lung, breast, urothelial and ovarian cancers [71]. The let-7 family plays an important role in controlling cellular growth by regulating the Kirsten rat sarcoma viral oncogene homolog (KRAS) [141], and the specific downregulation of family members is a frequent event associated with a poor prognosis in lung cancer and the presence of lymph node metastases in breast cancer [71]. Moreover, in both types of tumor, polymorphisms in the 3′ UTR of KRAS mRNA are able to impair the let-7 regulating action and predict a worse prognosis [142, 143]. Along with KRAS, let-7 targets comprise LIN28, HMGA2 [54], MYC [144], and cell-cycle-related factors such as cyclinD2, CDK6, and CDC25A [145], allocating let-7 to every process involved in tumor growth and progression.

miR-34 Family The miR-34 family includes three miRNAs – miR-34a, miR-34b and miR-34c – under the direct transcriptional control of p53 [146]. In particular, p53 activation promotes a positive feedback loop in which an increased expression of miR-34 results in the downregulation of E2F, MET and SIRT1 and an indirect activation of p53 [141]. The promoter region of miR-34 family genes also contains CpG islands, and aberrant CpG methylation has shown to reduce miR-34 expression in multiple cancer types [80]. Other studies have progressively identified a number of targets, including BCL2, MYCN, GMNN and HDAC1 of the apoptotic cascade, cell-cycle crucial effectors such as CDC25A, CDK4 and c-MYC, or even WNT, HMGA2, and SNAIL1 that modulate migration and the invasion of cancer cells. For instance, a lower miR-34b expression in NSCLC has been correlated with a higher incidence of lymph node metastasis and a poor prognosis [54].

The miR-200 Family The miR-200 family members (miR-200a, miR-200b, miR-200c, miR-141, and miR-429) are downregulated in human tumors, where they play a critical role in the suppression of EMT, the cellular process through which epithelial cells lose their polarity and cell–cell adhesion, by targeting the key factors ZEB1 and ZEB2 [147] and promoting invasion. In lung cancer cells, an induced overexpression of the miR-200 family promotes mesenchymal-to-epithelial transition by repressing ZEB1 and increasing E-cadherin [148]. Accordingly, a loss of miR-200 family members correlates with a lack of E-cadherin expression in invasive breast cancer cell lines and breast tumor specimens, supporting an *in-vivo* role for the miR-200 family in EMT repression. Interestingly, ZEB1 and SIP1 are likely to recognize an E-box proximal minimal promoter element, generating a potential double-negative feedback loop in which the repression of a primary transcript and mature miR-200 expression allows the maintenance of high ZEB1/SIP1 levels that favor the mesenchymal phenotype of human breast cancer cells [149]. In addition, *in-vitro* studies have described a type of Akt–miR-200–E-cadherin axis, where Akt1 and Akt2 are able to control the abundance of miR-200 and E-cadherin and thus regulate breast cancer metastasis. This has been confirmed by the ratio of Akt1 to Akt2, and the abundance of miR-200 family members and E-cadherin in a set of primary and metastatic human breast cancers [150]. The deregulation of miR-220c/141 has also been linked to aberrant DNA methylation [151]. In conclusion, the miR-200 family prevalently assumes a tumor-suppressor role, and downregulation of its family members is significantly associated with an aggressive cancer cell phenotype.

When taken together, these data and the novel information that is continuously collected from the increasing number of molecular profiling studies in human cancers, highlight "miRNA addiction" as a key feature of cancer cells for developing and maintaining the malignant phenotype. Investigating miRNAs and the molecular circuitries that mediate their regulatory function in cancer can provide new clues about the nature of the cancer itself, and the molecules or mechanisms which may be targeted therapeutically to improve the clinical management of cancer patients.

4 Current Prospects for RNA Interference-Based Therapy for Cancer

To date, it is well documented that cancers can build sophisticated biological networks supporting their ability to progress and, moreover, to evade treatment. The promise of personalized medicine is to streamline clinical decision-making, according to each patient's genetic and physiological profile, distinguishing those who can benefit from a treatment from those who can experience side effects without relief. Indeed, distinct gene expression patterns may affect the body's response to medications, and an improved comprehension of the signaling cascades regulating tumor survival has determined the transition from a more promiscuously chemotherapeutic approach to highly selective targeted pharmaceutics. Initially, the antibody-based trastuzumab (Herceptin®) represented the first targeted therapy specifically for HER2-positive metastatic breast cancer (in 1997). Later, the small molecule imatinib mesylate (Gleevec™) became the first approved kinase inhibitor for targeting BCR/ABL in chronic myeloid leukemia (CML), and then for the treatment of gastrointestinal stromal tumors (GISTs), targeting c-kit.

A number of gain/loss-of-function experiments, in combination with target prediction analyses, have demonstrated that perturbation in miRNAs expression can affect not only every step of the tumorigenic process, but also the sensibility to drugs. Hence, the targeting of miRNAs, and the mechanisms orchestrating the endogenous RNAi, may undoubtedly represent a rich soil for the development of novel anti-cancer targeted therapies. RNAi-induced gene silencing mirrors the inhibitory effects of conventional drugs, such as protein-based compounds (e.g., antibodies and vaccines) and small molecules, whose inhibitory effect involves blocking the function of their targets. However, some disease-related molecules do not have enzymatic function, or they have a conformation that is not accessible to conventional drugs or small molecules, and so they are considered "undruggable." Conversely, RNAi technologies can be designed to virtually target any gene, including such undruggable molecules [152–154], with an exclusively allele-specific gene silencing [155]. Furthermore, antagonizing a target gene expression rather than its function has a more powerful downstream effect, as a single mRNA molecule is translated in multiple copies of a protein. Finally, compared to traditional drugs, the synthesis and production of RNAi modulators is more rapid, does not need the use of cellular expression systems or refolding steps, and can guarantee a higher selectivity and potency.

Nevertheless, to date, there are still some pending issues regarding the efficiency of the delivery systems, the choice of

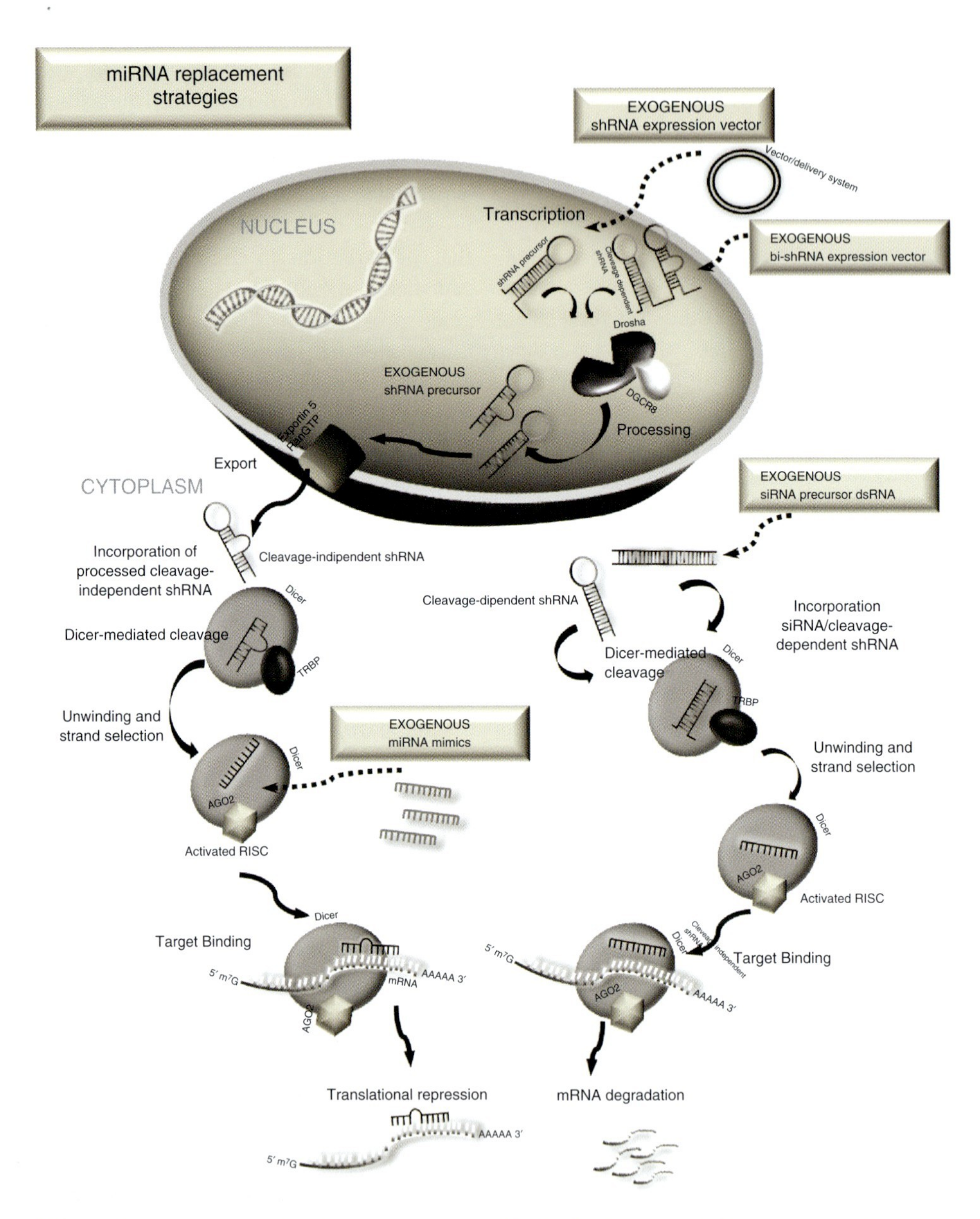

Fig. 2 MicroRNAs as targeted therapeutics. When aiming to repress oncogenic proteins, the expression levels of tumor suppressor microRNAs can be restored by using different strategies including siRNAs, shRNAs (short hairpin RNAs), bi-shRNAs (bifunctional siRNA), targeting by perfect base-complementarity oncogenic mRNAs, or miRNA mimics, synthetic analogs of endogenous miRNAs.

RNAi targets and the safety profile, that need to be addressed before allowing the widespread use of RNAi strategy *in vivo* [156].

4.1
Harnessing the Endogenous RNAi Apparatus for Clinical Purposes

The endogenous RNAi machinery can be engaged by designing exogenous triggers that enter the pathway at different points. Several synthetic modulators and strategies have been implemented to boost and optimize the effects on cancer-related pathways.

Modulating the RNAi machinery consists of either replenishing the levels of downregulated tumor-suppressor miRNAs to block tumor-promoting mRNAs, or sequestering oncogenic, overexpressed miRNAs to allow the transcription of tumor-suppressor proteins. All of the examples described below are simplified in Figures 2 and 3.

4.1.1 Restoring miRNA's Endogenous Regulation of Transcription: siRNAs, shRNAs, and Derivatives

The replacement of tumor suppressor miRNAs (let-7 family, miR-34a, miR-24, miR-26) [51, 74] finds its rationale in the global miRNA downregulation that characterizes the neoplastic phenotype. This strategy would permit the potential inhibition of oncogene transcripts that are aberrantly increased in the cancer state.

Short-interfering RNAs (siRNAs) are synthetic, short (usually 19–23 bp) dsRNAs which enter miRNA biogenesis at the cytoplasmic stage, when the pre-miRNA is chopped by Dicer, to be directly incorporated into the RISC complex. The siRNA guide-strand binds to and cleaves the complementary mRNA by a perfect-complementary match. Importantly, the guide-strand–RISC complex has the ability to be recycled, allowing the translational repression of several mRNA molecules, and propagation of the induced gene-silencing activity.

However, as the half-life of a siRNA is relatively short [157], shRNAs have been designed to enter the nucleus and be transcribed by RNA polymerase II or III from an external expression vector bearing a pre-designed, short, double-stranded DNA sequence. This means that shRNAs are constantly synthesized in the host cells, and achieve a longer gene-silencing effect. The primary transcript generated from the RNA polymerase II promoter contains a hairpin-like stem–loop structure that is regularly processed by Drosha, and translocated into the cytoplasm as pre-shRNA, to be cleaved by Dicer and incorporated into RISC, followed by the same cytoplasmic RNAi process as in siRNA [157].

A new class of bi-functional short hairpin RNAs (bi-shRNAs) has been developed for a more efficient and prolonged target-gene knockdown [158]. These consist of a vector construct for the expression of two (bi) stem–loop

Fig. 3 MicroRNAs as therapeutic targets. Aberrantly overexpressed microRNAs (oncomiRs) can conversely be inhibited by using: antimiRs (ASO, antisense oligonucleotides; LNAs, locked nucleic acids), designed to be complementary to the endogenous miRNAs; miRNA sponges, which contain multiple binding sites for microRNAs and prevent their loading onto the RISC complex; miRNA masks, small antisense oligonucleotides, which bind the target mRNA at the 3′-UTR miRNA-target sites, avoiding the interaction between mRNA and endogenous miRNAs, and transcriptional repression.

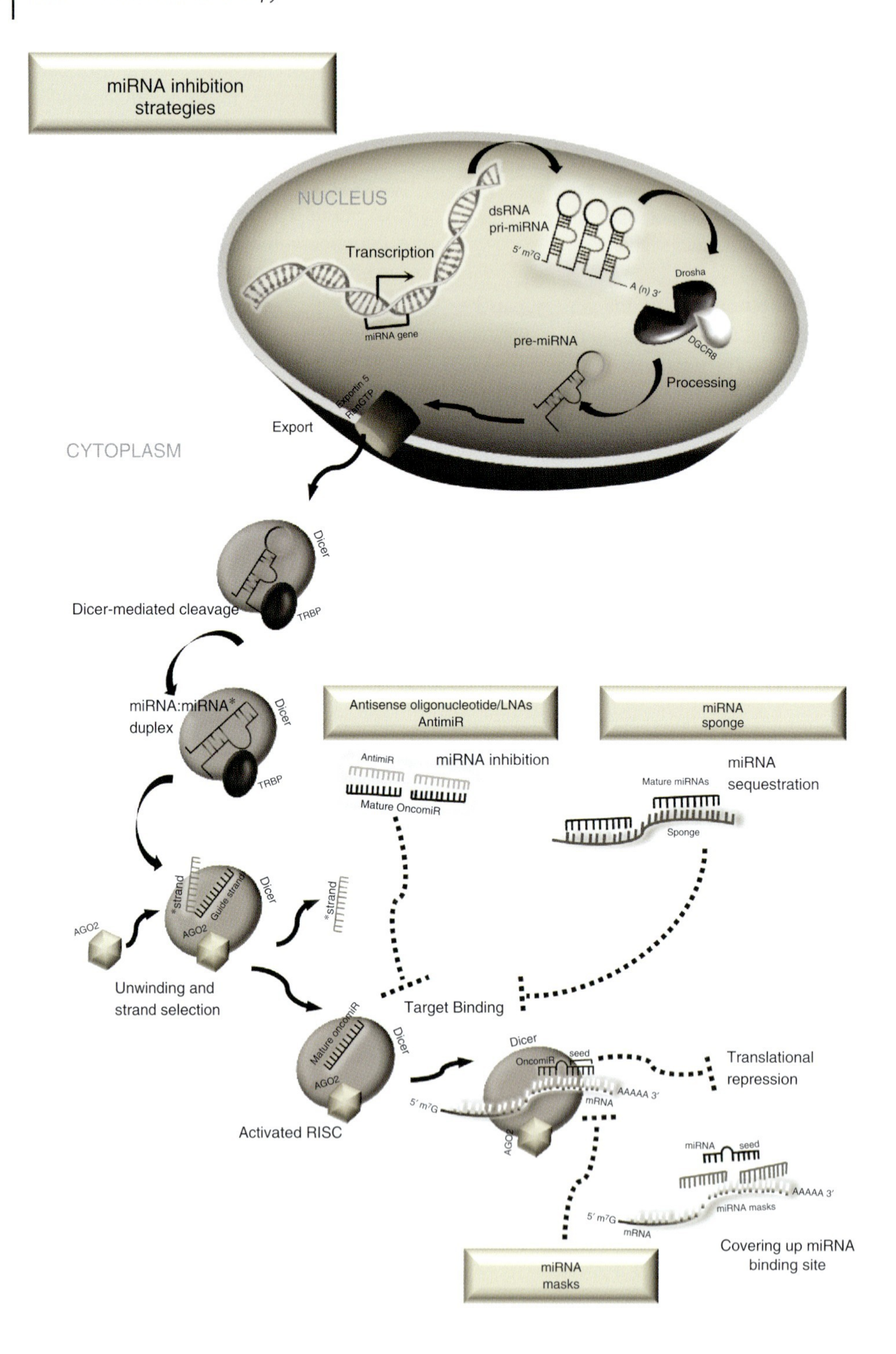
miRNA inhibition strategies
NUCLEUS
Transcription
miRNA gene
dsRNA
pri-miRNA
Drosha
DGCR8
pre-miRNA
Processing
Exportin 5
RanGTP
Export
CYTOPLASM
Dicer
Dicer-mediated cleavage
TRBP
miRNA:miRNA*
duplex
Antisense oligonucleotide/LNAs
AntimiR
miRNA inhibition
Mature OncomiR
miRNA sponge
miRNA
sequestration
Mature miRNAs
Sponge
*strand
Guide strand
AGO2
Unwinding and
strand selection
Target Binding
Mature oncomiR
Activated RISC
OncomiR
seed
mRNA
Translational
repression
miRNA masks
Covering up miRNA
binding site
miRNA
masks

shRNAs for each targeted mRNA: one with perfect-matching stem sequences (as siRNAs), and a second with mismatches at the central location (bases 9–12) and additional locations of the stem. Compared to siRNAs, this strategy would promote the loading of mature shRNAs onto both cleavage-dependent (AGO2-mediated) and cleavage-independent (blocked-AGO2 by the mismatch) RISCs, thus achieving a higher and more effective shut down of the target mRNA by the concurrent involvement of both degradation and translation inhibition mechanisms [158].

ssRNAs may also be used as RNAi triggers but, although they have a greater structural flexibility (ssRNA versus dsRNA) and a more amphiphilic nature [159], they are not efficiently recognized by the miRNA biogenesis apparatus (~100-fold less efficient) as siRNAs and shRNAs.

Moreover, a new method to restore the expression levels of endogenous, downregulated miRNAs (single or cluster) employs double-stranded, short artificial miRNAs, termed miRNA *mimics*, which are analogs of endogenous miRNAs and repress mRNA molecules by an imperfect binding to the 3′-UTR following the same mechanism of action [51]. Indeed, the ectopic expression of synthetic miRNAs mimics with tumor suppressor function as miR-15-a and miR-29 in cancer cells has shown to induce apoptosis in prostate and AML cell lines, respectively [160]. The first *in-vivo* evidence derives from Kota and colleagues, who recently cloned tumor suppressor miR-26 into an adeno-associated virus (AAV) vector for intravenous injection in an established MYC-dependent liver cancer mouse model. As expected, the restoration of miR-26 expression, which generally was reduced in liver cancer cells compared to normal cells, resulted in a suppression of tumorigenicity through the downregulation of cyclins D2 and E2 and a consequent suppression of tumor growth, without signs of toxicity [161].

Of note, it is likely that the mimic strategy would prevent the development of resistance to an RNAi approach by cancer cells better than siRNA formulations. Indeed, by mimicking endogenous miRNAs, mimics would impair the downstream function of multiple genes at the same time; thus, it would be more difficult for cancer cells to rearrange the sequence of several genes simultaneously to escape the mimic targeting and inhibition.

4.1.2 Inhibition of oncomiRs: AntimiRs, miRNA Sponges, and miR-Masks

The inverse approach consists of the inhibition of overexpressed miRNAs with oncogenic roles as miR-17~92 cluster, miR-155, by using small, single-stranded ASO drugs or the new generation of locked nucleic acids (LNAs) [51, 74]. These compounds belong to the antimiRs category, the properties of which are complementary to a target miRNA, with an ability to bind with high affinity and specificity and to produce a functional inhibition, thereby derepressing the mRNA targets of that miRNA [162].

A novel technology for managing miRNAs involves the use of expressing vector-carrying tandem miRNA target sites, to saturate an endogenous miRNA, and prevent the binding to its natural targets. This approach, known as "decoy," "sponge," or "eraser," employs different systems of gene delivery such as plasmids and adenovirus-, lentivirus- or retrovirus-based vectors that are normally used for gene knock-down studies. Since the inhibition of one miRNA can be compensated by the expression of other miRNA genes or family members, the tandem target sites built on these vectors can be assembled to catch different miRNAs sharing the same binding site, thus amplifying the effect of miRNA inhibition. Moreover, the sponges have been shown to be stable, they do not require repetitive rounds of administration, and they assure cell-type specificity compared to ASOs. Notably, the problem of a safety profile (i.e., toxicity), the partial interference that can occur, and the unknown targets that still must be identified for each putative miRNA, has led to the use of these constructs being a major challenge that is still in progress [163].

As an alternative to miRNA sponges, MiRNA-Masking Antisense oligonucleotides (miR-Masks) are single-stranded 2′-*O*-methyl-modified ASOs, that are fully complementary to predicted miRNA binding sites in the 3′-UTR of the target mRNA [164]. They compete with an endogenous miRNA by covering up the miRNA binding site and de-repressing its target mRNA. As this approach has the advantage of being gene-specific, it can significantly reduce any off-target effects [164]; however, it does not allow the targeting of multiple pathways.

5 Rational Design of an RNAi-Based Therapeutic Approach in Cancer

5.1 RNAi Target Selection

A successful RNAi clinical pipeline first relies on the rational selection of targets, according to parameters such as tumor specificity, tumor survival relevance, and the predictivity of tumor biological and clinical responses.

Whole-genome RNAi screening has been applied to human research for the first time for the identification of host genes affecting influenza A virus replication [165] and West Nile virus infection [166]. Using this approach, Tiedemann and colleagues identified crucial survival genes for the non-curative myeloma cancer after transfection of siRNA constructs of myeloma human cell [167]. To date, *in-vitro* whole-genome RNAi screening, combined with mutational signatures from different types of human cancer [156], has provided valuable clues on putative molecules to conveniently hit for disturbing oncogenic pathways and inducing tumor regression. For instance, GATA2 and BRCA1 have been identified as potential therapeutic targets in tumors with *KRAS* or *CCNE-1* mutations [168, 169], respectively. Furthermore, RNAi screening in animal models can provide information about the normal physiological functions of the putative RNAi-targets, and predict toxic effects that can derive from their incorrect targeting, compromising the clinical outcome. In the near future, the implementation of RNAi screening with "omics" data from clinical samples will provide a robust framework for prioritizing targets according to their translational potentials (Fig. 4).

5.2 *In-Vivo* RNAi Challenges

Overall, RNAi has important advantages over traditional pharmaceuticals such as protein-based drugs and small molecules for treating cancer, as it can be used to silence any target gene involved in tumor initiation, growth, angiogenesis, metastases formation, and chemoresistance [170]. The design of specific RNAi-effectors can

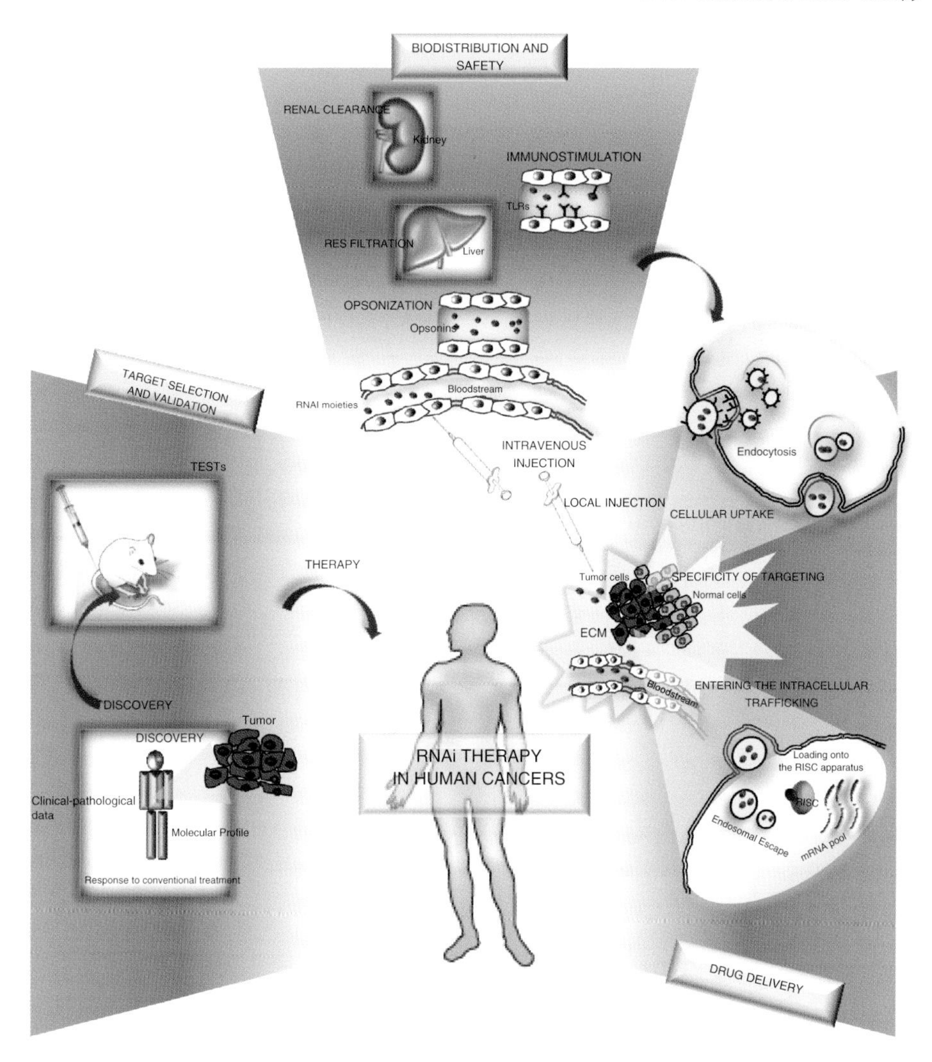

Fig. 4 Rationale design and challenges of RNAi therapeutics for human cancers. Schematic showing the critical issues to be taken into account when designing an RNAi-based strategy for cancer treatment: selection of the targets according to the tumor characteristics and patients' information from the clinic; the biological barriers affecting biostability and safety profile of RNAi therapeutics; and an efficient drug delivery culminating in the correct cellular uptake by tumor cells.

rely on the growing amount of information that is continuously being collected on potential mRNA targets, in terms of corresponding sequences and mechanisms of action. Moreover, their robustness and specificity makes them more efficient and less toxic than traditional drugs. However, despite a number of optimistic preclinical studies having been conducted and ongoing clinical trials, several barriers must first be overcome before RNAi molecules can be moved to the clinical setting. For instance, the efficient intratumoral delivery and correct cellular uptake of RNAi molecules remain critical issues; in addition, despite their sequence-specificity, the potential to knock-down non-targeted genes is an off-targeting effect that must be estimated before *in-vivo* administration. Finally, RNAi has also been shown to induce an innate immune response and to saturate the endogenous miRNA pathway, leading to cytokine-associated damage and cytotoxicity, respectively [170].

5.2.1 Stability and Bioavailability Issues

When systemic delivery methods are adopted, and before reaching their target tissue and cells, small RNA drugs need to overcome several biological barriers that derive from: (i) their intrinsic nature (superficial negative charge, and high molecular weight for crossing cellular membranes); (ii) the bloodstream, as renal clearance, nucleases digestion, opsonization, and consequent reticuloendothelial system (RES)-mediated elimination; and (iii) the target tissue itself, as the extracellular matrix (ECM), cellular uptake and endosome internalization [171] (Fig. 4).

Double-stranded RNAs are more stable compared to other oligonucleotides (ssRNA or ssDNA) [172] but once they reach the circulation, unmodified (naked) siRNAs have shown to be susceptible to nuclease degradation. Some studies have also shown that when administered, naked siRNAs accumulate preferentially in the kidney to achieve levels 40-fold higher than in other organs, and are excreted into the urine within 1 h, due to their relatively small size and effective glomerular pore size. Hence, the correct target-site accumulation of an effective dosage of RNAi agents may be one of several major concerns for their therapeutic use [173].

To this end, the insertion of chemical modifications has been explored as a strategy to increase siRNA stability without compromising the siRNA silencing potency. Most of the chemical modifications have been carried out on the 2′-OH group of the ribose moiety as it is not essential for siRNA silencing activity [174]. Such modifications are commonly known as 2′-modifications, and mainly referred to as 2′-O, 2′-Methyl-nucleoside (2′-OMe) and 2′-deoxyflouridine (2′-F) modifications [175, 176]. Another modification of the 2′ position of the ribose consisting of a methylene linkage between the 2′ and 4′ positions generated LNAs [177]. Indeed, LNA-based oligonucleotides, as miRNA mimics or inhibitors, appear to be highly resistant to nuclease degradation and displays low toxicity in biological systems. An additional type of chemical modification is the phosphoromonothioate (PS) linkage to the 3′ terminus of siRNA's phosphate backbone [178]. AntagomiRs are among the first antimiRs with proven *in vivo* efficacy and, similar to siRNAs, are 2′-O-Me oligonucleotide PS-substitution-modified constructs with the addition of a 3′-cholesteryl moiety [179].

5.2.2 Multiple Off-Target Effects

Although the mRNA cleavage induced by siRNAs and the translational repression by miRNAs are distinct pathways with different requirements [9], the two RNAi mechanisms share common cofactors and entry points, such as the RISC processing machinery. A "miRNA-like off-target gene silencing" can occur when siRNAs or shRNAs are loaded onto the RISC complex in a miRNA-like conformation, due to similarities between the core region at the 5′ end of a siRNA and the *seed* region of miRNAs, both of which guide recognition of the complementary 3′ UTR of the target mRNA and stimulate the expression of often hundreds of off-target genes [180, 181]. Such adverse side effects are a safety concern, especially when RNAi therapeutics are administered over extended periods of time [180]. A substantial reduction in off-targeting may be achieved by combining and implementing different strategies of rational drug design. Bioinformatics, which considers potential interactions between the siRNA *seed* region and the genome-wide repertoire of 3′-UTRs during the *in-silico* stage of RNAi agent selection, can be considered a rational approach to bypass such off-targeting. In addition, chemical and structural modifications can reduce guide strand-mediated off-targeting, by penalizing *seed*-only interactions via more extensive base-pairing (e.g., by including 2′-O-methyl, unlocked nucleic acid-modifications in the *seed*), and eliminate passenger strand-mediated off-targeting by 5′ blocking or by shortening the length of the passenger strand, without compromising the specific gene knockdown [180, 182].

The risk of saturation of the endogenous miRNA machinery by exogenous small RNAs [183] is another type of potential RNAi toxicity. The saturation mechanism involves several components, such as the Exportin-5 [184], and the RISC complex [183], which become less accessible to native miRNAs, especially in the case of a sustained expression of shRNAs [183]. Currently, the best strategy to avoid such complications refers to the use of lower concentrations of siRNA or, even better, to siRNA pools rather than single siRNAs [185, 186].

5.2.3 Immune Response

Besides delivery, acute immune responses from activating innate immune receptors or the complement system are critical "checkpoints" when assessing the safety profile of RNAi compounds. Both, RNAi-related "flu-like symptoms" with inflammatory cytokine elevations and the activation of alternative complement pathways have been observed as adverse effects in different clinical trials [187].

Indeed, synthetic, exogenous, RNAi modulators can induce an innate immune response via interaction with RNA-binding proteins such as Toll-like receptors (TLRs) and protein kinase receptors (PKRs) located on the cell surface or within the cell (in the cytoplasm or endosome compartments) [188] (Fig. 4). The innate immunity activation leads to a release of type I interferon (IFN) [189] and pro-inflammatory cytokines, such as interleukin (IL)-6 and tumor necrosis factor-alpha (TNFα).

It is now known that siRNA immune-stimulation can occur in a sequence-independent manner, as in the induction of TLR3 [190] and IFN pathways [189], as well as a sequence-dependent manner, when shRNAs or siRNAs contain a 5′-GUCCUUCAA-3′ motif or similar GU-rich sequences, and meet TLR7 and TLR8 [191, 192]. Moreover, the length of dsRNAs seems to influence the reactivity of the immune system independently of their nucleotide sequence; that is, a longer dsRNA will markedly increase cytokine production.

Adjustments of the RNAi trigger structure by chemical modifications, or reducing the concentration of siRNA preparation to mitigate immunogenicity, have been applied with positive results [193], although occasionally activation of the immune system can be partly beneficial as an adjuvant effect and improve the efficacy of a treatment against viral infections and some cancers.

5.2.4 Delivery Issues

The appropriate delivery of RNAi therapeutics to the target cells represents the major challenge in the development and *in-vivo* application of RNAi-based approaches. Once they have evaded the main systemic barriers (renal clearance, protein binding, and RES filtration), the RNAi particles must pass through the polysaccharide and fibrous network of the ECM to reach the cell surface of the inner tissues. Although the sinusoidal vasculature supporting organs such as liver, spleen, lymph nodes and bone marrow can facilitate the passive passage of particles up to 100 nm in size from the blood, continuous capillaries are more tightly interwoven (brain, muscle) and require active vascular escape strategies. Some of these strategies take advantage of the natural transvascular transportation that occurs within the endothelial cells, by conjugating specific ligands or bifunctional antibodies for receptors and molecular shuttles to RNAi moieties [187]. Subsequently, in order to allow the localization of RNAi compounds within close proximity of the target cells, some systemic delivery systems for the treatment of solid cancers benefit from an enhanced permeability and retention (EPR) effect, but an incorrect cellular uptake can still affect the success of the intervention. For example, RNAi drugs such as naked moieties do not readily cross the cell membranes due to their negative charge and large size [171]. Compared to direct cell penetration by lytic chemistries that risk inducing cell damage, receptor-mediated endocytosis is likely to be the more convenient and efficient pathway to be exploited. ApoE/LDL (low-density lipoprotein)- or folate/folate–receptor interactions in the case of stable nucleic acid particle lipid-based nanoparticle (SNALP LNP; see below) have demonstrated very productive internalizing pathways because of the wide distribution and elevated turnover of the receptors involved. On the other hand, cationic polyplexes/lipoplexes generally enter the cells by non-receptor-mediated physical association with the outer membrane and macropinocytosis [187]. Finally, internalization through endocytosis is normally followed by trapping in the endosome trafficking, culminating in an enzyme-mediated degradation into the lysosome. The RNAi agents must escape into the cytoplasm to reach and be loaded onto the RISC. To this end, delivery strategies including ionizable liposomes (long cationic polymers) are designed to enable endosomal release before maturation into lysosomes. Most of these strategies exploit the acidification that occurs naturally through endosomal maturation because of proton influx. The neutral charge of some compounds at physiologic pH in the blood becomes positive in the endosomes, facilitating disruption of the endosomal membrane. The low pH can also unmask membrano-lytic groups conjugated to the backbone of the RNAi compounds. Moreover, enzymes in the early endosomes can be harnessed to shed any chemistries linked to the RNAi trigger and that might otherwise interfere with RISC incorporation (e.g., ligands) [187].

In conclusion, to allow the implementation of RNAi-based therapy into the clinic, delivery strategies must overcome several technical and biological challenges. Herein are reported disparate systems that have been developed so far in attempts to satisfy not only efficacy but also safety requirements to support and expand the therapeutic application of RNAi molecules.

5.3 RNAi Implementation with Nanomedicine

To summarize, RNAi regulators of downstream gene expression include miRNA mimics and (mostly employed) siRNAs/shRNAs, whose purpose is to restore downregulated tumor suppressor miRNAs. Alternatively, antimiRs can be used for specific miRNA-silencing when the target miRNAs show oncogenic activity in the tumor.

The rational design of RNAi therapeutics for *in-vivo* applications aims to provide safety and effectiveness, by reducing renal clearance and immunogenicity, and optimizing stability, biocompatibility and site-specific delivery of the synthetic agents. These criteria are imperative not only for naked RNAi compounds but also (as explained later) carrier-formulated systems. In the latter case a low tendency for aggregation, which otherwise could induce toxicity and impair delivery activity, solubility in water and endosomal escape ability, are additional features of adequate drug design.

5.3.1 Chemical Modifications

As noted above, chemical modifications to siRNA strands (ribose, nucleotide bases, phosphorous acid, strand termini or backbone of both sense and antisense strands) can significantly improve the performance of naked RNAi molecules, preventing nuclease-based degradation and prolonging the siRNA half-life in the serum from a few minutes to two to three days [175]. Accordingly, conjugation to the strand ends of an siRNA of small molecules (such as cholesterol, folic acid, and galactose), polymers such as polyethylenimine (PEI) and polyethyleneglycol (PEG), antibodies, peptides [e.g., cell-penetrating peptides (CPPs) and transcriptional activator (TAT) peptide] or aptamers is able to increase cellular uptake by promoting caveolin or clathrin (glycoreceptor-mediated) endocytosis [175], thus enhancing the potency of siRNAs. For instance, the intravenous injection of cholesterol-siRNA conjugates targeting apolipoprotein B (ApoB) in transgenic mice led to a successful downregulation of plasma ApoB protein levels, as well as a total reduction of cholesterol levels in the liver and jejunum [194, 195]. Furthermore, siRNA–PEG conjugates have shown active function, prolonged circulation times, and reduced siRNA urinary excretion following intravenous injection [196]. Collectively, these chemical alterations, along with their combination [197], have been shown to ameliorate the *in-vitro* and *in-vivo* stability, bioavailability and duration of action of RNAi molecules, while remaining associated with a discrete effect of gene silencing.

5.3.2 Nanocarriers

Carrier-mediated transport represents a more robust strategy to overcome biological barriers and optimize uptake by the target cells, correctly locating the RNAi compounds into the expected site of action.

Carrier-mediated delivery strategies can be categorized into viral and nonviral vectors. The viral category includes retroviruses, lentiviruses, baculoviruses, adenoviruses, and AAVs, which are mostly known for their high delivery and transfection efficiency, and are distinguished by size and transgene capacity, target, performance, and duration of required expression (Table 3).

Nevertheless, critical limitations that include the immunogenicity and safety profile of viral carriers have boosted the development of alternative technologies based on nonviral vectors which fulfill the safety, because of a low risk of infection, and

Tab. 3 Viral-delivery systems.

Vectors	*Description*	*Target cells*	*Integration into the host genome*	*Expression status*	
Lentiviral vectors (LVs)	Retroviral-based vector, modified to infect several types of cells. The new generation include replication defective and self-inactivating viruses	Dividing cells	Yes	Transient or stable	***Pros:*** Effective in a wide variety of cell lines and tissues
		Nondividing cells			***Cons***: Immunogenicity and oncogenic potentiality following integration
Retrovirus vectors (RVs)	Enveloped viruses containing a single-stranded RNA molecule converted into a double-stranded DNA once in the infected cell	Dividing cells	Yes	Stable	***Pros:*** Long-lasting gene expression
					Cons: Limited to dividing cells, receptor-mediated internalization, and oncogenic potentiality due to random integration
Adenoviral vectors (AVs)	Non-enveloped virus (d = 60–90 nm) containing a linear double- stranded DNA genome	Dividing cells	No	Transient	***Pros:*** Efficient transfection in several cell types
		Nondividing cells			***Cons:*** High immunogenicity, strong inflammatory stimulation, and low loading capacity (7.5 kb)
Adenovirus-associated vectors (AVVs)	Single-stranded DNA human parvoviruses	Dividing cells	No	Transient or stable	***Pros:*** Nonpathogenic
		Nondividing cells			***Cons:*** Small packaging size (4.5 kb)

large-scale production requirements for their use. Promising nanocarrier systems for the clinical applications of RNAi include metallic, cationic lipid-based and polymer-based systems, collectively referred as nanoparticles (NPs). Nanoparticles are nanoscaled devices (generally in the range of 1 to 1000 nm), the size, shape and chemical surface configuration of which can be modified according to the purpose of their use [171]. To date, several types of NPs have been engineered to protect RNAi agents from serum nuclease degradation and the immune system, and to allow their uptake and cytosolic release by target cells. The most extensively studied systems are detailed in Table 4.

Additional intriguing alternatives for the targeted and intracellular delivery of RNAi compounds are represented by extracellular vesicles (EVs) and minicells. The EVs function as natural carriers of cellular components, including miRNAs, which mediate cell-to-cell communication and systemic trafficking of molecules [198]. They include exosomes, activation- or apoptosis-induced microvesicles (MVs)/microparticles, and apoptotic bodies. Although exosomes and microvesicles differ in size and biogenesis, both are produced by different cell types such as dendritic cells, intestinal epithelial cells, T cells and B cells, as well as cancer cells [199]. The detection of miRNAs in body fluids such as plasma, saliva and breast milk (despite the presence of RNase in the circulation) suggests that they can be (selectively) released by the cells into EVs in order to prevent degradation and reach distant tissues. Thus, compared to other conventional devices, EVs possess natural transporting properties that may be relevant when searching for a biocompatible and efficient system of delivery. Moreover, they are amenable to surface and content manipulation, for improving stability, reducing potential immunogenicity, and favoring the release of the RNAi moieties within the cells. An example of this was reported by Ohno and colleagues, who demonstrated that HEK293-derived EVs, expressing GE11 or EGF on their surface, were able to release let-7a to EGFR-expressing xenografted breast cancer tissue in immune-deficient mice. Similar results were obtained when a siRNA was used as alternative to miRNA [198]. Hence, despite some pending clues – such as stability in the circulation, purification protocols, mechanisms of internalization, composition, and immunogenicity in host organisms – the first experimental data seemed to fully encourage the application of EV formulations to the design of delivery systems for RNAi therapeutics.

Minicells are non-living, anucleate, bacteria-derived cells that are approximately 400 nm in diameter and are produced by mutants in which division is uncoupled from chromosomal replication [200]. They can be emptied of their endogenous RNA and protein contents, and packaged with therapeutic amounts of drugs that include chemotherapeutics, siRNAs and shRNAs. Compared to other carriers (such as liposomes), minicells have the properties of a simple drug packaging with different compounds (independent of their structure, charge, hydrophobicity and solubility) following a one-step coincubation, a higher payload capacity ($1-10 \times 10^6$ molecules versus 10 000 within each liposome), an ability to deliver the drug within a tumor cell without leakage during systemic circulation, an ease of conjugation to affinity ligands for tumor targeting, and no off-target effects [201]. Importantly, as minicells are of bacterial origin, purification procedures to eliminate free endotoxin and free bacterial components must be carried out to minimize

Tab. 4 Non-viral delivery systems: NPs-based technologies.

Delivery system	*Description*	*Common formulations*	*Targeting mechanisms*	*Pros and Cons*
Lipid-based delivery				
Lipid-based nanoparticles (LNPs)/lipoplexes (LPPs)	Lipid-based neutral micelles (d = 100 nm), consisting of lipid bilayer surrounding a hydrophilic aqueous core	• DOPC liposomes • DOPE liposomes • DOTAP liposomes	EPR	***Pros:*** • Protection of nucleic acids from enzymatic degradation • Low immunogenicity • Low toxicity • Biocompatibility • Ease of preparation
Cationic LNPs	"Stealth-coated" cationic lipid-based micelles (d = 100–150 nm) able to interact with the negative charge of RNAi moieties. The net surface charge, due to high risk of opsonization and immunogenicity, is generally neutralized by coating with PEG polymer	• SNALP • SLN	MPS	***Pros:*** • Increased half-life • Enhanced cellular uptake • Enhanced endosomal escape ***Cons:*** • Opsonization (SNALP) • Potential immunogenicity (SNALP)
Lipidoids	Non-glycerol cationic lipid-based micelles derived from conjugation with amines to acrylate and acrylamide	• Epoxide-based lipidoids	MPS	***Pros:*** • Suitability for siRNAs pool
Bubble liposomes (BLs)	PEG-modified liposomes containing echo-contrast gas, which can function as a gene and siRNA delivery tool with ultrasound exposure	• DOTAP liposomes • DSDAP liposomes	EPR	***Pros:*** • Ultrasound-enhanced delivery • Stability and long half-life • Biocompatibility

Polymer-based delivery				
Polyethylenimines (PEIs)	High cationic charge density polymers, presenting linear or branched structures with variable molecular weights (MWs)	• PEG-PEI • DA-PEI	Mucoadhesion	***Pros:*** • High transfection efficiency (high MW and branches forms > low MW and linear forms) • Endosomal pH resistance (proton sponge effect). • Physical protection of nucleic acid from degradation. ***Cons:*** • Proven cytotoxicity: cell apoptosis (high MW and branches forms > low MW and linear forms) • Activation of the complement system
Chitosan	High positive-charged mucopolysaccharide. The degree of deacetylation and MW can influence the capability to interact with nucleic acid. Both low degree of acetylation and MW decrease the interaction with RNAi moieties and the stability of nanocomplexes	Chitosan	Mucoadhesion	***Pros:*** • Biodegradability • Low immunogenicity

(*continued overleaf*)

Tab. 4 (*Continued*)

Delivery system	*Description*	*Common formulations*	*Targeting mechanisms*	*Pros and Cons*
Poly($_{D,L}$-lactic-*co*-glycolic acid) (PLGA)	Small lactic and glycolic acids polymer	• PLGA • PLGA-PEG	Mucoadhesion	***Pros:*** • Biodegradability • Rapid tissue penetration (small size) • Physical protection of RNAi molecules from RNase degradation
Dendrimers and dendritic polymers	Highly branched, globular, nanoscaled macromolecules consisting of a core, an interior dendritic backbone (the branches), and an exterior surface with functional groups	• PAMAM • PPI • Polyglycerols • Peptide dendritic polymers	EPR	***Pros:*** • Solubility, and very low intrinsic viscosities • High transfection efficiency, and drug delivery (endosomal escape) • Ease of preparation and commercial availability ***Cons:*** Low degradation rate and toxicity
Magnetic nanoparticles (MNPs)	A class of NP consisting of a magnetic core of a pure metal (Co, Mn, Ni, Fe), or their alloys and oxides usually coated by a shell, isolating the core against the environment	Coated MNPs (silane shell, gold shell, polymeric shell, dendrimeric shell): SPIONs, CPMNs	EPR	***Pros:*** • Tumor targeting by the aid of an external magnetic field • Visualization opportunity by MRI • Enhanced cellular uptake

				Cons: • Tendency to aggregation • Opsonization • Liver, spleen, and brain accumulation
Others				
Carbon nanotubes (CNTs)	Nonspherical nano-cylindrical particles, 50–100 nm in length (d = 1–100 nm)	• SWCNT • MWCNT	MPS	***Pros:*** • Ease of functionalization (PEGylation, Tween-20 coating, Ab-conjugation) • Good biodistribution (ease of shape manipulation to escape renal clearance) ***Cons:*** Non-biodegradable (pending issue)

DOPC, 1,2-oleoyl-*sn*-glycero-3-phosphatidylcholine; DOPE, dioleoylphosphatidylethanolamine; DOTAP, *N*-[1-(2,3-dioleoyloxy)-propyl]-*N*,*N*,*N*-trimethylammoniummethylsulfate; EPR, enhanced permeability and retention; MPS, mononuclear phagocytic system; SNALP, stable lipid-nucleic acid particle; SLNs, solid lipid nanoparticles; PEG, polyethyleneglycol; DSDAP, 1,2-distearoyl-3-dimethylammonium-propane; PEI, polyethyleneimine; DA, deoxycholic acid; PAMAM, polyamidoamine; PPI, poly(propyleneimine); SPIONs, superparamagnetic iron oxide nanoparticles; CPMNs, cell-penetrating magnetic nanoparticles; MRI, magnetic resonance imaging; SWCNTs, one-dimensional single-walled carbon nanotubes; MWCNTs, multiwalled carbon nanotubes.

the potential for toxic side effects that can derive from the inflammatory responses activated by TLRs.

In their earlier report, MacDiarmid *et al.* successfully demonstrated the *in-vivo* efficacy of minicells as targeting vehicles for an array of chemotherapeutic drugs, including 5-fluoracil, carboplatin, cisplatin, doxorubicin (DOX), irinotecan, paclitaxel (PTX), and vinblastine. Targeting was mediated by bispecific antibody (BsAb) conjugates that recognized both O-polysaccharide and a tumor antigen. As a consequence, targeted minicell-mediated drug delivery was shown to produce a highly significant inhibition and even a regression of tumor growth in different xenograft models of human breast, ovarian and lung cancer, and also leukemia [201]. More recently, the same authors demonstrated the feasibility of using minicells for the siRNA-mediated treatment of mouse xenograft models. In an attempt to treat drug-resistant tumors, they applied a dual sequential treatment that consisted of an initial administration of si/shRNA-containing minicells linked to BsAb that targeted a known drug resistance mechanism. After allowing a sufficient knockdown of the drug resistance-mediating protein, BsAb-targeted minicells packaged with cytotoxic drug were then administered intravenously. This approach, which required a markedly small amount of drug, siRNA and antibody than was needed for a conventional systemic administration of free therapeutics, led to the complete survival of mice with tumor xenografts, with no signs of toxicity [202]. Overall, similar to EVs (caution must be taken here regarding the immunogenic profile and *in-vivo* efficacy of this system), the data acquired supported the potential of a minicell-based anticancer therapeutic for transfer to the clinical setting.

5.4
Administration

Both, local and systemic modalities of RNAi administration have been assessed in animal models, and are currently under investigation in several clinical trials [202]. Local delivery through intraperitoneal, intranasal, intramuscular, intrahepatic, intraretinal, intratesticular or subcutaneous injection, and through inhalation certainly offers the advantage of site-specificity and high bioavailability with small amounts of even naked RNAi drugs in the target tissue [203]. For instance, effective gene silencing and tumor suppression have been achieved by the direct subcutaneous and intraperitoneal injection of siRNA complexed with a commercial transfection reagent to induce specific gene silencing of human papillomavirus (HPV) 16 E6 oncogene in a model of cervical cancer [172]. Nonetheless, such a methodology is manageable only for certain technically accessible sites, such as mucosal and subcutaneous tissues. Indeed, the RNAi molecules must be distributed throughout a sufficiently large interstitial space, as occurs in the skin. Moreover, a local delivery and passive diffusion of the drugs can be achieved if energy-dependent mechanisms such as pressure, local heat gradients or muscle-related pressure gradients are present. More importantly, local delivery cannot be successfully applied to hematological malignancies or solid tumors that have already metastasized.

Alternatively, a less well-documented modality of administration is represented by electropermeabilization (EP), which involves the application of a calibrated electric field pulse to target cells or tissues, which in turn allows the RNAi molecules to rapidly "catch" their targets. To date, this approach has been used for various *in-vitro* and *in-vivo* evaluations of drug and nucleic acid deliveries. Similar to local administration, it allows the cellular uptake to bypass external barriers such as clearance, and internal barriers such as endosomes/lysosome internalization. EP-mediated siRNA or the introduction of LNA/DNA oligomers has been shown not only to be effective in silencing the gene expression [204] but also to allow a homogeneous distribution of the siRNA in the cytoplasm, facilitating its entry to the RISC complex as it is directly localized in the cytoplasm of the recipient cell [205, 206]. Successful results have also been obtained by *in-vivo* direct siRNA delivery into the tissues of different animal models [207]. However, a lack of knowledge of the biophysical mechanisms that support reorganization of the cellular membranes after induced permeabilization can limit the wide application of this methodology.

In this regard, the systemic delivery of RNAi molecules (mainly by intravenous infusion) is the best-suited and multifaceted approach, and disparate formulations and targeting systems have been explored in this regard. Along with the above-mentioned chemical modifications, strategies such as the physical encapsulation of RNAi molecules, including nanocarrier functionalization with target moieties for intracellular diffusion, have been assessed as a means of breaking through the biological barriers and maintaining the targeting of diseased tissues. But even in this context, precautionary measures must be taken into account; indeed, as with other small molecules the intravenous route of administration of RNAi therapeutics can be accompanied by infusion reactions. Experience obtained from clinical trials has shown that these effects are dose-dependent and are especially related to the first infusion; consequently, a slowing rate of infusion, coupled with one-monthly optimized formulations (such as SNALP LNPs), have been shown to make adverse reactions significantly more manageable and clinically endurable [187].

5.5 Targeting

5.5.1 Passive Targeting

The unique abnormal architecture of the tumor vasculature can be exploited to enable nanodrugs to specifically accumulate in tumor tissues. Typically, tumor vessels are poorly aligned, with defective endothelial cells and wide fenestrations among them, together with a compromised lymphatic drainage. The “leaky” vascularization, which relates to the EPR effect, allows the extravasation and accumulation of macromolecules up to 400 nm greater in diameter in tumor tissue than in normal tissue [208]. The coating of cationic or hydrophobic nanocarriers with hydrophilic polymers allows them to elude opsonization and serum nuclease degradation, and also to localize in the tumor vessels. PEGylated liposomes (as SNALPs), which represent advanced liposome vehicles for passive siRNA delivery, show a much longer half-life in plasma than do traditional cationic liposomes [208].

As noted above, once the RNAi carriers access in the ECM they need to reach the target cells. The higher expression level of $\alpha_v\beta_3$ integrin in neoplastic cells compared to normal epithelial cells can be exploited by linking RGD (Arg-Gly-Asp) peptides to the NP surface to facilitate its binding to the tumor ECM, with consequent accumulation [168]. Lastly, because of the high metabolic rate of rapidly growing tumor cells, the microenvironment surrounding the tumor tissue is characterized by a lower pH; hence, some liposome preparations have been designed to be stable at physiological pH (7.4), but will degrade to release drug molecules at an acidic pH [209].

Despite these promising manipulations, the passive delivery process for RNAi therapeutics is difficult to control, and the lack of EPR in certain tumors as well as the heterogeneous permeability of vessels throughout a single tumor make passive delivery difficult to be applied extensively [209].

5.5.2 Active Targeting: Antibodies, Aptamers, Small Molecules

This strategy involves the attachment of affinity ligands (antibodies, peptides, aptamers, or small molecules) to naked RNAi or NPs–RNAi complex molecules [205–213], for molecules that are highly expressed on the cell surface of the diseased tissue. This would improve the bioavailability of the exogenous compounds, minimize nonspecific delivery, and induce cellular uptake with minimal doses of compounds.

Immuno-NP with monoclonal antibodies (mAbs), and engineered fragment antibodies (Fabs) or single chain fragment variable antibodies (scFv) has been widely used, and represents an effective method for targeting specific cells HIV-1 gp160 and HER2 represent peculiar examples. A protamine–antibody fusion protein (F105-P) against the HIV-1 envelope specifically delivered a fluorescein isothiocyanate (FITC)-silencing siRNA to HIV-1 infected and HIV-1 envelope-transfected cells, and induced silencing only in those cells enriched in gp160 (+) [205]. Similarly, a Polo-like Kinase 1 (PLK1)-targeting siRNA was injected intravenously into HER2(+) breast cancer xenografts by using an anti-Her2 ScFv-protamine fusion protein. When using this approach, the complex demonstrated a correct delivery of the siRNA to the HER-positive cells, without releasing to bystander tissues, while the siRNA reached a full PLK1 silencing, culminating in tumor growth retardation, metastasis reduction, and prolonged survival of mice [214].

Aptamers are synthetic, short, single-strand oligonucleotides with several advantages compared to full-size or single-chain antibodies that include a small size that allows for a better tissue internalization, a low immunogenicity and a high stability, as well as an amenability for modification and thus a capability to bind specific and different ligands with secondary or tertiary structures (proteins, receptors) [174, 208]. Aptamers can be easily linked covalently or noncovalently to both naked RNAi molecules (i.e., Aptamer siRNA chimera) and NPs-formulated constructs. For example, in a xenograft model of human prostate cancer, an A10 aptamer–NP complex was shown to recognize and bind the prostate membrane-specific antigen (PMSA) with high specificity, allowing its conjugated siRNAs into the cells to induce BCL2 and PLK1 silencing [215]. In another study, a dual inhibitory anti-gp120 aptamer/siRNA chimera, in which both the aptamer and the siRNA were directed against HIV genes, demonstrated effective silencing.

Cell-penetrating peptides (CPPs) are cationic oligopeptides, derived from natural sequences of viral, insect or mammalian proteins [216], that can be linked to RNAi moieties or NPs by covalent bonds or by electrostatic interactions, in order to accelerate their internalization by increasing cellular plasma membrane permeability. A typical example is the HIV–TAT family, penetratin and the chimeric peptide transportan, and oligoarginines (arginine-repeated unit) [217], whose ability to mediate siRNA silencing has been proven in the central nervous system of mice.

Small molecules can also function as effective ligands for increasing siRNA stability and cellular uptake. Cholesterol and its derivatives increase the binding of RNAi compounds to serum albumin, but reduce the interaction with LDLs and high-density lipoproteins (HDLs), with a consequent improved biodistribution in certain target tissues, such as liver [194, 197, 218]. The folate receptor is a membrane glycoprotein that is highly overexpressed in a number of human tumors, including ovarian, colorectal and breast cancer, but is almost absent in all other normal tissues, except kidney [197, 219]. Similarly, transferrin receptor-mediated delivery has been extensively studied in a variety of targets, including tumors, brain and endothelial cells [220]. Their several advantages include small size, a lack of immunogenicity, convenient availability, and easy chemical conjugation [218, 219].

5.5.3 Alternative Targeting

Alternative approaches include magnetic targeting and ultrasound (US)-enhanced delivery. Magnetic targeting utilizes the magnetic properties of metallic NPs (e.g., silica or gold NPs) to induce their specific accumulation in a target tissue by applying external, local, strong magnetic fields following systemic injection [221]. Once in the circulation, magnetic nanoparticles (MNPs) can be easily monitored using real-time magnetic resonance imaging (MRI) to follow their distribution and cellular uptake [222]. Disparate formulations of siRNA nanovectors exploiting magnetic targeting have been developed for the *in-vivo* imaging of delivered siRNA and gene silencing in tumors, and many others are currently under investigation. For instance, cell-penetrating magnetic nanoparticles (CPMNs) containing CPP-coated MNPs, conjugated with a low-molecular-weight protamine peptide, have consistently been shown to improve transfection efficiency and intracellular siRNA release *in vitro* [223]. In addition, functionalized superparamagnetic iron oxide nanoparticles (SPIONs), combined with chitosan and siRNAs, have been used to couple the SPION suitability for MRI and/or thermal therapy to the ability of chitosan to increase the stability of associated siRNA [224].

A combination of microbubbles and US has been initially proposed as a less-invasive and tissue-specific method for gene delivery. Following infusion, the application of US produces transient changes in the permeability of the cell membrane, due to pressure oscillations, and allows for the site-specific intracellular delivery of combined molecules. Due to the limits of the microbubbles, in terms of their size, stability and targeting functionality, Suzuki and colleagues have developed different formulations of so-called "bubble liposomes"; these are PEG-modified liposomes in which an US imaging gas is trapped and can function as a novel plasmid DNA (pDNA), siRNA and, more recently, as a miRNA delivery tool when used with US exposure both *in vitro* and *in vivo* [225–228].

5.6 RNAi-Based Drug Combination Strategy (coRNAi)

Initial studies in viral infection-related diseases [hepatitis B virus, hepatitis C virus (HCV), human immunodeficiency virus, and other human viruses] have prompted the evaluation of a coRNAi technology for the treatment initially of metabolic or blood disorders, and then of cancer [229].

Combination therapy has been recommended in cancer treatment because it allows the use of lower doses of each drug, compared to single-drug administration, but reaches a higher efficacy based on additive or synergistic anticancer effects. This would prevent the toxic side effects that are often related to common chemotherapeutics, as well as the development of multidrug resistance (MDR) within the tumor. In fact, as human tumors tend to accumulate mutations and acquire resistance towards single-agent-based therapies, the concurrent targeting of multiple key factors and pathways would impair the development of secondary resistance. The *MDR-1* gene is primarily responsible for activating mechanisms for the expulsion of chemotherapeutics upon expression of the components of the P-gp pump (as MRP) within the cell membrane, which is likely to be induced following the repeated administration of drugs to oppose the cytotoxic action [230]. In addition, cancer cells can activate anti-apoptotic proteins such as Bcl2 in further protective attempts at survival over chemotherapy [230].

To date, one of the more widely investigated coRNAi strategies to combat chemoresistance in cancer has employed siRNAs targeting key components of the MDR pathways, in combination with standard chemotherapeutics. As an example, an *in-vitro* sensitization of cervical cancer cells has been obtained following the administration of silica NPs containing both an siRNA against the *MRP* gene and DOX, to compromise assembly of the P-gp pump and allow DOX to cross the cell membrane and inhibit cellular growth [231]. Similarly, cationic micelles carrying anti-BCL2 siRNA and docetaxel (DTX) have been used to target the anti-apoptotic pathway in an *in-vivo* model of human breast cancer [232]. A synergistic effect of RNAi triggers, when combined with PTX, has also been documented [233]; in this case, the coadministration of PTX and an siRNA against midkine (MKsiRNA) in human prostate cancer xenografts led to a significant augmentation of the antitumor therapeutic effect.

A more promising and recently emerging coRNAi approach relies on the coadministration of siRNAs and miRNAs (siRNA/miRNA-based coRNAi). Several studies have reported the successful delivery of both molecules into the tumor tissue, with surprising synergistic anticancer effects. Nishimura *et al.* showed that the dual inhibition of EphA2 (which is overexpressed in ovarian cancer cells and patients) by using an anti-EphA2 siRNA, combined with a miR-520d-3p mimic, targeting EphA2 and with high expression levels, correlated with a better prognosis and significantly augmented the therapeutic effect in inhibiting cell proliferation, migration and invasion *in vitro*, and of tumor growth *in vivo*, compared to monotherapy with each of the drugs [234].

Similar results with lung cancer have confirmed the effectiveness of this approach, as well as the relevance of nanotechnology in supporting such experimental and therapeutic design. Indeed, a new class of polymer-based NPs, 7C1, has been used to encapsulate an equal molar ratio of anti-KRAS siRNAs and miR-34a mimic in a single NP formulation, suggesting an easier and more flexible approach for coRNAi application. Both, *in-vitro* and *in-vivo* assessments confirmed the synergistic anticancer effect derived from the RNAi compound combination, and this was even more enhanced after coadministration with cisplatin (CP) [235].

Although miRNA-based combination therapy is still in its infancy, initial experiments seem to have demonstrated great potential. Kasinski *et al.* assessed the efficacy and safety of miRNA mimics coadministration in an aggressive *KRAS*; *TP53*NSCLC mouse model for replacing the expression of two lung tumor suppressor miRNAs, let-7 and miR-34. The systemic delivery of miR-34a and let-7 by lung-targeting NPs – either neutral lipid emulsion (NLE) or NOV34 (a miRNA delivery agent already undergoing clinical trials) – resulted in tumor growth suppression and a survival advantage [236].

In conclusion, although delivery issues and the risk of reciprocal competition between codelivered RNAi drugs still needs to be fully resolved [237], the above-mentioned results and ongoing clinical trials have provided great encouragement for coRNAi translation in the clinical setting of human cancer. In particular, it is a less-toxic and more direct method for targeting multiple biologically pathways that are relevant to the genesis and progression of tumors.

5.7 RNAi in the Clinical Trials

The first RNAi-based clinical trial began in 2004, with an intravitreal injection of naked siRNA targeting the vascular endothelial growth factor (VEGF) pathway in patients with age-related macular degeneration (AMD) and diabetic macular edema (DME). Most of the naked siRNAs were locally administered to treat topical diseases that included AMD, DME and non-arteritic anterior ischemic optic neuropathy (NAION) or virus infections such as respiratory syncytial virus and HCV. The clinical was trial terminated due to low efficacy, off-target effects, or company issues.

Following the application of nanotechnology to the design of siRNA delivery vehicles, new RNAi agents based on systemic delivery entered clinical trials. To date, the terminated siRNA agents in clinical trials include naked delivered bevasiranib, AGN-745, and PF-655 (NCT0030694; NCT00363714; NCT1445899; www:ClinicalTrials.gov), for the treatment of eye-related disorders, and the SNALP-based delivery agent TKM–ApoB (NCT00927459) for cardiovascular diseases. Bevasiranib administration was terminated because of low efficacy at Phase III in a clinical trial to treat for AMD/DME, while AGN-745 was terminated during the treatment of AMD due to off-target effects at Phase II of clinical testing. TKM–ApoB was terminated because only a transient reduction of cholesterol level was observed at Phase I in the clinical study. One common adverse effect among these agents was the activation of TLRs; however, although most terminated RNAi agents were naked siRNAs, those to be administered locally will still have the chance to enter further testing as long as they demonstrate appropriate safety and efficacy.

CALAA-01 (Calando Pharmaceuticals; NCT006889065) has been the first active targeted delivery agent to enter clinical trials for the treatment of solid tumors. The agent consisted of an unmodified siRNA targeting Ribonucleotide Reductase M2 (RRM2) into cyclodextrin-containing polymer (CDP) NPs, decorated with transferrin, and administered intravenously to patients with metastatic melanoma. Despite the promising data obtained, the CALAA-01 trial was recently terminated, without any clear explanations.

Tab. 5 siRNA-based RNAi cancer interventions in clinical trials.

Drug	*Formulation*	*siRNA-target*	*Cancer disease*	*Status*	*Identifier at: ClinicalTrials.gov*
CALAA-01	Systemic CDP-NPs	RRM2: cell proliferation	Advanced solid tumors	Phase I: terminated	NCT00689065
ALN-VSP02	Systemic SNALP liposomes (short circulating lipid). coRNAi with two siRNAs	VEGF: angiogenesis; KSP: cell proliferation	Liver cancer and advanced solid tumors with liver metastases	Phase I: completed	NCT01158079
iP siRNA	*Ex vivo* transfection of mature dendritic cells and vaccination transfected mature dendritic cells	LMP2, LMP7, MECL1: proteasome-mediated antigen-presenting processing	Metastatic melanoma	Phase I: completed	NCT00672542
TKM-PLK1	Systemic SNALP liposomes (long-circulating lipid)	PLK1: cell proliferation	Solid tumors, lymphomas, primary and secondary liver metastases	Phase I/II: recruiting	NCT01262235
Atu027	Systemic PEGylated lipoplex (Atuplex)	PKN3: angiogenesis and metastasis	Advanced solid tumors	Phase I: completed	NCT00938574
Atu 027 + Gemcitabine			Advanced pancreatic adenocarcinoma	Phase I: recruiting	NCT01808638
SIG12D LODER	Local LODER polymer (PLGA-based slow-release formulation)	Mutated KRAS	Advanced pancreatic adenocarcinoma	Phase I: ongoing	NCT01188785
				Phase II: not yet recruiting	NCT01676259
EPHARNA *siRNA-EphA2-DOPC*	Non-PEGylated DOPC liposome	EPHA2 cell proliferation	Advanced solid tumors	Phase I: not yet recruiting	NCT01591356
DCR-MYC	Systemic liposome	MYC: cell proliferation	Solid tumors	Phase I: recruiting	NCT02110563
			Multiple myeloma Non-Hodgkins lymphoma		

Tab. 6 miRNA-based RNAi cancer interventions in clinical trials.

Drug	*Formulation*	*Target*	*Cancer disease*	*Status*	*Identifier at: ClinicalTrials.gov*
MRX34	Systemic NLE (neutral lipid emulsion) liposomal miR-34 mimic	BCL2: cell apoptosis	Primary liver cancer or metastatic cancer with liver involvement. lymphoma, leukemia, multiple myeloma	Phase I: recruiting	NCT01829971
SPC3649 (Miravirsen)	Systemic LNA-ASO (antisense oligonucleotide)	miR-122: inhibition of miR-122–HCV complex formation essential to the stability and propagation of HCV RNA	HCV and hepatocellular carcinoma	Phase I/II: completed	NCT01200420
EZN2968 *anti-HIF-1α LNA AS-ODN*	Systemic LNA-ASO (antisense oligonucleotide)	HIF-1α: hypoxia response	Advanced solid tumors or lymphoma	Phase I: completed	NCT00466583
SPC2996	LNA-ASO	BCL2: cell apoptosis	CLL	Phase I/II: completed	NCT00285103

Atu027 (Silence Therapeutic; NCT00938574), another siRNA-containing PEG-nanoparticle (AtuPLEX) was the first siRNA to be applied for the inhibition of metastases. No drug-related side effects were observed at Phase I of the clinical trial, in which toxicity was evaluated in 27 of 33 patients with advanced solid tumor. A Phase Ib/IIa trial (NCT01808638), in which patients with locally advanced or metastatic pancreatic adenocarcinoma were recruited has since been undertaken, to evaluate the combination of Atu027 with gemcitabine.

ALN-VSP02 (Alnylam/Tekmira; NCT01158079) is the first dual-targeted RNAi formulation combining two siRNAs into a lipid-NP carrier system (SNALP) targeting both VEGF and kinesin spindle protein (KSP), respectively. A Phase I clinical evaluation (in 28 patients) reported that ALN-VSP02 was well tolerated, with neither hepatotoxicity nor any elevation of complement proteins. Although increases in cytokine levels were observed these were considered due to higher doses, and immediately resolved within 24 h after administration.

On-going and future trials include the TKM-PLK1 compound (NCT01262235) for the treatment of primary liver cancer or liver metastases, and two promising coRNAi trials for siG12D LODER (Silenceed Ltd; NCT01676259), a siRNA targeting KRAS combined with gemcitabine to treat advanced pancreatic cancer, and siRNA-EphA2-DOPC (1,2-oleoyl-*sn*-glycero-3-phosphatidylcholine) (EPHARNA, MD Anderson Cancer Center; NCT01591356). In this case, nonPEGylated liposomes will be administered to patients with advanced and recurrent solid tumors. Details of other, similar, studies are provided in Tables 5 and 6.

6 Summary

In combining the characterization of RNAi in humans with increasing knowledge of the molecular alterations that support the genesis of human malignancies, RNAi therapeutics represent the most promising approach towards innovative and personalized medical interventions on cancer, and a clear alternative to the limited effects of conventional treatments (e.g., chemotherapy, surgery, or radiotherapy).

During the past decades, gene expression profiling studies of human malignancies have revealed that the global dysregulation of small, noncoding RNAs – that is, miRNAs – which are the genuine users of the endogenous RNAi apparatus for fine-tuning gene expression, has a significant effect on every step of tumorigenesis. Thus, the ability to harness the RNAi machinery by restoring the regulatory function of downregulated miRNAs using synthetic miRNA mimics, siRNAs, viral expression constructs, or inhibiting an oncomiR function by chemically modified antimiR oligonucleotides, miRNA sponges, or miR-masks, offers the chance to target the expression of any gene linked to neoplastic phenomena, including those which have proved so far to be “undruggable.”

Although a number of RNAi therapeutics have emerged as putative, effective anticancer drugs, the major drawback in their translation into the clinic remains a lack of safe and efficient delivery systems, and the ability to release the RNAi moieties at a therapeutic concentration specifically to the target cells, without affecting healthy tissues. In this context, during the past few years remarkable achievements have been made by designing NPs-based carriers (cationic lipids/polymers) with high transfection abilities, biocompatibilities, stabilities, and a lack of immunogenicity. Today, encouraging results appear to be emerging from ongoing clinical trials regarding the safety and efficacy of RNAi therapeutics, even when repeated routes of administration are used. However, seminal investigations into the real therapeutic potential of these strategies are still required before entering the clinical setting and supporting patient management.

References

1. Hannon, G.J. (2002) RNA interference. *Nature*, **418**, 244–251.
2. Couzin, J. (2002) Breakthrough of the year. Small RNAs make big splash. *Science*, **298**, 2296–2297.
3. Fire, A., Xu, S., Montgomery, M.K., Kostas, S.A. *et al.* (1998) Potent and specific genetic interference by double-stranded RNA in *Caenorhabditis elegans*. *Nature*, **391**, 806–811.
4. Elbashir, S.M., Harborth, J., Lendeckel, W., Yalcin, A. *et al.* (2001) Duplexes of 21-nucleotide RNAs mediate RNA interference in cultured mammalian cells. *Nature*, **411** (6836), 494–498.
5. Elbashir, S.M., Lendeckel, W., Tuschl, T. (2001) RNA interference is mediated by 21-and 22-nucleotide RNAs. *Genes Dev.*, **15**, 188–200.
6. Jorgensen, R. (1990) Altered gene expression in plants due to trans interactions between homologous genes. *Trends Biotechnol.*, **8**, 340–344.
7. Romano, N., Macino, G. (1992) Quelling: transient inactivation of gene expression in *Neurospora crassa* by transformation with homologous sequences. *Mol. Microbiol.*, **6**, 3343–3353.
8. Hammond, S.M., Bernstein, E., Beach, D., Hannon, G.J. (2000) An RNA-directed nuclease mediates posttranscriptional gene silencing in *Drosophila* cells. *Nature*, **404**, 293–296.
9. Fire, A., Albertson, D., Harrison, S.W., Moerman, D.G. (1991) Production of antisense RNA leads to effective and specific inhibition of gene expression in *C. elegans* muscle. *Development*, **113**, 503–514.
10. Dernburg, A.F., Zalevsky, J., Colaiacovo, M.P., Villeneuve, A.M. (2000) Transgene-mediated cosuppression in the *C. elegans* germ line. *Genes Dev.*, **14**, 1578–1583.
11. Guo, S., Kemphues, K.J. (1995) par-1, a gene required for establishing polarity in *C. elegans* embryos, encodes a putative Ser/Thr kinase that is asymmetrically distributed. *Cell*, **81**, 611–620.
12. de Carvalho, F., Gheysen, G., Kushnir, S., Van Montagu, M. *et al.* (1992) Suppression of beta-1,3-glucanase transgene expression in homozygous plants. *EMBO J.*, **11**, 2595–2602.
13. Waterhouse, P.M., Graham, M.W., Wang, M.B. (1998) Virus resistance and gene silencing in plants can be induced by simultaneous expression of sense and antisense RNA. *Proc. Natl Acad. Sci. USA*, **A95** (23), 13959–13964.
14. Hamilton, A.J., Baulcombe, D.C. (1999) A species of small antisense RNA in post-transcriptional gene silencing in plants. *Science*, **286**, 950–952.
15. Bernstein, E., Caudy, A.A., Hammond, S.M., Hannon, G.J. (2001) Role for a bidentate ribonuclease in the initiation step of RNA interference. *Nature*, **409**, 363–366.
16. Reinhart, B.J., Slack, F.J., Basson, M., Pasquinelli, A.E. *et al.* (2000) The 21-nucleotide let-7 RNA regulates developmental timing in *Caenorhabditis elegans*. *Nature*, **403**, 901–906.
17. Lee, R.C., Feinbaum, R.L., Ambros, V. (1993) The *C. elegans* heterochromic gene lin-4 encodes small RNAs with antisense complementarity to lin-14. *Cell*, **75**, 843–854.
18. Hutvágner, G., McLachlan, J., Pasquinelli, A.E., Bálint, E. *et al.* (2001) A cellular function for the RNA-interference enzyme Dicer in the maturation of the let-7 small temporal RNA. *Science*, **293**, 834–838.
19. Grishok, A., Pasquinelli, A.E., Conte, D., Li, N. *et al.* (2001) Genes and mechanisms related to RNA interference regulate expression of the small temporal RNAs that control *C. elegans* developmental timing. *Cell*, **106**, 23–34.
20. Volpe, T.A., Kidner, C., Hall, I.M., Teng, G. *et al.* (2002) Regulation of heterochromatic

silencing and histone H3 lysine-9 methylation by RNAi. *Science*, **297**, 1833–1837.

21. Mochizuki, K., Fine, N.A., Fujisawa, T., Gorovsky, M.A. (2002) Analysis of a piwi-related gene implicates small RNAs in genome rearrangement in *Tetrahymena*. *Cell*, **110**, 689–699.
22. Farazi, T.A., Spitzer, J.I., Morozov, P., Tuschl, T. (2011) miRNAs in human cancer. *J. Pathol.*, **223** (2), 102–115.
23. Siolas, D., Lerner, C., Burchard, J., Ge, W. *et al.* (2005) Synthetic shRNAs as potent RNAi triggers. *Nat. Biotechnol.*, **23** (2), 227–231.
24. Carmell, M.A., Hannon, G.J. (2004) RNase III enzymes and the initiation of gene silencing. *Nat. Struct. Mol. Biol.*, **11**, 214–218.
25. Kim, V.N. (2005) MicroRNA biogenesis: coordinated cropping and dicing. *Nat. Rev. Mol. Cell Biol.*, **6**, 376–385.
26. Wilson, R.C., Doudna, J.A. (2013) Molecular mechanisms of RNA interference. *Annu. Rev. Biophys.*, **42**, 217–239.
27. Denli, A.M., Tops, B.B, Plasterk, R.H, Ketting, R.F. *et al.* (2004) Processing of primary microRNAs by the Microprocessor complex. *Nature*, **432**, 231–235.
28. Han, J., Lee, Y., Yeom, K.H., Kim, Y.K. *et al.* (2004). The Drosha–DGCR8 complex in primary microRNA processing. *Genes Dev.*, **18**, 3016–3027.
29. Landthaler, M., Yalcin, A., Tuschl, T. (2004) The human Di George syndrome critical region gene 8 and its *D. melanogaster* homolog are required for miRNA biogenesis. *Curr. Biol.*, **14**, 2162–2167.
30. Lund, E., Guttinger, S., Calado, A., Dahlberg, J.E. *et al.* (2004) Nuclear export of microRNA precursors. *Science*, **303**, 95–98.
31. Gregory, R.I., Chendrimada, T.P., Cooch, N., Shiekhattar, R. (2005) Human RISC couples microRNA biogenesis and post-transcriptional gene silencing. *Cell*, **123**, 631–640.
32. Ambros, V., Lee, R.C., Lavanway, A., Williams, P.T. *et al.* (2003) MicroRNAs and other tiny endogenous RNAs in *C. elegans*. *Curr. Biol.*, **13**, 807–818.
33. Aravin, A.A., Lagos-Quintana, M., Yalcin, A., Zavolan, M. *et al.* (2003) The small RNA profile during *Drosophila melanogaster* development. *Dev. Cell*, **5**, 337–350.
34. Xie, Z., Johansen, L.K., Gustafson, A.M., Kasschau, K.D. *et al.* (2004) Genetic and functional diversification of small RNA pathways in plants. *PLoS Biol.*, **2**, e104.
35. Vazquez, F., Vaucheret, H., Rajagopalan, R., Lepers, C. *et al.* (2004) Endogenous trans-acting siRNAs regulate the accumulation of *Arabidopsis* mRNAs. *Mol. Cell*, **16**, 69–79.
36. Tomari, Y., Matranga, C., Haley, B., Martinez, N. *et al.* (2004). A protein sensor for siRNA asymmetry. *Science*, **306**, 1377–1380.
37. Carmell, M.A., Xuan, Z., Zhang, M.Q., Hannon, G.J. (2002) The Argonaute family: tentacles that reach into RNAi, developmental control, stem cell maintenance, and tumorigenesis. *Genes Dev.*, **16**, 2733–2742.
38. Hutvagner, G., Simard, M.J. (2008) Argonaute proteins: key players in RNA silencing. *Nat. Rev. Mol. Cell Biol.*, **9**, 22–32.
39. Liu, J., Carmell, M.A., Rivas, F.V., Marsden, C.G. *et al.* (2004) Argonaute2 is the catalytic engine of mammalian RNAi. *Science*, **305**, 1437–1441.
40. Caudy, A.A., Ketting, R.F., Hammond, S.M., Denli, A.M. *et al.* (2003) A micrococcal nuclease homologue in RNAi effector complexes. *Nature*, **425**, 411–414.
41. Ishizuka, A., Siomi, M.C., Siomi, H.A. (2002) *Drosophila* fragile X protein interacts with components of RNAi and ribosomal proteins. *Genes Dev.*, **16**, 2497–2508.
42. Mourelatos, Z., Dostie, J., Paushkin, S., Sharma, A. *et al.* (2002) miRNPs: a novel class of ribonucleoproteins containing numerous microRNAs. *Genes Dev.*, **16**, 720–728.
43. Valencia-Sanchez, M.A., Liu, J., Hannon, G.J., Parker, R. (2006) Control of translation and mRNA degradation by miRNAs and siRNAs. *Genes Dev.*, **20**, 515–524.
44. Friedman, R.C., Farh, K.K., Burge, C.B., Bartel, D.P. (2009) Most mammalian mRNAs are conserved targets of microRNAs. *Genome Res.*, **19**, 92–105.
45. Bartel, D.P. (2004) MicroRNAs: genomics, biogenesis, mechanism, and function. *Cell*, **116**, 281–297.
46. Ørom, U.A., Nielsen, F.C., Lund, A.H. (2008) MicroRNA-10a binds the 5'UTR of ribosomal protein mRNAs and enhances their translation. *Mol. Cell*, **30**, 460–471.

47. Lytle, J.R., Yario, T.A., Steitz, J.A. (2007) Target mRNAs are repressed as efficiently by microRNA-binding sites in the 5′ UTR as in the 3′ UTR. *Proc. Natl Acad. Sci. USA*, **104**, 9667–9672.
48. Moretti, F., Thermann, R., Hentze, M.W. (2010) Mechanism of translational regulation by miR-2 from sites in the 5′ untranslated region or the open reading frame. *RNA*, **16**, 2493–2502.
49. Qin, W. Shi, Y., Zhao, B., Yao, C. *et al.* (2010) miR-24 regulates apoptosis by targeting the open reading frame (ORF) region of FAF1 in cancer cells. *PLoS One*, **5**, e9429.
50. Vasudevan, S., Tong, Y., Steitz, J.A. (2007) Switching from repression to activation: microRNAs can up-regulate translation. *Science*, **318**, 1931–1934.
51. Filipowicz, W., Bhattacharyya, S.N., Sonenberg, N. (2008) Mechanisms of post-transcriptional regulation by microRNAs: are the answers in sight? *Nature*, **9**, 102–114.
52. Hwang, H.W., Wentzel, E.A., Mendell, J.T. (2007) A hexanucleotide element directs microRNA nuclear import. *Science*, **315**, 97–100.
53. Valadi, H., Ekström, K., Bossios, A., Sjöstrand, M. *et al.* (2007) Exosome-mediated transfer of mRNAs and microRNAs is a novel mechanism of genetic exchange between cells. *Nat. Cell Biol.*, **9**, 654–659.
54. Berindan-Neagoe, I., Moroig, P.d.C., Pasculli, B., Calin, G.A. (2014) MicroRNAome genome: a treasure for Cancer Diagnosis and therapy. *CA Cancer J. Clin.*, **64** (5), 311–336.
55. Selcuklu, S.D., Donoghue, M.T., Spillane, C. (2009) miR-21 as a key regulator of oncogenic processes. *Biochem. Soc. Trans.*, **37**, 918–925.
56. Medina, P.P., Nolde, M., Slack, F.J. (2010) OncomiR addiction in an in vivo model of miRNA-21-induced pre-B-cell lymphoma. *Nature*, **467**, 86–90.
57. Calin, G.A., Dumitru, C.D., Shimizu, M., Bichi, R. *et al.* (2002) Frequent deletions and down-regulation of micro-RNA genes miR15 and miR16 at 13q14 in chronic lymphocytic leukemia. *Proc. Natl Acad. Sci. USA*, **99**, 15524–15529.
58. Cimmino, A., Calin, G.A., Fabbri, M., Iorio, M.V. *et al.* (2005) miR-15 and miR-16 induce apoptosis by targeting BCL2. *Proc. Natl Acad. Sci. USA*, **102**, 13944–13949.
59. Klein, U., Lia, M., Crespo, M., Siegel, R. *et al.* (2010) The DLEU2/miR-15a/16-1 cluster controls B cell proliferation and its deletion leads to chronic lymphocytic leukemia. *Cancer Cell*, **17**, 28–40.
60. Chen, R.W., Bemis, L.T., Amato, C.M., Myint, H. *et al.* (2008) Truncation of CCND1 mRNA alters miR-16-1 regulation in mantle cell lymphoma. *Blood*, **112**, 822–829.
61. Calin, G.A., Croce, C.M. (2006) MicroRNA signatures in human cancers. *Nat. Rev. Cancer*, **6**, 857–866.
62. Calin, G.A., Ferracin, M., Cimmino, A., Di Leva, G. *et al.* (2005) A MiRNA signature associated with prognosis and progression in chronic lymphocytic leukemia. *N. Engl. J. Med.*, **353**, 1793–1801.
63. Iorio, M.V., Ferracin, M., Liu, C.G., Veronese, A. *et al.* (2005) MiRNA gene expression deregulation in human breast cancer. *Cancer Res.*, **65**, 7065–7070.
64. He, H., Jazdzewski, K., Li, W., Liyanarachchi, S. *et al.* (2005) The role of miRNA genes in papillary thyroid carcinoma. *Proc. Natl Acad. Sci. USA*, **102**, 19075–19080.
65. Yanaihara, N., Caplen, N., Bowman, E., Seike, M. *et al.* (2006) Unique miRNA molecular profiles in lung cancer diagnosis and prognosis. *Cancer Cell*, **9**, 189–198.
66. Ciafré, S.A., Galardi, S., Mangiola, A., Ferracin, M. *et al.* (2005) Extensive modulation of a set of miRNAs in primary glioblastoma. *Biochem. Biophys. Res. Commun.*, **334**, 1351–1358.
67. Roldo, C., Missiaglia, E., Hagan, J.P., Falconi, M. *et al.* (2006) MiRNA expression abnormalities in pancreatic endocrine and acinar tumors are associated with distinctive pathologic features and clinical behavior. *J. Clin. Oncol.*, **24**, 4677–4684.
68. Murakami, Y., Yasuda, T., Saigo, K., Urashima, T. *et al.* (2006) Comprehensive analysis of miRNA expression patterns in hepatocellular carcinoma and non-tumorous tissues. *Oncogene*, **25**, 2537–2545.
69. Porkka, K.P., Pfeiffer, M.J., Waltering, K.K., Vessella, R.L. *et al.* (2007) MiRNA expression profiling in prostate cancer. *Cancer Res.*, **67**, 6130–6135.

70. Chen, W., Tang, Z., Sun, Y., Zhang, Y. *et al.* (2012) miRNA expression profile in primary gastric cancers and paired lymph node metastases indicates that miR-10a plays a role in metastasis from primary gastric cancer to lymph nodes. *Exp. Ther. Med.*, **3**, 351–356.
71. Di Leva, G., Garofalo, M., Croce, C.M. (2014) MicroRNAs in cancer. *Annu. Rev. Pathol.*, **9**, 287–314.
72. Lee, E.J., Baek, M., Gusev, Y., Brackett, D.J *et al.* (2008) Systematic evaluation of microRNA processing patterns in tissues, cell lines, and tumors. *RNA*, **14**, 35–42.
73. Esquela-Kerscher, A., Slack, F.J. (2006) Oncomirs: miRNAs with a role in cancer. *Nat. Rev. Cancer*, **6**, 259–269.
74. Petrocca, F., Lieberman, J. (2011) Promise and challenge of RNA interference-based therapy for cancer. *J. Clin. Oncol.*, **29** (6), 747–754.
75. O'Donnell, K.A., Wentzel, E.A., Zeller, K.I., Dang, C.V. *et al.* (2005) c-Myc-regulated microRNAs modulate E2F1 expression. *Nature*, **435**, 839–843.
76. Lopez-Serra, P., Esteller, M. (2012) DNA methylation-associated silencing of tumor suppressor microRNAs in cancer. *Oncogene*, **31**, 1609–1622.
77. Weber, B., Stresemann, C., Brueckner, B., Lyko, F. (2007) Methylation of human microRNA genes in normal and neoplastic cells. *Cell Cycle*, **6**, 1001–1005.
78. Saito, Y., Liang, G., Egger, G., Friedman, J.M. *et al.* (2006) Specific activation of microRNA-127 with downregulation of the proto-oncogene BCL6 by chromatin-modifying drugs in human cancer cells. *Cancer Cell*, **9**, 435–443.
79. Lehmann, U., Hasemeier, B., Christgen, M., Müller, M. *et al.* (2008) Epigenetic inactivation of microRNA gene has-miR- 9-1 in human breast cancer. *J. Pathol.*, **214**, 17–24.
80. Toyota, M. Suzuki, H., Sasaki, Y., Maruyama, R. *et al.* (2008) Epigenetic silencing of microRNA-34b/c and B-cell translocation gene 4 is associated with CpG island methylation in colorectal cancer. *Cancer Res.*, **68**, 4123–4132.
81. Lujambio, A., Ropero, S., Ballestar, E., Fraga, M.F. *et al.* (2007) Genetic unmasking of an epigenetically silenced microRNA in human cancer cells. *Cancer Res.*, **67**, 1424–1429.
82. Iorio, M.V., Visone, R., Di Leva, G., Donati, V. *et al.* (2007) MicroRNA signatures in human ovarian cancer. *Cancer Res.*, **67**, 8699–8707.
83. Scott, G.K., Mattie, M.D., Berger, C.E., Benz, S.C. *et al.* (2006) Rapid alteration of microRNA levels by histone deacetylase inhibition. *Cancer Res.*, **66**, 1277–1281.
84. Saito, Y., Suzuki, H., Tsugawa, H., Nakagawa, I. *et al.* (2009) Chromatin remodeling at Alu repeats by epigenetic treatment activates silenced microRNA-512-5p with downregulation of Mcl-1 in human gastric cancer cells. *Oncogene*, **28**, 2738–2744.
85. Rhodes, L.V., Nitschke, A.M., Collins-Burow, B.M. (2012) The histone deacetylase inhibitor trichostatin A alters microRNA expression profiles in apoptosis-resistant breast cancer cells. *Oncol. Rep.*, **27**, 10–16.
86. Raveche, E.S., Salerno, E., Scaglione, B.J., Manohar, V. *et al.* (2007) Abnormal microRNA-16 locus with synteny to human 13q14 linked to CLL in NZB mice. *Blood*, **109**, 5079–5086.
87. Hu, Z., Liang, J., Wang, Z., Tian, T. *et al.* (2009) Common genetic variants in pre-microRNAs were associated with increased risk of breast cancer in Chinese women. *Hum. Mutat.*, **30**, 79–84.
88. Jazdzewski, K., Murray, E.L., Franssila, K., Jarzab, B. *et al.* (2008) Common SNP in pre-miR-146a decreases mature miR expression and predisposes to papillary thyroid carcinoma. *Proc. Natl Acad. Sci. USA*, **105**, 7269–7274.
89. Hu, Z., Chen, J., Tian, T., Zhou, X. *et al.* (2008) Genetic variants of miRNA sequences and non-small cell lung cancer survival. *J. Clin. Invest.*, **118**, 2600–2608.
90. Mishra, P.J., Humeniuk, R., Longo-Sorbello, G.S., Banerjee, D. *et al.* (2007) A miR-24 microRNA binding-site polymorphism in dihydrofolate reductase gene leads to methotrexate resistance. *Proc. Natl Acad. Sci. USA*, **104**, 13513–13518.
91. Tan, Z., Randall, G., Fan, J., Camoretti-Mercado, B. *et al.* (2007) Allele-specific targeting of microRNAs to HLA-G and risk of asthma. *Am. J. Hum. Genet.*, **81**, 829–834.
92. Karube, Y., Tanaka, H, Osada, H., Tomida, S. *et al.* (2005) Reduced expression of Dicer

associated with poor prognosis in lung cancer patients. *Cancer Sci.*, **96**, 111–115.
93. Merritt, W.M., Lin, Y.G., Han, L.Y., Kamat, A.A. *et al.* (2008) Dicer, Drosha, and outcomes in patients with ovarian cancer. *N. Engl. J. Med.*, **359**, 2641–2650.
94. Guo, X., Liao, Q., Chen, P., Li, X. *et al.* (2012) The microRNA-processing enzymes: Drosha and Dicer can predict prognosis of nasopharyngeal carcinoma. *J. Cancer Res. Clin. Oncol.*, **138**, 49–56.
95. Lin, R.J., Lin, Y.C., Chen, J., Kuo, H.H. *et al.* (2010) microRNA signature and expression of Dicer and Drosha can predict prognosis and delineate risk groups in neuroblastoma. *Cancer Res.*, **70**, 7841–7850.
96. Grelier, G., Voirin, N., Ay, A.S., Cox, D.G. *et al.* (2009) Prognostic value of Dicer expression in human breast cancers and association with the mesenchymal phenotype. *Br. J. Cancer*, **101**, 673–683.
97. Faber, C., Jelezcova, E., Chandran, U., Acquafondata, M. *et al.* (2011) Overexpression of Dicer predicts poor survival in colorectal cancer. *Eur. J. Cancer*, **47**, 1414–1419.
98. Chiosea, S., Jelezcova, E., Chandran, U., Acquafondata, M. *et al.* (2006) Upregulation of dicer, a component of the MicroRNA machinery, in prostate adenocarcinoma. *Am. J. Pathol.*, **169**, 1812–1820.
99. Muralidhar, B., Winder, D., Murray, M., Palmer, R. *et al.* (2011) Functional evidence that Drosha overexpression in cervical squamous cell carcinoma affects cell phenotype and microRNA profiles. *J. Pathol.*, **224**, 496–507.
100. Martello, G., Rosato, A., Ferrari, F., Manfrin, A. *et al.* (2010) A MicroRNA targeting dicer for metastasis control. *Cell*, **141**, 1195–1207.
101. Lambertz, I., Nittner, D., Mestdagh, P., Denecker, G. *et al.* (2010) Monoallelic but not biallelic loss of Dicer1 promotes tumorigenesis in vivo. *Cell Death Differ.*, **17**, 633–641.
102. Heravi-Moussavi, A., Anglesio, M.S., Cheng, S.W., Senz, J. *et al.* (2012) Recurrent somatic DICER1 mutations in nonepithelial ovarian cancers. *N. Engl. J. Med.*, **366**, 234–242.
103. Melo, S.A., Ropero, S., Moutinho, C., Aaltonen, L.A. *et al.* (2009) A TARBP2 mutation in human cancer impairs microRNA processing and DICER1 function. *Nat. Genet.*, **41**, 365–370.
104. Melo, S.A., Moutinho, C., Ropero, S., Calin, G.A. *et al.* (2010) A genetic defect in Exportin-5 traps precursor microRNAs in the nucleus of cancer cells. *Cancer Cell*, **18**, 303–315.
105. Adams, B.D., Kasinski, A.L., Slack, F.J. (2014) Aberrant regulation and function of MicroRNAs in cancer. *Curr. Biol.*, **24**, R762–R776.
106. Tang, R., Li, L., Zhu, D., Hou, D. *et al.* (2012) Mouse miRNA-709 directly regulates miRNA- 15a/16-1 biogenesis at the posttranscriptional level in the nucleus: evidence for a microRNA hierarchy system. *Cell Res.*, **22**, 504–515.
107. Chang, T.C., Wentzel, E.A., Kent, O.A., Ramachandran, K. *et al.* (2007) Transactivation of miR-34a by p53 broadly influences gene expression and promotes apoptosis. *Mol. Cell*, **26**, 745–752.
108. Piovan, C., Palmieri, D., Di Leva, G., Braccioli, L. *et al.* (2012) Oncosuppressive role of p53-induced miR-205 in triple negative breast cancer. *Mol. Oncol.*, **6** (4), 458–472.
109. Chang, T.C., Zeitels, L.R., Hwang, H.W., Chivukula, R.R. *et al.* (2009) Lin-28B transactivation is necessary for Myc mediated let-7 repression and proliferation. *Proc. Natl Acad. Sci. USA*, **106**, 3384–3389.
110. Mott, J.L., Kurita, S., Cazanave, S.C., Bronk, S.F. *et al.* (2010) Transcriptional suppression of mir-29b-1/mir-29a promoter by c-Myc, hedgehog, and NF-kappaB. *J. Cell. Biochem.*, **110**, 1155–1164.
111. Kulshretha, R., Davuluri, R.V., Calin, G.A., Ivan, M. (2006) A microRNA component of the hypoxic response. *Cell Death Differ.*, **15**, 667–671.
112. Burk, U., Schubert, J., Wellner, U., Schmalhofer, O. *et al.* (2008) A reciprocal repression between ZEB1 and members of the miR-200 family promotes EMT and invasion in cancer cells. *EMBO Rep.*, **9**, 582–589.
113. Felli, N., Fontana, L., Pelosi, E., Botta, R. *et al.* (2005) MicroRNAs 221 and 222 inhibit normal erythropoiesis and erythroleukemic cell growth via kit receptor down-modulation. *Proc. Natl Acad. Sci. USA*, **102**, 18081–18086.

114. Garofalo, M., Quintavalle, C., Romano, G., Croce, C.M. *et al.* (2012) miR221/222 in cancer: their role in tumor progression and response to therapy. *Curr. Mol. Med.*, **12**, 27–33
115. Babar, I.A., Czochor, J., Steinmetz, A., Weidhaas, J.B. *et al.* (2011) Inhibition of hypoxia-induced miR-155 radiosensitizes hypoxic lung cancer cells. *Cancer Biol. Ther.*, **12**, 908–914.
116. Kong, W., He, L., Coppola, M., Guo, J. *et al.* (2010) MicroRNA-155 regulates cell survival, growth, and chemosensitivity by targeting FOXO3a in breast cancer. *J. Biol. Chem.*, **285**, 17869–17879.
117. Kong, W., Yang, H., He, L., Zhao, J.J. *et al.* (2008) MicroRNA-155 is regulated by the transforming growth factor beta/Smad pathway and contributes to epithelial cell plasticity by targeting RhoA. *Mol. Cell Biol.*, **28**, 6773–6784.
118. Mendell, J.T. (2008) miRiad roles for the miR-17-92 cluster in development and disease. *Cell*, **133**, 217–222.
119. Volinia, S., Calin, G.A., Liu, C.G., Ambs, S. *et al.* (2006) A microRNA expression signature of human solid tumors defines cancer gene targets. *Proc. Natl Acad. Sci. USA*, **103**, 2257–2261.
120. Taguchi, A., Yanagisawa, K., Tanaka, M., Cao, K. *et al.* (2008) Identification of hypoxia-inducible factor-1 alpha as a novel target for miR-17-92 microRNA cluster. *Cancer Res.*, **68**, 5540–5545.
121. Dews, M., Homayouni, A., Yu, D., Murphy, D. *et al.* (2006) Augmentation of tumor angiogenesis by a Myc-activated microRNA cluster. *Nat. Genet.*, **38**, 1060–1065.
122. Petrocca, F., Visone, R., Onelli, M.R., Shah, M.H. *et al.* (2008) E2F1-regulated microRNAs impair TGFβ-dependent cell-cycle arrest and apoptosis in gastric cancer. *Cancer Cell*, **13**, 272–286.
123. Fornari, F., Gramantieri, L., Ferracin, M., Veronese, A. *et al.* (2008) miR-221 controls CDKN1C/p57 and CDKN1B/p27 expression in human hepatocellular carcinoma. *Oncogene*, **27**, 5651–5661.
124. Pallante, P., Visone, R., Ferracin, M., Ferraro, A. *et al.* (2006) MicroRNA deregulation in human thyroid papillary carcinomas. *Endocr. Relat. Cancer*, **13**, 497–508.
125. Felicetti, F., Errico, M.C., Bottero, L., Segnalini, P. *et al.* (2008) The promyelocytic leukemia zinc finger–microRNA-221/-222 pathway controls melanoma progression through multiple oncogenic mechanisms. *Cancer Res.*, **68**, 2745–2754.
126. Di Leva, G., Gasparini, P., Piovan, C., Ngankeu, A. *et al.* (2010) MicroRNA cluster 221-222 and estrogen receptor αinteractions in breast cancer. *J. Natl Cancer Inst.*, **102**, 706–721.
127. Conti, A., Aguennouz, M., La Torre, D., Tomasello, C. *et al.* (2009) miR-21 and 221 upregulation and miR-181b downregulation in human grade II-IV astrocytic tumors. *J. Neurooncol.*, **93**, 325–332.
128. Garofalo, M., Di Leva, G., Romano, G., Nuovo, G. *et al.* (2009) miR-221&222 regulate TRAIL resistance and enhance tumorigenicity through PTEN and TIMP3 downregulation. *Cancer Cell*, **16**, 498–509.
129. Stinson, S., Lackner, M.R., Adai, A.T., Yu, N. *et al.* (2011) miR-221/222 targeting of trichorhinophalangeal (TRPS1) promotes epithelial-to-mesenchymal transition in breast cancer. *Sci. Signal.*, **4**, ra41.
130. Garofalo, M., Romano, G., Di Leva, G., Nuovo, G. *et al.* (2011) EGFR and MET receptor tyrosine kinase-altered microRNA expression induces tumorigenesis and gefitinib resistance in lung cancers. *Nat. Med.*, **18**, 74–82.
131. Zhang, J.G., Wang, J.J., Zhao, F., Liu, Q. *et al.* (2010) MicroRNA-21 (miR-21) represses tumor suppressor PTEN and promotes growth and invasion in non-small cell lung cancer (NSCLC). *Clin. Chim. Acta*, **411**, 846–852.
132. Zhu, S., Si, M.L., Wu, H., Mo, Y.Y. (2007) MicroRNA-21 targets the tumor suppressor gene tropomyosin 1 (TPM1). *J. Biol. Chem.*, **282**, 14328–14336.
133. Hatley, M.E., Patrick, D.M., Garcia, M.R., Richardson, J.A. *et al.* (2010) Modulation of K-Ras-dependent lung tumorigenesis by MicroRNA-21. *Cancer Cell*, **18**, 282–293.
134. Gabriely, G., Wurdinger, T., Kesari, S., Esau, C.C. *et al.* (2008) MicroRNA 21 promotes glioma invasion by targeting matrix metalloproteinase regulators. *Mol. Cell Biol.*, **28**, 5369–5380.
135. Seike, M., Goto, A., Okano, T., Bowman, E.D. *et al.* (2009) MiR-21 is an EGFR-regulated anti-apoptotic factor in lung

cancer in neversmokers. *Proc. Natl Acad. Sci. USA*, **106**, 12085–12090.

136. Chan, J.A., Krichevsky, A.M., Kosik, K.S. (2005) MicroRNA-21 is an antiapoptotic factor in human glioblastoma cells. *Cancer Res.*, **65**, 6029–6033.
137. Ambs, S., Prueitt, R.L., Yi, M., Hudson, R.S. *et al.* (2008) Genomic profiling of microRNA and messenger RNA reveals deregulated microRNA expression in prostate cancer. *Cancer Res.*, **68**, 6162–6170.
138. Bandi, N., Zbinden, S., Gugger, M., Arnold, M. *et al.* (2009) miR-15a and miR-16 are implicated in cell cycle regulation in a Rb-dependent manner and are frequently deleted or down-regulated in non-small cell lung cancer. *Cancer Res.*, **69**, 5553–5559.
139. Roccaro, A.M., Sacco, A., Thompson, B., Leleu, X. *et al.* (2009) MicroRNAs 15a and 16 regulate tumor proliferation in multiple myeloma. *Blood*, **26**, 6669–6680.
140. Pouliot, L.M., Chen, Y.C., Bai, J., Guha, R. *et al.* (2012) Cisplatin sensitivity mediated by WEE1 and CHK1 is mediated by miR-155 and the miR-15 family. *Cancer Res.*, **72**, 5945–5955.
141. Johnson, S.M., Grosshans, H., Shingara, J., Byrom, M. *et al.* (2005) RAS is regulated by the let-7 microRNA family. *Cell*, **120**, 635–647.
142. Chin, L.J., Ratner, E., Leng, S., Zhai, R. *et al.* (2008) A SNP in a let-7 microRNA complementary site in the KRAS 3′ untranslated region increases non-small cell lung cancer risk. *Cancer Res.*, **68**, 8535–8540.
143. Paranjape, T., Heneghan, H., Lindner, R., Keane, F.K. *et al.* (2011) A 3′-untranslated region KRAS variant and triple-negative breast cancer: a case-control and genetic analysis. *Lancet Oncol.*, **12**, 377–386.
144. Sampson, V.B., Rong, N.H., Han, J., Yang, Q. *et al.* (2007) MicroRNA let-7a down-regulates MYC and reverts MYC-induced growth in Burkitt lymphoma cells. *Cancer Res.*, **67**, 9762–9770.
145. Johnson, C.D., Esquela-Kerscher, A., Stefani, G., Byrom, M. *et al.* (2007) The let-7 microRNA represses cell proliferation pathways in human cells. *Cancer Res.*, **67**, 7713–7722.
146. Corney, D.C., Flesken-Nikitin, A., Godwin, A.K., Wang, W. *et al.* (2007) MicroRNA-34b and microRNA-34c are targets of p53 and cooperate in control of cell proliferation and adhesion-independent growth. *Cancer Res.*, **67**, 8433–8438.
147. Gregory, P.A., Bert, A.G., Paterson, E.L., Barry, S.C. *et al.* (2008) The miR-200 family and miR-205 regulate epithelial to mesenchymal transition by targeting ZEB1 and SIP1. *Nat. Cell Biol.*, **10**, 593–601.
148. Hurteau, G.J., Carlson, J.A., Spivack, S.D., Brock, G.J. (2007) Overexpression of the microRNA hsa-miR-200c leads to reduced expression of transcription factor 8 and increased expression of E-cadherin. *Cancer Res.*, **67**, 7972–7976.
149. Bracken, C.P., Gregory, P.A., Kolesnikoff, N., Bert *et al.* (2008) A double-negative feedback loop between ZEB1-SIP1 and the microRNA-200 family regulates epithelial-mesenchymal transition. *Cancer Res.*, **68**, 7846–7854.
150. Iliopoulos, D., Polytarchou, C., Hatziapostolou, M., Kottakis, F. *et al.* (2009) MicroRNAs differentially regulated by Akt isoforms control EMT and stem cell renewal in cancer cells. *Sci. Signal.*, **2**, ra62.
151. Vrba, L., Jensen, T.J., Garbe, J.C., Heimark, R.L. *et al.* (2010) Role for DNA methylation in the regulation of miR-200c and miR-141 expression in normal and cancer cells. *PLoS One*, **5**, e8697.
152. Shen, J., Samul, R., Silva, R.L., Akiyama, H. *et al.* (2006) Suppression of ocular neovascularization with siRNA targeting VEGF receptor 1. *Gene Ther.*, **13** (3), 225–234.
153. Filleur, S., Courtin, A., Ait-Si-Ali, S., Guglielmi, J. *et al.* (2003) SiRNA-mediated inhibition of vascular endothelial growth factor severely limits tumor resistance to antiangiogenic thrombospondin-1 and slows tumor vascularization and growth. *Cancer Res.*, **63** (14), 3919–3922.
154. Gaudilliere, B., Shi, Y., Bonni, A. (2002) RNA interference reveals a requirement for myocyte enhancer factor 2A in activity-dependent neuronal survival. *J. Biol. Chem.*, **277** (48), 46442–46446.
155. Miller, V.M., Xia, H., Marrs, G.L., Gouvion, C.M. *et al.* (2003) Allele-specific silencing of dominant disease genes. *Proc. Natl Acad. Sci. USA*, **100** (12), 7195–7200.
156. Wu, S.Y., Lopez-Berestein, G., Calin, G.A., Sood, A.K. (2014) RNAi therapies: drugging

the undruggable. *Sci. Transl. Med.*, **6** (240), 240ps7.

157. Rao, D.D., Vorhies, J.S., Senzer, N., Nemunaitis, J. (2009) siRNA vs. shRNA: similarities and differences. *Adv. Drug Deliv. Rev.*, **61**, 746–759.
158. Lima, W.F., Prakash, T.P., Murray, H.M., Kinberger, G.A. *et al.* (2012) Single-stranded siRNAs activate RNAi in animals. *Cell*, **150**, 883–894.
159. Maples, P.B., Senzer, N., Kumar, P., Wang, Z. *et al.* (2010) Enhanced target gene knockdown by a bifunctional shRNA: a novel approach of RNA interference. *Cancer Gene Ther.*, **17** (11), 780–791.
160. Garzon, R., Marcucci, G., Croce, C.M. (2010) Targeting microRNAs in cancer: rationale, strategies and challenges. *Nat. Rev. Drug Discov.*, **9** (10), 775–789.
161. Kota, J., Chivukula, R.R., O'Donnell, K.A., Wentzel, E.A. *et al.* (2009) Therapeutic microRNA delivery suppresses tumorigenesis in a murine liver cancer model. *Cell*, **137**, 1005–1017.
162. Stenvang, J., Petri, A., Lindow, M., Obad, S. *et al.* (2012) Inhibition of microRNA function by antimiR oligonucleotides. *Silence*, **3** (1), 1.
163. Brown, B.D., Naldini, L. (2009) Exploiting and antagonizing microRNA regulation for therapeutic and experimental applications. *Nat. Rev.*, **10** (8), 578–585.
164. Choi, W.Y., Giraldez, A.J., Schier, A.F. (2007) Target protectors reveal dampening and balancing of Nodal agonist and antagonist by miR-430. *Science*, **318**, 271–274.
165. Karlas, A., Machuy, N., Shin, Y., Pleissner, K.P. *et al.* (2010) Genome-wide RNAi screen identifies human host factors crucial for influenza virus replication. *Nature*, **463** (7282), 818–822.
166. Krishnan, M.N., Ng, A., Sukumaran, B., Gilfoy, F.D. *et al.* (2008) RNA interference screen for human genes associated with West Nile virus infection. *Nature*, **455** (7210), 242–245.
167. Tiedemann, R.E., Zhu, Y.X., Schmidt, J., Shi, C.X. *et al.* (2012) Identification of molecular vulnerabilities in human multiple myeloma cells by RNA interference lethality screening of the druggable genome. *Cancer Res.*, **72** (3), 757–768.
168. Steckel, M., Molina-Arcas, M., Weigelt, B., Marani, M. *et al.* (2012) Determination of synthetic lethal interactions in KRAS oncogene-dependent cancer cells reveals novel therapeutic targeting strategies. *Cell Res.*, **22**, 1227–1245.
169. Etemadmoghadam, D., Weir, B.A., Au-Yeung, G., Alsop, K. *et al.*, Australian Ovarian Cancer Study Group (2013) Synthetic lethality between CCNE1 amplification and loss of BRCA1. *Proc. Natl Acad. Sci. USA*, **110**, 19489–19494.
170. Phalon, C., Rao, D.D., Nemunaitis, J. (2010) Potential use of RNA interference in cancer therapy. *Expert Rev. Mol. Med.*, **12**, e26.
171. Daka, A., Peer, D. (2012) RNAi-based nanomedicines for targeted personalized therapy. *Adv. Drug Deliv. Rev.*, **64**, 1508–1521.
172. Niu, X.Y., Peng, Z.L., Duan, W.Q., Wang, H. *et al.* (2006) Inhibition of HPV 16 E6 oncogene expression by RNA interference in vitro and in vivo. *Int. J. Gynecol. Cancer*, **16** (2), 743–751.
173. Guo, P., Coban, O., Snead, N.M., Trebley, J. *et al.* (2010) Engineering RNA for targeted siRNA delivery and medical application. *Adv. Drug Deliv. Rev.*, **62**, 650–666.
174. Chiu, Y.L., Rana, T.M. (2003) siRNA function in RNAi: a chemical modification analysis. *RNA*, **9** (9), 1034–1048.
175. Morrissey, D.V., Lockridge, J.A., Shaw, L., Blanchard, K. *et al.* (2005) Potent and persistent in vivo anti-HBV activity of chemically modified siRNAs. *Nat. Biotechnol.*, **23** (8), 1002–1007.
176. Morrissey, D.V., Blanchard, K., Shaw, L., Jensen, K. *et al.* (2005) Activity of stabilized short interfering RNA in a mouse model of hepatitis B virus replication. *Hepatology*, **41** (6), 1349–1356.
177. Elmen, J., Thonberg, H., Ljungberg, K., Frieden, M. *et al.* (2005) Locked nucleic acid (LNA) mediated improvements in siRNA stability and functionality. *Nucleic Acids Res.*, **33** (1), 439–447.
178. Bumcrot, D., Manoharan, M., Koteliansky, V., Sah, D.W. (2006) RNAi therapeutics: a potential new class of pharmaceutical drugs. *Nat. Chem. Biol.*, **2** (12), 711–719.
179. Krutzfeldt, J., Rajewsky, N., Braich, R., Rajeev, K.G *et al.* (2005) Silencing of microRNAs in vivo with 'antagomiRs'. *Nature*, **438**, 685–689.
180. Jackson, A.L., Burchard, J., Schelter, J., Chau, B.N. *et al.* (2006) Widespread siRNA

"off-target" transcript silencing mediated by seed region sequence complementarity. *RNA*, **12** (7), 1179–1187.

181. Doench, J.G., Sharp, P.A. (2004) Specificity of microRNA target selection in translational repression. *Genes Dev.*, **18** (5), 504–511.
182. Dykxhoorn, D.M., Lieberman, J. (2006) Knocking down disease with siRNAs. *Cell*, **126** (2), 231–235.
183. Grimm, D., Streetz, K.L., Jopling, C.L., Storm, T.A. *et al.* (2006) Fatality in mice due to oversaturation of cellular microRNA/short hairpin RNA pathways. *Nature*, **441** (7092), 537–541.
184. Wang, Z., Rao, D.D., Senzer, N., Nemunaitis, J. (2011) RNA interference and cancer therapy. *Pharm. Res.*, **28**, 2983–2995.
185. Persengiev, S.P., Zhu, X.C., Green, M.R. (2004) Nonspecific, concentration-dependent stimulation and repression of mammalian gene expression by small interfering RNAs (siRNAs). *RNA*, **10** (1), 12–18.
186. Jackson, A.L., Bartz, S.R., Schelter, J., Kobayashi, S.V. *et al.* (2003) Expression profiling reveals off-target gene regulation by RNAi. *Nat. Biotechnol.*, **21** (6), 635–637.
187. Haussecker, D. (2014) Current issues of RNAi therapeutics delivery and development. *J. Control. Release*, **195**, 49–54.
188. Agrawal, S., Kandimalla, E.R. (2004) Antisense and siRNA as agonists of Toll-like receptors. *Nat. Biotechnol.*, **22** (12), 1533–1537.
189. Sledz, C.A., Holko, M., de Veer, M.J., Silverman, R.H. *et al.* (2003) Activation of the interferon system by short-interfering RNAs. *Nat. Cell Biol.*, **5** (9), 834–839.
190. Kariko, K., Bhuyan, P., Capodici, J., Weissman, D. (2004) Small interfering RNAs mediate sequence-independent gene suppression and induce immune activation by signaling through toll-like receptor. *J. Immunol.*, **172** (11), 6545–6549.
191. Judge, A.D., Sood, V., Shaw, J.R., Fang, D. *et al.* (2005) Sequence-dependent stimulation of the mammalian innate immune response by synthetic siRNA. *Nat. Biotechnol.*, **23** (4), 457–462.
192. Hornung, V., Nemorin, J.G., Montino, C., Müller, C. *et al.* (2005) Sequence-specific potent induction of IFN-alpha by short interfering RNA in plasmacytoid dendritic cells through TLR7. *Nat. Med.*, **11** (3), 263–270.
193. Forsbach, A., Nemorin, J.G., Montino, C., Müller, C. *et al.* (2008) Identification of RNA sequence motifs stimulating sequence-specific TLR8-dependent immune responses. *J. Immunol.*, **180** (6), 3729–3738.
194. Bramsen, J.B., Kjems, J. (2011) Chemical modification of small interfering RNA. *Methods Mol. Biol.*, **721**, 77–103.
195. Soutschek, J., Akinc, A., Bramlage, B., Charisse, K. *et al.* (2004) Therapeutic silencing of an endogenous gene by systemic administration of modified siRNAs. *Nature*, **432**, 173–178.
196. Di Figlia, M., Sena-Esteves, M., Chase, K., Sapp, E. *et al.* (2007) Therapeutic silencing of mutant huntingtin with siRNA attenuates striatal and cortical neuropathology and behavioral deficits. *Proc. Natl Acad. Sci. USA*, **104**, 17204–17209.
197. Iversen, F., Yang, C.X., Dagnaes-Hansen, F., Schaffert, D.H. *et al.* (2013) Optimized siRNA–PEG conjugates for extended blood circulation and reduced urine excretion in mice. *Theranostics*, **3**, 201–209.
198. Hagiwara, K., Ochiya, T., Kosaka, N. (2014) A paradigm shift for extracellular vesicles as small RNA carriers: from cellular waste elimination to therapeutic applications. *Drug Deliv. Transl. Res.*, **4** (1), 31–37.
199. György, B., Szabó, T.G., Pásztói, M., Pál, Z. *et al.* (2011) Membrane vesicles, current state-of-the-art: emerging role of extracellular vesicles. *Cell. Mol. Life Sci.*, **68** (16), 2667–2688.
200. Adler, H.I., Fisher, W.D., Cohen, A., Hardigree, A.A. (1967) Miniature *Escherichia coli* cells deficient in DNA. *Proc. Natl Acad. Sci. USA*, **57**, 321–326.
201. MacDiarmid, J.A., Mugridge, N.B., Weiss, J.C., Phillips, L. *et al.* (2007) Bacterially derived 400 nm particles for encapsulation and cancer cell targeting of chemotherapeutics. *Cancer Cell*, **11**, 431–445.
202. Deng, Y., Wang, C.C., Choy, K.W., Du, Q. *et al.* (2014) Therapeutic potentials of gene silencing by RNA interference: principles, challenges, and new strategies. *Gene*, **538**, 217–227.
203. Wu, S.Y., Yang, X., Gharpure, K.M., Hatakeyama, H. *et al.* (2014) 2′-OMe-phosphorodithioato-modified siRNAs show increased loading into the RISC complex

and enhanced anti-tumour activity. *Nat. Commun.*, **5**, 3459.

204. Wilson, J.A., Jayasena, S., Khvorova, A., Sabatinos, S. *et al.* (2003) RNA interference blocks gene expression and RNA synthesis from hepatitis C replicons propagated in human liver cells. *Proc. Natl Acad. Sci. USA*, **100**, 2783–2788.
205. Paganin-Gioanni, A., Bellard, E., Escoffre, J.M., Rols, M.P. *et al.* (2011) Direct visualization at the single-cell level of siRNA electrotransfer into cancer cells. *Proc. Natl Acad. Sci. USA*, **108**, 10443–10447.
206. Chabot, S., Orio, J., Castanier, R., Bellard, E. *et al.* (2012) LNA-based oligonucleotide electrotransfer for miRNA inhibition. *Mol. Ther.*, **20**, 1590–1598.
207. Chabot, S., Teissié, J., Golzio, M. (2015) Targeted electro-delivery of oligonucleotides for RNA interference: siRNA and antimiR. *Adv. Drug Deliv. Rev.*, **81**, 161–168.
208. Zhou, Y., Zhang, C., Liang, W. (2014) Development of RNAi technology for targeted therapy – A track of siRNA based agents to RNAi therapeutics. *J. Control. Release*, **10** (193), 270–281.
209. Bamrungsap, S., Zhao, Z., Chen, T., Wang, L. *et al.* (2012) Nanotechnology in therapeutics: a focus on nanoparticles as a drug delivery system. *Nanomedicine (Lond.)*, **7** (8), 1253–1271.
210. Song, E., Zhu, P., Lee, S.K., Chowdhury, D. *et al.* (2005) Antibody mediated in vivo delivery of small interfering RNAs via cell-surface receptors. *Nat. Biotechnol.*, **23** (6), 709–717.
211. Arap, W., Pasqualini, R., Ruoslahti, E. (1998) Cancer treatment by targeted drug delivery to tumor vasculature in a mouse model. *Science*, **279** (5349), 377–380.
212. Wu, Y., Sefah, K., Liu, H., Wang, R. *et al.* (2010) DNA aptamer-micelle as an efficient detection/delivery vehicle toward cancer cells. *Proc. Natl Acad. Sci. USA*, **107**, 5–10.
213. Leamon, C.P., Reddy, J.A. (2004) Folate-targeted chemotherapy. *Adv. Drug Deliv. Rev.*, **56**, 1127–1141.
214. Yao, Y.D., Sun, T.M., Huang, S.Y., Dou, S. *et al.* (2012) Targeted delivery of PLK1–siRNA by ScFv suppresses Her2+ breast cancer growth and metastasis. *Sci. Transl. Med.*, **4**, 130ra148.
215. McNamara, J.O. II, Andrechek, E.R., Wang, Y., Viles, K.D. *et al.* (2006) Cell type-specific delivery of siRNAs with aptamer-siRNA chimeras. *Nat. Biotechnol.*, **24** (8), 1005–1015.
216. Jones, A.T., Sayers, E.J. (2012) Cell entry of cell penetrating peptides: tales of tails wagging dogs. *J. Control. Release*, **161**, 582–591.
217. Gupta, B.,. Torchilin, V.P. (2006) Transactivating transcriptional activator-mediated drug delivery. *Expert Opin. Drug Deliv.*, **3** 177–190.
218. Lorenz, C., Hadwiger, P., John, M., Vornlocher, H.P. *et al.* (2004) Steroid and lipid conjugates of siRNAs to enhance cellular uptake and gene silencing in liver cells. *Bioorg. Med. Chem. Lett.*, **14**, 4975–4977.
219. Yu, B., Zhao, X., Lee, L.J., Lee, R.J. (2009) Targeted delivery systems for oligonucleotide therapeutics. *AAPS J.*, **11**, 195–203.
220. Wilner, S.E., Wengerter, B., Maier, K., Borba Magalhães, M.L. *et al.* (2012) An RNA alternative to human transferrin: a new tool for targeting human cells. *Mol. Ther. Nucleic Acids*, **1**, e21.
221. Scherer, F., Anton, M., Schillinger, U., Henke, J. *et al.* (2002) Magnetofection: enhancing and targeting gene delivery by magnetic force in vitro and in vivo. *Gene Ther.*, **9**, 102–109.
222. Medarova, Z., Pham, W., Farrar, C., Petkova, V. *et al.* (2007) In vivo imaging of siRNA delivery and silencing in tumors. *Nat. Med.*, **13** (3), 372–377.
223. Qi, L., Wu, L., Zheng, S., Wang, Y. *et al.* (2012) Cell-penetrating magnetic nanoparticles for highly efficient delivery and intracellular imaging of siRNA. *Biomacromolecules*, **13** (9), 2723–2730.
224. David, S., Marchais, H., Bedin, D., Chourpa, I. (2014) Modelling the response surface to predict the hydrodynamic diameters of theranostic magnetic siRNA nanovectors. *Int. J. Pharm.*, **478** (1), 409–415.
225. Suzuki, R., Takizawa, T., Negishi, Y., Hagisawa, K. *et al.* (2007) Gene delivery by combination of novel liposomal bubbles with perfluoropropane and ultrasound. *J. Control. Release*, **117**, 130–136.
226. Suzuki, R., Takizawa, T., Negishi, Y., Utoguchi, N. *et al.* (2008) Tumor specific ultrasound enhanced gene transfer in vivo with novel liposomal bubbles. *J. Control. Release*, **125**, 137–144.

227. Suzuki, R., Takizawa, T., Negishi, Y., Utoguchi, N. *et al.* (2008) Effective gene delivery with novel liposomal bubbles and ultrasonic destruction technology. *Int. J. Pharm.*, **354**, 49–55.
228. Endo-Takahashi, Y., Negishi, Y., Nakamura, A., Ukai, S. *et al.* (2014) Systemic delivery of miR-126 by miRNA-loaded Bubble liposomes for the treatment of hindlimb ischemia. *Sci. Rep.*, **24** (4), 3883.
229. Grimm, D., Kay, M.A. (2007) Combinatorial RNAi: a winning strategy for the race against evolving targets? *Mol. Ther.*, **15** (5), 878–888.
230. Gandhi, N.S., Tekade, R.K., Chougule, M.B. (2014) Nanocarrier mediated delivery of siRNA/miRNA in combination with chemotherapeutic agents for cancer therapy: current progress and advances. *J. Control. Release*, **194C**, 238–256.
231. Meng, H., Liong, M., Xia, T., Li, Z. *et al.* (2010) Engineered design of mesoporous silica nanoparticles to deliver doxorubicin and P-glycoprotein siRNA to overcome drug resistance in a cancer cell line. *ACS Nano*, **4**, 4539–4550.
232. Zheng, C., Zheng, M., Gong, P., Deng, J. *et al.* (2013) Polypeptide cationic micelles mediated co-delivery of docetaxel and siRNA for synergistic tumor. *Biomaterials*, **34** (13), 3431–3438.
233. Takei, Y., Kadomatsu, K., Goto, T., Muramatsu, T. (2006) Combinational antitumor effect of siRNA against midkine and paclitaxel on growth of human prostate cancer xenografts. *Cancer*, **107**, 864–873.
234. Nishimura, M., Jung, E.J., Shah, M.Y., Lu, C. *et al.* (2013) Therapeutic synergy between microRNA and siRNA in ovarian cancer treatment. *Cancer Discov.*, **3** (11), 1302–1315.
235. Xue, W., Dahlman, J.E., Tammela, T., Khan, O.F. *et al.* (2014) Small RNA combination therapy for lung cancer. *Proc. Natl Acad. Sci. USA*, **111** (34), E3553–E3561.
236. Kasinski, A.L., Kelnar, K., Stahlhut, C., Orellana, E. *et al.* (2014) A combinatorial microRNA therapeutics approach to suppressing non-small cell lung cancer. *Oncogene*. doi: 10.1038/onc.2014.282.
237. Castanotto, D., Sakurai, K., Lingeman, R., Li, H. *et al.* (2007) Combinatorial delivery of small interfering RNAs reduces RNAi efficacy by selective incorporation into RISC. *Nucleic Acids Res.*, **35**, 5154–5164.

13
Smart Nanoparticles in Brain Cancer Therapy

Yinhao Wu[1], Tao Sun[1], Lisha Liu[1], Xi He[1], Yifei Lu[1], Sai An[1], and Chen Jiang[1,2]
[1]Fudan University, Ministry of Education, Department of Pharmaceutics, Key Laboratory of Smart Drug Delivery, School of Pharmacy, 826 Zhangheng Road, Shanghai 201203, China
[2]Fudan University, State Key Laboratory of Medical Neurobiology, 138 Yixueyuan Road, Shanghai 200032, China

Keywords

Nanoparticles
Nanoparticles are particles between 1 and 100 nm in size. In nanotechnology, a particle is defined as a small object that behaves as a whole unit with respect to its transport and properties.

Translational Nanomedicine. First Edition. Edited by Robert A. Meyers.

Active-targeting
Active-targeting nanoparticles utilize cell-specific ligands or the capability to be activated by a trigger that is specific to the target site to enhance the effects of passive targeting to make the nanoparticle more specific to a target site.

Stimuli-responsive release
Stimuli-responsive release depends on the sensitivity to a number of factors, such as temperature, humidity, pH, the wavelength, or intensity of light, or an electrical or magnetic field to make the nanoparticle more specific to a target site.

Brain cancer
Brain cancer is the malignant form of brain tumors. The lesions can be subdivided into primary tumors that start within the brain, and secondary tumors that have spread from elsewhere the latter are known as brain metastasis tumors.

Brain cancer therapy has become a huge challenge compared with peripheral cancers because of the physiological characteristic of the brain-blood barrier (BBB), which prevents most therapeutical drugs from reaching the cancer tissues. For years, efforts have been made in nanotechnology, especially employing nanoparticles (NPs), to overcome this limitation. Currently, research groups are mainly focusing on three elements, namely functionalization, targeting, and imaging, to fabricate smart nanoparticles that are capable of crossing the BBB, can respond to multiple internal or external stimuli, and deliver therapeutic or diagnostic agents to cancer cells through systemic administration. In this review, a brief discussion is provided of how these 'smart' nanoparticles are designed, and the advantages they present in the treatment of brain cancer.

1 Introduction

In 2015, an estimated 22 850 adults were diagnosed with primary cancerous tumors of the brain and spinal cord, and a number of 15 320 adults died from them in the United States, according to the statistics provided by the American Cancer Society [1]. Brain tumors can be classified into malignant/cancerous tumors and benign tumors based on the severity of the tumors, or they can be subdivided into primary tumors and secondary (metastatic) tumors based on the origin of the tumor cells. Among all malignant primary brain tumors in adults, the most common malignant tumor is glioblastoma multiforme (GBM). In adolescents aged 15–19 years, gliomas accounts for approximately 35% of tumors [2]. Metastatic brain tumors have become prevalent as new detection methods have been developed, the main sources of metastatic brain tumors being lung cancer, breast cancer, and melanoma [3].

Optimal surgical resection is an important therapy for primary brain cancers, and surgery is typically the first option when there are only a few brain metastases present [4]. Since it is impossible to thoroughly remove all cancer cells through surgical resection, radiation therapy and chemotherapy are usually combined as conventional therapeutical approaches. Compared to chemotherapy, radiation therapy is costly and is often accompanied by many complications. Thus, chemotherapy plays an important role in the treatment of malignant brain tumors (primary glioma and metastatic cancers).

Temozolomide (TMZ) is regarded as the priority choice for brain cancer chemotherapy. The main problem is that, after surgical resection, the majority of available therapeutic agents are unable to reach cancer cells that reside behind an intact blood–brain barrier (BBB). Another problem is the fact that chemotherapeutic agents circulating in the blood may cause severe systemic toxicity. Even more frustrating are the cases where, although a very small proportion of chemotherapeutic agents are able to pass through the BBB, the drug resistance of cancer cells and the existence of a high interstitial fluid pressure (IFP) in cancer tissues prevents these agents from playing an effective therapeutic role [5, 6]. Clearly, drug-delivery systems are urgently required to provide a means of crossing the BBB and improving drug accumulation in cancer cells within the brain, while simultaneously reducing the peripheral toxicity of these compounds.

Due to the rapid proliferation of cancer cells, tumor angiogenesis occurs in abnormal pathological states such as pericyte deficiency and aberrant basement membrane formation, leading to enhanced vascular permeability [7]. However, the absence of lymphatic drainage in cancers leads to particle retention at the tumor sites. This phenomenon, termed the "enhanced permeability and retention (EPR)" effect, has been employed in the design of passive tumor-targeting nanoparticles. While the BBB remains intact during the early stages of brain cancers, brain tumors remain beyond the reach of passive targeting nanoparticles. However, as brain cancers develop the permeability of new blood vessels in the region of the cancer is usually increased, so as to demonstrate the EPR effect. In the past few years, well-designed smart nanoparticles have been created. In possessing properties of tissue selectivity and the controlled release of therapeutic agents, these nanoparticles showed enhanced accumulation in brain cancers and improved therapeutic effects, providing promising chemotherapy and gene therapy for brain.

2
Barriers and Problems

The BBB is a dynamic barrier that protects the brain from invading organisms and unwanted substances, except for necessary nutrients such as certain ions and glucose for normal brain function. The integral BBB supporting system consists of brain capillary endothelial cells (ECs), the extracellular base membrane, adjoining pericytes, astrocytes, and microglia [8]. The presence of a multilayer membrane around the brain capillary and a lack of fenestration means that the BBB has a low and selective permeability to molecules. In addition, resistance towards chemotherapeutic agents attributes to the high expression of efflux transporters such as p-glycoprotein (P-gp) and multi-drug resistance related proteins (MRPs) at the BBB [5, 9]. Together, these features impede drug transport into the brain via the blood circulation.

In addition to the obstacle of the BBB, the existence of a high IFP and drug resistance in brain cancer tissues result in a more difficult approach to chemotherapy. It is now well established that the IFP of most solid tumors is increased, and although the mechanisms that determine such increases are not fully understood they are most likely related to blood vessel leakiness, lymph vessel abnormalities, interstitial fibrosis, and a contraction of the interstitial space mediated by stromal fibroblasts. The increase in IFP contributes to a decreased transcapillary transport into cancers, and represents an obstacle to treatment by causing a reduced uptake of drugs or therapeutic antibodies into the cancer [6].

P-gp and MRPs are present not only at the BBB but also in cancer cells [10], which further decreases the accumulation of anti-tumor agents in cancer cells. Notably, the existence of P-gp and MRPs is crucial to the healthy brain tissues. Although the inhibition of these proteins may help therapeutic agents to cross the BBB and accumulate in cancer cells, the inhibition of P-gp function could also result in toxicity to the brain. Moreover, an inhibition of the transport of proteins such as P-gp and MRPs to healthy tissues may increase the distribution of chemotherapeutic agents into other organs, with an accompanying risk of increasing systemic drug toxicity [5]. Hence, the suggestion to increase the accumulation of anti-tumor agents in brain cancers by inhibiting P-gp and MRPs is not viable.

Recent developments in nanotechnology have created exciting opportunities for the diagnosis and therapy of brain tumors. Nanoparticles may include polymeric nanoparticles, lipid nanocapsules, albumin nanoparticles, liposomes, micelles, dendrimers, and nanoemulsions. Subsequently, when therapeutic agents are loaded onto or into nanoparticles, the properties of the therapeutic agent are changed, so as to demonstrate improved biodistribution and pharmacokinetic properties. Indeed, advanced designs of the nanoparticles may provide outstanding functions in cancer therapies.

In the present review, attention is focused on the application of 'smart' nanoparticles to brain cancer therapy. Smart nanoparticles incorporate two main aspects, namely the targeting of the vehicles and the smart release of their cargoes. The major aim of smart targeting is to provide an enhanced accumulation of therapeutic agents in cancer tissues, while normal cells within the same region can avoid the embedded 'cargoes' and avoid being killed simultaneously with tumor cells. Combined with the design of ingenious release mechanisms based on differences between the intracellular and extracellular environments, smart nanoparticles can prevent therapeutic agents from leaking and also provide a guaranteed effect within the tumor cells. In general, the design of smart nanoparticles can provide an amplified therapeutic effect while simultaneously reducing the toxicity of a chemotherapeutic agent. In this chapter a specific classification of smart nanoparticles will be introduced, and several examples provided to illustrate how such strategies are used in practice.

3 Nanoparticles with Smart Targeting Strategies

As an essential prerequisite, the designed nanoparticles must be able to cross the BBB and enter cancer cells. Since the BBB provides the brain with nutrients necessary for normal brain function, large numbers of receptors and transporters are expressed at the BBB. As the tumor cells proliferate rapidly they demonstrate an enormous requirement for nutriments, and the rapid generation of neovascular tissue also results in the over-expression of specific receptors and transporters in tumor cells. These include transferrin receptors (TfRs), low-density lipoprotein receptor-related proteins (LRPs), facilitative glucose transporters, and choline transporters [11]. The abnormal gene amplification and over-expression in tumor cells also increases the numbers of some receptors at tumor cells [12].

Based on the over-expression of abundant receptors, transporters at the BBB and brain cancers, multifarious active targeting nanoparticles are designed to cross the BBB and to accumulate in cancer cells. For brain cancer targeting, dual targeting strategies and two-order targeting strategies are commonly employed in designing the drug-delivery or diagnosis systems to overcome the obstacle of BBB and to enhance accumulation in cancer cells (Fig. 1) [13, 14].

3.1 The Dual Targeting Strategies

The benefits of dual targeting to BBB and glioma with one ligand are mainly focused on a convenient modification and a higher biocompatibility due to less-modified ligands. Generally speaking, dual-targeting ligands can be classified as receptor-mediated targeting, transporter-mediated targeting, and adsorptive-mediated transcytosis.

3.1.1 Receptor-Mediated Targeting

A wide range of receptors such as TfRs [15, 16], low-density LRP receptors [17], epidermal growth factor receptors (EGFRs) [18], opioid receptors [19], and leptin receptors

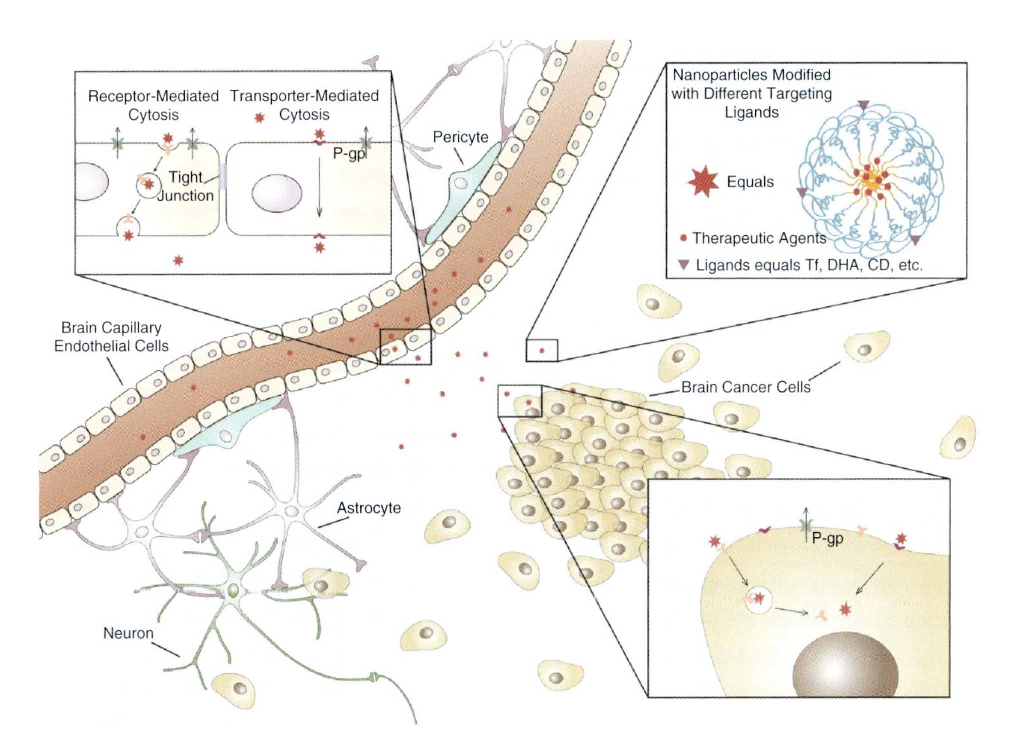

Fig. 1 Nanoparticles with dual-targeting strategies or two-order targeting strategies in brain cancer therapy. The surfaces of nanoparticles could be modified with only one type of targeting ligands, such as Tf, DHA, and CD, or two different types of targeting ligand.

[20] are over-expressed on both brain capillary endothelial cells and brain cancer cells. Thus, nanoparticles are able to dual-target the BBB and cancer cells via a modification with one single type of targeting ligand.

TfRs are engaged in regulating the balance of iron, which is important to the growth and metabolism of normal cells or cancer cells. Gao *et al.* prepared polyamidoamine dendrimer (PAMAM) modified with transferrin (Tf) to deliver plasmid pORF-hTRAIL for glioma gene therapy [21]. Zheng *et al.* combined transferrin and cell-penetrating peptide to enhance the doxorubicin delivery into the brain [22]. Kang *et al.* found a novel iron-mimic peptide which targeted the Tf/TfR complex and enhanced the penetration of nanoparticles [23]. Although Tf can be modified on the surface of nanoparticles to cross both brain capillary endothelia obstacles and the tumor cell membranes, high concentrations of endogenous Tf may cause a competitive inhibition of the Tf-modified nanoparticles [24]. Under these circumstances, peptide T7 (sequence His-Ala-Ile-Tyr-Pro-Arg-His) has been discovered which has the similar affinity to target Tf receptors [25]. Hence, DGL-PEG-T7/DNA nanoparticles were designed as a dual-targeting gene delivery system [26]. These nanoparticles, when modified with T7, showed an increased cellular uptake in both brain capillary endothelial cells (BCECs) and U87 glioma cells, and increased distribution in tumor sites. In particular, the amplification of Tf on the targeting of T7-modified nanoparticles was discovered due to the different binding sites of Tf and T7.

Low-density LRP plays an active role in transporting numerous ligands across the BBB, such as lipoproteins and extracellular matrix proteins [27]. The over-expression of LRP in both BBB and human glioblastoma cancer cells has been demonstrated using Western blotting [14]. Based on the fact that lactoferrin is transported by LRP, Miao *et al.* designed a dual-targeting nanoparticle modified with lactoferrin for anti-glioblastoma therapy [28], while Fang *et al.* developed nanoparticles for brain tumor therapy by co-delivering doxorubicin and curcumin [29]. Not only for therapy purposes, nanoparticles can also be modified with lactoferrin for improved diagnosis of brain cancers. Xie *et al.* explored lactoferrin-conjugated superparamagnetic iron oxide (SPIO) nanoparticles as a magnetic resonance imaging (MRI) contrast agent for the detection of brain gliomas *in vivo.* The nanoparticles could be bound to brain glioma with high selectivity and sensitivity, and showed a greater accumulation in glioma [30]. Angiopep-2 is a peptide derived from the common peptide sequence of the LRP protein ligands, and demonstrates a much higher BBB transcytosis efficacy than transferrin and its parent molecule aprotinin. Thus, PAMAM was modified with angiopep-2 to deliver tumor necrosis factor-related apoptosis-inducing ligand (TRAIL) into glioma for gene therapy. The biodistribution of these nanoparticles showed that the uptake of angiopep-2-modified nanoparticles into the brain – especially at tumor sites – was more conspicuous than that of other, unmodified nanoparticles [31].

EGFRs belong to the ErbB family of receptor tyrosine kinases that are normally expressed in epithelial, mesenchymal, and neuronal tissues [32, 33]. In addition to the prevalence of EGFRs in malignant gliomas, therapeutic approaches that target EGFRs have been proposed. Liao *et al.* conjugated a series of EGFR antibodies to the microbubble for glioma treatment, and the microbubble was further combined with pulsed-mode ultrasound for temporarily enhanced BBB permeability [34].

Pulsed-mode ultrasound treatment prolonged the circulatory half-life of targeting microbubbles and enhanced the anti-tumor effect of EGFR antibodies. Chong *et al.* found that TMZ-resistant glioma was susceptible to the combined treatment of nimotuzumab (a EGFR monoclonal antibody) and rapamycin [35]. Likewise, EGFRs can be targeted for enhancing MRI. Mu *et al.* synthesized EGFR monoclonal antibody-conjugated SPIO nanoparticles for glioma imaging [36].

Apart from endogenous ligands, various infectious agents are able to cross the BBB, including prions and certain neurotropic viruses [37]. Liu *et al.* utilized a bacteria-derived peptide with the sequence of EPRNEEK which binds to laminin receptors for dual-targeting glioma gene delivery [38].

Receptor-mediated transcytosis possesses the characteristics of high specificity, selectivity, and affinity. Nonetheless, homeostasis and endogenous ligands may compete with the targeting ligands to reduce the targeting efficiency. Hence, the uptake of nanoparticles by the BBB and cancer cells can be restricted to a certain degree. Fortunately, receptors are not the only ways to be used in designing targeting nanoparticles.

3.1.2 Transporter-Mediated Targeting

Since numerous transporters are strongly expressed on both brain capillary endothelial cells and brain tumor cells, aiding the transportation of nutriments into the brain [8], nanoparticles can take advantage of these transport systems to overcome the obstacle of the BBB and deliver therapeutic agents into cancer cells. These transporters include glucose transporter isoform 1 (GLUT1) [39, 40], L-type amino acid transporter 1 LAT1 [41, 42], and choline transporters [43]. Compared to the receptor-mediated pathway, the transporter-mediated pathway is faster, with a higher efficiency, and is also less impacted by endogenous substrates [44, 45].

Facilitative GLUT1 is a representative member of the GLUT family. This transporter is widely distributed in normal tissues, such as the BBB, and over-expressed on brain tumor cells [46]. 2-Deoxy-D-glucose, galactose, mannose, D-glucose, and dehydroascorbic acid (DHA) are all endogenous substrates of GLUT1. Jiang *et al.* modified nanoparticles with 2-deoxy-D-glucose to deliver paclitaxel (PTX) into glioma [47]. However, the bidirectional transport of GLUT1 may reverse-pump the nanoparticles out of the brain before they can release their cargoes. Li *et al.* designed and synthesized a new ligand, L-TDS-G, to give the nanoparticles a ‘lock-in’ ability [48]. The thiamine disulfide (TDS) unit of L-TDS-G could be reduced by a relative reductase and then ring-closed to form a thiazolium, after which the lipophobic thiazolium could be locked into the BBB. DHA is also reported to be a ‘one-direction’ substrate of GLUT1. Once transported into the cell, DHA is rapidly reduced to ascorbate, being trapped within the cell and causing a continuous accumulation of DHA in the brain [49]. Meanwhile, owing to the virtual absence of sodium ascorbate co-transporters, which mainly transport vitamin C, DHA is the primary substitute of vitamin C for tumor cells [50]. DHA was introduced onto the surface of the micelles to achieve a 2.84-fold higher targeting efficiency [51].

The central nervous system (CNS) requires choline for the synthesis of membrane phospholipids and acetylcholine. Because the brain cannot synthesize choline *de novo*, choline uptake from extracellular fluids by choline transporters at the BBB is essential [52]. Due to the high malignancy of glioma, choline transporters are over-expressed on glioma cells. Thus, a novel choline derivate (CD) was designed and synthesized as a BBB and glioma dual-targeting ligand for doxorubicin delivery. By modifying the CD on the nanoparticles, drug accumulation was significantly enhanced in tumors, showing a 2.37-fold higher drug concentration in tumors than for unmodified nanoparticles [53]. A brain-targeting gene delivery nanoparticle modified with the CD was also developed.

3.1.3 Adsorptive-Mediated Transcytosis

At systemic physiological pH, the luminal surface and abluminal surface of cerebral endothelial cells (ECs) and the basement membrane contiguous to the basolateral side of the cereal ECs present an overall negative charge. All three of these consecutive electrostatic barriers provide both the potential for the binding of cationic molecules to the luminal surface of ECs, and subsequently for exocytosis at the abluminal surface [54].

Cationic proteins and cell-penetrating peptides (CPPs) are normally utilized for brain drug delivery. When nanoparticles modified with cationic proteins had crossed the BBB, they further accumulated in tumor cells due to the higher expression levels of negatively charged glycoproteins on brain tumor cells. Lu *et al.* reported the synthesis of cationic albumin-conjugated PEGylated nanoparticles (CBSA-NPs) and further incorporated plasmid pORF-hTRAIL (pDNA) into CBSA-NP as a nonverbal vector for the gene therapy of gliomas. CBSA-NP-hTRAIL was found to be far more extensively distributed in tumors than NP-hTRAIL [55].

Rao *et al.* conjugated TAT peptide to the nanoparticles for delivery to the CNS of an anti-virus agent [56]. Although the TAT peptide helped the nanoparticles to cross the BBB, it showed no trait for selective tumor cell delivery.

Whilst cationic proteins and CPPs can enhance the permeability of nanoparticles through the BBB, various limitations and pitfalls restrict their utilization in drug-delivery systems. One limitation in using cationic proteins and CPPs is their low selectivity, which is a disadvantage for the specific targeting of brain tumor tissues. The toxicity and immunogenicity of these materials also restrict their use in drug-delivery systems.

3.2 The Two-Order Targeting Strategies

In two-order targeting strategies, two types of ligand are modified on the nanoparticles. One ligand with binding affinity to the receptors or transporters expressed on the brain capillary endothelial cells could facilitate transcytosis across the BBB, while a second ligand with high affinity to the receptors over-expressed at brain cancers could enhance accumulation into cancer cells.

Lactoferrin (Lf) and folic acid were used by Kuo and Chen to crosslink on poly(lactide-*co*-glycolide) (PLGA) nanoparticles for treating human brain malignant glioblastoma [57]. Since folic acid is crucial to the biosynthesis of nucleic acids and amino acids that are necessary for cell proliferation, the folate receptor has shown to be over-expressed in a wide variety of human tumors, and folate receptor density appears to increase as the stage or grade of the cancer worsens [58]. Nanoparticles modified with folic acid have been widely developed to enhance targeting efficiency. Gao *et al.* built a doxorubicin (Dox) liposome and modified the surface with both folic acid and Tf to improve the transport of doxorubicin across the BBB, and also entry into the brain glioma. The results showed a significantly improved transportation of Dox across the BBB and its accumulation in the brain glioma [59].

$\alpha_V\beta_3$ integrin as a receptor for extracellular matrix proteins is over-expressed on the activated endothelial cells of tumor neovasculatures, brain tumor cells, but not on the normal vasculatures [60]. Therefore, $\alpha_V\beta_3$ integrin could be used as a target for tumor diagnosis and therapy. The cyclic arginine-glycine-aspartic (RGD) tripeptide shows high binding affinity to $\alpha_V\beta_3$ integrin, which is a potential ligand for targeting tumor cells [61]. Yan *et al.* designed nanoparticles that had been modified with both angiopep-2 and c[RGDyK] for brain tumor imaging [14], and showed improved imaging results compared to nanoparticles modified with only angiopep-2 or c[RGDyK].

Yue *et al.* combined TfR monoclonal antibody OX26 with the substrate of matrix metalloproteinase-2 (MMP-2) and chlorotoxin (CTX) to accomplish gene delivery in glioma treatment [62]. MMPs play a key role in the invasion process of cancers, and their expression levels correlate with an histological grade of malignancy [63]. CTX is a peptide that originates from *Leiurus quinquestriatus* venom and specifically binds to MMP-2, the levels of which are increased in gliomas [64].

4 Nanoparticles with Smart Release Strategies

In brain cancer therapy, the release of a therapeutic agent is equally important as its targeting to the brain. The ideal release process should be efficient, and spatially-temporally controlled, and the released cargoes should retain their original effects. In some cases, only a very slight change in the structure of a chemotherapeutic agent or a gene could cause the loss of any therapeutic effect.

An abnormal metabolism in cancer cells can result in a higher level of reactive oxygen species (ROS) -related stress, which is considered fatal to all cells. Thus, accordingly a higher glutathione (GSH) concentration typically arises in cancer cells as a self-defense strategy. Similarly, the 'Warburg effect' suggests that there is a lower pH environment in cancer tissues. At this point, details of how smart release strategies are achieved, based on these abnormal microenvironments, and the concept of external stimuli-responsive strategies will be introduced.

4.1 GSH-Sensitive Release

The GSH pool has been shown to maintain cellular redox homeostasis. According to previous reports, in tumor tissues there exists a GSH concentration gradient between extracellular conditions and intracellular conditions. Several intracellular compartments such as the cytosol, mitochondria and cell nucleus maintain a high concentration of GSH (about 2–10 mM), which is 100- to 1000-fold higher than that in the extracellular fluids and circulation (about 2–20 μM) [65]. Thus, GSH has been recognized as an ideal internal stimulus for the rapid destabilization of nanoparticles in cells to accomplish an efficient intracellular drug release [66]. It should be noted, however, that both cancer cells and normal cells have a high intracellular GSH concentration, and that of cancer cells is only slightly higher [67]. Hence, cancer-targeting strategies are the premise of employing GSH-sensitive release.

Stephen *et al.* used O^6-benzylguanine (BG)-linked chitosan to develop a GSH-sensitive-based nanoparticle [68]. In GBM

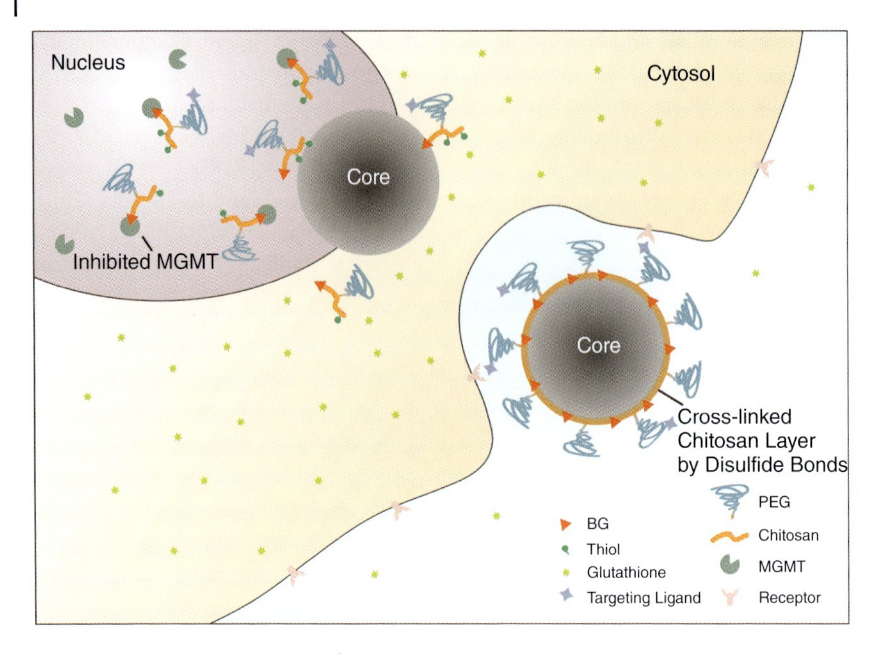

Fig. 2 GSH-sensitive release of O^6-benzylguanine (BG) in cancer cells. When the nanoparticles are exposed to a high concentration of GSH in the cyotosol, crosslinked chitosan falls off the core and the drugs are free to inhibit MGMT.

chemotherapy, upregulation of the DNA repair protein O^6-methylguanine-DNA methyltransferase (MGMT) caused the GBM to be resistant to TMZ-based regimens [69]. Although GBM resistance to TMZ can be overcome by inhibiting MGMT with DNA repair inhibitors such as BG, the poor permeation of BG, as well as other shortcomings in term of pharmacokinetics, remain to be improved [70]. In these studies, chitosan was modified with several thiols and further crosslinked by disulfide bonds in order to coat the iron oxide core of the nanoparticles, while BG was further linked to the chitosan layer. Following uptake of the nanoparticles, the high concentration of GSH in cancer cells led to cleavage of the disulfide bonds of the crosslinked chitosan, after which BG-linked chitosan was free to inhibit the MGMT with BG ligand, thus overcoming the resistance of TMZ-based chemotherapy (Fig. 2). The nanoparticles were released rapidly under conditions which mimicked the intracellular environment, and achieved maximum drug release at 1 h. However, under conditions which mimicked blood, maximum drug release was not achieved until 24 h. Moreover, a prominent prolongation of median survival indicated that the GSH-sensitive strategy was a viable approach for smart drug release.

The two arms of lipooligomer 49 were terminated with cysteines which could stabilize siRNA complexes by disulfide crosslinks. Then intracellular glutathione then reduced the disulfide bonds such that small interfering RNA) (siRNA) could be released, resulting in an outstanding gene down-regulation in glioma [71]. Shao *et al.* used 3,3′-dithiobis(sulfosuccinimidyl propionate) (DTSSP) to anchor micelle materials together, using disulfide bonds, endowing the micelles with a higher stability in an extracellular environment [51]. The aim for both of these nanoparticles was to increase nanoparticle stability in the blood circulation while maintaining the capability of releasing their cargoes in target cells.

4.2 pH-Sensitive Release

According to the Warburg effect, compared to normal tissues, tumor cells can take up 12-fold glucose and have an increased aerobic glycolysis. However, they have the tendency to limit themselves to glycolysis, even when provided with adequate oxygen, thus generating more lactate [72]. Following glycolysis, lactic acid is exported from cells, mediating an increased amount of protons in the tumor microenvironment [73]. Besides lactic acid, large amounts of CO_2, which is produced by tumor cells via oxidative metabolism, is another major source of tumor extracellular acidity [74]. Unfortunately, due mainly to poor lymphatic drainage and the elevated interstitial pressure of tumors, metabolic acids exported from cancer cells cannot be exported to the blood rapidly [75]. This results in an increased proton production and poor proton clearance, such that the extracellular pH may have a low-value range of approximately 6–7 [76, 77].

Other than a mild acidic extracellular pH in cancers, an even greater pH decrease can be found in intracellular compartments compared to normal tissues (pH 7.4), and this reduction can also be used for the controlled release of drugs. The pH values of the early endosome and late lysosome are about pH 5–6 and pH 4–5, respectively, which is much more acidic than normal tissues [78]. A more advanced approach to efficient cytoplasmic drug release to cancer cells is to use drug vehicles that have lysosome pH-triggered drug-release properties. Acid-cleavable chemical bonds and lysosome pH-sensitive materials to release drugs have been used extensively in intracellular pH-triggered drug-delivery systems for tumor therapy [77].

It must be noted that the low pH environment in intracellular compartments is not tumor-specific. The pH values of endosomes and lysosomes are also acidic in normal cells; that is to say, intracellular pH-sensitive strategies are also based on the cancer-targeting strategies.

Cheng *et al.* utilized an acid-labile hydrazine bond to load doxorubicin onto the surface of TAT-Au nanoparticles [79]. The ketone group of doxorubicin was conjugated to the hydrazine-modified TAT-Au nanoparticles to form the pH-sensitive bond. With the help of acidic extracellular pH in tumor tissues and acidic intracellular pH in specific compartments, the hydrazine bonds could be cleaved and doxorubicin released in its active form. Zhao *et al.* selected a pH-responsive peptide H7K(R2)2 as a targeting ligand to modify the liposome, which could respond to the acidic pH environment in gliomas, possessing cell-penetrating characteristics. The pH-sensitive liposomes could also respond to the acidic pH environment in gliomas and release encapsulated drugs [80]. Xu *et al.* synthesized two new amphiphilic pH-sensitive carriers which can form stable nanoparticles with plasmid DNA and siRNA for therapeutic gene delivery [81]. Qiao *et al.* also synthesized a novel copolymer exhibiting pH-sensitive properties to release hydrophobic drugs from the nanoparticles upon a change in pH [82]. At a low pH, protons combined with the tertiary amine of the di(ethylene glycol)diacrylate (DEPA) units in the copolymer and impelled them to be ionized, leading to a dissociation of the nanoparticles and corresponding drug release.

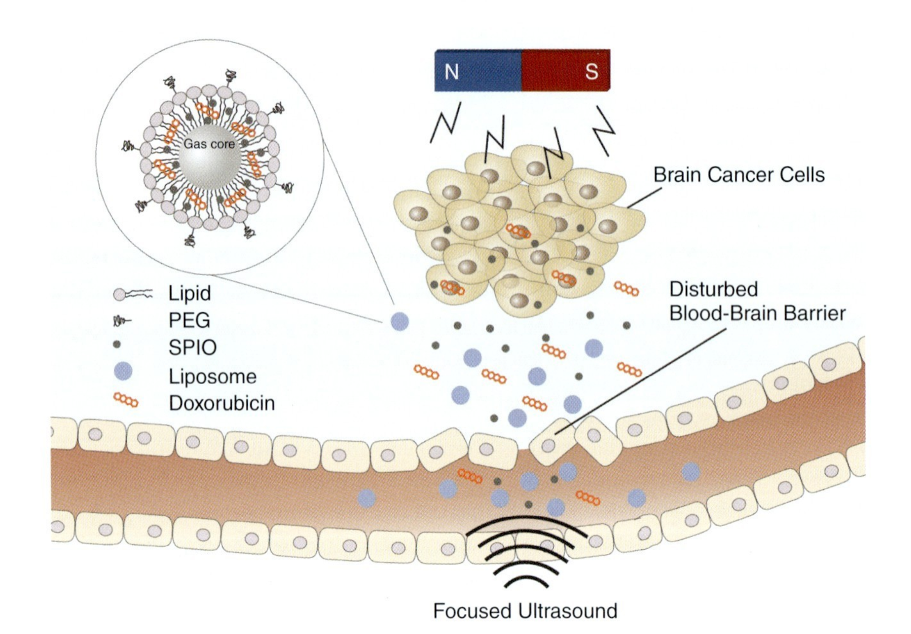

Fig. 3 The release of doxorubicin (Dox) and superparamagnetic iron oxide (SPIO) with the help of focused ultrasound (FUS) and an external magnetic field. The liposome with a gas core serves as the vehicle and microbubbles. FUS disturbs both the liposome and the BBB in the presence of microbubbles. The external magnetic field enhances the accumulation of SPIO in cancer cells to improve the MRI imaging.

4.3 External Stimuli-Sensitive Release

As both GSH-sensitive and pH-sensitive release strategies are less specific, external stimuli are used to realize a more spatially specific drug release in cancer cells.

Lin *et al.* developed a cationic liposome nanoparticle, DOX-CLs, performing drug release upon exposure to focused ultrasound (FUS) [83]. The utilization of FUS was aimed to release the loaded doxorubicin and also to open the BBB temporarily in the local presence of lipid- or polymer-shelled microbubbles [84–87]. Fan *et al.* combined microbubbles with liposomes to deliver SPIO and doxorubicin for both brain cancer chemotherapy and MRI imaging (Fig. 3) [88].

Therapeutic approaches employing light include photodynamic therapy (PDT) and photothermal therapy (PTT). In PDT, light can be used to activate photosensitizers to generate ROS, whereas in PTT the light can trigger a photothermal effect to achieve tissue ablation [89, 90]. Metal nanoparticles such as gold nanostructures [91] can absorb light, serving as photothermal agents to improve the efficiency of near-infrared (NIR) penetration depth. Visible light excitation of metal nanoparticles also induces singlet O_2 formation and PDT [92].

Febvay *et al.* presented a light-triggered mesoporous silica nanoparticle, whereby the fluorescent dye Alexa546 was delivered as a model drug [93]. Following endocytosis of the nanoparticles the cargo was released into the endosome. Exposure to light at the dye's excitation wavelength (546 nm) then promoted ROS-mediated membrane damage of the endosome and promoted drug release into the cytosol.

Glioma contains numerous tumor-associated macrophages which are reported to have the capability of bypassing the BBB, and have been used for brain cancer therapy as nanoparticle vehicles [94, 95]. Madsen *et al.* innovatively used macrophages to deliver gold nanoparticles into glioma for PTT [96]. Cell-based vehicles could maintain an elevated concentration of nanoparticles at the tumor site and prevent them from spreading into healthy tissues.

5 Conclusions

Dual-targeting and two-order targeting strategies are widely used when designing active targeting nanoparticles for relieving long-standing issues in brain cancer treatments. When modified with ligands that bind specifically to highly expressed receptors and transporters at the BBB and cancer cells, nanoparticles have the ability to accumulate in both brain and cancer cells. Depending on the external stimuli or different microenvironmental factors between cancer tissues and normal tissues, such as GSH and pH, therapeutic agents are thus enabled for release sufficiently within cancer cells rather than leaking into the blood circulation or accumulating in normal brain tissues. Together, active-targeting and stimuli-responding constitute smart nanoparticles in brain cancer chemotherapy and gene therapy, and have shown promising therapeutic effects. Clinical data acquired from smart nanoparticles should also be evaluated in the near future.

6 Perspective

Apart from brain cancer therapy, smart nanoparticles have also been applied for curing other diseases and, indeed, some may provide inspiration in designing novel strategies for brain cancer therapy.

As LAT1 are overexpressed at both the BBB and in some brain tumor cells [97], the LAT1 substrate may serve as a dual-targeting ligand in active-targeting nanoparticles. Vyas *et al.* modified an amino acid micronutrient to nanoparticles that were delivered with a non-nucleoside reverse transcriptase inhibitor into brain for the treatment of AIDS-related encephalopathy. Wu *et al.* found that an aspartate-modified doxorubicin was accumulated to a greater extent in LAT1-overexpressing tumors, while the treatment of tumor-bearing mice with the same agent demonstrated a significant inhibition of tumor growth [98].

ROS are chemically reactive molecules containing oxygen which, in a biological context, are formed as natural byproducts of the normal metabolism of oxygen and have important roles in cell signaling and homeostasis [99]. Cancer cells exhibit a greater ROS stress level than normal cells, due partly to the oncogenic stimulation, increased metabolic activity, and mitochondrial malfunction [100]. More importantly, unlike the GSH environment, high ROS stress is specific to cancer cells. For instance, Wang *et al.* established a nanoparticle with both GSH-sensitive and ROS-sensitive release mechanisms for cancer chemotherapy [101].

High levels of cathepsin B are found in a wide range of human cancers. Cathepsin B is a lysosomal cysteine protease which degrades extracellular matrix components during the invasion and metastasis of cancer cells [102, 103]. Lee *et al.* conjugated doxorubicin and nanoparticles with a cathepsin B-cleavable peptide for enzyme-responsive release in colorectal cancer cells [104]. Similarly, Tian *et al.* conjugated a photosensitizer through a cathepsin B-cleavable peptide to a nanoprobe for the *in-situ* therapeutic monitoring of photosensitive cell death. Following uptake into cancer cells, the peptide can be cleaved to release drug cargoes in the presence of cancer-associated cathepsin B.

Taken together, these findings suggest that there still exists a variety of unexploited strategies for designing smart nanoparticles in brain cancer therapy. Many of these are used to treat other types of cancers, or are still at the basic research stage, and this inspires the application of such novel strategies in brain cancer therapy. A great matter of priority, however, is that smart nanoparticles should be investigated in clinical trials in the future, rather than designing new smart nanoparticles in endless fashion.

References

1. Siegel, R.L., Miller, K.D., Jemal, A. (2015) Cancer statistics, 2015. *CA Cancer J. Clin.*, **65**, 5–29.
2. Ostrom, Q.T., Gittleman, H., Liao, P., Rouse, C., Chen, Y., Dowling, J., Wolinsky, Y., Kruchko, C., Barnholtz-Sloan, J. (2014) CBTRUS statistical report: primary brain and central nervous system tumors diagnosed in the United States in 2007–2011. *NeuroOncology*, **16**, iv1–iv63.
3. Pérez-Larraya, J.G., Hildebrand, J. (2014) Brain metastases. *Handb. Clin. Neurol.*, **121**, 1143–1157.
4. Parrish, K.E., Sarkaria, J.N., Elmquist, W.F. (2015) Improving drug delivery to primary and metastatic brain tumors: strategies to overcome the blood–brain barrier. *Clin. Pharmacol. Ther.*, **97**, 336–346.
5. Bredel, M., Zentner, J. (2002) Brain-tumour drug resistance: the bare essentials. *Lancet Oncol.*, **3**, 397–406.
6. Heldin, C.-H., Rubin, K., Pietras, K., Östman, A. (2004) High interstitial fluid pressure – an obstacle in cancer therapy. *Nat. Rev. Cancer*, **4**, 806–813.
7. Danhier, F., Feron, O., Préat, V. (2010) To exploit the tumor microenvironment: passive and active tumor targeting of nanocarriers for anti-cancer drug delivery. *J. Controlled Release*, **148**, 135–146.
8. Chen, Y., Liu, L. (2012) Modern methods for delivery of drugs across the blood–brain barrier. *Adv. Drug Delivery Rev.*, **64**, 640–665.
9. Uchida, Y., Ohtsuki, S., Katsukura, Y., Ikeda, C., *et al.* (2011) Quantitative targeted absolute proteomics of human blood–brain barrier transporters and receptors, *J. Neurochem.*, **117** 333–345.
10. Su, W., Pasternak, G.W. (2013) The role of multidrug resistance-associated protein in the blood–brain barrier and opioid analgesia. *Synapse*, **67**, 609–619.
11. Wohlfart, S., Gelperina, S., Kreuter, J. (2012) Transport of drugs across the blood–brain barrier by nanoparticles. *J. Controlled Release*, **161**, 264–273.
12. Hynes, N.E., MacDonald, G. (2009) ErbB receptors and signaling pathways in cancer. *Curr. Opin. Cell Biol.*, **21**, 177–184.
13. Ying, X., Wen, H., Lu, W.-L., Du, J., *et al.* (2010) Dual-targeting daunorubicin liposomes improve the therapeutic efficacy of brain glioma in animals. *J. Controlled Release*, **141**, 183–192.
14. Yan, H., Wang, L., Wang, J., Weng, X., *et al.* (2011) Two-order targeted brain tumor imaging by using an optical/paramagnetic nanoprobe across the blood brain barrier. *ACS Nano*, **6**, 410–420.
15. Ji, B., Maeda, J., Higuchi, M., Inoue, K., *et al.* (2006) Pharmacokinetics and brain uptake of lactoferrin in rats. *Life Sci.*, **78**, 851–855.
16. Zhang, P., Hu, L., Yin, Q., Zhang, Z., *et al.* (2012) Transferrin-conjugated polyphosphoester hybrid micelle loading paclitaxel for

brain-targeting delivery: synthesis, preparation and in vivo evaluation. *J. Controlled Release*, **159**, 429–434.

17. Gao, X., Qian, J., Zheng, S., Xiong, Y., *et al.* (2013) Up-regulating blood brain barrier permeability of nanoparticles via multivalent effect. *Pharm. Res.*, **30**, 2538–2548.
18. Hatanpaa, K.J., Burma, S., Zhao, D., Habib, A.A. (2010) Epidermal growth factor receptor in glioma: signal transduction, neuropathology, imaging, and radioresistance. *Neoplasia*, **12**, 675–684.
19. Janecka, A., Fichna, J., Janecki, T. (2004) Opioid receptors and their ligands. *Curr. Top. Med. Chem.*, **4**, 1–17.
20. Riolfi, M., Ferla, R., Valle, L.D., Piña-Oviedo, S., *et al.* (2010) Leptin and its receptor are overexpressed in brain tumors and correlate with the degree of malignancy. *Brain Pathol.*, **20**, 481–489.
21. Gao, S., Li, J., Jiang, C., Hong, B., *et al.* (2016) Plasmid pORF-hTRAIL targeting to glioma using transferrin-modified polyamidoamine dendrimer. *Drug Des. Dev. Ther.*, **10**, 1–11.
22. Zheng, C., Ma, C., Bai, E., Yang, K., *et al.* (2015) Transferrin and cell-penetrating peptide dual-functioned liposome for targeted drug delivery to glioma. *Int. J. Clin. Exp. Med.*, **8**, 1658–1668.
23. Kang, T., Jiang, M., Jiang, D., Feng, X., *et al.* (2015) Enhancing glioblastoma-specific penetration by functionalization of nanoparticles with an iron-mimic peptide targeting transferrin/transferrin receptor complex. *Mol. Pharm.*, **12**, 2947–2961.
24. Ulbrich, K., Hekmatara, T., Herbert, E., Kreuter, J. (2009) Transferrin-and transferrin-receptor-antibody-modified nanoparticles enable drug delivery across the blood–brain barrier (BBB). *Eur. J. Pharm. Biopharm.*, **71**, 251–256.
25. Lee, J.H., Engler, J.A., Collawn, J.F., Moore, B.A. (2001) Receptor mediated uptake of peptides that bind the human transferrin receptor. *Eur. J. Biochem.*, **268**, 2004–2012.
26. Kuang, Y., An, S., Guo, Y., Huang, S., *et al.* (2013) T7 peptide-functionalized nanoparticles utilizing RNA interference for glioma dual targeting. *Int. J. Pharm.*, **454**, 11–20.
27. Shibata, M., Yamada, S., Kumar, S.R., Calero, M., *et al.* (2000) Clearance of Alzheimer's amyloid-β1-40 peptide from brain by LDL receptor-related protein-1 at the blood–brain barrier. *J. Clin. Invest.*, **106**, 1489–1499.
28. Miao, D., Jiang, M., Liu, Z., Gu, G., *et al.* (2013) Co-administration of dual-targeting nanoparticles with penetration enhancement peptide for antiglioblastoma therapy. *Mol. Pharm.*, **11**, 90–101.
29. Fang, J.H., Lai, Y.H., Chiu, T.L., Chen, Y.Y., *et al.* (2014) Magnetic core–shell nanocapsules with dual-targeting capabilities and co-delivery of multiple drugs to treat brain gliomas. *Adv. Healthcare Mater.*, **3**, 1250–1260.
30. Xie, H., Zhu, Y., Jiang, W., Zhou, Q., *et al.* (2011) Lactoferrin-conjugated superparamagnetic iron oxide nanoparticles as a specific MRI contrast agent for detection of brain glioma in vivo. *Biomaterials*, **32**, 495–502.
31. Huang, S., Li, J., Han, L., Liu, S., *et al.* (2011) Dual targeting effect of angiopep-2-modified, *DNA-loaded nanoparticles for glioma. Biomaterials*, **32**, 6832–6838.
32. Bublil, E.M., Yarden, Y. (2007) The EGF receptor family: spearheading a merger of signaling and therapeutics. *Curr. Opin. Cell Biol.*, **19**, 124–134.
33. Ohgaki, H., Kleihues, P. (2007) Genetic pathways to primary and secondary glioblastoma. *Am. J. Pathol.*, **170**, 1445–1453.
34. Liao, A.-H., Chou, H.-Y., Hsieh, Y.-L., Hsu, S.-C., *et al.* (2015) Enhanced therapeutic epidermal growth factor receptor (EGFR) antibody delivery via pulsed ultrasound with targeting microbubbles for glioma treatment. *J. Med. Biol. Eng.*, **35**, 156–164.
35. Chong, D.Q., Toh, X.Y., Ho, I.A., Sia, K.C., *et al.* (2015) Combined treatment of Nimotuzumab and rapamycin is effective against temozolomide-resistant human gliomas regardless of the EGFR mutation status. *BMC Cancer*, **15**, 255.
36. Mu, K., Zhang, S., Ai, T., Jiang, J., *et al.* (2015) Monoclonal antibody-conjugated superparamagnetic iron oxide nanoparticles for imaging of epidermal growth factor receptor-targeted cells and gliomas. *Mol. Imaging*, **14**, 1–12.
37. Mbazima, V., Da Costa Dias, B., Omar, A., Jovanovic, K., *et al.* (2010) Interactions between PrP (c) and other ligands with the 37-kDa/67-kDa laminin receptor. *Front. Biosci.*, **15**, 3667.

38. Liu, Y., He, X., Kuang, Y., An, S., *et al.* (2014) A bacteria deriving peptide modified dendrigraft poly-l-lysines (DGL) self-assembling nanoplatform for targeted gene delivery. *Mol. Pharm.*, **11**, 3330–3341.
39. Nishioka, T., Oda, Y., Seino, Y., Yamamoto, T., *et al.* (1992) Distribution of the glucose transporters in human brain tumors. *Cancer Res.*, **52**, 3972–3979.
40. Boado, R.J., Black, K.L., Pardridge, W.M. (1994) Gene expression of GLUT3 and GLUT1 glucose transporters in human brain tumors. *Mol. Brain. Res.*, **27**, 51–57.
41. Yanagida, O., Kanai, Y., Chairoungdua, A., Kim, D.K., *et al.* (2001) Human L-type amino acid transporter 1 (LAT1): characterization of function and expression in tumor cell lines. *Biochim. Biophys. Acta Biomembr.*, **1514**, 291–302.
42. Chen, J., Liang, L., Liu, Y., Zhang, L. (2015) Development of a cell-based screening method for compounds that inhibit or are transported by large neutral amino acid transporter 1, a key transporter at the blood–brain barrier. *Anal. Biochem.*, **486**, 81–85.
43. Li, J., Guo, Y., Kuang, Y., An, S., *et al.* (2013) Choline transporter-targeting and co-delivery system for glioma therapy. *Biomaterials*, **34**, 9142–9148.
44. De Boer, A., Van Der Sandt, I., Gaillard, P. (2003) The role of drug transporters at the blood–brain barrier. *Annu. Rev. Pharmacol. Toxicol.*, **43**, 629–656.
45. Pardridge, W.M. (2007) Blood–brain barrier delivery. *Drug Discovery Today*, **12**, 54–61.
46. Wood, I.S., Trayhurn, P. (2003) Glucose transporters (GLUT and SGLT): expanded families of sugar transport proteins. *Br. J. Nutr.*, **89**, 3–9.
47. Jiang, X., Xin, H., Ren, Q., Gu, J., *et al.* (2014) Nanoparticles of 2-deoxy-D-glucose functionalized poly (ethylene glycol)-*co*-poly (trimethylene carbonate) for dual-targeted drug delivery in glioma treatment. *Biomaterials*, **35**, 518–529.
48. Li, X., Qu, B., Jin, X., Hai, L., *et al.* (2014) Design, synthesis and biological evaluation for docetaxel-loaded brain targeting liposome with "lock-in" function. *J. Drug Targeting*, **22**, 251–261.
49. Agus, D.B., Gambhir, S.S., Pardridge, W.M., Spielholz, C., *et al.* (1997) Vitamin C crosses the blood–brain barrier in the oxidized form through the glucose transporters. *J. Clin. Invest.*, **100**, 2842–2848.
50. Airley, R.E., Mobasheri, A. (2007) Hypoxic regulation of glucose transport, anaerobic metabolism and angiogenesis in cancer: novel pathways and targets for anticancer therapeutics. *Chemotherapy*, **53**, 233–256.
51. Shao, K., Ding, N., Huang, S., Ren, S., *et al.* (2014) Smart nanodevice combined tumor-specific vector with cellular microenvironment-triggered property for highly effective antiglioma therapy. *ACS Nano*, **8**, 1191–1203.
52. Michel, V., Yuan, Z., Ramsubir, S., Bakovic, M. (2006) Choline transport for phospholipid synthesis. *Exp. Biol. Med.*, **231**, 490–504.
53. Li, J., Yang, H., Zhang, Y., Jiang, X., *et al.* (2015) Choline derivate-modified doxorubicin-loaded micelle for glioma therapy. *ACS Appl. Mater. Interfaces*, **7**, 21589–21601.
54. Vorbrodt, A. (1989) Ultracytochemical characterization of anionic sites in the wall of brain capillaries. *J. Neurocytol.*, **18**, 359–368.
55. Lu, W., Sun, Q., Wan, J., She, Z., *et al.* (2006) Cationic albumin-conjugated pegylated nanoparticles allow gene delivery into brain tumors via intravenous administration. *Cancer Res.*, **66**, 11878–11887.
56. Rao, K.S., Reddy, M.K., Horning, J.L., Labhasetwar, V. (2008) TAT-conjugated nanoparticles for the CNS delivery of anti-HIV drugs. *Biomaterials*, **29**, 4429–4438.
57. Kuo, Y.-C., Chen, Y.-C. (2015) Targeting delivery of etoposide to inhibit the growth of human glioblastoma multiforme using lactoferrin-and folic acid-grafted poly (lactide-*co*-glycolide) nanoparticles. *Int. J. Pharm.*, **479**, 138–149.
58. Parker, N., Turk, M.J., Westrick, E., Lewis, J.D., *et al.* (2005) Folate receptor expression in carcinomas and normal tissues determined by a quantitative radioligand binding assay. *Anal. Biochem.*, **338**, 284–293.
59. Gao, J.-Q., Lv, Q., Li, L.-M., Tang, X.-J., *et al.* (2013) Glioma targeting and blood–brain barrier penetration by dual-targeting doxorubincin liposomes. *Biomaterials*, **34**, 5628–5639.
60. Brooks, P.C., Clark, R., Cheresh, D.A. (1994) Requirement of vascular integrin alpha

v beta 3 for angiogenesis. *Science*, **264**, 569–571.

61. Schottelius, M., Laufer, B., Kessler, H., Wester, H.-J. (2009) Ligands for mapping $\alpha_v\beta_3$-integrin expression in vivo. *Acc. Chem. Res.*, **42**, 969–980.
62. Yue, P.-J., He, L., Qiu, S.-W., Li, Y., *et al.* (2014) OX26/CTX-conjugated PEGylated liposome as a dual-targeting gene delivery system for brain glioma. *Mol. Cancer*, **13**, 1–13.
63. McCawley, L.J., Matrisian, L.M. (2000) Matrix metalloproteinases: multifunctional contributors to tumor progression. *Mol. Med. Today*, **6**, 149–156.
64. Deshane, J., Garner, C.C., Sontheimer, H. (2003) Chlorotoxin inhibits glioma cell invasion via matrix metalloproteinase-2. *J. Biol. Chem.*, **278**, 4135–4144.
65. Schafer, F.Q., Buettner, G.R. (2001) Redox environment of the cell as viewed through the redox state of the glutathione disulfide/glutathione couple. *Free Radical Biol. Med.*, **30**, 1191–1212.
66. Cheng, R., Feng, F., Meng, F., Deng, C., *et al.* (2011) Glutathione-responsive nano-vehicles as a promising platform for targeted intracellular drug and gene delivery. *J. Controlled Release*, **152**, 2–12.
67. Lee, F., Vessey, A., Rofstad, E., Siemann, D., *et al.* (1989) Heterogeneity of glutathione content in human ovarian cancer. *Cancer Res.*, **49**, 5244–5248.
68. Stephen, Z.R., Kievit, F.M., Veiseh, O., Chiarelli, P.A., *et al.* (2014) Redox-responsive magnetic nanoparticle for targeted convection-enhanced delivery of O^6-benzylguanine to brain tumors. *ACS Nano*, **8**, 10383–10395.
69. Bobola, M.S., Tseng, S.H., Blank, A., Berger, M.S., *et al.* (1996) Role of O^6-methylguanine-DNA methyltransferase in resistance of human brain tumor cell lines to the clinically relevant methylating agents temozolomide and streptozotocin. *Clin. Cancer Res.*, **2**, 735–741.
70. Quinn, J.A., Jiang, S.X., Reardon, D.A., Desjardins, A., *et al.* (2009) Phase II trial of temozolomide plus O^6-benzylguanine in adults with recurrent, temozolomide-resistant malignant glioma. *J. Clin. Oncol.*, **27**, 1262–1267.
71. An, S., He, D., Wagner, E., Jiang, C. (2015) Peptide-like polymers exerting effective glioma-targeted siRNA delivery and release for therapeutic application. *Small*, **11**, 5142–5150.
72. Weinberg, R. (2013) *The Biology of Cancer*, Garland Science.
73. Stubbs, M., McSheehy, P.M., Griffiths, J.R., Bashford, C.L. (2000) Causes and consequences of tumour acidity and implications for treatment. *Mol. Med. Today*, **6**, 15–19.
74. Helmlinger, G., Sckell, A., Dellian, M., Forbes, N.S., *et al.* (2002) Acid production in glycolysis-impaired tumors provides new insights into tumor metabolism. *Clin. Cancer Res.*, **8**, 1284–1291.
75. Vaupel, P. (2004) Tumor microenvironmental physiology and its implications for radiation oncology. *Semin. Radiat. Oncol.*, **14**(3), 198–206.
76. Engin, K., Leeper, D., Cater, J., Thistlethwaite, A., *et al.* (1995) Extracellular pH distribution in human tumours. *Int. J. Hyperthermia*, **11**, 211–216.
77. He, X., Li, J., An, S., Jiang, C. (2013) pH-sensitive drug-delivery systems for tumor targeting. *Ther. Delivery*, **4**, 1499–1510.
78. Fleige, E., Quadir, M.A., Haag, R. (2012) Stimuli-responsive polymeric nanocarriers for the controlled transport of active compounds: concepts and applications. *Adv. Drug Delivery Rev.*, **64**, 866–884.
79. Cheng, Y., Dai, Q., Morshed, R.A., Fan, X., *et al.* (2014) Blood–brain barrier-permeable gold nanoparticles: an efficient delivery platform for enhanced malignant glioma therapy and imaging. *Small*, **10**, 5137–5150.
80. Zhao, Y., Ren, W., Zhong, T., Zhang, S., *et al.* (2016) Tumor-specific pH-responsive peptide-modified pH-sensitive liposomes containing doxorubicin for enhancing glioma targeting and anti-tumor activity. *J. Controlled Release*, **222**, 56–66.
81. Xu, R., Wang, X.-L., Lu, Z.-R. (2010) New amphiphilic carriers forming pH-sensitive nanoparticles for nucleic acid delivery. *Langmuir*, **26**, 13874–13882.
82. Qiao, Z.-Y., Qiao, S.-L., Fan, G., Fan, Y.-S., *et al.* (2014) One-pot synthesis of pH-sensitive poly (RGD-*co*-β-amino ester) s for targeted intracellular drug delivery. *Polym. Chem.*, **5**, 844–853.
83. Lin, Q., Mao, K.-L., Tian, F.-R., Yang, J.-J., *et al.* (2015) Brain tumor-targeted delivery and therapy by focused ultrasound introduced doxorubicin-loaded cationic

liposomes. *Cancer Chemother. Pharmacol.*, 1–12.

84. Hynynen, K., McDannold, N., Sheikov, N.A., Jolesz, F.A., *et al.* (2005) Local and reversible blood–brain barrier disruption by noninvasive focused ultrasound at frequencies suitable for trans-skull sonications. *Neuroimage*, **24**, 12–20.
85. O'Reilly, M.A., Waspe, A.C., Ganguly, M., Hynynen, K. (2011) Focused-ultrasound disruption of the blood–brain barrier using closely-timed short pulses: influence of sonication parameters and injection rate. *Ultrasound Med. Biol.*, **37**, 587–594.
86. Vlachos, F., Tung, Y.S., Konofagou, E. (2011) Permeability dependence study of the focused ultrasound-induced blood–brain barrier opening at distinct pressures and microbubble diameters using DCE-MRI. *Magn. Reson. Med.*, **66**, 821–830.
87. Hosseinkhah, N., Goertz, D.E., Hynynen, K. (2015) Microbubbles and blood–brain barrier opening: a numerical study on acoustic emissions and wall stress predictions. *IEEE Trans. Biomed. Eng.*, **62**, 1293–1304.
88. Fan, C.-H., Ting, C.-Y., Lin, H.-J., Wang, C.-H., *et al.* (2013) SPIO-conjugated, doxorubicin-loaded microbubbles for concurrent MRI and focused-ultrasound enhanced brain-tumor drug delivery. *Biomaterials*, **34**, 3706–3715.
89. Huang, X., El-Sayed, M.A. (2011) Plasmonic photo-thermal therapy (PPTT). *Alexandria J. Med.*, **47**, 1–9.
90. Master, A., Livingston, M., Gupta, A.S. (2013) Photodynamic nanomedicine in the treatment of solid tumors: perspectives and challenges. *J. Controlled Release*, **168**, 88–102.
91. O'Neal, D.P., Hirsch, L.R., Halas, N.J., Payne, J.D., *et al.* (2004) Photo-thermal tumor ablation in mice using near infrared-absorbing nanoparticles. *Cancer Lett.*, **209**, 171–176.
92. Cheng, Y., Meyers, J.D., Broome, A.-M., Kenney, M.E., *et al.* (2011) Deep penetration of a PDT drug into tumors by noncovalent drug–gold nanoparticle conjugates. *J. Am. Chem. Soc.*, **133**, 2583–2591.
93. Febvay, S., Marini, D.M., Belcher, A.M., Clapham, D.E. (2010) Targeted cytosolic delivery of cell-impermeable compounds by nanoparticle-mediated, light-triggered endosome disruption. *Nano Lett.*, **10**, 2211–2219.
94. Hirschberg, H., Baek, S.-K., Kwon, Y.J., Sun, C.-H., *et al.* (2010) Bypassing the blood–brain barrier: delivery of therapeutic agents by macrophages. *BiOS, International Society for Optics and Photonics*, pp. 75483Z–75485Z.
95. Madsen, S.J., Baek, S.-K., Makkouk, A.R., Krasieva, T., *et al.* (2012) Macrophages as cell-based delivery systems for nanoshells in photothermal therapy. *Ann. Biomed. Eng.*, **40**, 507–515.
96. Madsen, S.J., Christie, C., Hong, S.J., Trinidad, A., *et al.* (2015) Nanoparticle-loaded macrophage-mediated photothermal therapy: potential for glioma treatment. *Lasers Med. Sci.*, **30**, 1357–1365.
97. Youland, R.S., Kitange, G.J., Peterson, T.E., Pafundi, D.H., *et al.* (2013) The role of LAT1 in ^{18}F-DOPA uptake in malignant gliomas. *J. Neuro-oncol.*, **111**, 11–18.
98. Wu, W., Dong, Y., Gao, J., Gong, M., *et al.* (2015) Aspartate-modified doxorubicin on its N-terminal increases drug accumulation in LAT1-overexpressing tumors. *Cancer Sci.*, **106**(6), 747–756.
99. Devasagayam, T.P., Tilak, J.C., Boloor, K.K., Sane, K.S., *et al.* (2004) Free radicals and antioxidants in human health: current status and future prospects. *J. Assoc. Phys. India*, **52**, 794–804.
100. Liou, G.-Y., Storz, P. (2010) Reactive oxygen species in cancer. *Free Radical Res.*, **44**, 479–496.
101. Wang, J., Sun, X., Mao, W., Sun, W., *et al.* (2013) Tumor redox heterogeneity-responsive prodrug nanocapsules for cancer chemotherapy. *Adv. Mater.*, **25**, 3670–3676.
102. Bervar, A., Zajc, I., Sever, N., Katunuma, N., *et al.* (2003) Invasiveness of transformed human breast epithelial cell lines is related to cathepsin B and inhibited by cysteine proteinase inhibitors. *Biol. Chem.*, **384**, 447–455.
103. Ruan, H., Hao, S., Young, P., Zhang, H. (2015) Targeting cathepsin B for cancer therapies. *Horiz. Cancer Res.*, **56**, 23–40.
104. Lee, S.J., Jeong, Y.-I., Park, H.-K., Kang, D.H., *et al.* (2015) Enzyme-responsive doxorubicin release from dendrimer nanoparticles for anticancer drug delivery. *Int. J. Nanomed.*, **10**, 5489–5503.

Part V
Tissue Engineering and Regeneration

Translational Nanomedicine. First Edition. Edited by Robert A. Meyers.

14
Bone Tissue Engineering: Nanomedicine Approaches

Michael E. Frohbergh[1], Peter Newman[2], Calogera M. Simonaro[1], and Hala Zreiqat[2]
[1]Icahn School of Medicine at Mount Sinai, Department of Genetics and Genomic Sciences, 1425 Madison Avenue, Rm 14-26A, New York, NY 10128, USA
[2]University of Sydney, Biomaterials and Tissue Engineering Research Unit, School of Aeronautical Mechanical and Mechatronics Engineering, Building J07, Sydney NSW 2006, Australia

Keywords

Bone tissue engineering
The field of bone tissue engineering aims to regenerate damaged bone tissue by using therapeutics, including biochemicals, cells, and biomaterials.

Translational Nanomedicine. First Edition. Edited by Robert A. Meyers.

Nanomedicine
The application of nanotechnologies in the field of medicine, including the diagnosis, treatment, and prevention of disease.

Regenerative medicine
The process of synthetically inducing the regeneration of diseased or damaged human tissue or cells.

Biomaterials
Biomaterials combines the fields of biology and materials science. It includes the study of naturally occurring biological materials as well as synthetic materials designed to interact with biological systems.

Surface modification
Within the field of nanomedicine, surface modification involves nanoscale alterations to the dimensional, chemical, optical or electrical surface properties of a material. Such alterations are used to enhance the interface between a material and the biological system with which it interacts.

Tissue engineering focuses on the use of cellular and material-based therapies aimed at targeted tissue regeneration caused by traumatic, degenerative, and genetic disorders. Current treatments for bone injuries and defects that will not spontaneously heal employ replacement rather than regeneration, which is accompanied by long-term complications. In this chapter, attention is focused on state-of-the-art research techniques and materials that are aimed at alleviating these complications by inducing bone-healing rather than bone-substitution; specifically, the use of nanoscale materials and surface modifications in order to closely mimic the microenvironment of bone. Today, these nanoscale technologies are coming to the forefront in medicine because of their biocompatibility, tissue-specificity, and integration and ability to act as therapeutic carriers.

1 Introduction

Extensive bone defects, resulting from traumatic injuries, genetic defects, degenerative diseases and cancers, remain major challenges in orthopedic surgery. The worldwide prevalence of such defects leads to billions of dollars in healthcare costs and a decreased quality of life for those afflicted. Whilst minor fractures where casting and self-healing can produce desirable and longlasting effects, large bone defects do not heal spontaneously and surgical intervention is required.

Current treatment options are limited, and are associated with a high incidence of complications, often leading to non-unions or re-fractures. Filling of the defect with autologous bone grafts harvested from the patient's iliac crest – the current "gold standard" – leads to donor site morbidity in 20–30% of cases [1]. If the autologous bone graft material harvested is insufficient – which often is the case for extensive bone defects – allografts from bone banks may be used, but these are plagued by variable bioactivity and the risk of transmitting infections. Another treatment option – bone segment transport – requires the patient to use an external fixation system for up to one-and-a-half years, and the process is very laborious and painful.

In light of these problems with current treatment modalities, combined with an increasing demand for bone graft materials due to the aging population and increasing patient numbers, there is a great and growing need for alternative techniques to replace, restore, or regenerate bone. Surgical methods use either inert metals or materials [2–9], but lack the appropriate integrative properties to induce native tissue ingrowth and promote self-healing; rather, they serve simply as "filler" materials to provide a temporary stability that decreases over time.

Issues with the current treatments have led to intense investigations to develop new and novel therapies for bone repair and regeneration. These therapies include materials that interact with the native tissue to induce self-healing, cell and growth factor therapies to induce regenerative signaling, and *in-vitro* systems to model the natural processes of bone healing and development outside of the body. All of these tools allow for the diagnosis and evaluation of unique and regenerative strategies to address clinical needs.

A particularly promising alternative is synthetic bone substitutes, the global market for which is around US$ 2 billion each year. The essential requirements for successful bone repair and regeneration using a synthetic-based tissue engineering approach are: (i) mechanical stability; (ii) bioactivity; (iii) a biodegradable scaffold with high porosity and interconnectivity that will allow for the efficient migration of bone cell precursors and the vascularization necessary for bone formation [10]. Otherwise, bone formation and vascularization would be largely limited to the periphery of the scaffold [11–14].

In the case of large, critical-size bone defects, however, the synthetic materials currently available fall short of meeting the *combined* requirements for high porosity and interconnectivity while maintaining mechanical properties and bioactivity; they are also unable to produce a scaffold that maintains architectural support throughout the duration of the healing phase under load-bearing conditions. The result is a suboptimal regeneration of bone within the defect, leading to significant morbidity and considerable economic cost [15]. There is, therefore, an unmet clinical need for a material that could strike the correct balance between mechanical properties, implant architecture, and bioactivity.

1.1 Bone Structure, Development, and Healing

Bone is a hierarchically structured connective tissue responsible for support, motion, and internal organ protection (Fig. 1) [16]. The skeletal system develops through two processes:

- Endochondral ossification, where chondrocyte hypertrophy results in mineralized tissue formation.
- Intramembranous ossification, where mesenchymal cells differentiate into osteoprogenitor cells and osteoblasts which directly synthesize mineralized tissue.

Long bones, such as the femur, generally undergo endochondral ossification, which

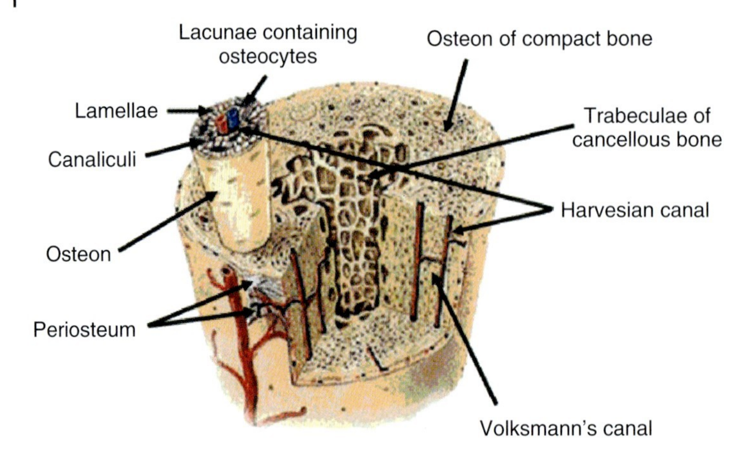

Fig. 1 The hierarchical structure of bone, revealing its macroscale, microscale, and nanoscale components. Nanomedicines aim to closely mimic all of these regulated and unique features. (*http://training.seer.cancer.gov/anatomy/skeletal/tissue.html*)

can be observed by an elongating growth plate of chondrocytes that progresses during skeletal development and eventually fuses at maturity. Flat and irregular bones, such as the bones of the skull, generally undergo intramembranous ossification. The hierarchical structure of bone can be broken down into macro-, micro-, and nanoscale structures [17].

An example of macroscale bone is the lamellar structures that compose the osteons which make up cortical bone. Within osteons, porous microscale layers of bone are composed of osteoid matrix surrounding lacunae and internetworking canals for vascularization and innervation. The nanoscale levels of bone can be found within the osteoid matrix and lacunae, namely the extracellular matrix (ECM) molecules and cellular components that are responsible for depositing, maintaining, and resorbing bone during homeostasis [17].

The nanoscale components of bone can be described as either cellular or ECM. The latter is composed of a biphasic environment consisting of hydroxyapatite nanoparticles (inorganic phase) and collagen type I fibers (organic phase), as well as small amounts of proteoglycans and glycoproteins. Osteoblasts, osteocytes and osteoclasts are the three main cell types found in bone tissue, and all reside within the ECM. Osteoblasts are responsible for the mineralization and deposition of new bone ECM during development and healing. Once surrounded by a dense, mineralized ECM (known as osteoid), the cells are embedded in the lacunae and are referred to as osteocytes, which are responsible for maintaining bone. In a dynamic process known as bone homeostasis, osteoclasts resorb bone tissue, allowing osteoblasts to continuously be recruited for remodeling. This process keeps bone tissue strong and healthy over a person's lifetime, but if and when osteoblast/osteoclast homeostasis is disrupted and the ratio of resorption to deposition increases, then bone diseases such as osteoporosis will become manifest [17].

The healing process for long bones is similar to the process of intramembranous ossification. Initially, a wound callus forms, which is highly populated by inflammatory cells and cytokines. Specific cellular signaling recruits osteoprogenitor cells from the periosteum to the tissue surrounding the site of injury where they begin to differentiate into osteoblasts, mineralize and deposit bone ECM. Over time, mesenchymal stem cells from the endosteum and bone marrow are recruited to aid in and complete the wound healing [18].

2 Nanotechnologies in Bone Tissue Engineering: Importance, Advances, Challenges, and Future Directions

2.1 Strategies and Criteria

One strategy to bone tissue regeneration involves creating materials which support cell life and tissue regrowth by mimicking the macro-, micro-, and nano-environments of the ECM. Because cells tend to dedifferentiate *in vitro* and lose their specific phenotypes and functionalities, much attention has been paid to synthesizing materials that maintain relevant environmental cues to support healthy cell function [19–22]. These technologies are being designed to enhance the standards for bone regeneration. As mentioned above, the major drawback of clinically used materials is that they do not induce regenerative behavior; rather, they act only to support the void left when defected/degenerated areas are removed and do not suitably mimic natural bone in terms of mechanical strength and bioactivity. For example, when metal implants are used to support major femur fractures they will not heal, as the loads crucial for bone homeostasis are transferred from the bone to the implant, leading to stress shielding. Bone requires a day-to-day loading in order to maintain deposition/resorption homeostasis, and when these forces are removed then degenerative diseases begin to develop over time [23]. Bearing these points in mind, many tissue-engineering remedies have been aimed at developing technologies that can aid in the regeneration of bone tissue rather than its replacement.

Currently, different technologies and fabrication methods for developing bone-regenerating materials are under investigation, many of which offer their own unique properties for tissue regeneration. Three main characteristics must be applied when developing a new bone technology:

- An appropriate material must be biocompatible and nontoxic, it must meet the mechanical criteria for bone, and must also mimic the macro-, micro-, and nanostructural details of the bone ECM.
- An appropriate cell source must be used that can differentiate into functional osteoblasts.
- Growth factors, cytokines and biomolecules must be present that can induce native tissue migration, promote tissue mineralization and ECM production as well as act as chemoattractants to form attractive gradients from the tissue onto the implanted material [24].

A number of different nanotechnologies currently being investigated for bone tissue engineering, and how they implement the above criteria, are outlined in the following section.

2.2 Nanomaterials

By the simplest definitions, nanomaterials are organic or inorganic materials that have been intentionally created with at least one dimension less than 100 nm. These materials usually have discrete or functional parts, and are created for their unique properties that are not available to the same material at larger scales. These unique size-dependent properties range from diverse chemistries [25] to distinct optical properties [26] and high strengths and conductivity [27]. Generally, their use is considered "value-added", and their extraordinary properties are included with larger macroscale materials and applications. In biology, nanomaterials are currently being investigated for use in bioimaging, drug and DNA delivery, cancer therapies, biosensing, and tissue engineering, which form the subject of this review.

2.2.1 Hydrogels

The intrinsic biomimetic properties of hydrogels make them an interesting material for bone tissue engineering. They are composed of microscale and nanoscale fibers that resemble those of the ECM, they are hydrophilic, and their structure, mechanical properties and bioactivity can be controlled using a number of fabrication techniques. Unfortunately, the use of hydrogels in bone tissue engineering is limited to nonstructural applications as a result of the materials' poor mechanical properties, whilst trade-offs also exist when optimizing for architecture, mechanical properties, degradation, and bioactivity. This has led to problems in identifying a hydrogel that is suited to many functions simultaneously [28]. The image in Fig. 2

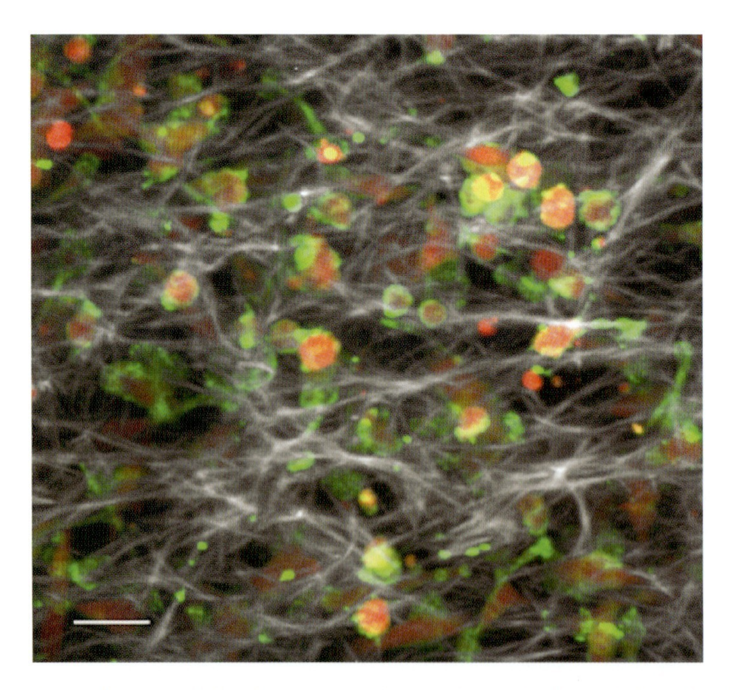

Fig. 2 Maximum intensity projection of a type I rat tail collagen hydrogel with embedded mouse peritoneal macrophages stained for cytoplasm (red) and F-actin (phalloidin, green). Scale bar = 30 μm. Image courtesy of Maté Biro of the Centenary Institute of Cancer Medicine and Cell Biology, Australia.

shows cells proliferating through a generic collagen hydrogel, where the ECM-mimetic nature of the material is evident in the macroscale and nanoscale fibers.

Hydrogels are hydrophilic polymer materials that have the ability to retain a high fraction of water within their structure, without dissolving; indeed, the water content may be up to many thousand-fold the polymer's dehydrated weight [29]. A hydrophilic polymer or gel backbone is a large nanoscale macromolecule that can be created from a variety of sources, both natural and synthetic. The natural sources include collagen, hyaluronic acid, elastin, chondroitin sulfate, Matrigel, alginate, and gelatin, all of which are appealing due to their inherent biocompatible properties. Alternatively, synthetically derived sources may be used, such as poly(vinyl alcohol) (PVA), poly(acrylic acid) (PAA), poly(acrylamide) (PAAm), poly(ethylene glycol) (PEG), and poly(lactic-*co*-glycolic acid) (PLGA).

Hydrogels represent a promising biomaterial due to their ability to manipulate a broad range of physico-chemical properties, such as the ability to vary bioactivity by dispersing biochemicals throughout their water phase and/or biochemical conjugation to (or within) the hydrophilic polymer/gel backbone. This can include small peptides motifs such as the cell adhesion ligand RGD [30] or larger growth factors and genes [31]. The chemical charge can be neutral, cationic or anionic, or even ampolytic as a result of differences in the gel backbone [32]. Hydrogels can be synthesized with a wide range of mechanical properties that vary due to changes in the chemical composition of the polymer, the degree to which it is crosslinked, the water content, and multiphasic inclusions. The gel backbone of hydrogels also forms large three-dimensional (3D) mesh structures composed of nanoscale macromolecules that bear an architectural resemblance to the ECM. Depending on the gel, these dimensional parameters can be varied across the nanoscale.

While it is intuitive to dismiss the use of hydrogels in bone tissue engineering due to their low degree of stiffness and an inability to hold the loads required of bone, they do show promise when load-bearing is not required. This includes bone repair in craniofacial defects or for smaller defects where loads could be supported by other means. The low degree of stiffness and easy deformation of hydrogels also makes them perfect for use in minimally invasive surgeries, as they can easily be injected.

The continued development of hydrogels has yielded a variety of novel technologies. As noted above, the bioactivity of hydrogels can be improved by dispersing biochemicals through the hydrogel network, or by biochemical conjugation to the gel backbone. In general, such biochemical enhancement of hydrogels for bone tissue engineering utilizes the cell-binding adhesive peptide RGD, and Burdick *et al.* first created an RGD-modified PEG hydrogel for bone tissue engineering in 2002 [30]. Because, in this case, a photopolymerization technique was used to crosslink the hydrogel, a homogeneous dispersion of osteoblastic cells could be incorporated into the hydrogel such that degradation of the scaffold was not required for cell migration. Taken together, these findings indicated that increases in the concentration of RGD would result in a greater degree of mineralization, and this point was confirmed by Yang *et al.* in 2005 [33].

Subsequently, hydrogels incorporating RGD have become more complex. In one study, when the effects of adding multiple adhesive peptides [RGD (Arg-Gly-Asp) and PHSRN (Pro-His-Ser-Arg-Asn)] into a hydrogel simultaneously were examined, increasing concentrations of RGD were shown to favor osteogenic differentiation and to improve cell proliferation [34]. This effect was also achieved with a hydrogel conjugation of both RGD and the growth factor bone morphogenetic protein-2 (BMP-2) [35]. When using bone marrow stromal cells, RGD and BMP-2 were each shown to increase bone mineral formation separately, but if they were applied together they acted synergistically. This demonstrated the importance of a multitude of cell signals in tissue scaffolds, and thus the development of more complex hydrogel systems. Since then, a variety of larger-sized biochemicals have been incorporated into hydrogels, including growth factors such as vascular endothelial growth factor (VEGF) [36], BMP-2 [37], and transforming growth factor β1 (TGF-β1). In addition to growth factors, hydrogels have been loaded with pharmaceuticals that include anticancer drugs and steroids [38], as well as with DNA for cell transfection studies [39].

Other investigators have employed techniques to achieve nanoscale spatial control over RGD domains in a gel [40, 41]. For example, Comisar *et al.* used a nanoscale patterning of RGD throughout an alginate hydrogel to analyze focal adhesion behavior, cell spreading, proliferation, and differentiation [40]. The results obtained supported the findings of Burdick [30] and Yang *et al.* [33], whereby RGD increased spreading, proliferation and osteogenic differentiation, but also managed to decouple these behaviors; notably, while cell spreading was related to RGD patterning, the proliferative behavior was dependent only on the overall RGD concentration. Subsequently, the positive adhesive behaviors of RGD nanopatterning were shown only to be influential when the RGD spacing was <70 nm [41].

Hydrogel composites represent another popular approach to bone tissue engineering, whereby the hydrogels are commonly mixed with inorganic materials such as tricalcium phosphate ($Ca_3(PO_4)_2$) or hydroxyapatite (HA). These materials are appealing as they not only mimic the organic–inorganic/collagen–HA composition of bone but also have relatively greater mechanical properties.

Photolithography is another method used to fabricate hydrogels with nanoscale details. Such hydrogels may contain catalysts which, on irradiation, cause the hydrogel to be crosslinked; hence, by controlling the area of irradiation, hydrogels can be created with nanoscale patterning. This approach has also been used to conjugate bioactive molecules including RGD [42] and VEGF [43] within a hydrogel.

2.2.2 Self-Assembled Structures

Peptide amphiphiles (PAs) are a class of peptide hydrogels that assemble spontaneously as a result of hydrogen bonds and peptide interactions. These materials have the advantage that they are made from peptides and thus have an inherent biocompatibility. Once formed, they are mimetic of biological hierarchical structures, ranging from protein primary structures through to microscale and macroscale organizations. In the same manner that small changes to protein primary structures can have profound effects on those at a higher level, fine-tuning the order and number of peptides in a PA can specify noncovalent interactions that create a diverse range of highly resolved nanostructures [44]. These include structures commonly found in biological systems, such as fibers, spheres and membrane/bilayer structures. The structure of a PA is shown in Fig. 3; one end of the molecule is hydrophobic, while the other hydrophilic end is charged and contains

beta sheet forming domain charged and bioactive end

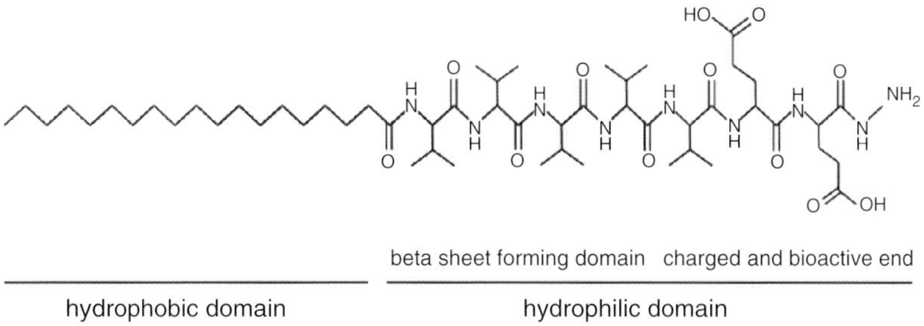

hydrophobic domain hydrophilic domain

Fig. 3 A self-assembling supramolecular peptide amphiphile. The amphiphilic nature of the molecule can be seen with hydrophilic and lipophilic/hydrophobic domains at either end. Alterations can be made to the beta-sheet-forming subregion of the hydrophilic domain to further change the supramolecular structure. Likewise, functional and biochemical groups can be made to the end of the hydrophilic domain to increase the bioactivity or delivery of pharmaceuticals.

the peptide sequence that specifies the larger-scale structure through noncovalent interaction.

Because their order and length can be easily altered, PAs can be used for the rational bottom-up design of nanostructured biomaterials, an approach which has been used previously to control the dimensional parameters of self-assembled nanofibers. By changing the number of repeats in the charged peptide end of a PA, Moyer *et al.* were able to change the width of the fibers from 10 to 100 nm [45], while Sur and coworkers altered the mechanical properties of PAs by making slight variations to the structure's hydrophobic end. In doing this, it could be shown that the spreading of neural cells over stiffer substrates was slowed [46].

Block copolymers represent another type of nanostructured self-assembling protein. As a result of immiscibility between two polymer types, block copolymers undergo a phase separation, thus creating nanostructures. Although investigations into block copolymers have been limited particularly to use with bone tissue engineering, various other studies have been conducted to examine the use of block copolymers to change the topology of a material and to enhance cell attachment and spreading [47, 48].

Self-assembled structures offer an intriguing approach to tissue engineering, as their formation mirrors that seen in the creation of biological systems such as membranes, the ECM, and in DNA transcription. Moreover, because the chemical interactions that determine their formation can be controlled, these materials can be used in bottom-up design approaches to tissue engineering, allowing alterations to be made to the materials' mechanical, structural, biochemical, and degradation properties. Based on these points, self-assembled structures have shown promise for the production of complex biomaterial systems with physico-chemical properties similar to those of the ECM. Unfortunately, as a class of hydrogels self-assembled structures suffer from similar problems, in that they lack sufficient mechanical strength for load-bearing applications. In addition, because their structure, mechanical properties and bioactivity are intrinsically linked, the independent optimization of these properties would be problematic.

2.2.3 Nanoparticles, Nanotubes, Nanofibers, and their Composites

Recently, much research has been directed towards investigating the potential use of nanoparticles (NPs) as biomaterials – whether alone or as composites – for bone regeneration. The most common NPs and nanotubes used for bone tissue replacements include ceramics such as nanoscale HA, titanium nanotubes, graphene, carbon nanotubes (CNTs), and iron oxide [26, 49, 50]. The main benefits of using these nanomaterials are the increased surface-to-volume ratios that result from the particles' decreased sizes; the main downside to their use relates to the difficulties encountered during their processing. Because NPs are often highly hydrophobic they have a propensity to agglomerate and, as a consequence, this can lead to poor dispersions that in fact *reverse* the desired effect of their inclusion. For example, when NPs are poorly dispersed in nanocomposite materials they can facilitate stress concentration and crack formation, thereby lowering a material's mechanical properties [25]. In other situations, agglomeration will reduce the available surface area of NPs so as to render any attached biochemicals unavailable. Clearly, efforts are required to maintain a homogeneous dispersion, and this is most commonly achieved by sonicating the materials during the fabrication process [27].

Due to its chemical similarity with the inorganic phase of bone, HA is a common choice as a synthetic bone substitute. In natural bone, HA exists as platelets with nanoscale dimensions of 2–5 nm thickness, 10–80 nm width, and 200 nm length [49]. As per the biomimetic paradigm, it is to be expected that cells interacting with nanosized HA would better recreate natural cell behavior and, indeed, biomaterials made from HA have good biocompatibility, osteoconductivity and osteoinductivity, and can also be templated and then sintered into most shapes, porosities, and architectures. Unfortunately, HA-based materials suffer from being brittle and having a poor fatigue resistance, which means they can only be used in applications where there are lower loads. This problem is further exaggerated when the scaffold porosity is increased in order to encourage nutrient exchange, vascularization, and bone ingrowth. These disadvantages have led to HA-based materials being used only in non-load-bearing applications, as fillers and coatings.

These problems can be overcome, however, by using NP HA powders to create bone scaffolds, and as a result of the increased grain boundaries and improved crack-arresting characteristics, nanosized ceramics are now available that exhibit increased straining and toughness [51]. For example, when scaffolds created from NP HA were compared to those made from larger, submicron-particle HA, the former scaffolds were seen to be superior at repairing segmental bone defects in New Zealand White rabbits [52]; these results were supported by numerous cell-based studies which showed an increased adhesion and proliferation of osteoblasts on nanosized ceramics [53–55]. Some research groups have attributed this benefit to the increased roughness provided by sintering the NP HA, as well as the consequential hydrophilicity [51], increased surface area, and improved adsorption of proteins [56].

In bone tissue engineering, metal NPs are less commonly used than those made from ceramics or polymers, due mainly to the high toxicity and oxidizing effects of metals. Typically, metal NPs are used in their oxide form and conjugated with biochemicals, the most common metals including silver, gold, and iron oxide. The same NPs have been used to label bone cells for imaging and to deliver drugs for bone regeneration and/or magnetotransfection, though their use in orthopedic applications is relatively unexplored to date [57]. In one recent application, Skaat *et al.* used iron oxide NPs conjugated with thrombin to create a 3D scaffold that is bioactive with growth factors and can be visualized using magnetic resonance imaging (MRI). Moreover, as such scaffolds can be easily visualized after implantation, magnetic scaffolds can be reseeded with both cells and biochemicals [58].

Polymeric NPs are popular due to their versatility, as they can be easily functionalized, they display a wide range of mechanical properties, and they have a relatively high biodegradability. Similar to metal NPs, the use of polymeric NPs in orthopedic applications has been limited; however, whilst being rarely used in composites they are commonly used alone for DNA/drug delivery, bioimaging, and cancer therapies [59]. The creation of composite materials represents a strategy whereby the advantages of different materials can be combined. For example, the use of NPs in a composite can provide a bulk material with access to the extraordinary properties of NPs, and this combination has yielded many promising materials with relatively enhanced mechanical and bioactive properties.

Some of these composites have attempted to mimic the organic–inorganic nanoscale structure of bone by combining nanosized HA into polymer materials [60]. Such materials show that there is a more favorable response than using either of the components separately [61, 62], and *in-vivo* studies with them have demonstrated bone formation within critical-sized defects in rats [63].

Recently, Zreiqat and coworkers developed a composite ceramic, comprising strontium-doped Hardystonite ($Ca_2ZnSi_2O_7$) (85 wt%) and Gahnite ($ZnAl_2O_4$) (15 wt%), called Sr-HT Gahnite (patent #2011902160). The novel Sr-HT Gahnite ceramic is a three-phase micro-nano crystal glass ceramic composite that mimics bone microstructures with (80%) porosity, interconnectivity, and (300–500 μm) pore size. This micro- and nanoscale composite displays high mechanical strength and toughness [64], and was the first to demonstrate a capability of bone regeneration under active loads [65].

Another interesting use of NPs is to control the biodegradability of a material. Hydrophobic NPs can be used to decrease degradation, while hydrophilic NPs will increase degradation. In this process, the hydrophilicity of the NP will cause changes to the diffusion of water throughout the scaffold, thereby increasing or decreasing water exchange and the breakdown of the scaffold [66]. Another appealing nanomaterial for use in bone tissue engineering is created by the anodization of titanium, during which process titanium is oxidized and then etched away to form porous nanotube structures over its surface. The various parameters employed in this process can be altered to change the length and diameter of the nanotubes. This process is particularly relevant in orthopedics, where many of the devices used are created from titanium alloys. Further, TiO_2 nanotubes are biocompatible [67], have been shown to have an increased ability for protein adsorption and calcium deposition [68, 69], can be created with different chemical functionalization [70], intercalated with pharmaceuticals for drug delivery [71], and controlled to optimize the proliferation and differentiation of osteoblastic cells [72]. As result of this versatility and pre-existence of titanium orthopedic devices, TiO_2 nanotubes represent a promising nanotechnology for orthopedic devices.

The carbon nanomaterials graphene, graphene oxide (GO), and CNTs are also of interest for bone tissue engineering. These materials share many properties, including biocompatibility in osteogenic applications [50, 73], exceptional mechanical properties [27, 74], they are easily functionalized with either simple or complex chemicals, growth factors, and pharmaceuticals [49, 75], and they are highly conductive. Graphene and CNTs are highly hydrophobic, however, and therefore have a tendency to agglomerate; accordingly, these materials require dispersion prior to their use. Although sonication is a common strategy, it is often combined with chemical treatments to functionalize with carboxyl groups. Carboxylation increases the material's hydrophilic character and also aids dispersion.

Previously, carbon nanomaterials have been used in nanocomposites to provide improved mechanical properties and bioactivity [76–79]. In 2011, Depan *et al.* created a chitosan–GO nanocomposite which demonstrated an enhanced cellular attachment as well as an increased elastic modulus and hardness when compared to chitosan alone. Ogihara *et al.* were able to achieve a similar result, creating a nanocomposite made from CNTs and alumina; when this composite was compared to noncomposite alumina scaffolds, a 120% increase in fracture toughness was demonstrated, as well as a superior proliferative ability. In another study conducted by Verdejo *et al.*, CNTs were successfully coated over polyurethane foams by using a chemical vapor deposition (CVD) process, and shown to have no detrimental effect on osteoblast viability or adhesion; however, the wettability of the scaffold was increased, showing an incorporation of osteoblasts around the pores of the scaffolds [80]. Although these authors failed to add sufficient CNTs to cause mechanical changes to the foams, it is realistic to expect this strategy could be used to increase the mechanical stability and perhaps also act to improve fixative devices.

Graphene has demonstrated promising results as an osteogenic substrate. When comparing Si/SiO_2, polyethylene terephthalate (PET), and polydimethyl siloxane (PDMS) substrates coated with graphene to uncoated substrates, Nayak *et al.* showed the osteogenic differentiation over graphene substrates to be less than for those treated with the osteogenic growth factor, BMP-2 [50].

Toxicologically, CNTs have proven to be problematic, with the relevant mechanisms believed to be threefold: (i) they can create reactive oxidative stresses; (ii) they can rupture membranes due to their needle-like dimensionality; and/or (iii) they can disrupt biochemical processes. Many experiments have reported a wide variability in CNT toxicity, where the differences reflect the wide range of dimensional and chemical properties that CNTs can have. To further complicate these findings, CNTs in bulk, within a composite, in solution, or attached to a substrate surface, have also shown mixed biological responses [49], and it is for this reason that the use of CNTs in tissue engineering should be approached with caution.

The similarities between graphene, GO and CNTs suggest that caution should be exercised when studying the toxicology of graphene and GO. In 2011, Liao *et al.* demonstrated both size-and dose-dependent toxicological effects of graphene and GO NPs [81], with the rupture of red blood cells or hemolytic activity being shown to increase as the graphene and GO NPs decreased in size. A similar outcome was noted when the NPs were allowed to aggregate into larger particles, with larger aggregates showing a lesser toxicity. Additional studies have demonstrated a dose-dependent effect between toxicity and GO concentration [82, 83]. For example, granulation formation occurred when GO was injected directly into the lungs of mice *in vivo* [84], while subsequent reports of the intravenous injection of PEG-GO sheets resulted in PEG-GO accumulation in the liver and spleen of mice and subsequent excretion, but with no further damage to the animals [85]. Similar to CNTs, the study-to-study variation in the toxicity of graphene and GO may reflect the many different chemical, dimensional, and environmental factors attributable to versatile NPs during exposure. Perhaps most importantly, it stresses the need for great caution when using these materials in tissue engineering.

Electrospinning Electrospinning is used to develop microscale and nanoscale fibrous scaffolds whose origins derive from the textile industry [86, 87]. Due to the large number of different polymers that can be electrospun, it is a versatile fabrication method that can be aimed at many different tissues in the body.

The use of electrospun scaffolds imparts several advantages. The nanoscale ECM of most tissues is fibrous in nature, and the fibrous scaffolds produced will closely mimic these structural details [88–90]. As per the biomimetic paradigm, to mimic the fibrous structure of the ECM will promote the natural, healthy behaviors of cells *in vivo*. Moreover, due to the controllability of electrospun scaffolds, the mechanical properties, alignment and dimensionality of the fibers can easily be altered to suit the appropriate tissue type [91, 92]. In tissues such as the heart, the aligned fibers can offer cells a similar environment to native cardiac muscle and promote anisotropy [93]. The mechanics can be altered by polymer choice, crosslinking, or by creating composite material scaffolds [94–97].

The main drawback to electrospun scaffolds is the lack of substantial porosity. While the natural porous structure of electrospun scaffolds allows for the passage of growth factors, biomolecules and cytokines, the scaffolds actually produced are too densely packed with fibers to allow appropriate cell infiltration, vascularization, and tissue formation *in vivo* [98, 99]. Numerous attempts have been made to enhance the porosity of these scaffolds, including the use of sacrificial fibers [100, 101], etching [102], and salt leaching [103, 104].

Currently, the many materials being investigated with the aim of using electrospinning to guide bone regeneration can be classified as either natural or synthetic polymers. Natural polymers are derived from either the human body or other organic life, and have an inherent biocompatibility that can include natural ECM proteins, growth factors, ligands, and polysaccharides [105–108]. Unfortunately, however, natural polymers lack the processability of synthetic polymers which, because they are man-made, can easily be adjusted in terms of fiber size, mechanical strength, and degradation rates. In contrast, natural polymers tend to maintain their inherent properties [109–111].

Because of the unique mechanical requirements of bone, many studies have employed either composite scaffolds or scaffolds containing a mixture of materials, in order to generate all of the necessary properties for creating a bone scaffold. Composite scaffolds can contain either

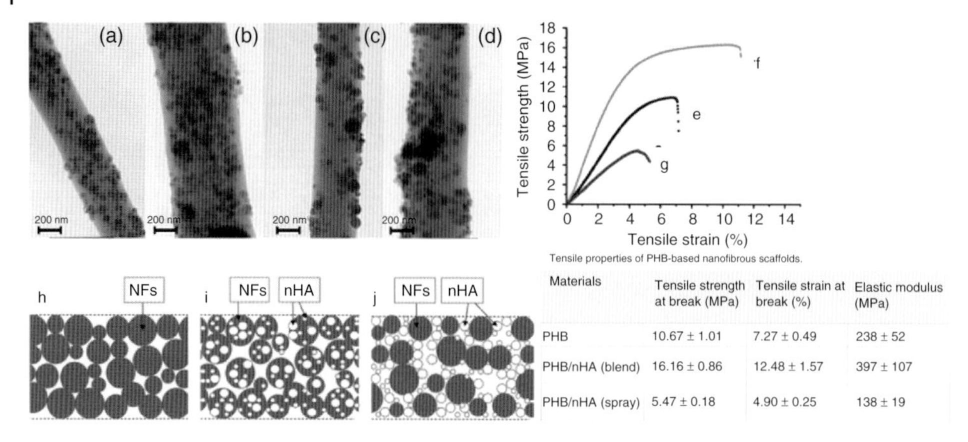

Tensile properties of PHB-based nanofibrous scaffolds.

Materials	Tensile strength at break (MPa)	Tensile strain at break (%)	Elastic modulus (MPa)
PHB	10.67 ± 1.01	7.27 ± 0.49	238 ± 52
PHB/nHA (blend)	16.16 ± 0.86	12.48 ± 1.57	397 ± 107
PHB/nHA (spray)	5.47 ± 0.18	4.90 ± 0.25	138 ± 19

Fig. 4 Composite nanofibers (NFs) synthesized via co-electrospinning procedures. Transmission electron microscopy reveals that hydroxyapatite nanoparticles (nHA) can either be incorporated into (a and b) or deposited onto (c and d) poly(3-hydroxybutyrate) nanofibers by creating blended solutions, or spraying them from heterogeneous solutions, respectively. Incorporating nHA changes the tensile strength of the fibrous scaffolds showing an increase in strength when using the blended solution (f) and a decrease in strength when spraying (g) when compared to scaffolds without HA (e). Tensile strength, strain at break and elastic modulus are also presented in the table. These changes in mechanical properties are described in the schematic to the lower left where, as opposed to nonfibers without nHA (h), it is predicted that increases from nHA incorporation with blending is due to strengthening of the fibers (i) whereas decreases from spraying is due to disruption of the interfibrous space (j).

natural and/or synthetic blends of polymers in order to include the mechanical properties and biocompatibility required; alternatively, a polymer and inorganic material(s) can be blended to more closely mimic the biphasic nature of bone ECM. Li *et al.* showed that the incorporation of gelatin/nanohydroxyapatite into PLGA electrospun fibers had a significant effect on osteoblast adhesion and proliferation, as well as an increase in the osteogenic differentiation of bone-marrow derived mesenchymal stem cells (bMSCs), as evidenced by increases in osteogenic markers and mineral deposition. The same group also demonstrated a superior performance in an inflammatory subcutaneous implant model [112]. Similarly, Ramier *et al.* demonstrated an electrospinning/electrospraying technique to develop novel scaffolds for enhanced bioactivity with bMSCs. In this case, when poly(3-hydroxybutyrate) (PHB) nanofibers were compared as-spun, blended with HA, or sprayed with HA, although the mechanics of the blends were much higher, an enhanced porosity resulting from NPs not being trapped in the fibers but rather being present on the surface and in between the fibers, led to a better cell attachment and an enhanced potential for *in-vivo* tissue ingrowth (Fig. 4) [113]. Cheng *et al.* showed that coating polycaprolactone (PCL) and chitosan blended fibers with collagen significantly enhanced the adhesion and spreading of rat bMSCs as well as osteogenic differentiation [114]. Collagen I is the primary fibrous protein in the ECM of bone, and these results indicate the importance of using naturally occurring factors to functionalize materials towards specific tissues. In another pioneering study, Frohbergh *et al.*

showed that a composite of the natural polymer, chitosan, and HA crosslinked with genipin could be used to enhance the mechanical strength of chitosan fibers. In this case, the materials exhibited a Young's modulus similar to that of callus formation during bone healing. This confirmed an enhanced effect on the osteogenic maturation of pre-osteoblast cells, as indicated by an increase in alkaline phosphatase (ALP) activity and bone-specific gene expression markers [94].

2.2.4 3D Printing and Rapid Prototyping

Despite the progress made in the engineering of two-dimensional (2D) bionanomaterials, translating the techniques into 3D materials remains a significant challenge. Consequently, the concepts involved in 3D printing and rapid prototyping have become a general "hot-spot" for biomedical devices, organ templates, and surgical tools [115–118]. Originally referred to as stereolithography, 3D printing involves the layer-by-layer deposition of a material to form a 3D structure. Today, two methods are available to import the information to print:

- The use of computer design programs (Solid Works, Cad programs, etc.) which send the template files directly to the printer, which then prints the object.
- To scan an object and import the data to the printer, which can then generate an infinite number of copies of the scanned object.

The two methods of 3D printing currently available are laser sintering and dry-powder printing. Laser sintering uses a laser beam to sinter a block or other shaped material into the detailed 3D object that is programmed into the printer. Although useful, this is mainly applicable to objects where the only important feature is the outer surface, such as models or toys. Dry-powder printing uses polymer or ceramic powders that are passed through an extrusion tip and subsequently cooled or dried to solidify. This is performed layer-by-layer in the specific template that is exported to the printer, and is much more applicable for medical devices and tissue engineering purposes.

The two main focuses for orthopedics involving 3D printing is the development of organ models for use in surgical training [119, 120] and, potentially in the future, the rapid screening of therapeutics *in vitro*, and to create tissue engineering constructs for *in-vitro* analysis and *in-vivo* evaluation. With 3D printing, porous structures with enhanced interconnectivity and variable sizes can be synthesized from bone-specific materials, such as tricalcium phosphate [121, 122]. Specific printers can directly print organic materials in the forms of hydrogels [123], within which microparticles containing drugs, growth factors or other therapeutics can be simultaneously printed through a dual-nozzle extrusion to template the particles throughout the scaffold and control both temporal and spatial release.

2.3 Surface Modification Techniques

2.3.1 Micropatterning, Nanopatterning, and Lithography

The microenvironment plays a crucial role in cell fate determination and functionality [21, 124–126]. In particular, bone tissue has a unique organic–inorganic nanoenvironment that osteoblasts deposit during development, remodeling, and repair. Nanopatterning by different lithography techniques allows the manipulation of material surfaces at the nanoscale. Currently, several different methods are employed to modify surface characteristics by using nanopatterning.

Both, microlithography and nanolithography techniques have become prominent forms of surface modification in the field of biotechnology during the past 15–20 years. The most commonly used form is photolithography, in which thin films called photomasks and light (commonly ultraviolet) are used to generate micropatterns on photosensitive materials called photoresists [127]. Unfortunately, this process has limitations due to the fact that it can only print on the submicron level (>100 nm), can only be used on photosensitive materials, and can only create 2D surfaces. Soft lithography and probing lithography techniques were later developed that could create surface modifications on the nanoscale, thus enhancing resolution and applicability [127, 128]. Today, many different methods of lithographic techniques are used in bone tissue engineering, one of the more common soft lithography methods being a PDMS stamping or casting technique developed by Whitesides in 1994 [129].

In this technique, microscale or nanoscale structures can be imprinted onto the surface of a material to change the surface topology and cell-substrate interaction behaviors. For example, Kim *et al.* showed that stamping microposts of varying diameters and different inter-channel distances had a significant effect on cell behavior [130]. Higher posts yielded different cell morphologies, while mid-size posts yielded relatively high cell numbers, and smaller posts yielded the highest mechanical strength of the materials. Wang *et al.* also demonstrated a potential use for treating bone cancers using nanoscale PDMS mold casting, where 27 nm patterns stamped onto PLGA scaffolds were able to significantly enhance the osteoblast-to-osteosarcoma cell ratio [131]. The use of such nanopatterned materials *in vivo* could help to suppress tumor progression and restore the healthy balance of factors secreted by healthy osteoblasts.

Two alternative forms of soft lithography are ion beam lithography and interference lithography. Although relatively new to tissue engineering, these techniques are capable of producing controlled high-resolution nanopattern modifications on surfaces. Lamers *et al.* used laser interference lithography (LIL) in order to fabricate nanogrooved surfaces that were capable of guiding osteoblast differentiation at early-stage time points, indicating their potential for osteogenic induction [132]. Different size grooves were created on silicon chips and observed for topographic characteristics, effects on cell spreading, and gene expression of osteoblasts. Nanogrooved patterns indicated the highest potential for cell alignment, focal adhesion, and osteogenic gene expression [132]. Although ion beam lithography is capable of producing arbitrary patterning much more effectively than LIL (which can only create template patterns), the latter technique is relatively quicker and more cost-effective, which makes its up-scalability much more attractive in terms of marketability.

Nanoimprint lithography (NIL) is a specific form of scanning probe lithography, which uses a mechanical stylus to create nanoscale chemical or physical modifications on the surface of a material. By coupling NIL with LIL, Prodanov *et al.* showed that LIL could be used to generate nanogrooved structures on silica, after which NIL could be used to imprint these structures onto titanium implants (Fig. 5); in this way, the authors demonstrated an *in-vivo* capacity to regenerate bone in critical size defects [133]. These results indicated that nanogrooves have a significant effect on inducing host-implant interactions and

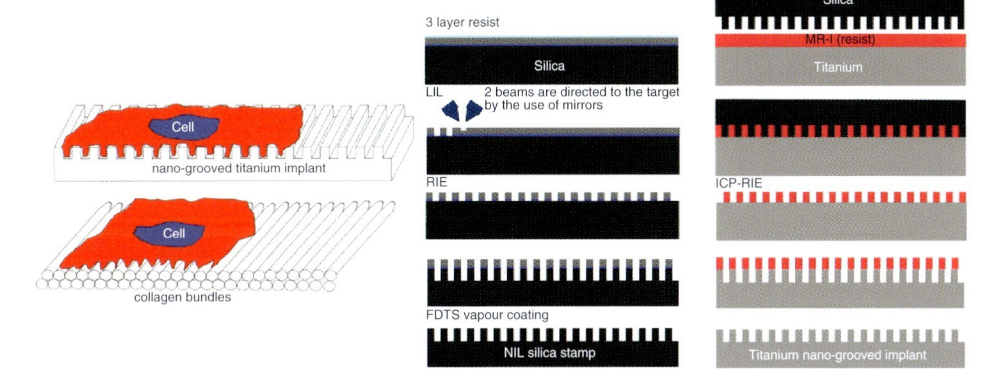

Fig. 5 Schematic representations of the structure and fabrication of nanogrooved titanium implants for improved osteocompatibility. The hypothesis states that using nanogrooved implants will more closely mimic how cells behave, organize the extracellular matrix (ECM) and deposit minerals on a substrate that is biomimetic to the natural collagen bundles that make up the bone ECM (left top and bottom). To synthesize these modified titanium implants, laser interference lithography (LIL) and reactive ion etching (RIE) were used to introduce nanogrooves on silica wafers. Perfluorodecyltrichlorosilane (FDTS) was then applied on the silica stamps before the nanoimprint lithography (NIL) step, followed by a magnetic resonance imaging resist onto the titanium surface. Inductive-coupled plasma reactive ion etching (ICP-RIE) was used to complete the imprint, resulting in nanogrooved modified titanium implants.

allowing superior tissue ingrowth. While these technologies are new to the field of tissue engineering, they show great promise for the development of smart, functionalized materials.

2.4 Chemical Etching and Vapor Deposition

Chemical etching (also known as chemical milling) is a subtractive fabrication technique in which temperature-responsive chemical baths are used to selectively remove specific materials and create a desired shape. For bone, chemical etching is also used for cortical imaging, where a chemical is used to etch the surface of the bone and to create a clearer lamellar image that is free from organic and inorganic debris after diamond or band-saw cutting. As recently discussed by Congiu *et al.*, this process is still being optimized in terms of etching material [134]. Stronger acids and bases tend to demineralize the surface and damage the collagen fibrils, whereas weak acids do not sufficiently clear the debris. In this study, dilute polyphosphate was shown to yield the best morphological representations and lamellar structure integrity [134].

Etching chemicals are also used to create and modify metal implants that are used for bone replacement. Chen *et al.* showed that, by using different surface-modification techniques on titanium (Ti) implants, different results were obtained when osteoprogenitor cells were cultured on them [135], due to the different resulting surfaces that had been produced by each technique. An electrochemical modification was observed through potentiostatic anodization (ECH), sandblasting and acid etching (SLA), sandblasting and hydrogen peroxide treatment with heating (SAOH), and sandblasting with alkali heating and etching (SMART). Upon evaluation, the rougher surfaces of SMART and SLA yielded the quickest cell adhesion and proliferation rates, as well as a quick initial expression of ALP and osteocalcin (OCN). However, in the long term the ECH and SMART surfaces showed the highest expression of ALP and OCN. These data provided an initial insight into how these modifications may cause superior interactions between the host and the implant material [135].

Acid etching is used primarily to modify Ti implants, as noted above. As Ti is used widely in dental implants and fillings, this technique can also be applied to enhance the osseointegrative properties of Ti implants, which is a common requirement in dental research. Kim *et al.* showed that the treatment of SLA titanium implants with magnesium ions led to a significant increase in cell attachment, proliferation, and ALP activity, indicating that further surface modification using plasma source ion implantation methods may further enhance these dental implants [136]. When Oliveria *et al.* conducted gene expression studies of a number of osteogenic markers after acid etching and acid/alkaline etching of titanium implants [137], an interesting observation made involved an increase in some early bone markers, such as osteopontin (OPN); this was most likely due to an enhanced cell adhesion and Runx2, an early marker transcription factor present in maturing osteoblasts. A lack of ALP and collagen I and III expression may indicate that, after two weeks in culture, these cells still have yet to reach a mature osteoblast state [137].

Due to its mechanical stability, carbon is a promising candidate for bone tissue regeneration. However, the main drawback of carbon is that it is bioinert and does induce tissue integration or cell adhesion well. Chemical etching to create a functionalized surface can be used to enhance this drawback. For example, Hoppe *et al.* studied the surface modification of carbon substrates using HCl/HNO_3 etching, followed by soaking in simulated body fluid (SBF) for HA deposition. The rough surface, coupled with the HA deposition, led to an enhanced cell adhesion, viability, proliferation, and ALP expression [138]. Chemical etching can also be used to remove sacrificial cores and create drug-delivery devices. The process begins with a selected removable material, which is subsequently coated with a desired material. After coating, the inner material is carefully removed to create a hollow shell that is capable of being loaded with drugs, growth factors, cytokines, and other deliverables. Previously, Guo *et al.* used such drug-delivery systems to observe the ability of carbonated HA coated on calcium carbonate microspheres to deliver bone-specific therapeutics [139]. In this case, chemical etching with citric acid was used to remove the calcium carbonate core, leaving behind a hollow carbonated HA shell that was subsequently loaded with vancomycin hydrochloride to observe the release kinetics of the carriers. This study was targeted at instances of bone infections, where the marrow swells and leads to necrosis of the surrounding bone tissue. In this case, a biomimetic scaffold would be only half of the solution as the infection would also need to be treated with antibiotics. Drug carriers, as presented here, have the potential to address both of these needs simultaneously [139].

Etching and surface modification is not only limited to metals and carbon, but can also be applied to polymeric scaffolds to increase roughness and cell–surface interactions. Lyndon *et al.* studied the effects of self-reinforced HA-coated poly(L-lactic acid) (PLLA) scaffolds for use as potential fixation devices [140]. PLLA is commonly used as a fixation device because of its biocompatibility and resistance to immune responses [141, 142]. In this study, the PLLA fibers were chemically etched using alkaline calcium hydroxide, followed by self-reinforcement by melting the top layer and cooling to crystallize it; the fibers were then soaked in SBF to create a HA layer for functionalization. Mechanical testing showed that the presence of HA increased the flexural modulus of the scaffolds with increasing concentrations from 0% (~8.3 GPa) to 15% (~9.7 GPa), indicating a potential application in fixation devices [140].

A further form of surface modification is that of CVD, where gas-phase chemical depositions are created as a thin film on the surface of a material. Several different CVD methods are available, and the process is widely used in nanotechnology surface modification for electronics, semi-conductor synthesis, and biomedical applications [143–145]. The process is performed in a pressure-controlled chamber where the substrates are exposed to the gases or precursors. The chamber is heated and exhausted appropriately such that, when the gas comes into contact with the substrate, the chemical(s) is (are) deposited onto the surface of the substrate as a thin film, usually on the picometer scale. Some of the major precursors deposited during the CVD process that could have potential applications for bone tissue engineering are diamond [146], carbon [147], and graphene [148].

3 Conclusions

Nanomaterials are rapidly approaching the forefront of medical research, and have already begun to show great promise. Indeed, the future surely holds a wide array of nanoscale materials, fabrication and modification techniques, drug-delivery methods, and a number of other exciting improvements.

Through new and exciting techniques, research groups have begun to utilize the power of these nanoscale technologies by fabricating hydrogels, nanoparticles, nanotubes, nanofibers, and self-assembling peptide scaffolds. The robust applications of 3D printing are also aimed at the rapid prototyping of tissue scaffolds and cell/scaffold constructs for *in-vitro* tissue models and potential organ transplants. New and improved nanoscale surface-modification techniques also allow the functionalization and manipulation of already-existing materials to enhance their osteogenic nature and osseointegrative capacity.

The use of these nanoscale developments holds much promise for promoting the regeneration of bone defects, and may help to improve the quality of life in people suffering from traumatic injuries, degenerative diseases such as osteoporosis, and also genetic disorders that hinder skeletogenesis, such as osteogenesis imperfecta.

References

1. Schwartz, C.E., Martha, J.F., Kowalski, P., Wang, D.A., *et al.* (2009) Prospective evaluation of chronic pain associated with posterior autologous iliac crest bone graft harvest and its effect on postoperative outcome. *Health Qual. Life Outcomes*, 7, 49.
2. Camera, A., Cattaneo, G., Tedino, R. (2010) Use of tantalum trabecular metal in total knee arthroplasty. *Minerva Ortop. Traumatol.*, **61** (3), 159–164.
3. Mohan, V., Inacio, M.C.S., Namba, R.S., Sheth, D., *et al.* (2013) Monoblock all-polyethylene tibial components have a lower risk of early revision than metal-backed modular components. A registry study of

27,657 primary total knee arthroplasties. *Acta Orthop.*, **84** (6), 530–536.
4. Robinson, Y., Tschoeke, S.K., Kayser, R., Boehm, H. (2009) Reconstruction of large defects in vertebral osteomyelitis with expandable titanium cages. *Int. Orthop.*, **33** (3), 745–749.
5. van Manen, C.J., Dekker, M.L., van Eerten, P.V., Rhemrev, S.J., *et al.* (2008) Bioresorbable versus metal implants in wrist fractures: a randomised trial. *Arch. Orthop. Trauma Surg.*, **128** (12), 1413–1417.
6. Cameron, H.U. (1986) Six-year results with a microporous-coated metal hip prosthesis. *Clin. Orthop. Relat. Res.*, **208**, 81–83.
7. Ghanaati, S., Lorenz, J., Obreja, K., Choukroun, J. (2014) Nanocrystalline hydroxyapatite-based material already contributes to implant stability after 3 months: a clinical and radiologic 3-year follow-up investigation. *J. Oral Implantol.*, **40** (1), 103–109.
8. Hahn, Y., Bojrab, D.I. (2013) Outcomes following ossicular chain reconstruction with composite prostheses: hydroxyapatite-polyethylene vs. hydroxyapatite-titanium. *Ear Nose Throat J.*, **92** (6), 250, 252, 254 passim.
9. Schroder, J., Grosse-Dresselhaus, F., Schul, C., Wassmann, H. (2007) PMMA versus titanium cage after anterior cervical discectomy – a prospective randomized trial. *Zentralbl. Neurochir.*, **68** (1), 2–7.
10. Kanczler, J.M., Oreffo, R.O.C. (2008) Osteogenesis and angiogenesis: the potential for engineering bone. *Eur. Cell. Mater.*, **15**, 100–114.
11. Roohani-Esfahani, S.I., Nouri-Khorasani, S., Lu, Z.F., Appleyard, R. (2010) The influence of hydroxyapatite nanoparticle shape and size on the properties of biphasic calcium phosphate scaffolds coated with hydroxyapatite-PCL composites. *Biomaterials*, **31** (21), 5498–5509.
12. Roohani-Esfahani, S.I., Nouri-Khorasani, S., Lu, Z.F., Appleyard, R.C. (2011) Effects of bioactive glass nanoparticles on the mechanical and biological behavior of composite coated scaffolds. *Acta Biomater.*, **7** (3), 1307–1318.
13. Karageorgiou, V., Kaplan, D. (2005) Porosity of 3D biomaterial scaffolds and osteogenesis. *Biomaterials*, **26** (27), 5474–5491.
14. Suarez-Gonzalez, D., Barnhart, K., Saito, E., Vanderby, R., Jr (2010) Controlled nucleation of hydroxyapatite on alginate scaffolds for stem cell-based bone tissue engineering. *J. Biomed. Mater. Res. A*, **95** (1), 222–234.
15. Willie, B.M., Petersen, A., Schmidt-Bleek, K., Cipitria, A., *et al.* (2010) Designing biomimetic scaffolds for bone regeneration: why aim for a copy of mature tissue properties if nature uses a different approach? *Soft Matter*, **6** (20), 4976–4987.
16. Steele, D.G., Bramblett, C.A. (1988) *The Anatomy and Biology of the Human Skeleton*. Texas A&M University Press, College Station, viii, 291 pp.
17. Boskey, A.L., Posner, A.S. (1984) Bone structure, composition, and mineralization. *Orthop. Clin. North Am.*, **15** (4), 597–612.
18. Maxson, S., Lopez, E.A., Yoo, D., Danilkovitch-Miagkova, A. (2012) Concise review: role of mesenchymal stem cells in wound repair. *Stem Cells Transl. Med.*, **1** (2), 142–149.
19. Nishimura, I., Garrell, R.L., Hedrick, M., Iida, K. (2003) Precursor tissue analogs as a tissue-engineering strategy. *Tissue Eng.*, **9**, S77–S89.
20. Larsen, M., Tremblay, M.L., Yamada, K.M. (2003) Phosphatases in cell-matrix adhesion and migration. *Nat. Rev. Mol. Cell Biol.*, **4** (9), 700–711.
21. Suzuki, K., Saito, J., Yanai, R., Yamada, N. (2003) Cell-matrix, and cell-cell interactions during corneal epithelial wound healing. *Prog. Retin. Eye Res.*, **22** (2), 113–133.
22. Nguyen, L.L., D'Amore, P.A. (2001) Cellular interactions in vascular growth and differentiation. *Int. Rev. Cytol. – Surv. Cell Biol.*, **204**, 1–48.
23. Huiskes, R., Weinans, H., Vanrietbergen, B. (1992) The relationship between stress shielding and bone-resorption around total hip stems and the effects of flexible materials. *Clin. Orthop. Relat. Res.*, **274**, 124–134.
24. Daculsi, G., Fellah, B.H., Miramond, T., Durand, M. (2013) Osteoconduction, osteogenicity, osteoinduction, what are the fundamental properties for a smart bone substitutes. *IRBM*, **34** (4-5), 346–348.
25. Sapsford, K.E., Algar, W.R., Berti, L., Gemmill, K.B., *et al.* (2013) Functionalizing nanoparticles with biological molecules: developing chemistries that

facilitate nanotechnology. *Chem. Rev.*, **113** (3), 1904–2074.
26. Gupta, A.K., Gupta, M. (2005) Synthesis and surface engineering of iron oxide nanoparticles for biomedical applications. *Biomaterials*, **26** (18), 3995–4021.
27. Coleman, J.N., Khan, U., Blau, W.J., Gun'ko, Y.K. (2006) Small but strong: a review of the mechanical properties of carbon nanotube-polymer composites. *Carbon*, **44** (9), 1624–1652.
28. Annabi, N., Tamayol, A., Uquillas, J.A., Akbari, M., *et al.* (2014) 25th Anniversary Article: Rational design and applications of hydrogels in regenerative medicine. *Adv. Mater.*, **26** (1), 85–124.
29. Peppas, N.A. (2006) New intelligent and targetted drug delivery systems. Pharmaceutical and biomedical applications. *Ann. Pharm. Fr.*, **64** (4), 260–275.
30. Burdick, J.A., Anseth, K.S. (2002) Photoencapsulation of osteoblasts in injectable RGD-modified PEG hydrogels for bone tissue engineering. *Biomaterials*, **23** (22), 4315–4323.
31. Lee, K.Y., Peters, M.C., Anderson, K.W., Mooney, D.J. (2000) Controlled growth factor release from synthetic extracellular matrices. *Nature*, **408** (6815), 998–1000.
32. Slaughter, B.V., Khurshid, S.S., Fisher, O.Z., Khademhosseini, A., *et al.* (2009) Hydrogels in regenerative medicine. *Adv. Mater.*, **21** (32-33), 3307–3329.
33. Yang, F., Williams, C.G., Wang, D.A., Lee, H. (2005) The effect of incorporating RGD adhesive peptide in polyethylene glycol diacrylate hydrogel on osteogenesis of bone marrow stromal cells. *Biomaterials*, **26** (30), 5991–5998.
34. Benoit, D.S.W., Anseth, K.S. (2005) The effect on osteoblast function of colocalized RGD and PHSRN epitopes on PEG surfaces. *Biomaterials*, **26**, 5209–5220.
35. He, X., Ma, J., Jabbari, E. (2008) Effect of grafting RGD and BMP-2 protein-derived peptides to a hydrogel substrate on osteogenic differentiation of marrow stromal cells. *Langmuir*, **24** (21), 12508–12516.
36. Porter, A.M., Klinge, C.M., Gobin, A.S. (2011) Biomimetic hydrogels with VEGF induce angiogenic processes in both hUVEC and hMEC. *Biomacromolecules*, **12** (1), 242–246.
37. Lee, S.S., Hsu, E.L., Mendoza, M., Ghodasra, J., *et al.* (2008) Gel scaffolds of BMP-2-binding peptide amphiphile nanofibers for spinal arthrodesis. *Adv. Healthcare Mater.* (online publication).
38. Hoare, T.R., Kohane, D.S. (2008) Hydrogels in drug delivery: progress and challenges. *Polymer*, **49** (8), 1993–2007.
39. Segura, T., Chung, P.H., Shea, L.D. (2005) DNA delivery from hyaluronic acid-collagen hydrogels via a substrate-mediated approach. *Biomaterials*, **26** (13), 1575–1584.
40. Comisar, W.A., Kazmers, N.H., Mooney, D.J., Linderman, J.J. (2007) Engineering RGD nanopatterned hydrogels to control preosteoblast behavior: a combined computational and experimental approach. *Biomaterials*, **28** (30), 4409–4417.
41. Huang, J.H., Grater, S.V., Corbellinl, F., Rinck, S., *et al.* (2009) Impact of order and disorder in RGD nanopatterns on cell adhesion. *Nano Lett.*, **9** (3), 1111–1116.
42. Luo, Y., Shoichet, M.S. (2004) A photolabile hydrogel for guided three-dimensional cell growth and migration. *Nat. Mater.*, **3** (4), 249–253.
43. Mosiewicz, K.A., Kolb, L., van der Vlies, A.J., Martino, M.M., *et al.* (2013) In situ cell manipulation through enzymatic hydrogel photopatterning. *Nat. Mater.*, **12** (11), 1071–1077.
44. Matson, J.B., Stupp, S.I. (2012) Self-assembling peptide scaffolds for regenerative medicine. *Chem. Commun.*, **48** (1), 26–33.
45. Moyer, T.J., Cui, H., Stupp, S.I. (2013) Tuning nanostructure dimensions with supramolecular twisting. *J. Phys. Chem. B*, **117** (16), 4604–4610.
46. Sur, S., Newcomb, C.J., Webber, M.J., Stupp, S.I. (2013) Tuning supramolecular mechanics to guide neuron development. *Biomaterials*, **34** (20), 4749–4757.
47. Otsuka, H., Nagasaki, Y., Kataoka, K. (2001) Self-assembly of poly(ethylene glycol)-based block copolymers for biomedical applications. *Curr. Opin. Colloid Interface Sci.*, **6** (1), 3–10.
48. Tessmar, J.K., Gopferich, A.M. (2007) Customized PEG-derived copolymers for tissue-engineering applications. *Macromol. Biosci.*, **7** (1), 23–39.
49. Newman, P., Minett, A., Ellis-Behnke, R., Zreiqat, H. (2013) Carbon nanotubes: their potential and pitfalls for bone tissue

regeneration and engineering. *Nanomed.: Nanotechnol. Biol. Med.*, **9** (8), 1139–1158.
50. Nayak, T.R., Andersen, H., Makam, V.S., Khaw, C., *et al.* (2011) Graphene for controlled and accelerated osteogenic differentiation of human mesenchymal stem cells. *ACS Nano*, **5** (6), 4670–4678.
51. Dorozhkin, S.V. (2010) Nanosized and nanocrystalline calcium orthophosphates. *Acta Biomater.*, **6** (3), 715–734.
52. Zhou, H., Lee, J. (2011) Nanoscale hydroxyapatite particles for bone tissue engineering. *Acta Biomater.*, **7** (7), 2769–2781.
53. Webster, T.J., Ergun, C., Doremus, R.H., Siegel, R.W. (2000) Enhanced functions of osteoblasts on nanophase ceramics. *Biomaterials*, **21** (17), 1803–1810.
54. Smith, I.O., McCabe, L.R., Baumann, M.J. (2006) MC3T3-E1 osteoblast attachment and proliferation on porous hydroxyapatite scaffolds fabricated with nanophase powder. *Int. J. Nanomedicine*, **1** (2), 189–194.
55. Shi, Z., Huang, X., Cai, Y., Tang, R., *et al.* (2009) Size effect of hydroxyapatite nanoparticles on proliferation and apoptosis of osteoblast-like cells. *Acta Biomater.*, **5** (1), 338–345.
56. Chan, C.K., Kumar, T.S., Liao, S., Murugan, R., *et al.* (2006) Biomimetic nanocomposites for bone graft applications. *Nanomedicine (Lond.)*, **1** (2), 177–188.
57. Tran, N., Webster, T.J. (2009) Nanotechnology for bone materials. *Wiley Interdiscip. Rev. Nanomed. Nanobiotechnol.*, **1** (3), 336–351.
58. Skaat, H., Ziv-Polat, O., Shahar, A., Last, D., *et al.* (2012) Magnetic scaffolds enriched with bioactive nanoparticles for tissue engineering. *Adv. Healthcare Mater.*, **1** (2), 168–171.
59. Elsabahy, M., Wooley, K.L. (2012) Design of polymeric nanoparticles for biomedical delivery applications. *Chem. Soc. Rev.*, **41** (7), 2545–2561.
60. Kikuchi, M., Itoh, S., Ichinose, S., Shinomiya, K. (2001) Self-organization mechanism in a bone-like hydroxyapatite/collagen nanocomposite synthesized in vitro and its biological reaction in vivo. *Biomaterials*, **22** (13), 1705–1711.
61. Marra, K.G., Szem, J.W., Kumta, P.N., DiMilla, P.A. (1999) In vitro analysis of biodegradable polymer blend/hydroxyapatite composites for bone tissue engineering. *J. Biomed. Mater. Res.*, **47** (3), 324–335.
62. Blaker, J.J., Gough, J.E., Maquet, V., Notingher, I., *et al.* (2003) In vitro evaluation of novel bioactive composites based on bioglass-filled polylactide foams for bone tissue engineering scaffolds. *J. Biomed. Mater. Res. A*, **67** (4), 1401–1411.
63. Kim, S.S., Ahn, K.M., Park, M.S., Lee, J.H. (2007) A poly(lactide-*co*-glycolide)/hydroxyapatite composite scaffold with enhanced osteoconductivity. *J. Biomed. Mater. Res. A*, **80** (1), 206–215.
64. Roohani-Esfahani, S.I., Chen, Y.J., Shi, J., Zreiqat, H. (2013) Fabrication and characterization of a new, strong and bioactive ceramic scaffold for bone regeneration. *Mater. Lett.*, **107**, 378–381.
65. Roohani-Esfahani, S.I., Dunstan, C.R., Li, J.J., Lu, Z. (2013) Unique microstructural design of ceramic scaffolds for bone regeneration under load. *Acta Biomater.*, **9** (6), 7014–7024.
66. Liu, H., Slamovich, E.B., Webster, T.J. (2006) Less harmful acidic degradation of poly(lactic-*co*-glycolic acid) bone tissue engineering scaffolds through titania nanoparticle addition. *Int. J. Nanomedicine*, **1** (4), 541–545.
67. Popat, K.C., Leoni, L., Grimes, C.A., Desai, T.A. (2007) Influence of engineered titania nanotubular surfaces on bone cells. *Biomaterials*, **28** (21), 3188–3197.
68. Gao, L., Feng, B., Wang, J.X., Lu, X., *et al.* (2009) Micro/Nanostructural porous surface on titanium and bioactivity. *J. Biomed. Mater. Res. B Appl. Biomater.*, **89B** (2), 335–341.
69. Zhao, L.Z., Mei, S.L., Chu, P.K., Zhang, Y.M., *et al.* (2010) The influence of hierarchical hybrid micro/nano-textured titanium surface with titania nanotubes on osteoblast functions. *Biomaterials*, **31** (19), 5072–5082.
70. Vasilev, K., Poh, Z., Kant, K., Chan, J. (2010) Tailoring the surface functionalities of titania nanotube arrays. *Biomaterials*, **31** (3), 532–540.
71. Shokuhfar, T., Sinha-Ray, S., Sukotjo, C., Yarin, A.L. (2013) Intercalation of anti-inflammatory drug molecules within TiO_2 nanotubes. *RSC Adv.*, **3** (38), 17380–17386.
72. Oh, S., Brammer, K.S., Li, Y.S.J., Teng, D., *et al.* (2009) Stem cell fate dictated solely

by altered nanotube dimension. *Proc. Natl. Acad. Sci. USA*, **106** (7), 2130–2135.

73. Newman, P., Roohani-Esfahani, S.I., Zreiqat, H., Minett, A. (2014) See the extracellular forest for the nanotrees. Exploring three-dimensional bone scaffolds. *Mater. Today*, **17** (1), 43–44.
74. Lee, C., Wei, X.D., Kysar, J.W., Hone, J. (2008) Measurement of the elastic properties and intrinsic strength of monolayer graphene. *Science*, **321** (5887), 385–388.
75. Yang, K., Feng, L.Z., Hong, H., Cai, W.B., *et al.* (2013) Preparation and functionalization of graphene nanocomposites for biomedical applications. *Nat. Protoc.*, **8** (12), 2392–2403.
76. Depan, D., Girase, B., Shah, J.S., Misra, R.D.K. (2012) Structure–process–property relationship of the polar graphene oxide-mediated cellular response and stimulated growth of osteoblasts on hybrid chitosan network structure nanocomposite scaffolds. *Acta Biomater.*, **7** (9), 3432–3445.
77. Ogihara, N., Usui, Y., Aoki, K., Shimizu, M., *et al.* (2012) Biocompatibility and bone tissue compatibility of alumina ceramics reinforced with carbon nanotubes. *Nanomedicine*, **7** (7), 981–993.
78. Jell, G., Verdejo, R., Safinia, L., Shaffer, MSP. (2008) Carbon nanotube-enhanced polyurethane scaffolds fabricated by thermally induced phase separation. *J. Mater. Chem.*, **18** (16), 1865–1872.
79. Lin, C.L., Wang, Y.F., Lai, Y.Q., Yang, W. (2011) Incorporation of carboxylation multi-walled carbon nanotubes into biodegradable poly(lactic-*co*-glycolic acid) for bone tissue engineering. *Colloids Surf., B - Biointerfaces*, **83** (2), 367–375.
80. Verdejo, R., Jell, G., Safinia, L., Bismarck, A., *et al.* (2009) Reactive polyurethane carbon nanotube foams and their interactions with osteoblasts. *J. Biomed. Mater. Res. A*, **88A** (1), 65–73.
81. Liao, K.H., Lin, Y.S., Macosko, C.W., Haynes, C.L. (2011) Cytotoxicity of graphene oxide and graphene in human erythrocytes and skin fibroblasts. *ACS Appl. Mater. Interfaces*, **3** (7), 2607–2615.
82. Chang, Y.L., Yang, S.T., Liu, J.H., Dong, E. (2011) In vitro toxicity evaluation of graphene oxide on A549 cells. *Toxicol. Lett.*, **200** (3), 201–210.
83. Singh, S.K., Singh, M.K., Nayak, M.K., Kumari, S. (2011) Thrombus inducing property of atomically thin graphene oxide sheets. *ACS Nano* **5** (6):4987–4996.
84. Zhang, X.Y., Yin, J.L., Peng, C., Hu, W.Q. (2011) Distribution and biocompatibility studies of graphene oxide in mice after intravenous administration. *Carbon*, **49** (3), 986–995.
85. Yang, K., Wan, J.M., Zhang, S.A., Zhang, Y.J., *et al.* (2011) In vivo pharmacokinetics, long-term biodistribution, and toxicology of PEGylated graphene in mice. *ACS Nano*, **5** (1), 516–522.
86. Bergshoef, M.M., Vancso, G.J. (1999) Transparent nanocomposites with ultrathin, electrospun nylon-4,6 fiber reinforcement. *Adv. Mater.*, **11** (16), 1362–1365.
87. Jin, H.J., Fridrikh, S.V., Rutledge, G.C., Kaplan, D.L. (2002) Electrospinning *Bombyx mori* silk with poly(ethylene oxide). *Biomacromolecules*, **3** (6), 1233–1239.
88. Li, W.J., Laurencin, C.T., Caterson, E.J., Tuan, R.S., *et al.* (2002) Electrospun nanofibrous structure: a novel scaffold for tissue engineering. *J. Biomed. Mater. Res.*, **60** (4), 613–621.
89. Liao, S., Li, B., Ma, Z., Wei, H. (2006) Biomimetic electrospun nanofibers for tissue regeneration. *Biomed. Mater.*, **1** (3), R45–R53.
90. Chew, S.Y., Wen, Y., Dzenis, Y., Leong, K.W. (2006) The role of electrospinning in the emerging field of nanomedicine. *Curr. Pharm. Des.*, **12** (36), 4751–4770.
91. Baker, B.M., Gee, A.O., Metter, R.B., Nathan, A.S. (2008) The potential to improve cell infiltration in composite fiber-aligned electrospun scaffolds by the selective removal of sacrificial fibers. *Biomaterials*, **29** (15), 2348–2358.
92. Uttayarat, P., Perets, A., Li, M., Pimton, P., *et al.* (2010) Micropatterning of three-dimensional electrospun polyurethane vascular grafts. *Acta Biomater.*, **6** (11), 4229–4237.
93. Masoumi, N., Larson, B.L., Annabi, N., Kharaziha, M., *et al.* (2014) Electrospun PGS:PCL microfibers align human valvular interstitial cells and provide tunable scaffold anisotropy. *Adv. Healthcare Mater.*, **2** (6), 929–939

94. Frohbergh, M.E., Katsman, A., Botta, G.P., Lazarovici, P., *et al.* 2012 Electrospun hydroxyapatite-containing chitosan nanofibers crosslinked with genipin for bone tissue engineering. *Biomaterials*, **33** (36), 9167–9178
95. Thomas, V., Dean, D.R., Jose, M.V., Mathew, B., *et al.* (2007) Nanostructured biocomposite scaffolds based on collagen coelectrospun with nanohydroxyapatite. *Biomacromolecules*, **8** (2), 631–637.
96. Han, J., Lazarovici, P., Pomerantz, C., Chen, X., *et al.* (2010) Co-electrospun blends of PLGA, gelatin, and elastin as potential nonthrombogenic scaffolds for vascular tissue engineering. *Biomacromolecules*, **12** (2), 399–408
97. Kim, H., Che, L., Ha, Y., Ryu, W. (2014) Mechanically-reinforced electrospun composite silk fibroin nanofibers containing hydroxyapatite nanoparticles. *Mater. Sci. Eng., C: Mater. Biol. Appl.*, **40**, 324–335.
98. Eichhorn, S.J., Sampson, W.W. (2005) Statistical geometry of pores and statistics of porous nanofibrous assemblies. *J. R. Soc. Interface*, **2** (4), 309–318.
99. Joshi, V.S., Lei, N.Y., Walthers, C.M., Wu, B. (2013) Macroporosity enhances vascularization of electrospun scaffolds. *J. Surg. Res.*, **183** (1), 18–26.
100. Phipps, M.C., Clem, W.C., Grunda, J.M., Clines, G.A. (2012) Increasing the pore sizes of bone-mimetic electrospun scaffolds comprised of polycaprolactone, collagen I and hydroxyapatite to enhance cell infiltration. *Biomaterials*, **33** (2), 524–534.
101. Skotak, M., Ragusa, J., Gonzalez, D., Subramanian, A. (2011) Improved cellular infiltration into nanofibrous electrospun cross-linked gelatin scaffolds templated with micrometer-sized polyethylene glycol fibers. *Biomed. Mater.*, **6** (5), 055012.
102. Cheng, Q., Lee, B.L., Komvopoulos, K., Li, S. (2013) Engineering the microstructure of electrospun fibrous scaffolds by microtopography. *Biomacromolecules*, **14** (5), 1349–1360.
103. Kim, T.G., Chung, H.J., Park, T.G. (2008) Macroporous and nanofibrous hyaluronic acid/collagen hybrid scaffold fabricated by concurrent electrospinning and deposition/leaching of salt particles. *Acta Biomater.*, **4** (6), 1611–1619.
104. Scaglione, S., Guarino, V., Sandri, M., Tampieri, A. (2012) In vivo lamellar bone formation in fibre-coated MgCHA-PCL-composite scaffolds. *J. Mater. Sci. Mater. Med.*, **23** (1), 117–128.
105. Li, M., Mondrinos, M.J., Gandhi, M.R., Ko, F.K. (2005) Electrospun protein fibers as matrices for tissue engineering. *Biomaterials*, **26** (30), 5999–6008.
106. Zhang, Y., Venugopal, J.R., El-Turki, A., Ramakrishna, S. (2008) Electrospun biomimetic nanocomposite nanofibers of hydroxyapatite/chitosan for bone tissue engineering. *Biomaterials*, **29** (32), 4314–4322.
107. Dhandayuthapani, B., Krishnan, U.M., Sethuraman, S. (2010) Fabrication and characterization of chitosan-gelatin blend nanofibers for skin tissue engineering. *J. Biomed. Mater. Res. B Appl. Biomater.*, **94** (1), 264–272.
108. Cai, Z.X., Mo, X.M., Zhang, K.H., Fan, L.P., *et al.* (2010) Fabrication of chitosan/silk fibroin composite nanofibers for wound-dressing applications. *Int. J. Mol. Sci.*, **11** (9), 3529–3539.
109. Xie, D., Huang, H., Blackwood, K., MacNeil, S. (2010) A novel route for the production of chitosan/poly(lactide-*co*-glycolide) graft copolymers for electrospinning. *Biomed. Mater.*, **5** (6), 065016.
110. Ba Linh, N.T., Min, Y.K., Lee, B.T. (2013) Hybrid hydroxyapatite nanoparticles-loaded PCL/GE blend fibers for bone tissue engineering. *J. Biomater. Sci. Polym. Ed.*, **24** (5), 520–538.
111. Torricelli, P., Gioffre, M., Fiorani, A., Panzavolta, S., *et al.* (2014) Co-electrospun gelatin-poly(L-lactic acid) scaffolds: modulation of mechanical properties and chondrocyte response as a function of composition. *Mater. Sci. Eng., C-Mater. Biol. Appl.*, **36**, 130–138.
112. Li, D., Sun, H., Jiang, L., Zhang, K., *et al.* (2014) Enhanced biocompatibility of PLGA nanofibers with gelatin/nano-hydroxyapatite bone biomimetics incorporation. *ACS Appl. Mater. Interfaces*, **6** (12), 9402–9410
113. Ramier, J., Bouderlique, T., Stoilova, O., Manolova, N., *et al.* (2014) Biocomposite scaffolds based on electrospun poly(3-hydroxybutyrate) nanofibers and electrosprayed hydroxyapatite nanoparticles for bone tissue engineering applications.

Mater. Sci. Eng., C-Mater. Biol. Appl., **38**, 161–169.

114. Cheng, Y., Ramos, D., Lee, P., Liang, D., *et al.* (2014) Collagen functionalized bioactive nanofiber matrices for osteogenic differentiation of mesenchymal stem cells: bone tissue engineering. *J. Biomed. Nanotechnol.*, **10** (2), 287–298.
115. Igawa, K., Chung, U.I., Tei, Y. (2008) Custom-made artificial bones fabricated by an inkjet printing technology. *Clin. Calcium*, **18** (12), 1737–1743.
116. Saijo, H., Igawa, K., Kanno, Y., Mori, Y., *et al.* (2009) Maxillofacial reconstruction using custom-made artificial bones fabricated by inkjet printing technology. *J. Artif. Organs*, **12** (3), 200–205.
117. Fedorovich, N.E., Alblas, J., Hennink, W.E., Oner, F.C., *et al.* (2011) Organ printing: the future of bone regeneration? *Trends Biotechnol.*, **29** (12), 601–606.
118. Schubert, C., van Langeveld, M.C., Donoso, L.A. (2014) Innovations in 3D printing: a 3D overview from optics to organs. *Br. J. Ophthalmol.*, **98** (2), 159–161.
119. Windisch, G., Salaberger, D., Rosmarin, W., Kastner, J., *et al.* (2007) A model for clubfoot based on micro-CT data. *J. Anat.*, **210** (6), 761–766.
120. Liu, Y.F., Xu, L.W., Zhu, H.Y., Liu, S.S. (2014) Technical procedures for template-guided surgery for mandibular reconstruction based on digital design and manufacturing. *Biomed. Eng. Online*, **13** (1), 63.
121. Tarafder, S., Davies, N.M., Bandyopadhyay, A., Bose, S. (2013) 3D printed tricalcium phosphate scaffolds: effect of SrO and MgO doping on osteogenesis in a rat distal femoral defect model. *Biomater. Sci.*, **1** (12), 1250–1259.
122. Tarafder, S., Bose, S. (2014) Polycaprolactone coated 3D printed tricalcium phosphate scaffolds: in vitro alendronate release behavior and local delivery effect on in vivo osteogenesis. *ACS Appl. Mater. Interfaces*, **6** (13), 9955–9965
123. Poldervaart, M.T., Gremmels, H., van Deventer, K., Fledderus, J.O., *et al.* (2014) Prolonged presence of VEGF promotes vascularization in 3D bioprinted scaffolds with defined architecture. *J. Control. Release*, **184**, 58–66.
124. Salomon, D.S. (2000) Cell-cell and cell-extracellular matrix adhesion molecules communicate with growth factor receptors: an interactive signaling Web. *Cancer Invest.*, **18** (6), 591–593.
125. Werle, M.J. (2008) Cell-to-cell signaling at the neuromuscular junction: the dynamic role of the extracellular matrix. *Ann. NY Acad. Sci.*, **1132**, 13–18.
126. Gentili, C., Cancedda, R. (2009) Cartilage and bone extracellular matrix. *Curr. Pharm. Des.*, **15** (12), 1334–1348.
127. Xia, Y.N., Whitesides, G.M. (1998) Soft lithography. *Annu. Rev. Mater. Sci.*, **28**, 153–184.
128. Brittain, S., Paul, K., Zhao, X.M., Whitesides, G. (1998) Soft lithography and microfabrication. *Phys. World*, **11** (5), 31–36.
129. Wilbur, J.L., Kumar, A., Kim, E., Whitesides, G.M. (1994) Microfabrication by microcontact printing of self-assembled monolayers. *Adv. Mater.*, **6** (7-8), 600–604.
130. Kim, E.J., Fleischman, A.J., Muschler, G.F., Roy, S. (2013) Response of bone marrow-derived connective tissue progenitor cell morphology and proliferation on geometrically modulated microtextured substrates. *Biomed. Microdevices*, **15** (3), 385–396.
131. Wang, Y.C., Zhang, L.J., Sun, L.L., Webster, T.J. (2013) Increased healthy osteoblast to osteosarcoma density ratios on specific PLGA nanopatterns. *Int. J. Nanomedicine*, **8**, 159–166.
132. Lamers, E., Walboomers, X.F., Domanski, M., te Riet, J. (2010) The influence of nanoscale grooved substrates on osteoblast behavior and extracellular matrix deposition. *Biomaterials*, **31** (12), 3307–3316.
133. Prodanov, L., Lamers, E., Domanski, M., Luttge, R. (2013) The effect of nanometric surface texture on bone contact to titanium implants in rabbit tibia. *Biomaterials*, **34** (12), 2920–2927.
134. Congiu, T., Pazzaglia, U.E., Basso, P., Quacci, D. (2014) Chemical etching in processing cortical bone specimens for scanning electron microscopy. *Microsc. Res. Tech.*, **77** (9), 653–660.
135. Chen, W.C., Chen, Y.S., Ko, C.L., Lin, Y., *et al.* (2014) Interaction of progenitor bone cells with different surface modifications of titanium implant. *Mater. Sci. Eng., C-Mater. Biol. Appl.*, **37**, 305–313.

136. Kim, B.S., Kim, J.S., Park, Y.M., Choi, B.Y., *et al.* (2013) Mg ion implantation on SLA-treated titanium surface and its effects on the behavior of mesenchymal stem cell. *Mater. Sci. Eng., C-Mater. Biol. Appl.*, **33** (3), 1554–1560.
137. Oliveira, D.P., Palmieri, A., Carinci, F., Bolfarini, C. (2014) Osteoblasts behavior on chemically treated commercially pure titanium surfaces. *J. Biomed. Mater. Res. A*, **102** (6), 1816–1822.
138. Hoppe, A., Will, J., Detsch, R., Boccaccini, A.R., *et al.* (2014) Formation and in vitro biocompatibility of biomimetic hydroxyapatite coatings on chemically treated carbon substrates. *J. Biomed. Mater. Res. A*, **102** (1), 193–203.
139. Guo, Y.J., Wang, Y.Y., Chen, T., Wei, Y.T., *et al.* (2013) Hollow carbonated hydroxyapatite microspheres with mesoporous structure: hydrothermal fabrication and drug delivery property. *Mater. Sci. Eng., C: Mater. Biol. Appl.*, **33** (6), 3166–3172.
140. Charles, L.F., Kramer, E.R., Shaw, M.T., Olson, J.R., *et al.* 2013 Self-reinforced composites of hydroxyapatite-coated PLLA fibers: fabrication and mechanical characterization. *J. Mech. Behav. Biomed. Mater.*, **17**, 269–277.
141. Macarini, L., Milillo, P., Mocci, A., Vinci, R., *et al.* (2008) Poly-L-lactic acid–hydroxyapatite (PLLA-HA) bioabsorbable interference screws for tibial graft fixation in anterior cruciate ligament (ACL) reconstruction surgery: MR evaluation of osteointegration and degradation features. *Radiol. Med.*, **113** (8), 1185–1197.
142. Thiele, A., Bilkenroth, U., Bloching, M., Knipping, S. (2008) Foreign body reaction to materials implanted as biocompatible for internal fixation. *HNO*, **56** (5), 545–548.
143. Tavares, J., Swanson, E.J., Coulombe, S. (2008) Plasma synthesis of coated metal nanoparticles with surface properties tailored for dispersion. *Plasma Processes Polym.*, **5** (8), 759–769.
144. Vollath, D., Szabo, D.V. (1999) Coated nanoparticles: a new way to improve nanocomposites. *J. Nanopart. Res.*, **1** (2), 235–242.
145. Qin, C., Coulombe, S. (2007) Organic layer-coated metal nanoparticles prepared by a combined arc evaporation/condensation and plasma polymerization process. *Plasma Sources Sci. Technol.*, **16** (2), 240–249.
146. Butler, J.E., Mankelevich, Y.A., Cheesman, A., Ma, J., *et al.* (2009) Understanding the chemical vapor deposition of diamond: recent progress. *J. Phys. Condens. Matter*, **21** (36), 364201.
147. Kumar, M., Ando, Y. (2010) Chemical vapor deposition of carbon nanotubes: a review on growth mechanism and mass production. *J. Nanosci. Nanotechnol.*, **10** (6), 3739–3758.
148. Oh, J.S., Kim, K.N., Yeom, G.Y. (2014) Graphene doping methods and device applications. *J. Nanosci. Nanotechnol.*, **14** (2), 1120–1133.

Index

Translational Nanomedicine. First Edition. Edited by Robert A. Meyers.

d

e

h

q

r

u